Springer Series in Optical Sciences

Volume 190

Springer Series in Optical Sciences

The Springer Series in Optical Sciences, under the leadership of Editor-in-Chief William T. Rhodes, Georgia Institute of Technology, USA, provides an expanding selection of research monographs in all major areas of optics: lasers and quantum optics, ultrafast phenomena, optical spectroscopy techniques, optoelectronics, quantum information, information optics, applied laser technology, industrial applications, and other topics of contemporary interest.

With this broad coverage of topics, the series is of use to all research scientists and engineers who need up-to-date reference books.

The editors encourage prospective authors to correspond with them in advance of submitting a manuscript. Submission of manuscripts should be made to the Editor-in-Chief or one of the Editors. See also www.springer.com/series/624

More information about this series at http://www.springer.com/series/624

Claudia S. Schnohr · Mark C. Ridgway
Editors

X-Ray Absorption Spectroscopy of Semiconductors

Springer

Editors
Claudia S. Schnohr
Institut für Festkörperphysik
Friedrich-Schiller-Universität Jena
Jena
Germany

Mark C. Ridgway
Department of Electronic Materials Engineering
Australian National University
Canberra, ACT
Australia

ISSN 0342-4111 ISSN 1556-1534 (electronic)
ISBN 978-3-662-44361-3 ISBN 978-3-662-44362-0 (eBook)
DOI 10.1007/978-3-662-44362-0

Library of Congress Control Number: 2014951156

Springer Heidelberg New York Dordrecht London

Printed on acid-free paper

Springer is part of Springer Science+Business Media (www.springer.com)

Preface

X-ray Absorption Spectroscopy (XAS) is a powerful technique with which to probe the properties of matter, equally applicable to the solid, liquid and gas phases. Its unique characteristics, including element-specificity and nanometer range, make it a versatile probe that provides structural information distinctly different and complementary to that obtained by other common techniques such as X-ray diffraction or electron microscopy. Since the pioneering works in the early 1970s, XAS has progressed tremendously with respect to both experimental techniques and theoretical understanding. Modern synchrotron lightsources not only enable standard XAS measurements with extremely high data quality, they also facilitate studies on the subsecond or nanometer scale. This provides a large variety of new applications such as time-resolved measurements of dynamic processes or structural characterization of single nanostructures. The theoretical understanding of XAS has progressed at a similar pace and several computer codes capable of calculating the X-ray absorption fine structure to within the experimental uncertainty are now readily available. It therefore seems a mere consequence that XAS is these days widely used in a large number of fields including physics, chemistry, material science, geology, biology and environmental science.

Semiconductors form the basis of an ever-growing variety of electronic and photonic devices that permeate almost every aspect of today's society. From mobile phones to cars, from washing machines to artificial light, semiconductor technology is at the bottom of nearly all modern appliances. Advanced telecommunications, the key to a global world, is utterly unthinkable without the achievements made in the semiconductor industry over the last decades. These developments, however, are far from completed. Currently, the whole new world of nanomaterials is being explored extensively and first concepts to utilize the unique properties thus discovered are being implemented. Semiconductor materials also play a vital role in the quest for a sustainable energy supply, one of the big global challenges of the twenty-first century. By directly converting sunlight to electricity, photovoltaic devices such as solar cells provide a versatile and renewable energy source. The growing and changing demands of future technology in nearly all aspects of modern life therefore continuously require improving current and developing new semiconductor devices.

The most effective utilization of these materials, today and tomorrow, necessitates a detailed knowledge of their structural properties as they determine other electrical, optical or magnetic properties crucial for device performance. XAS has provided unique and valuable insight into these relations for a large number of semiconductor systems. It is therefore the aim of this book to present a comprehensive overview of past and present research activities in this ever growing field.

Chapter 1 is dedicated to a short introduction to XAS and is aimed primarily at newcomers to the technique. It presents all the basic information necessary to follow the subsequent chapters and provides references for further reading. The following chapters are dedicated to XAS research of distinct groups of materials. Part I comprises Chaps. 2–6 and is dedicated to crystalline semiconductors spanning topics such as alloying, wide band gap materials, dopants and clusters and vibrational properties. Part II presents research on disordered semiconductors with amorphous materials covered in Chaps. 7 and 8 while phase changes due to extreme conditions such as high temperature and high pressure are discussed in Chap. 9. Part III consists of Chaps. 10–13 and is dedicated to semiconductor nanostructures such as quantum dots, nanoparticles and nanowires of various group IV, III–V and II–VI materials. The last section, Part IV, concerns the investigation of magnetic ions such as Mn, Co and Fe incorporated in different group IV, III–V and II–VI semiconductors discussed in Chaps. 14–16, respectively.

Each chapter summarizes the research activities of the respective field and highlights important experimental results thus demonstrating the capabilities and applications of the XAS technique. As such, this book provides a comprehensive review and valuable reference guide for both XAS newcomers and experts involved in semiconductor materials research.

Jena, Canberra

Claudia S. Schnohr
Mark C. Ridgway

Contents

Part III Semiconductor Nanostructures

Contributors

Federico Boscherini Department of Physics and Astronomy, University of Bologna, Bologna, Italy

Gianluca Ciatto Synchrotron SOLEIL, Saint-Aubin, Gif sur Yvette, France

Francesco d'Acapito CNR-IOM-OGG c/o European Synchrotron Radiation Facility-GILDA CRG, Grenoble, France

Andrea Di Cicco Physics Division, School of Science and Technology, Università di Camerino, Camerino, Italy

Adriano Filipponi Dipartimento di Scienze Fisiche e Chimiche, Università degli Studi dell'Aquila, Coppito, AQ, Italy

Paolo Fornasini Department of Physics, University of Trento, Povo (Trento), Italy

Alexander A. Guda Southern Federal University, Rostov-on-Don, Russia

Roberto Gunnella Scienze e Tecnologie, Università di Camerino, Camerino, MC, Italy

Steve M. Heald X-ray Science Divison, Advanced Photon Source, Argonne National Lab, Lemont, IL, USA

Maria Katsikini School of Physics, Section of Solid State Physics, Aristotle University of Thessaloniki, Thessaloniki, Greece

Alexander V. Kolobov Nanoelectronics Research Institute, National Institute of Advanced Industrial Science and Technology, Tsukuba, Ibaraki, Japan

Krystyna Lawniczak-Jablonska Institute of Physics, Polish Academy of Science, Warsaw, Poland

Mark C. Ridgway Department of Electronic Materials Engineering, Australian National University, Canberra, ACT, Australia

Claudia S. Schnohr Institut für Festkörperphysik, Friedrich-Schiller-Universität Jena, Jena, Germany

Tsun-Kong Sham Department of Chemistry, Soochow University-Western University Joint Centre for Synchrotron Radiation Research, University of Western Ontario, London, ON, Canada

Alexander V. Soldatov Southern Federal University, Rostov-on-Don, Russia

Mikhail A. Soldatov Southern Federal University, Rostov-on-Don, Russia

Xuhui Sun Soochow University-Western University Centre for Synchrotron Radiation Research, Institute of Functional Nano and Soft Materials (FUNSOM), Soochow University, Suzhou, Jiangsu, People's Republic of China

Chapter 1
Introduction to X-Ray Absorption Spectroscopy

Claudia S. Schnohr and Mark C. Ridgway

X-ray Absorption Spectroscopy (XAS) is a well-established analytical technique used extensively for the characterization of semiconductors in solid or liquid, crystalline or amorphous, bulk or nanoscale form. With this chapter, we provide a brief introduction to XAS, covering both theory and experiment, while we refer to more comprehensive texts for greater detail about this continually evolving technique. The chapter thus is a starting point upon which subsequent chapters build as they demonstrate the broad-ranging applications of XAS to semiconductors materials.

1.1 Basic Principle

X-ray absorption spectroscopy (XAS) measures the energy-dependent fine structure of the X-ray absorption coefficient near the absorption edge of a particular element. Detailed discussions of both theoretical and experimental aspects of XAS can be found in [1–5].

1.1.1 X-Ray Absorption

If X-rays of intensity I_0 are incident on a sample, as shown schematically in Fig. 1.1a, the extent of absorption depends on the photon energy E and sample thickness t.

C.S. Schnohr (✉)
Institut für Festkörperphysik, Friedrich-Schiller-Universität Jena,
Max-Wien-Platz 1, 07743 Jena, Germany
e-mail: c.schnohr@uni-jena.de

M.C. Ridgway
Department of Electronic Materials Engineering, Australian National University,
Canberra, ACT 0200, Australia
e-mail: mark.ridgway@anu.edu.au

C.S. Schnohr and M.C. Ridgway (eds.), *X-Ray Absorption Spectroscopy of Semiconductors*, Springer Series in Optical Sciences 190,
DOI 10.1007/978-3-662-44362-0_1

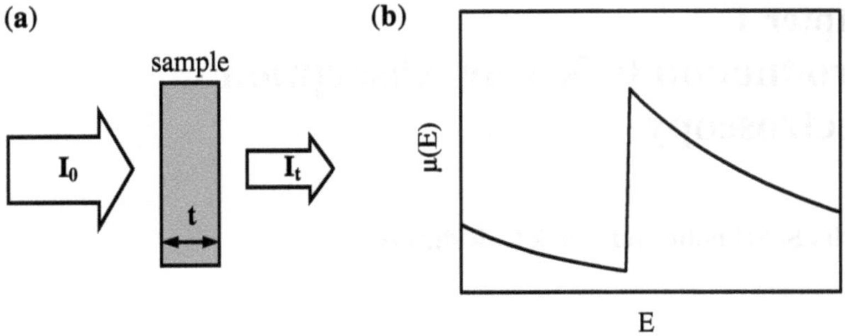

Fig. 1.1 **a** Schematic of incident and transmitted X-ray beam and **b** absorption coefficient $\mu(E)$ versus photon energy E around an absorption edge

According to Beer's Law, the transmitted intensity I_t is

$$I_t\,(t) = I_0 e^{-\mu(E)t} \tag{1.1}$$

where $\mu(E)$ is the energy-dependent X-ray absorption coefficient. Over large energy regions, $\mu(E)$ is a smooth function of the photon energy, varying approximately as $\mu(E) \sim dZ^4/mE^3$ [6]. Here d denotes the target density while Z and m are the atomic number and mass, respectively. Thus, $\mu(E)$ decreases with increasing photon energy. If the latter equals or exceeds the binding energy of a core electron, however, a new absorption channel is available in which the photon is annihilated thereby creating a photoelectron and a core-hole. This leads to a sharp increase in absorption coefficient as shown schematically in Fig. 1.1b. Above the absorption edge, the difference between the photon energy and the binding energy is converted into kinetic energy of the photoelectron and $\mu(E)$ continues to decrease with increasing photon energy. After a short time of the order of 10^{-15} s, the core-hole is filled by an electron from a higher energy state. The corresponding energy difference is released mainly via fluorescence X-ray or Auger electron emission [4].

1.1.2 Absorption Fine Structure

According to quantum mechanical perturbation theory, the transition rate between the core level and the final state is proportional to the product of the squared modulus of the matrix element M and the density of states ρ

$$\mu \propto |M|^2 \rho \propto |\langle f|H_p|i\rangle|^2 \rho \tag{1.2}$$

$|i\rangle$ and $|f\rangle$ denote the initial and final state, respectively, and H_p represents the interaction Hamiltonian that causes the transition, here the electromagnetic field of the X-ray photon [2]. Both factors can now cause a modulation of the absorption

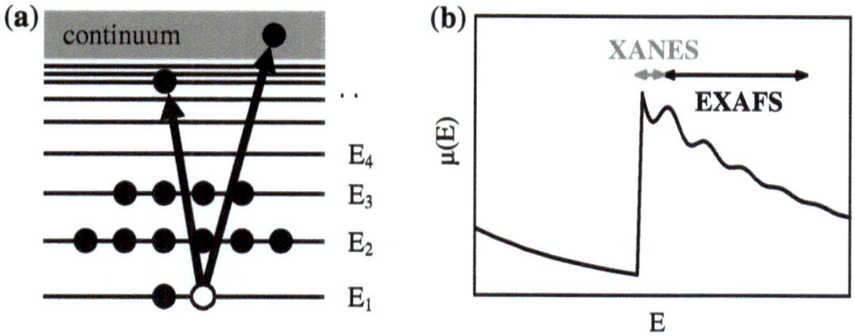

Fig. 1.2 **a** Schematic of the absorption process and **b** absorption coefficient $\mu(E)$ versus photon energy E including the fine structure above the edge divided into the XANES and EXAFS regions

Fig. 1.3 Schematic showing the absorbing atom (*yellow*) and its first nearest neighbors (*blue*). An interference pattern is created by the outgoing (*solid orange lines*) and reflected (*dashed blue lines*) photoelectron waves

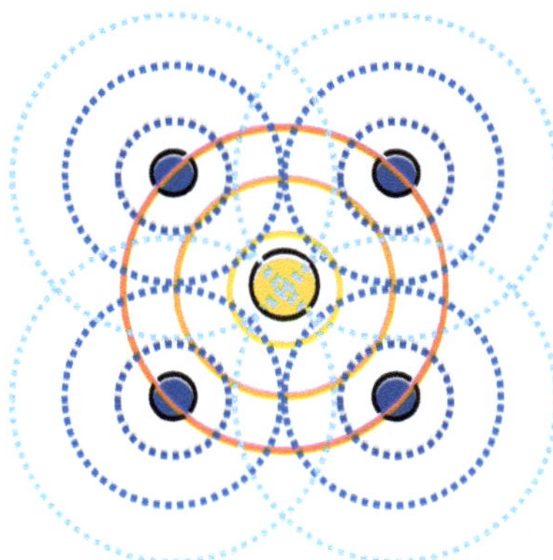

coefficient thus creating the X-ray absorption fine structure (XAFS). At the smallest X-ray energies for which the photon can be absorbed, the photoelectron will be excited to unoccupied bound states of the absorbing atom as shown schematically in Fig. 1.2a. This can lead to a strong increase of the absorption coefficient at particular X-ray energies corresponding to the energy difference between the core level and the unoccupied states. For higher X-ray energies, the photoelectron is promoted to a free or continuum state. The wave thus created propagates outwards and is scattered at neighboring atoms [7] as shown schematically in Fig. 1.3. The outgoing and scattered waves interfere in a manner that depends on the geometry of the absorber environment and on the photoelectron wavelength. The latter is inversely proportional to the photoelectron momentum and therefore changes with photon energy. Thus, the final state is an energy-dependent superposition of outgoing and scattered waves. Because the initial state is highly localized at the absorbing atom, the matrix element M in (1.2) depends on the magnitude of the final state wave function at the site of the absorbing atom. Constructive or destructive interference of outgoing and scattered waves thus increases or decreases the absorption probability, creating an energy-dependent fine structure of the absorption coefficient. Figure 1.2b schematically shows the $\mu(E)$ fine structure as a function of photon energy. Two regions are commonly distinguished, namely the X-ray absorption near edge structure (XANES) and the extended X-ray absorption fine structure (EXAFS).

1.1.2.1 XANES

The region very close to the absorption edge is characterized by transitions of the photoelectron to unoccupied bound states. XANES is therefore sensitive to the chemical bonding, exhibiting for example characteristic features for different oxidation states of the absorbing atom [4]. The XANES features are also influenced by strong multiple scattering effects which depend on the three-dimensional geometry of the crystal structure. This provides a means of distinguishing between different crystal phases [2]. Theoretical calculations of the fine structure in this region are complex and the accuracy of such simulations is still limited although significant progress has been made over recent years [8, 9]. Therefore, analysis typically compares the measured spectra to those of known standards and quantifies the ratios by which these standards are present in the sample using linear combination fitting. Often, the XANES region is also referred to as the near edge X-ray absorption fine structure (NEXAFS).

1.1.2.2 EXAFS

For photon energies higher than ~30 eV above the edge, the photoelectron is promoted to a free or continuum state. EXAFS is thus independent of chemical bonding and depends on the atomic arrangement around the absorber. It contains information about the coordination number, interatomic distances and structural and thermal disorder around a particular atomic species [7]. EXAFS does not require long-range order and is applicable to a wide range of ordered and disordered materials therefore providing a powerful tool for structural analysis. Theoretical calculations of the fine structure in the EXAFS region have also improved enormously during the last two decades and simulations with sufficient accuracy are now available [7, 9]. Nevertheless, the measurement of suitable standards still constitutes an important part of the experimental procedure.

1.2 Theoretical Description

1.2.1 Dipole Approximation

The Hamiltonian H_p in (1.2) describes the interaction of the electromagnetic field of the X-ray photon with the absorbing atom. It is proportional to the scalar product of the vector potential $\vec{A}$ of the X-ray field and the electron momentum operator

$\vec{p}$, $H_p \propto \vec{A} \cdot \vec{p}$. In principle, this is a many-body problem where all electrons of the absorbing atom would have to be considered. Practically, however, it is usually assumed that only one electron is involved in the transition and corrections due to many-body effects are added at a later stage. Using this one-electron approximation together with the dipole approximation for $\vec{A} \cdot \vec{p}$ yields

$$\mu \propto |\langle f|\hat{\epsilon} \cdot \vec{r}|i\rangle|^2 \rho \tag{1.3}$$

where $\hat{\epsilon}$ denotes the X-ray polarization vector [2, 5]. In most cases the dipole approximation is sufficient, however, quadrupole interactions may become important for high Z elements and L-edges [2].

Usually, synchrotron radiation is linearly polarized in the horizontal plane [2]. The matrix element in (1.3) therefore depends on the orientation of the line connecting absorber and scattering atom with respect to the X-ray polarization. In randomly oriented samples or in materials with cubic symmetry this angular dependence is averaged out. In contrast, the orientation dependence must be taken into account for single crystals or samples with a preferred particle or grain orientation. If unwanted, this X-ray linear dichroism can be averaged out experimentally by magic angle spinning of the sample [2]. It can, however, also be used intentionally as an additional source of information by performing systematic angle-dependent XAS measurements.

The matrix element in (1.3) is further subject to the well-known selection rules for transitions induced by electromagnetic radiation, i.e. $\Delta l = \pm 1$ and $\Delta m = 0, \pm 1$ for electric dipole interactions. Here, l and m denote the orbital angular momentum quantum number and its projection on the quantization axis, respectively [5]. The initial core state of the electron is to a good approximation given by an atomic-like state with well-defined quantum numbers l and m. In contrast, the final state is usually a superposition of wavefunctions with different values of l and m and only the fraction with the appropriate symmetry is of relevance for the transition [2]. Thus, for K- and L_1-edges (s states with $l = 0$) transitions occur only to final states containing p symmetry while for L_2- and L_3-edges (p states with $l = 1$) transitions are only allowed to final states containing s or d symmetry.

1.2.2 Quasi-Particle Model

While the initial state is well approximated by an atomic-like state, the final state is an excited state characterized by the presence of a core-hole ('final state rule'). In the quasi-particle model these final states Ψ_f are eigenstates of a Dyson equation[1]

$$h'\Psi_f = \left[\frac{p^2}{2m} + V' + \Sigma(E_f)\right]\Psi_f = E_f\Psi_f \tag{1.4}$$

[1] The analog of the Schrödinger equation for excited states.

where E_f denotes the energy of the photoelectron in the final state. The non-Hermitian Hamiltonian h' of the final state is characterized by the Coulomb potential V' calculated in the presence of a screened core-hole and by the complex valued and energy-dependent self-energy $\Sigma(E_f)$ which incorporates many-body effects and extrinsic inelastic losses [8]. The latter refer to losses during the propagation of the photoelectron and include excitations such as plasmons or electron-hole pairs and inelastic scattering in which the photoelectron loses energy [7]. The non-hermicity of h' corresponds to the complex nature of the eigenvalues E_f and is responsible for the finite lifetime of the final state [2]. Relativistic effects become important in the treatment of the initial atomic core states, especially for high elements, but have only weak effects on the propagation and scattering of the photoelectron in the final state [9].

1.2.3 Multiple Scattering Approach

The multiple scattering approach now separates the potential in (1.4) into individual contributions $v_{\vec{R}}$ localized at each atomic site $\vec{R}$ [8]

$$V' + \Sigma\left(E_f\right) = \sum_{\vec{R}} v_{\vec{R}}\left(\vec{r} - \vec{R}\right) \tag{1.5}$$

For electrons with energies of several eV or more above the threshold, the scattering depends mostly on the potential in the core of the neighboring atom which is approximately spherical [9]. The "muffin-tin" approximation therefore assumes spherically symmetric atomic potentials out to a finite radius and a constant potential in between the atoms. The approximation is a good description for close-packed structures but works less well for open structures. Deviations are most prominent for small anisotropic systems close to the absorption threshold [2, 9].

Despite this approximation, the calculation of final states turns out to be computationally demanding and very often impractical. The multiple scattering approach therefore makes use of the photoelectron Green's function or propagator $\boldsymbol{G}$ in real space. Applying the identity

$$-\frac{1}{\pi}\mathrm{Im}\,\boldsymbol{G} = \sum_f |f\rangle\, \delta\left(E + E_i - E_f\right)\, \langle f| \tag{1.6}$$

where E and E_i denote photon energy and electron energy in the initial state, respectively, (1.3) can be written as [2]

$$\begin{aligned}\mu &\propto \sum_f \langle i|\hat{\epsilon}\cdot\vec{r}'|f\rangle\, \delta\left(E + E_i - E_f\right) \langle f|\hat{\epsilon}\cdot\vec{r}|i\rangle \\ &\propto \mathrm{Im}\langle i|\hat{\epsilon}\cdot\vec{r}'\,\boldsymbol{G}\,\hat{\epsilon}\cdot\vec{r}|i\rangle\end{aligned} \tag{1.7}$$

The propagator $\boldsymbol{G}$ can be separated into a contribution $\boldsymbol{G}^c$ stemming from the central atom and a contribution $\boldsymbol{G}^{sc}$ due to multiple scattering from the environment, $\boldsymbol{G} = \boldsymbol{G}^c + \boldsymbol{G}^{sc}$. The nature of these contributions then allows expressing μ in terms of an atomic background μ_0 of the embedded absorber and the fine structure χ due to multiple scattering from the environment, $\mu = \mu_0(1+\chi)$ [7]. Within this framework, the fine structure component is now given by

$$\chi = \mathrm{Im}\, e^{i\delta} \left[\boldsymbol{1} - \boldsymbol{G}^0 \boldsymbol{T}\right]^{-1} \boldsymbol{G}^0 e^{i\delta'} \tag{1.8}$$

where $\boldsymbol{G}^0$ denotes the free particle propagator and $\boldsymbol{T}$ represents the scattering matrix while δ and δ' are partial-wave phase shifts [7, 8]. The matrix term in (1.8) can be written as a series expansion

$$\left[\boldsymbol{1} - \boldsymbol{G}^0 \boldsymbol{T}\right]^{-1} \boldsymbol{G}^0 = \boldsymbol{G}^0 \boldsymbol{T} \boldsymbol{G}^0 + \boldsymbol{G}^0 \boldsymbol{T} \boldsymbol{G}^0 \boldsymbol{T} \boldsymbol{G}^0 + \ldots \tag{1.9}$$

where the first term is missing due to the definition of $\boldsymbol{G}^0$ [7]. The fine structure contribution can thus be understood as the sum of individual scattering contributions arising from all possible paths of the photoelectron from the absorbing atom and back. The first, second, ... term in (1.9) correspond to single, double, ... scattering at surrounding atoms. The advantages of this multiple scattering Green's function formalism lie in the fact that it treats XANES and EXAFS within the same unified theory, that it avoids explicit calculation of the final state wave functions and that it naturally incorporates inelastic losses and other quasi-particle effects [7]. As an alternative to the path expansion, the fine structure contribution can also be expressed as the sum of irreducible n-body interactions which contain all scattering contributions due to a particular arrangement of n atoms including the absorber [10, 11]. This approach is directly related to the n-body distribution functions and is thus particularly suited for the study of highly disordered systems (see Chap. 9).

1.2.4 XANES

In the XANES region, the multiple scattering path expansion of (1.9) only converges satisfyingly for a few cases typically characterized by short core-hole lifetimes as given for the absorption by deep core electrons in high Z elements [8]. In most cases, however, convergence is poor and the multiple scattering expansion has to be carried out to very high or full order. In principle, this can be done by explicit matrix inversion of (1.8). Unfortunately, such a procedure is computationally very demanding and fast parallel Lanczos algorithms have been proposed and implemented to speed up calculations [8].

Another limitation of the current multiple scattering approach is given by the muffin-tin approach for the scattering potentials. This approximation usually works well for sufficiently high photoelectron energies as given in the EXAFS region. In contrast, the photoelectron energies in the XANES region are small enough for the scattering to become sensitive to the details of the surrounding potentials. To avoid this limitation, several full potential approaches have been reported ([2, 5, 8] and references therein). Band structure calculations based on ground state density functional theory can also predict the properties of low energy excited states, however, self-energy effects are typically neglected. Core-hole effects can be included by a super-cell approach leading to significant improvements of the calculated spectra. Several full multiple scattering cluster methods represent approaches intermediate between band structure calculations and path expansion and have been used for a variety of XANES calculations ([7] and references therein).

Comparison of experimentally determined spectra with ab initio calculations and even structural fitting of XANES data, especially for small molecules and clusters, have made tremendous progress in recent years. Nevertheless, theoretical calculations are still less mature and satisfying than in the EXAFS region. However, given that XANES is sensitive to both the three-dimensional atomic arrangement and the density of unoccupied states, improving its theoretical description is a field of much current effort and further progress can be expected in the near future.

1.2.5 EXAFS

1.2.5.1 EXAFS Equation

The EXAFS is expressed in terms of the fine structure contribution

$$\chi\left(E\right)=\frac{\mu\left(E\right)-\mu_0\left(E\right)}{\mu_0\left(E\right)}\sim\frac{\mu\left(E\right)-\mu_0\left(E\right)}{\Delta\mu_0} \tag{1.10}$$

where the energy-dependent denominator is approximated by a constant typically chosen as the height of the absorption edge, $\Delta\mu_0 = \mu_0(E_0)$ with E_0 being the energy of the absorption threshold. Instead of using $\chi(E)$, the fine structure is usually written as a function of the photoelectron wave number $k = \sqrt{2m_e(E - E_0)/\hbar^2}$, where m_e stands for the electron mass and $\hbar$ denotes Planck's constant divided by 2π. Using the *multiple scattering* path expansion described in Sect. 1.2.3, the fine structure contribution can be expressed as a sum over the scattering contributions arising from the various different paths

$$\chi\left(k\right)=\sum_j S_0^2 N_j\frac{\left|f_j(k)\right|}{kR_j^2}e^{-2R_j/\lambda(k)}e^{-2\sigma_j^2k^2} \times \sin\left[2kR_j+2\delta_c\left(k\right)+\delta_j(k)\right] \tag{1.11}$$

Paths with the same kinds of scattering atoms and a similar path length have been grouped under the index j. Equation (1.11) thus directly relates the EXAFS signal to the structural parameters N_j, R_j, and σ_j^2 which represent the number of such similar paths, the mean path length divided by two and the variation of all path lengths with index j, respectively. $f_j(k) = |f_j(k)|e^{i\delta_j(k)}$ represents the complex scattering amplitude while $\delta_c(k)$ stands for the phase shift experienced by the photoelectron wave in the potential of the absorbing atom. $\lambda(k)$ and S_0^2 denote the energy-dependent mean free path of the electron and the amplitude reduction factor, respectively.

Except for the factor S_0^2, (1.11) was first derived by Sayers, Stern, and Lytle for *single scattering* paths using the plane-wave approximation [12]. It assumes that the distance between the absorber-backscatter pair is sufficiently large to treat the outgoing spherical wave as a plane wave once it reaches the backscattering atom. For *single scattering* events, all paths involving the same kind of scattering atom in the same coordination shell around the absorber are grouped together. The structural parameters N_j, R_j, and σ_j^2 then represent the coordination number, the mean value, and the variance of the corresponding absorber-scatterer distance distribution, respectively. In case of the first nearest neighbor shell, absorbing and scattering atoms are usually connected by a real physical bond, and R_j and σ_j^2 signify the mean value and variance of the bond length distribution. Equation (1.11) has become known as the 'standard EXAFS equation' and has founded the application of XAS as a tool for structural analysis.

For an accurate calculation of the fine structure contribution, however, multiple scattering paths, curved-wave effects and many-body interactions must be taken into account [7]. Nevertheless, $\chi(k)$ can still be expressed in the same form as the original EXAFS equation. This provides a convenient parameterization of the absorber environment in terms of structural parameters for single and multiple scattering paths. The other quantities of (1.11) implicitly contain the curved-wave and many-body effects of modern XAS theory as discussed below. The key features of the EXAFS equation are as follows:

(i) As described in Sect. 1.1.2, the interference pattern depends on the photoelectron energy or wave number and on the distance between the absorbing and scattering atoms. This is given by the $\sin[2kR_j]$ term which causes the oscillatory nature of the fine structure contribution.

(ii) The strength of the scattering and thus the magnitude of the EXAFS depend on the number and type of the scattering atoms, represented by the coordination number or degeneracy of paths N_j and the modulus of the scattering amplitude $|f_j(k)|$, respectively. Modern XAS theory replaces the original plane-wave scattering amplitude by an effective curved-wave scattering amplitude for either single or multiple scattering events. Apart from the dependence on k, the effective scattering amplitudes are also characterized by a weak dependence on r [2].

(iii) The potential of the absorbing or scattering atom leads to a phase shift of the photoelectron wave represented by $\delta_c(k)$ and $\delta_j(k)$, respectively. The absorber potential acts twice on the photoelectron wave, once on the way out and once on the way back. The resulting term $2\delta_c(k)+\delta_j(k)$ appears in the sine function of (1.11).

(iv) The atoms in a particular coordination shell do not have exactly the same distance from the absorber. Differences are caused either by thermal vibrations (thermal disorder) or by structural variations in the interatomic distances (static disorder) and smear out the oscillations with increasing k. The phase difference in scattered waves due to a given difference in R_j increases with increasing k. This yields increased damping of the EXAFS at high wave numbers. In systems where the distance distributions exhibit only small asymmetry, Gaussian distributions with variance σ_j can be assumed. The term $e^{-2\sigma_j^2k^2}$ in (1.11) then accounts for the k-dependent damping of the EXAFS oscillations. In analogy to X-ray diffraction, σ_j^2 is often called the EXAFS Debye-Waller factor.

(v) The range that is probed by EXAFS is usually of the order of ten angstroms and is limited by the finite lifetime of the core hole and the finite mean free path of the photoelectron. The core-hole is eventually filled with an electron from a higher shell thereby emitting a fluorescence X-ray or an Auger electron while the photoelectron undergoes inelastic interactions with the surrounding material such as inelastic scattering and electron or plasmon excitation (extrinsic losses). The term $e^{-2R_j/\lambda(k)}$ in (1.11) accounts for the increasing decay of the photoelectron wave with increasing distance R_j. The mean free path is approximately given by $\lambda(k) \sim k/(|\mathrm{Im}\,\Sigma|+\Gamma/2)$ and thus comprises the finite core-hole lifetime Γ and extrinsic losses calculated in terms of the complex self-energy Σ [8]. The damping of the fine structure contribution by both $\lambda(k)$ and σ_j^2 makes EXAFS a local probe and ensures the convergence of the multiple scattering path expansion in this regime.

(vi) The one-electron approximation assumes that only a single electron participates in the absorption process. In reality, however, this is a many-body process and relaxation of the system in response to the sudden creation of the core-hole reduces the fine structure component. The corresponding amplitude reduction factor S_0^2 is, in principle, weakly energy dependent, particularly close to the absorption threshold. In the EXAFS region it can be taken as a constant to good approximation [8, 10].

1.2.5.2 Configurational Average

As already mentioned above, the sample contains an ensemble of slightly different atomic environments due to thermal and structural disorder. The EXAFS equation discussed above assumes Gaussian distance distributions with mean values R_j and standard deviations σ_j. For systems with significant asymmetry, a Gaussian function may no longer be an adequate approximation for the distance distribution and higher moments must be considered. EXAFS analysis based on a cumulant expansion was first proposed by Bunker [13]. Scattering from a coordination shell with a distance distribution $\rho(R)$ yields

$$\chi(k) = S_0^2 N \frac{|f(k)|}{k} \int_0^\infty \frac{e^{-2R/\lambda(k)}}{R^2} \sin\left[2kR + \delta(k)\right] \rho(R)\,\mathrm{d}R$$

$$= \mathrm{Re}\left[\frac{1}{i} S_0^2 N \frac{|f(k)|}{k} e^{i\delta(k)} \int_0^\infty \frac{e^{-2R/\lambda(k)}}{R^2} e^{2ikR} \rho(R)\,\mathrm{d}R\right] \tag{1.12}$$

for each term j in (1.11) with $\delta(k) = 2\delta_c(k) + \delta_j(k)$. Assuming a narrow distribution with the mean value $\langle R\rangle$, the integral can be approximated by

$$\int_0^\infty \frac{e^{-2R/\lambda(k)}}{R^2} e^{2ikR} \rho(R)\,\mathrm{d}R \sim \frac{e^{-2\langle R\rangle/\lambda(k)}}{\langle R\rangle^2} \int_0^\infty e^{2ikR} \rho(R)\,\mathrm{d}R \tag{1.13}$$

where the integral on the right hand side corresponds to a Fourier transformation of the distance distribution and is called the characteristic function. It can be expressed in terms of the raw moments p_n or the cumulants C_n of the distance distribution

$$\int_0^\infty e^{2ikR} \rho(R)\,\mathrm{d}R = \sum_{n=0}^\infty \frac{(2ik)^n}{n!} p_n = \exp\left[\sum_{n=0}^\infty \frac{(2ik)^n}{n!} C_n\right] \tag{1.14}$$

Although mathematically equivalent, the cumulant expansion converges much more rapidly than the expansion in terms of raw moments. Developing the sum in (1.14), using $C_0 = 0$ for normalized distributions, and neglecting cumulants higher than the fourth, (1.12) can be written as

$$\chi(k) = S_0^2 N \frac{|f(k)|}{k} \frac{e^{-2\langle R\rangle/\lambda(k)}}{\langle R\rangle^2} e^{\left(-2k^2 C_2 + \frac{2}{3} k^4 C_4\right)}$$
$$\times \sin\left[2kC_1 - \frac{4}{3} k^3 C_3 + \delta(k)\right] \tag{1.15}$$

This expression now allows the EXAFS to be analyzed in terms of the cumulants of the interatomic distance distribution: the first cumulant $C_1 = \langle R\rangle = R_j$ corresponding to the mean value, the second cumulant $C_2 = \sigma_j^2$ representing the variance and the third and fourth cumulants, C_3 and C_4, denoting asymmetric and symmetric deviations from a Gaussian profile, respectively. For very small or Gaussian disorder, only the first two cumulants have to be considered and (1.15) corresponds to (1.11).

EXAFS samples the one-dimensional distance distribution $\rho(R)$ that results from an angular average over the three-dimensional distribution $\rho(\vec{R})$. Therefore, care has to be taken when comparing the cumulants of $\rho(R)$ with the three-dimensional motion of the atoms or with structural parameters determined by other techniques such as X-ray diffraction (see Chap. 6).

For highly disordered systems, cumulants higher than the fourth may be necessary to adequately represent the distance distribution. In this case, the cumulant expansion is no longer a suitable description of the EXAFS and a different approach such as the one based on expressing the fine structure in terms of irreducible n-body interactions is more appropriate (see Chap. 9).

1.3 Experimental Aspects

1.3.1 Synchrotron Radiation

Most XAS experiments are performed at synchrotron sources due to the requirement of high X-ray intensities and a continuous energy spectrum [2]. Figure 1.4 shows the basic design of a modern synchrotron. Electrons produced in the source are first accelerated in a linear accelerator before their energy is further increased in the booster ring. From here they are transferred to the storage ring where they circulate over a million times each second, creating intensive electromagnetic radiation. Beamlines deliver the radiation to a number of end stations where it can be used for a variety of experimental techniques.

When a charged particle traverses a magnetic field it is forced to change its direction of motion thereby emitting electromagnetic radiation. In a synchrotron, the electrons are forced around the storage ring by a series of bending magnets [2]. The radiation thus created is characterized by a continuous energy spectrum over a wide range of wavelengths (from infrared to hard X-rays), high intensity, strong polarization and a pulsed nature. Modern synchrotron facilities also have additional elements, so-called insertion devices, placed in the straight sections between the

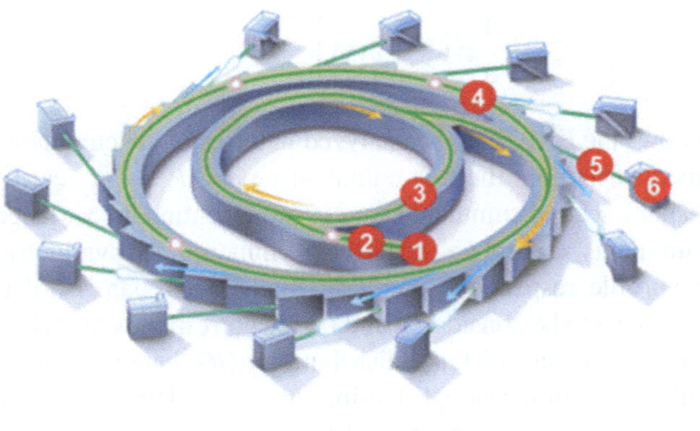

Fig. 1.4 Design of a modern synchrotron facility [14]

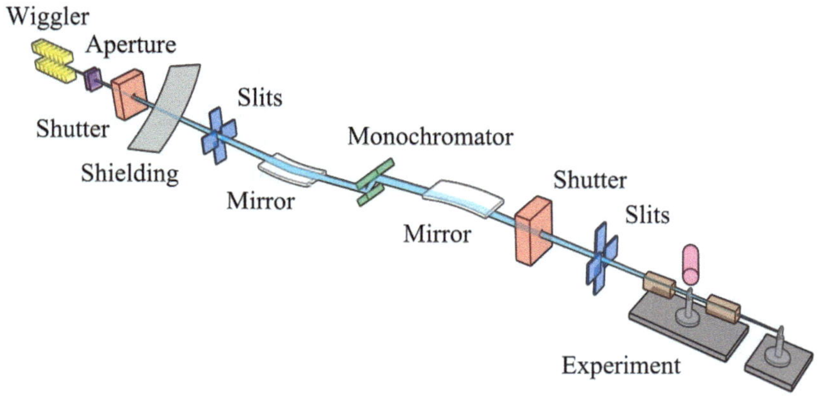

Fig. 1.5 Typical components of a modern XAS beamline [14]

bending magnets [2]. These devices constitute a series of alternating magnetic fields that force the electron beam to perform either strong (wiggler) or gentle (undulator) oscillations. The wiggler emits a broad beam of incoherent radiation characterized by increased intensity and a continuous energy spectrum extending to much higher X-ray energies compared to a bending magnet. The undulator emits a narrow beam of coherent radiation the intensity of which is amplified up to 10,000 times but only at certain energies. Based on these characteristics, the source (bending magnet, wiggler or undulator) best suited for a particular experimental technique is chosen.

Each beamline is usually configured to meet the requirements of a particular experimental technique [2]. Figure 1.5 depicts the various components of a typical modern XAS beamline. Mirrors are used to collimate and focus the beam while apertures and slits define its size. A double crystal monochromator is used to select X-rays of a very narrow energy band using the criterion for Bragg diffraction, $n\lambda = 2d \sin \theta$. Here, n is an integer, λ denotes the X-ray wavelength, d stands for the lattice spacing of the diffracting crystal and θ represents the angle under which the beam is incident on the crystal. Energies that satisfy the Bragg condition with $n \geq 2$ are called "harmonics" and have to be removed from the beam. This can be achieved by slightly detuning the monochromator which decreases the transmission of harmonics significantly more than that of the primary energy. Alternatively, X-ray mirrors can be used that only reflect energies below a critical value. With such an experimental arrangement, the absorption coefficient can be measured as a function of X-ray energy.

1.3.2 Experimental Setup

In general, the absorption coefficient can be detected either directly by measuring the intensities of incoming and transmitted beam (transmission mode) or indirectly by

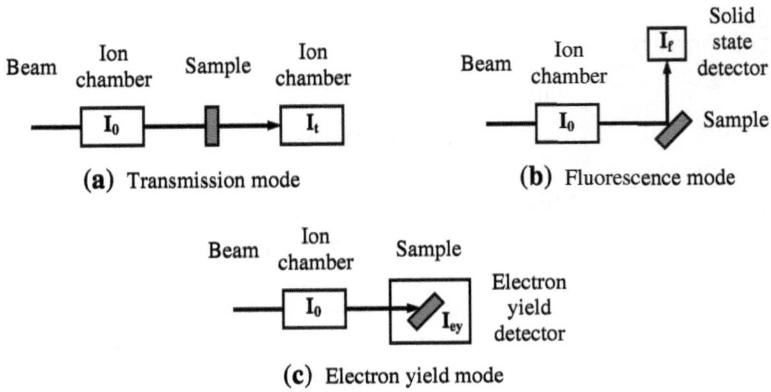

Fig. 1.6 Schematic of the experimental setup for the different XAS detection modes

measuring the intensity of the incoming beam and of the decay products such as fluorescent X-rays or Auger electrons (fluorescence or electron yield mode). The experimental setup for all three cases is shown schematically in Fig. 1.6 [2]. Alternatively, X-ray excited optical luminescence (XEOL) measures electromagnetic emissions in the optical range following X-ray absorption and is described in Chaps. 11 and 13.

1.3.2.1 Transmission Mode

In transmission mode both the incoming and the transmitted beam, I_0 and I_t, respectively, are measured by ion chambers and the absorption coefficient can be obtained according to (1.1) [2, 4]. Detecting the voltage or current generated in the ion chamber counting chain is inherently simpler than detecting single photons with a solid state detector often used for fluorescence measurements (see below). Using the same type of detector for I_0 and I_t also has the advantage of having a common energy dependence. However, transmission measurements require concentrated samples such that the difference between I_0 and I_t is significantly larger than the variation due to counting statistics. Furthermore, samples must be highly homogeneous, of constant thickness and free of pinholes. One means to prepare a sample that satisfies these requirements is to crush up an appropriate amount of material and mix it with a suitable binder such as boron nitride or cellulose. Once a fine, homogeneous powder is obtained, it is compacted into the small hole of a sample holder or pressed into a pellet and sealed on both sides with Kapton tape.

1.3.2.2 Fluorescence Mode

The intensity of the incoming beam I_0 is again measured by an ion chamber while the intensity of the characteristic fluorescence X-rays is usually detected by an energy-dispersive Si or Ge solid state detector [2]. The intensity of this fluorescence line is proportional to the absorption caused by the element under investigation, however, the relation is more complicated than for transmission measurements. So-called "self-absorption" effects must be taken into account especially for thick or concentrated samples [15]. Furthermore, detecting the fluorescence signal is also more complicated than measuring the transmitted intensity since the characteristic X-rays have to be isolated from other X-rays, particularly the elastically scattered beam itself. While Si and Ge solid state detectors are capable of the required energy discrimination they suffer from limited count rates. To improve the signal-to-noise ratio several independent detectors are coupled in an array to form multiple-element detectors with up to 100 channels [14]. Nevertheless, the big advantage of the fluorescence mode is the ability to study samples not suitable for measurement in transmission mode such as highly dilute and non-homogeneous samples.

1.3.2.3 Electron Yield Mode

Instead of detecting the fluorescent X-rays, one can also measure the electrons emitted from the sample such as the photoelectrons themselves, secondary electrons and Auger electrons. To that end the sample is situated inside the detector and the electrons are collected by suitable electrodes [2]. Given the relatively short mean free path of the electrons this technique is surface sensitive and does not suffer from the "self-absorption" effects mentioned above for the fluorescence mode [2, 4]. It is also beneficial for measuring samples in the soft X-ray regime where the filling of the core-hole is accompanied mainly by Auger electron production and only to a much lesser extent by fluorescence X-ray emission.

1.3.2.4 Specialized Experimental Techniques

A number of specialized experimental techniques have been developed to capitalize on additional physical effects or to accommodate the particular nature of a large variety of samples. Very thin films and nanostructures can be studied by grazing incidence measurements where the angle between the incoming X-ray beam and the sample surface is typically only a few degrees, equivalent to a few tens of milliradians [2, 5]. The penetration depth of the X-rays is then strongly reduced yielding a highly surface sensitive measurement. Usually, the emitted fluorescent X-rays are recorded with the detector placed at 45° with respect to the sample surface. By varying the angle of incidence, information about the absorption coefficient can be obtained for different depths. If the angle between the incoming beam and the sample surface is reduced below the critical angle of the material, typically a few milliradians, total external reflection will occur. The X-ray wave field is now confined to the immediate surface and the penetration depth is in the order of some nanometers. Detecting the intensity of the reflected beam then provides information about the absorption coefficient. However, care has to be taken when using this approach as the reflectivity depends on both the real and the imaginary part of the index of refraction yielding a complicated relation between the measured signal and the absorption coefficient (see also Chap. 12).

Atomic-scale structural information of crystalline samples can also be obtained from X-ray diffraction anomalous fine structure (DAFS) studies where the intensity of a particular Bragg reflection is measured as a function of X-ray energy [16, 17]. Due to the causal relationship between the real and imaginary parts of the atomic scattering amplitude, the energy-dependent variation of the scattering signal contains the same structural information as the X-ray absorption fine structure. In contrast to XAS, DAFS can provide site or spatial selectivity if the element of interest occupies inequivalent sites that have different structure factor contributions or if different spatial regions of the sample produce diffraction peaks at separate locations in reciprocal space [16]. The Bragg reflections of a strained $In_xGa_{1-x}As$ layer, for example, differ from those of the GaAs substrate thus allowing the study of the Ga and As environments in the ternary thin film without interference from the substrate [17]. Extracting structural information from the DAFS signal is, however, more complicated than in the case of XAS and usually requires very thin samples [16, 17]. Examples for the application of DAFS can be found in Chaps. 10 and 13.

As described in Sect. 1.2.1, the transition matrix element and hence the absorption coefficient depend on the orientation of the absorbing and scattering atom pair with respect to the X-ray polarization. For single crystal samples, such as strained epitaxial thin films or oriented nanostructures, the recorded signal therefore depends on the angle between the sample normal and the X-ray beam. Performing angle-dependent measurements then yields information about the structural parameters parallel and perpendicular to the sample normal as discussed in Chaps. 3, 5, 13 and 16.

For some applications, such as time-resolved studies (see e.g. Chap. 13), very short scanning times are required. During conventional XAS experiments, the monochromator is set at a certain angle corresponding to a specific X-ray energy and the intensities of interest are recorded before the monochromator is moved to the next setting. The scanning speed is thus limited by the time needed to move and settle the monochromator and to measure the signal. In contrast, scanning times can be dramatically reduced by using a monochromator moving with constant velocity and fast data acquisition systems [18, 19]. A full scan performed with such a quickXAS setup may take tens of seconds compared to 30–60 minutes for conventional scans. Even faster measurements in the order of one second or less can be realized by energy dispersive XAS where a polychromatic beam is incident on the sample and then diverges onto a position sensitive detector such as a photodiode array [18, 19]. The spectrum is thus recorded simultaneously and the time needed is limited only by the response time of the detector and the number of scans required for a sufficient signal to noise ratio.

Recent progress in the field of X-ray micro beams also opens up new possibilities, particularly for the study of inhomogeneous or heterogeneous materials and nanostructures (see e.g. Chap. 13). Brilliant synchrotron sources and specialized beam line optics such as compound refractive lenses, Fresnel zone plates and Kirkpatrick-Baez mirrors now provide X-ray beams with spot sizes of the order of tens of nanometers [20]. Structural parameters can thus be studied with a sub-micron spatial resolution. Often these microXAS measurements are combined with other techniques such as compositional analysis using micro X-ray fluorescence to obtain a more comprehensive picture of the material investigated.

1.4 Data Analysis

There exists a variety of ways to analyze XAS data and a large number of codes and programs are available. In general, data from the XANES and EXAFS region are analyzed separately. This is partly due to the different information contained in both spectral regions and partly due to the fact that theoretical modeling for XANES is not yet as advanced as it is for EXAFS. Even with the progress made over recent years, simultaneous calculation of the fine structure in both regions is numerically impractical thus favoring a separate analysis.

An important point common to both spectral regions, however, is the calibration of the energy scale and the alignment of different spectra. The beamline monochromator is calibrated with a known reference, often a thin metal foil, at each absorption edge. Ideally, this reference is then measured simultaneously with each sample of interest. In transmission mode this can be easily achieved by placing the reference between the second and a third ion chamber that detect the X-ray intensity incident on and transmitted through the reference, respectively. These reference spectra can then be used to align the energy scales of the samples of interest. In fluorescence mode this approach is often not feasible due to a thick absorbing substrate of the sample. It is then important to check the stability of the energy calibration at suitable intervals. Correct alignment of the sample spectra is crucial for the determination of edge shifts in the XANES and for bond length determination from EXAFS. Furthermore, thickness and "self-absorption" effects distort the amplitude of the fine structure and must be avoided or corrected for a meaningful analysis.

1.4.1 XANES

The absorption edge is often comprised of intense absorption peaks (white lines) and pre-edge features [2]. The white line results from strong transitions to final states confined to the near vicinity of the absorbing atom. These can be partially filled or empty bound states, sensitive to chemical bonding and the oxidation state of the absorbing atom, or low energy continuum states confined by strong multiple scattering, sensitive to the three dimensional structure surrounding the absorber. Pre-edge features are typically taken as a sign of broken inversion symmetry and thus also provide structural information. Small energy shifts of the absorption edge itself can be caused by charge transfer between the absorber and the neighboring atoms. More pronounced shifts can result from changes in the first nearest neighbor bond lengths [2].

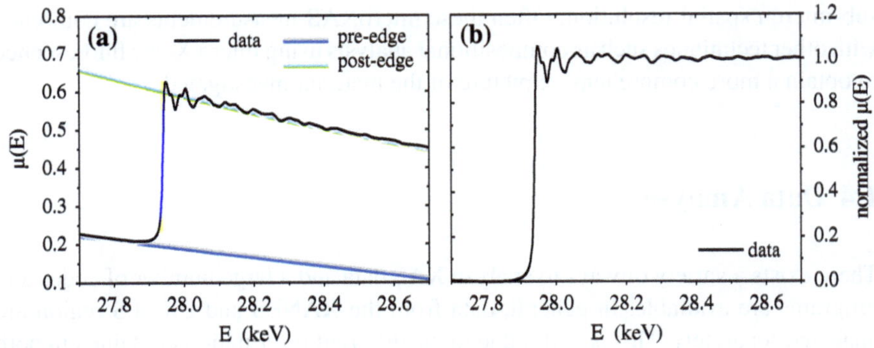

Fig. 1.7 **a** Raw $\mu(E)$ data measured at the In K-edge of crystalline InP together with the fitted pre-edge and post-edge lines. **b** Normalized $\mu(E)$ obtained from the spectrum in panel (*a*)

The first step in XANES analysis is the normalization of spectra [4]. To that end, the spectrum in the energy region below the absorption edge is fitted by a linear pre-edge line as shown in Fig. 1.7a. The spectrum well above the absorption edge is typically approximated by a quadratic post-edge line. The edge step $\Delta\mu_0$ is then obtained as the difference between pre-edge and post-edge lines at the absorption threshold E_0. A normalized XANES spectrum as shown in Fig. 1.7b is then obtained by subtracting the pre-edge line from the measured spectra over the whole energy range, dividing by the step height $\Delta\mu_0$ and flattening the spectra above threshold to account for the different slopes of pre-edge and post-edge lines. The resulting normalized spectrum equals zero below the edge, exhibits a step height of one and oscillates around this value for energies above E_0. Normalization thus removes effects of sample thickness and concentration and allows the direct comparison of different samples and measurements.

Comparing the spectra of the samples of interest with those of known standards already yields some qualitative assessment of the chemical and structural environment of the absorbing atom, provided the spectra of the standards are sufficiently different from each other. The analysis can be quantified by linear combination fitting. The spectrum of the sample of interest is modeled by weighting the spectrum of each known standard i with a factor f_i and adding them together

$$\mu_{calc} = \sum_i f_i \mu_i \tag{1.16}$$

The factors f_i represent the fraction with which each standard is present in the sample and are obtained by minimizing the difference between the calculated and measured spectrum. Ideally, they should sum to one. However, structural disorder can sometimes broaden the XANES features and the sum rule needs to be relaxed in these cases [4]. Linear combination fitting works well if the number of potential chemical and structural environments for the absorbing atom in the sample, and thus the number of standards, is small and if the spectra of these standards exhibit unique

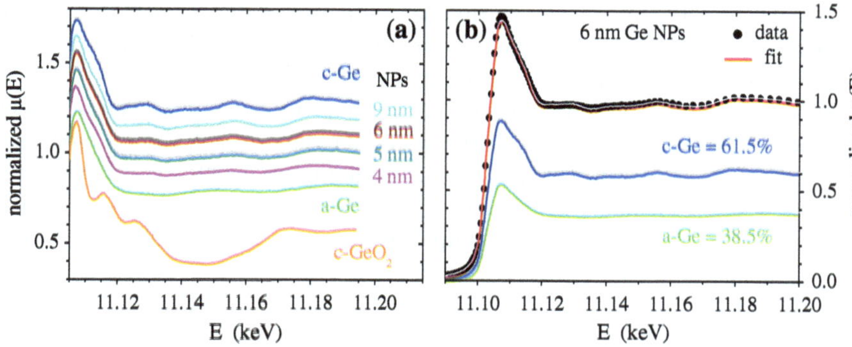

Fig. 1.8 **a** Ge K-edge XANES spectra for bulk crystalline Ge (c-Ge), amorphous Ge (a-Ge) and crystalline GeO_2 (c-GeO_2) compared with those for different sizes of Ge nanoparticles (NPs) embedded in a SiO_2 matrix. **b** Linear combination fitting of a nanoparticle spectrum with bulk crystalline and amorphous Ge standards [21]

features that allow to differentiate between them. Consequently, linear combination fitting is severely limited if a large number of possible standards has to be considered and if the spectra of different standards are very similar to each other.

As an example, Fig. 1.8a compares XANES spectra for bulk standards (crystalline Ge, amorphous Ge and crystalline GeO_2) with that for different sizes of Ge nanoparticles embedded in SiO_2 [21]. The features apparent in the spectra of the two crystalline standards stem from multiple scattering effects while, in contrast, the spectrum for the amorphous standard is effectively featureless. The latter demonstrates how the structural disorder inherent in the amorphous phase significantly impacts (diminishes) multiple scattering contributions. The nanoparticle spectra show no evidence of an oxide component and the crystalline-Ge-like features weaken as the nanoparticle size decreases. Figure 1.8b shows linear combination fitting of the spectrum for Ge nanoparticles of 6 nm diameter using the bulk crystalline and amorphous Ge standards. Clearly the XANES spectrum is well fitted with the given combination of standards. Within a Ge nanoparticle, Ge atoms are thus in either a crystalline or amorphous environment. This result was considered evidence for an amorphous-like Ge layer of constant thickness separating the crystalline Ge nanoparticle core and the amorphous SiO_2 matrix [21]. Ge nanoparticles are discussed in detail in Chap. 10.

A related technique is principal component analysis where a set of related samples of interest is analyzed in terms of characteristic principle components present in the samples. Interpretation of the results can, however, be difficult as the principal components are statistical abstractions and do not need to represent any physical standards [4].

Characteristic features such as white lines or pre-edge features can also be quantified by peak fitting procedures. To that end, the absorption edge itself is typically approximated by an arctangent function while the peaks are modeled by Lorentzian lines convoluted with a Gaussian function (Voigt function) to account for experimental broadening [2]. Such an analysis could yield, for example, the change in edge position related to a change in first nearest neighbor distances for a series of samples or the fraction of different oxidation states (characterized by different white line positions) of the absorbing atom within a sample [4].

With the progress of XAS theory and the improvement of numerical codes (see Sect. 1.2.4), comparison of calculated and measured spectra gains more and more importance as an analytical tool. Currently, the aim is mostly to refine the parameters of a single chemical and structural environment of the absorbing atom but characterization of multicomponent samples based on theoretical calculations may well become possible in the future.

1.4.2 EXAFS

The first step in EXAFS analysis is to isolate the fine structure $\chi(k)$ from the absorption background [4]. To that end, the spectrum is fitted with a pre-edge and post-edge line as already described in Sect. 1.4.1. The pre-edge line is again subtracted from the spectrum over the whole energy range. The absorption background $\mu_0(E)$ is then typically approximated by a spline function that approaches the post-edge line at energies well beyond the absorption edge as shown in Fig. 1.9a. The difference between the absorption coefficient $\mu(E)$ and the background $\mu_0(E)$ is normalized with respect to the step height $\Delta\mu_0$ yielding the fine structure $\chi(E)$. To convert the energy E to the photoelectron wave number $k = \sqrt{2m_e(E - E_0)/\hbar^2}$, the threshold energy E_0 is needed. Experimentally, E_0 is typically taken as the maximum of the derivative of $\mu(E)$ with respect to E or as the energy corresponding to the half of the step height [2]. This choice is somewhat arbitrary and the chosen E_0 may not be identical to the actual absorption threshold. This does not pose a major problem for the analysis, however, as long as E_0 is chosen consistently for all samples under investigation (see below). Figure 1.9b plots $\chi(k)$ isolated from the data shown in panel (a) and weighted with k^3 to emphasize the data at higher k. Depending on the sample and the absorber-backscatterer pair, different k-weights may be chosen.

The next step is usually to Fourier transform the data into R-space [2, 4]. Fourier transformation (FT) of the EXAFS provides a means to visualize different scattering contributions and is often used during analysis. The benefit of such a procedure was first shown by Sayers et al. [12]. When Fourier transformed, different scattering contributions with a large difference in R_j and small values for σ_j^2 produce well separated peaks with amplitudes approximately proportional to N_j/σ_j. However, as $\lambda(k)$, $|f_j(k)|$, and $\delta_j(k)$ are complicated functions of k, the FT of the EXAFS cannot be expressed in a simple analytical form and may differ significantly from the radial distribution function [1]. Furthermore, different k-weights might be used and a smooth window function is usually applied to account for the finite data range. Nevertheless, EXAFS results are typically presented as the magnitude of the FT due to the illustrative character of such a plot. Figure 1.9b shows a window function of the Hanning type and Fig. 1.9c plots the magnitude of the resulting FT. Distinct peaks due to different scattering contributions are readily apparent. The first peak,

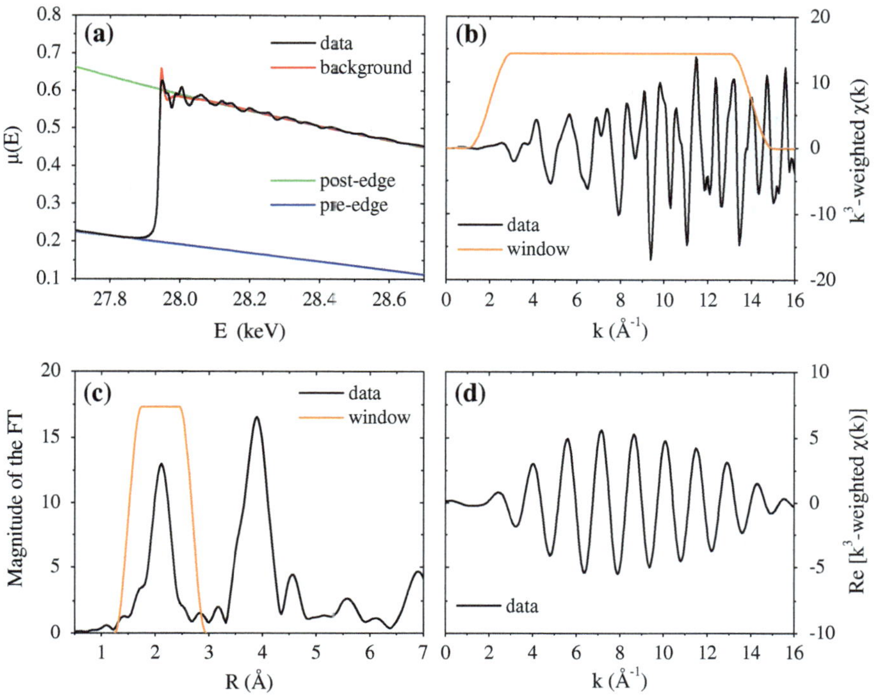

Fig. 1.9 **a** Raw $\mu(E)$ data measured at the In K-edge of crystalline InP together with the fitted pre-edge and post-edge lines and the background function. **b** k^3-weighted $\chi(k)$ obtained after background removal and conversion from energy scale to photoelectron wave number scale. **c** Magnitude of the Fourier transformation (FT) obtained from the k^3-weighted $\chi(k)$ using the smooth Hanning window plotted in panel (*b*). **d** Real part of the back-transformed data for the window plotted in panel (*c*) selecting the first nearest neighbor scattering peak

corresponding to scattering of the photoelectron wave at first nearest neighbor atoms, is usually well isolated. In contrast, the scattering contributions from higher coordination shells often overlap and a complicated peak structure may result. Furthermore, one should remember that the FT is a complex function and both magnitude and phase (or alternatively, real and imaginary part) have to be considered for the full information content.

A back-transformation can be used to isolate different scattering contributions if their signals are well separated in R-space. This methodology has been extensively used for the analysis of first nearest neighbor scattering by the Ratio Method (see below). However, it usually fails for higher coordination shells due to the overlap of different scattering contributions. Figure 1.9c shows a Hanning window selecting only the first nearest neighbor scattering peak. The real part of the resulting back-transformed k-spectra is plotted in Fig. 1.9d.

1.4.2.1 Ratio Method

Analysis using the Ratio Method is based on the EXAFS equation and the cumulant expansion and yields the differences in structural parameters between the sample of interest and a known reference. Expressing the fine structure of a single scattering contribution in terms of its amplitude $A(k)$ and phase $\Phi(k)$, $\chi(k) = A(k)\sin\Phi(k)$, (1.12) yields

$$A(k) = S_0^2 N \frac{|f(k)|}{k} \left| \int_0^\infty \frac{e^{-2R/\lambda(k)}}{R^2} e^{2ikR} \rho(R)\,\mathrm{d}R \right| \tag{1.17}$$

$$\Phi(k) = \delta(k) + \arg\left[\int_0^\infty \frac{e^{-2R/\lambda(k)}}{R^2} e^{2ikR} \rho(R)\,\mathrm{d}R \right] \tag{1.18}$$

The integral in both terms represents the FT of the *effective* distance distribution $\rho'(R) = \rho(R)e^{-2R/\lambda(k)}/R^2$ [13]. Using a cumulant expansion similar to (1.14), the amplitude and phase can be written as

$$A(k) = S_0^2 N \frac{|f(k)|}{k} \exp\left[C_0' - 2k^2 C_2' + \frac{2}{3} k^4 C_4' \right] \tag{1.19}$$

$$\Phi(k) = \delta(k) + 2kC_1' - \frac{4}{3} k^3 C_3' \tag{1.20}$$

where C_n' denotes the cumulants of the effective distribution and orders higher than $n = 4$ have been neglected. In contrast to C_0, C_0' does not vanish as the effective distribution is generally not normalized. It can be estimated by [22]

$$C_0' = -2\frac{C_1'}{\lambda(k)} - 2\ln C_1' \tag{1.21}$$

The relation between C_1' and C_1 is approximately given by [22]

$$C_1' = C_1 - 2\frac{C_2}{C_1}\left(1 + \frac{C_1}{\lambda(k)}\right) \tag{1.22}$$

Experimentally, $\lambda(k)$ is usually taken as a constant. For $n \geq 2$, the difference between the effective and the real cumulants is typically within the experimental uncertainty and $C_n' = C_n$ is assumed.

If the chemical and structural environment of the absorbing atom in the sample of interest (index s) and the reference (index r) is similar, such that $S_{0,s}^2 = S_{0,r}^2$ and

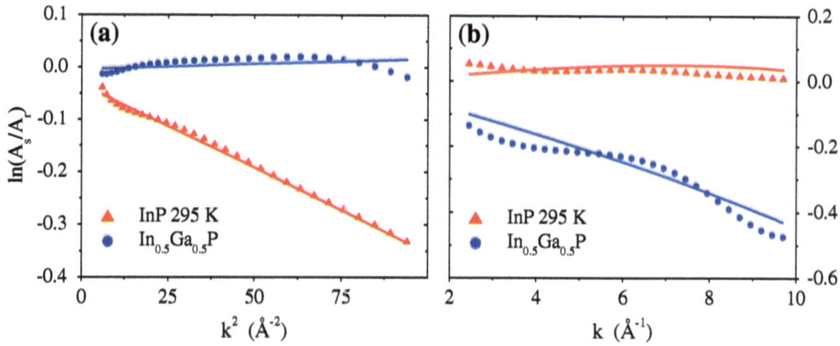

Fig. 1.10 Ratio method analysis of the samples InP measured at 295 K and $In_{0.5}Ga_{0.5}P$ measured at 20 K compared to the reference of InP measured at 20 K. **a** Logarithm of the amplitude ratio, in $|A_s(k)/A_r(k)|$, plotted versus k^2 and **b** phase difference $\Phi_s(k) - \Phi_r(k)$ plotted versus k where s and r denote the sample and the reference, respectively

$f_s(k) = f_r(k)$, comparison of amplitude and phase yields

$$\ln\left|\frac{A_s(k)}{A_r(k)}\right| = \ln\frac{N_s}{N_r} + \Delta C_0' - 2k^2\Delta C_2' + \frac{2}{3}k^4\Delta C_4' \tag{1.23}$$

$$\Phi_s(k) - \Phi_r(k) = 2k\Delta C_1' - \frac{4}{3}k^3\Delta C_3' \tag{1.24}$$

These differences $\Delta C_n' = C_{n,s}' - C_{n,r}'$ in the effective cumulants can then be determined from polynomial fits. As an example, Fig. 1.10a, b plot the logarithm of the amplitude ratio and the phase difference, respectively, obtained from a study of the In K-edge of InP measured at 295 K and $In_{0.5}Ga_{0.5}P$ measured at 20 K. The reference is in both cases InP measured at 20 K. The spectra were processed and Fourier transformed as described above yielding the amplitude and phase of the back-transformed first nearest neighbor scattering contribution. For InP measured at two different temperatures, the logarithm of the amplitude ratio is strongly sloped indicative of a significant increase of $C_2' = C_2$ for 295 K compared to 20 K. This increase is caused by increased thermal disorder yielding a larger variance of the distance distribution. In contrast, the slope is basically zero for $In_{0.5}Ga_{0.5}P$ demonstrating a similar width of the distance distribution in binary and ternary. Regarding the phase difference, only a small increase of C_1' is noted for InP measured at 295 K compared to InP measured at 20 K while the strong slope for $In_{0.5}Ga_{0.5}P$ results from a significant reduction of C_1' compared to InP. Increasing the measurement temperature thus only slightly increases the average In-P distance in InP whereas alloying with GaP significantly reduces the value of C_1 (see also Chaps. 2 and 6).

The advantage of the Ratio Method is that it is a model independent approach that does not assume any structure a priori. It analyzes the differences in structural parameters between a sample and a reference and thus avoids the need to explicitly know the scattering amplitudes and phase shifts. Due to this comparative nature, many experimental effects also cancel out. It does, however, assume that the chemical and structural environment of the absorbing atom is sufficiently similar in the sample and the reference. In many cases, this requirement is satisfied but there may be samples for which it is difficult to find a suitable reference. Another major limitation of the Ratio Method is the need to have scattering contributions well isolated in R-space which usually limits the analysis to the first coordination shell.

1.4.2.2 Path Fitting

Path fitting is another approach to EXAFS analysis provided for example by the IFEFFIT code [23] and the corresponding user interfaces ATHENA and ARTEMIS [24]. It is a model-dependent approach based on the cumulant expansion of the different single and multiple scattering paths and requires some pre-existing knowledge about the system under investigation. The analysis starts with a model structure that specifies the absorbing atom and the position and type of the surrounding atoms that are to be considered in the fitting procedure. Effective scattering amplitudes and phase shifts for the various single and multiple scattering paths are then calculated with the FEFF code [9]. The calculations are based on the multiple scattering approach described in Sect. 1.2.3 and include polarization dependence, curved-wave effects, core-hole effects and many-body interactions. Scattering paths are sorted with respect to their effective distance and their importance given by the scattering amplitude is listed. Usually, single scattering paths are stronger than multiple scattering paths of similar length. Nevertheless, some multiple scattering paths may be strong enough that an accurate representation of the experimental spectrum is only possible if they are considered in the fit. The selection of relevant paths to be included in the fitting procedure may vary depending on the system studied, the quality of the data and the aim of the investigation.

Figure 1.11a, b show the most prominent scattering paths calculated for absorption at the In K-edge of InP and $In_{0.5}Ga_{0.5}P$, respectively. For InP, the strongest multiple scattering (MS) path is also plotted leading from the In absorber to a second nearest neighbor In atom, then to a first nearest neighbor P atom and back to the absorber. For $In_{0.5}Ga_{0.5}P$, no multiple scattering paths were included in Fig. 1.11b for the sake of clarity. The complex Fourier transformed paths can add up constructively or destructively depending on their phases. As an example, the peak at $R \sim 4$ Å in Fig. 1.11a is smaller for the sum of all paths than for the second nearest neighbor In path alone, highlighting the importance of considering the full information content of the FT.

The structural parameters of the selected paths are refined in a least-squares fit to the processed experimental data. The first four cumulants $C_1 = \langle R \rangle$, $C_2 = \sigma^2$, C_3, and C_4 can be fixed, restrained or varied freely for each scattering path depending on the system studied, the quality of the data and the aim of the investigation. Fitting is often performed with multiple k-weights to minimize the correlation between the different parameters. Neither the E_0 value chosen during data processing nor the

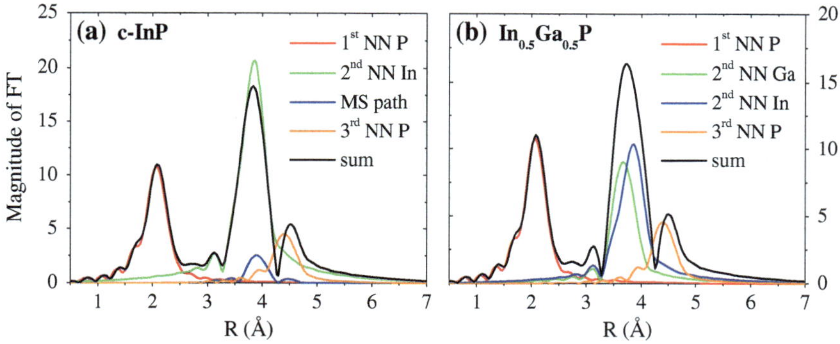

Fig. 1.11 Most important theoretical scattering paths for absorption at the In K-edge of **a** InP and **b** $In_{0.5}$ $Ga_{0.5}P$. The magnitude of the FT is plotted as a function of radial distance R for the different scattering path and their complex sum

theoretically calculated E_0 necessarily correspond to the actual absorption threshold. Therefore, an additional fitting parameter accounts for the difference between experimental effects such as data normalization. Often, E_0 and S_0^2 are determined from a suitable reference and are then kept constant during the analysis of the sample of interest.

The biggest advantage of the path fitting approach is the ability to analyze the structural parameters beyond the first coordination shell which often contain crucial information not available from the first nearest neighbor environment. Furthermore, the calculation of the scattering paths takes into account the full k-dependencies of all parameters and yields the cumulants of the *real* interatomic distance distribution. Path fitting analysis does not require the measurement of a reference with similar chemical and structural environment of the absorbing atom, although, it is advantageous for the determination of E_0 and S_0^2 if possible. A major drawback of the path fitting approach is the need for an initial structural model and thus the need for some pre-existing knowledge about the sample structure. Moreover, the quality of the results obviously depends strongly on the quality of the theoretical calculations. However, tremendous progress has been made in this area in recent years and, usually, shortcomings in the theory no longer pose a major problem.

1.4.2.3 Alternative Approaches

Other codes, with alternative approaches, have been developed for the analysis of XAS data and include, for example, EXCURVE [25] and GNXAS [10, 11]. Utilization of these and other packages can be advantageous for particular materials and forms thereof. The GNXAS code, for example, describes XAS data as a series of integrals over n-body distribution functions and is well suited to disordered systems including amorphous solids and liquids (see Chaps. 9 and 14).

1.5 Conclusions

We have sought to introduce and describe the fundamentals of XAS theory and experiment and, in doing so, lay the foundation for the more detailed chapters that now follow. As the reader will soon appreciate, these chapters demonstrate and reinforce the ever-expanding utilization of the XAS technique for the characterization of semiconductor materials.

References

1. D.C. Koningsberger, R. Prins, *X-ray Absorption: Principles, Applications, Techniques of EXAFS, SEXAFS and XANES* (Wiley, New York, 1988)
2. G. Bunker, *Introduction to XAFS* (Cambridge University Press, Cambridge, 2010)
3. S. Calvin, *XAFS for Everyone* (CRC Press, Taylor & Francis, Boca Raton, 2013)
4. S.D. Kelly, D. Hesterberg, B. Ravel, Analysis of soils and minerals using X-ray absorption spectroscopy, in *Methods of Soil Analysis—Part 5: Mineralogical Methods*, Soil Science Society of America Book Series No. 5, Madison (2008)
5. F. Boscherini, X-ray absorption fine structure in the study of semiconductor heterostructures and nanostructures, in *Characterization of Semiconductor Heterostructures and Nanostructures*, ed. by C. Lamberti (Elsevier, Amsterdam, 2008)
6. J. Als-Nielsen, D. McMorrow, *Elements of Modern X-ray Physics* (Wiley, Chichester, 2001)
7. J.J. Rehr, R.C. Albers, Rev. Mod. Phys. **72**, 621 (2000)
8. J.J. Rehr, A.L. Ankudinov, Coord. Chem. Rev. **249**, 131 (2005)
9. J.J. Rehr, J.J. Kas, F.D. Vila, M.P. Prange, K. Jorissen, Phys. Chem. Chem. Phys. **12**, 5503 (2010)
10. A. Filipponi, A. Di Cicco, C.R. Natoli, Phys. Rev. B **52**, 15122 (1995)
11. A. Filipponi, A. Di Cicco, Phys. Rev. B **52**, 15135 (1995)
12. D.E. Sayers, E.A. Stern, F.W. Lytle, Phys. Rev. Lett. **27**, 1204 (1971)
13. G. Bunker, Nucl. Instrum. Methods. **207**, 437 (1983)
14. Australian Synchrotron website (http://www.synchrotron.org.au/)
15. C.H. Booth, F. Bridges, Phys. Scripta **T115**, 202 (2005)
16. H. Stragier, J.O. Cross, J.J. Rehr, L.B. Sorensen, C.E. Bouldin, J.C. Woicik, Phys. Rev. Lett. **69**, 3064 (1992)
17. J.C. Woicik, J.O. Cross, C.E. Bouldin, B. Ravel, J.G. Pellegrino, B. Steiner, S.G. Bompadre, L.B. Sorensen, K.E. Miyano, J.P. Kirkland, Phys. Rev. B **58**, R4215 (1998)
18. A.J. Dent, Top. Catal. **18**, 27 (2002)
19. R. Frahm, J. Stötzel, D. Lützenkirchen-Hecht, Synchrotron Rad. News **22**, 6 (2009)
20. G.E. Ice, J.D. Budai, J.W.L. Pang, Science **334**, 1234 (2011)
21. L.L. Araujo, R. Giulian, D.J. Sprouster, C.S. Schnohr, D.J. Llewellyn, P. Kluth, D.J. Cookson, G.J. Foran, M.C. Ridgway, Phys. Rev. B **78**, 094112 (2008)
22. G. Dalba, P. Fornasini, R. Grisenti, D. Pasqualini, D. Diop, F. Monti, Phys. Rev. B **58**, 4793 (1998)
23. M. Newville, J. Synchrotron Radiat. **8**, 322 (2001)
24. B. Ravel, M. Newville, J. Synchrotron Radiat. **12**, 537 (2005)
25. N. Binstead, J.W. Campbell, S.J. Gurman, P.C. Stephenson, *The Excurve Programs* (Daresbury Laboratory, England, 1991)

Part I
Crystalline Semiconductors

Chapter 2
Binary and Ternary Random Alloys

Claudia S. Schnohr

Abstract The atomic-scale structure of binary and ternary bulk random alloys has been studied using extended X-ray absorption fine structure spectroscopy and a variety of theoretical models. The short-range structure of the alloys strongly deviates from the long-range crystallographic average and the element-specific bond lengths remain close to the values of the parent materials. The lattice mismatch is thus accommodated mostly by bond angle relaxation and only to a small extent by bond length relaxation.

2.1 Introduction

When alloying elemental or binary semiconductors, many important material properties, including lattice constant and band gap energy, can be tuned between the values of the parent materials by adjusting the alloy composition. Binary and ternary semiconductor alloys such as SiGe, InGaAs or GaAsP have therefore attracted much scientific and technological interest and are increasingly employed in electronic, optoelectronic and photonic devices. Their applications include light emitting diodes and lasers, infrared detectors, modulators, high electron mobility transistors, field effect transistors, power amplifiers, and high efficiency thin film solar cells [1–6].

Many group IV, III–V and II–VI semiconductors are completely miscible when mixed with another group IV, III–V and II–VI material, respectively. The resulting alloy typically has the same crystal structure as the two parent compounds. For $Si_{1-x}Ge_x$, the lattice sites of the diamond structure are occupied by both Si and Ge atoms as shown schematically in Fig. 2.1. The composition of the alloy is given by the Ge fraction x. For III–V and II–VI ternary alloys of the form $A_xB_{1-x}C$, the cation sublattice of the zincblende structure is populated with atoms of type A and B while the anion sublattice consists only of C atoms (see Fig. 2.1). Each A or B cation is thus bonded to four C anions while each C anion may have both A and B cations

C.S. Schnohr (✉)
Institut für Festkörperphysik, Friedrich-Schiller-Universität Jena, Max-Wien-Platz 1, 07743 Jena, Germany
e-mail: c.schnohr@uni-jena.de

C.S. Schnohr and M.C. Ridgway (eds.), *X-Ray Absorption Spectroscopy of Semiconductors*, Springer Series in Optical Sciences 190,
DOI 10.1007/978-3-662-44362-0_2

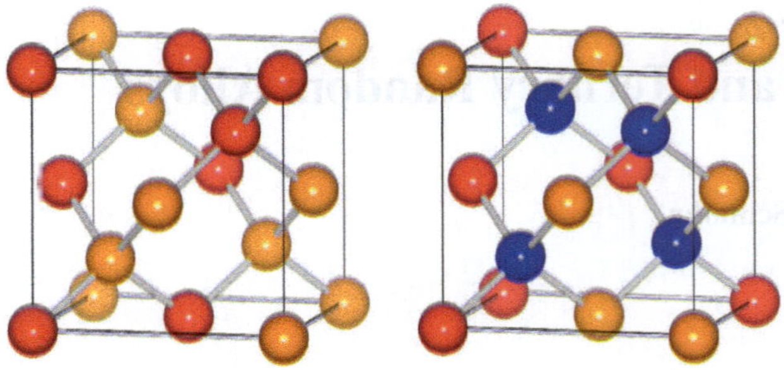

Fig. 2.1 *Left* Diamond structure of the binary SiGe alloy. The lattice sites are occupied by either Si (*red*) or Ge (*orange*) atoms. *Right* Zincblende structure of many ternary III–V or II–VI alloys. The common sublattice is populated by a single atomic species (*blue*) corresponding to the C anions or A cations for $A_xB_{1-x}C$ or AC_xD_{1-x}, respectively. The mixed sublattice is occupied by two different kinds of atoms (*red* and *orange*) corresponding to the A and B cations or C and D anions for $A_xB_{1-x}C$ or AC_xD_{1-x}, respectively. Images created with VESTA [7]

as first nearest neighbors (NN). Similarly, AC_xD_{1-x} alloys have a common cation sublattice occupied by atoms of type A and a mixed anion sublattice populated with C and D atoms. The arrangement of the two atomic species on the mixed sublattice can be either ordered or random. The former typically leads to a more complex crystal structure due to the increased long range order while the latter maintains the overall zincblende structure.

Systems are called lattice-matched when the two parent materials have the same or nearly the same lattice constant, as for example GaAs and AlAs [8]. Most materials, however, are characterized by a mismatch of their lattice constants. In case of InAs and GaAs, for example, the difference in lattice constants amounts to approximately 7 % [8]. For many of these lattice-mismatched systems, the lattice constant of the alloy is found to vary linearly with composition x in accordance with Vegard's Law [9]. The so-called virtual crystal approximation (VCA) hence assumes all atoms to occupy ideal lattice sites with the lattice constant determined by x [10]. All pairs of neighboring atoms have the same bond length and the bond angles remain unchanged. In contrast, Pauling and Huggins assume the conservation of atomic radii which results in different bond lengths for the different atomic pairs independent of x and identical to those of the parent materials [11]. The lattice mismatch in this case is accommodated by a change of the bond angles. Given that many material properties including the band gap directly depend on the short-range structure [12–14], detailed knowledge of the atomic-scale structural parameters is indispensable for a comprehensive understanding of these binary and ternary alloy systems.

In contrast to X-ray diffraction, which typically measures the long-range crystallographic structure, EXAFS is an ideal tool for studying the local atomic arrangement due to its short-range nature. Furthermore, its chemical sensitivity enables the determination of the element-specific first and second NN distances providing valuable information on the atomic-scale structure of binary and ternary alloys. Theoretical treatments range from simple spring models for the dilute limit to full first-principles pseudo-potential calculations. This chapter presents an overview of both experimental EXAFS studies and theoretical works focusing on random semiconductor alloys with diamond or zincblende structure.

2.2 $Si_{1-x}Ge_x$ Binary Alloys

2.2.1 First Shell

The lattice mismatch between Si and Ge amounts to approximately 4 % [8]. On the long-range crystallographic scale, the binary $Si_{1-x}Ge_x$ alloy exhibits the diamond structure similar to the parent materials. The lattice constant shows only a slight deviation from Vegard's Law in the form of a small downward bowing [15]. The *average* first NN distance thus changes almost linearly with composition x between the two elemental values.

On the short-range atomic scale, each atom can be bonded to both Si and Ge thus creating three types of first NN pairs: Si-Si, Si-Ge, and Ge-Ge (see Fig. 2.1). Si and Ge therefore both have a mixed first coordination shell and the EXAFS signal is a superposition of backscattering from both Si and Ge neighbors. Analysis of the first shell signal hence requires the fitting of two separate scattering paths. This leads to an increased uncertainty in the determination of the structural parameters compared to the analysis of a first shell that is comprised of only one type of neighboring atom. EXAFS analysis is further complicated by the fact that there is no experimental standard for the Si-Ge bond from which the scattering amplitude and phase shift could be inferred. For the Si-Si and Ge-Ge bonds these properties can be obtained from measuring the pure parent materials. For the mixed Si-Ge bond, the scattering amplitude and phase shift must be calculated theoretically using the FEFF code or approximated experimentally as the average of the values of the two adjacent isoelectronic compounds AlAs and GaP [16].

Early works on crystalline $Si_{1-x}Ge_x$ alloys unambiguously report different first NN distances for the different atomic pairs ([17] and references therein). However, there was some controversy as to whether or not these bond lengths change with composition x. In 1999, Aubry et al. [17] and Ridgway et al. [18] then clearly established that while the element-specific bond lengths of $Si_{1-x}Ge_x$ are much closer to the Pauling limit than to the VCA, they still change with varying Ge content. Most strikingly, this x-dependence is different for the different atomic pairs being largest for Ge-Ge and smallest for Si-Si. Figure 2.2 plots the results obtained by Aubry et al. at room temperature from total electron yield and fluorescence yield measurements at the Ge and Si K-edge, respectively [17]. Nearly identical results were observed by Ridgway et al. performing transmission measurements of the Ge K-edge at a temperature of 10 K [18] (see also Figs. 7.16 and 10.7). In contrast to these studies performed on $Si_{1-x}Ge_x$ alloys grown by molecular beam epitaxy, Yonenaga and Sakurai observed similar changes of the Si-Ge and Ge-Ge bond lengths with composition in Czochralski grown $Si_{1-x}Ge_x$ crystals [19].

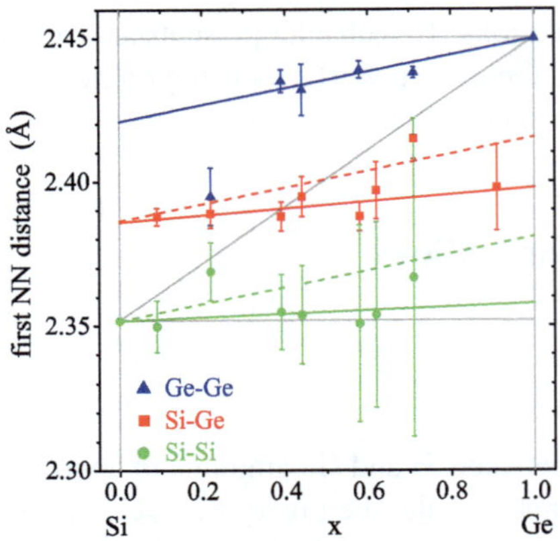

Fig. 2.2 Si-Si (*green*), Si-Ge (*red*), and Ge-Ge (*blue*) first NN distances as a function of composition x for binary $Si_{1-x}Ge_x$ alloys. The experimental values by Aubry et al. [17] (*symbols*) are shown together with the best linear fits (*solid lines*) and the theoretical calculations by Mousseau and Thorpe [20] (*dashed lines*). The *thin grey lines* represent the elemental values and the VCA

The lattice mismatch in $Si_{1-x}Ge_x$ alloys is thus accommodated mostly by bond angle distortions (bond bending) and only to a small extent by bond length distortions (bond stretching). The lesser composition dependence of the Si-Si distance compared to the Ge-Ge distance agrees with the fact that Si has a significantly higher Debye temperature than Ge [17]. The Si-Si bond is thus stiffer and harder to stretch or to compress than the Ge-Ge bond. Note that somewhat different results were observed for strained epitaxial $Si_{1-x}Ge_x$ layers where the lattice constant of the underlying substrate constrains the atomic-scale structure of the alloy (see Fig. 10.8).

2.2.2 Higher Shells

Sun et al. studied the Ge environment in a $Si_{0.994}Ge_{0.006}$ alloy by measuring the Ge K-edge fluorescence at several different temperatures [21]. Assuming that in the dilute limit the Ge impurity is surrounded only by Si atoms, they determined the first, second and third NN distances. The first NN bond length is considerably larger than that of pure Si and agrees well with the values reported by Aubry et al. and Ridgway et al. [17, 18]. In contrast, the second and third NN distances are almost identical to those of pure Si (see also Chap. 10). The distortions created in the Si lattice due to the introduction of a Ge impurity are thus mostly limited to the first NN shell. Doping therefore leads to a localized modification of the atomic-scale structure rather than to a change of the average crystallographic structure [21].

2.3 III–V and II–VI Ternary Alloys

2.3.1 First Shell

Many of the III–V and II–VI ternary alloys crystallize in the zincblende structure and their lattice constants closely follow Vegard's Law (e.g. [22, 23]). The *average* first NN distance thus changes linearly with composition x between the two binary values (VCA). For $A_xB_{1-x}C$ alloys, there are two different first NN pairs, namely A-C and B-C. The first coordination shell around the A and B cations thus consists only of C atoms while the C anions have a mixed first NN shell with both A and B neighbors (see Fig. 2.1). Similarly, AC_xD_{1-x} alloys feature A-C and A-D bonds. Here, the A cations are surrounded by a mixed first NN shell consisting of C and D anions while the anions have only A atoms as neighbors.

The first EXAFS study of the atomic-scale structure of semiconductor alloys was performed in the pioneering work on $In_xGa_{1-x}As$ by Mikkelsen and Boyce [22, 24]. The authors measured the In, Ga and As K-edge signal in transmission mode at a temperature of 77 K. Figure 2.3a plots the In-As and Ga-As first NN distances as a function of composition x. The two bond lengths are clearly different and much closer to the Pauling limit than to the VCA. Nevertheless, they exhibit a linear increase with increasing In content, although the change amounts to only 20 % of the difference between the two binary bond lengths.

Similar studies have since then been performed for a large number of III–V and II–VI ternary alloys including $In_xGa_{1-x}P$ [28, 29], $In_xGa_{1-x}Sb$ [28, 30], $InSb_xP_{1-x}$ [31], $GaAs_xP_{1-x}$ [23, 28], $Cd_xZn_{1-x}Te$ [32, 33], $Hg_xCd_{1-x}Te$ [32, 34],

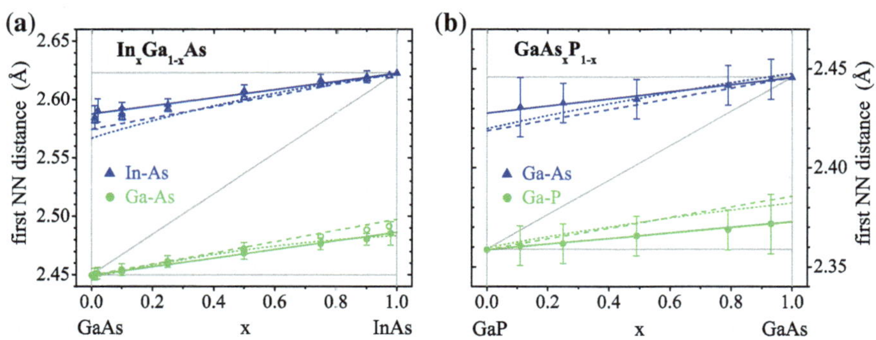

Fig. 2.3 Element-specific first NN distances as a function of composition x for **a** the mixed-cation alloy $In_xGa_{1-x}As$ and **b** the mixed-anion alloy $GaAs_xP_{1-x}$. The experimental values by Mikkelsen and Boyce [22] and by Wu et al. [23] (*full symbols*) are plotted together with the best linear fits (*solid lines*) and the theoretical calculations by Cai and Thorpe [25] (*dashed lines*) and by Schabel and Martins [26] (*dotted lines*). For $In_xGa_{1-x}As$, the model predictions by Balzarotti et al. [27] (*open symbols*) are also included for each composition studied experimentally. (Note that some of the points overlap with the experimental values.) The *thin grey lines* represent the binary values and the VCA

$ZnTe_xSe_{1-x}$ [28], and $HgTe_xSe_{1-x}$ [32]. As an example for a mixed-anion system, Fig. 2.3b plots the first NN distances of $GaAs_xP_{1-x}$ obtained by Wu et al. from transmission measurements of the Ga and As K-edge at a temperature of 80 K [23]. For all materials studied, the first NN distance distribution is bimodal and the values are closer to the Pauling limit than to the VCA. The latter is thus a poor description for the atomic-scale structure of semiconductor alloys. Most studies report a small linear change of the element-specific bond lengths with changing composition x. Only in case of $Cd_xZn_{1-x}Te$, Koteski et al. report a slightly nonlinear behavior, although the nonlinearity is of the same order as the uncertainty and scatter of the data [33]. Similar to $Si_{1-x}Ge_x$, a somewhat different behavior is found for strained epitaxial layers such as $In_xGa_{1-x}As$ thin films grown coherently on InP substrates [35].

A bimodal first NN distance distribution for ternary semiconductor alloys was also observed with other experimental techniques. Qualitatively, the difference between the A-C and B-C distances in $A_xB_{1-x}C$ alloys was referred from cross-section scanning tunneling microscopy studies for $In_xGa_{1-x}P$ and $In_xGa_{1-x}As$ [36, 37]. The element-specific bond lengths further have a strong influence on the phonon modes observed with Raman spectroscopy, particularly on the impurity modes of the dilute limit [38, 39]. As pointed out by Pagès et al. a full quantitative understanding of the relation between the local atomic arrangements and the various transverse and longitudinal optical modes is, however, far from trivial [39]. In contrast, Petkov et al. were able to quantitatively determine the element-specific bond lengths of $In_xGa_{1-x}As$ over the whole compositional range by analyzing high-resolution pair distribution functions obtained from X-ray diffraction measurements [40]. Their results are in excellent agreement with those reported by Mikkelsen and Boyce [22] and confirm the characteristic behavior discussed above.

This behavior, though, is not limited to III–V and II–VI ternary alloys with zincblende structure. The nitride alloys $In_xGa_{1-x}N$ and $Ga_xAl_{1-x}N$ crystallize in the wurtzite structure. EXAFS measurements of the In and Ga K-edge have revealed that the In-N and Ga-N first NN distances are again much closer to the binary values than to the VCA [41–44] (see also Chap. 3). Furthermore, they typically show a small change with changing composition similar to their zincblende counterparts. However, clustering and phase segregation, strain, and atomic ordering due to the specific preparation conditions may significantly influence the structural parameters of these nitride alloys [44, 45]. Numerous EXAFS studies were also performed on II–VI semiconductors with Mn substitution on the cation sublattice including $Zn_xMn_{1-x}Se$ [46], $Zn_xMn_{1-x}Te$ [47], $Cd_xMn_{1-x}Te$ [27], $Hg_xMn_{1-x}Te$ [34], and $Zn_xMn_{1-x}S$ [48]. All works report a bimodal first NN distance distribution with values closer to the Pauling limit than to the VCA (see also Chap. 16). A similar behavior was also observed for the I–VII ionic alloys $Rb_xK_{1-x}Br$ and RbI_xBr_{1-x} which crystallize in the rocksalt structure [49, 50]. The bimodal first NN distance distribution thus seems to be characteristic feature of III–V, II–VI, and I–VII ternary alloys independent of their crystal structure.

To assess and compare the amount of bond stretching for different alloy systems, the dimensionless relaxation parameter ϵ was introduced. It is defined for the dilute limit as the difference between the impurity bond length and the host bond length

Table 2.1 Relaxation parameter ϵ for the two different bonds of III–V and II–VI $A_xB_{1-x}C$ and AC_xD_{1-x} ternary alloys with zincblende structure

$A_xB_{1-x}C$	A-C	B-C	References	AC_xD_{1-x}	A-C	A-D	References
$In_xGa_{1-x}P$	0.76	0.80	[28]	$GaAs_xP_{1-x}$	0.75	0.76	[28]
	0.80	0.80	[29]		0.79	0.84	[23]
$In_xGa_{1-x}As$	0.80	0.79	[22]	$ZnTe_xSe_{1-x}$	0.78	0.80	[28]
$In_xGa_{1-x}Sb$	0.78	0.79	[28]	$HgTe_xSe_{1-x}$	0.96	0.74	[32]
	0.79	0.87	[30]				
$Cd_xZn_{1-x}Te$	0.70	0.76	[32]				
	0.86	0.92	[33]				

The uncertainty of the values is typically around ± 0.05

divided by the difference between the two binary values. Thus, $\epsilon = 0$ (no relaxation) corresponds to an impurity bond length equal to that of the host. In this case, the lattice is completely rigid, forcing the impurity first NN distance to adopt the value of the host lattice in accordance with the VCA and the lattice mismatch is accommodated solely by bond stretching. In contrast, $\epsilon = 1$ (full relaxation) corresponds to an impurity bond length equal to the value of its own binary compound. In this case, the lattice is completely floppy, allowing all first NN distances to retain their binary values and the lattice mismatch is accommodated solely by bond bending. Table 2.1 lists the experimentally determined ϵ values for III–V and II–VI ternary alloys with zincblende structure. As can be seen, the values are mostly between 0.75 and 0.80. While some works report a slightly smaller ϵ for the longer A-C bond compared to the shorter B-C or A-D bond, the differences are typically within the experimental uncertainty of ± 0.05. Bond bending is thus strongly favored over bond stretching for both first NN pairs.

Apart from the mean value, EXAFS analysis also yields the width of the element-specific distance distribution (see Chap. 1). In the binary parent compounds, thermal motion leads to a distribution of first NN bond lengths characterized by the standard deviation σ. Surprisingly, the A-C and B-C or A-C and A-D distance distributions of the ternary alloys show the same standard deviations as the binary materials within experimental uncertainty [22, 23, 29]. A broadening of the distributions due to an increase in structural disorder in the alloy is thus not apparent for the first NN bonds.

2.3.2 *Higher Shells*

In the zincblende structure, each atom has twelve second NN all occupying the same sublattice as the central atom (see Fig. 2.1). For $A_xB_{1-x}C$ alloys, however, each pair of second NN anions is linked by one cation that can be either A or B. This leads to two different types of second NN anion pairs, namely C-A-C and C-B-C. Second NN cations are always linked by one C anion but there are three possible cation combinations. Three different types of second NN cation pairs can thus be distinguished, namely A-C-A, A-C-B (B-C-A), and B-C-B. Similarly, AC_xD_{1-x} alloys feature two types of second NN cation pairs, namely A-C-A and A-D-A, and three types of second NN anion pairs, namely C-A-C, C-A-D (D-A-C), and D-A-D.

Figure 2.4 plots the second NN distances as a function of composition x for (a) the mixed-cation alloy $In_xGa_{1-x}As$ and (b) the mixed-anion alloy $GaAs_xP_{1-x}$ as determined by Mikkelsen and Boyce [22] and by Wu et al. [23], respectively. Regarding the common sublattice, Mikkelsen and Boyce found that the scattering contribution stemming from the second NN shell around the As atoms could not be fit with a single peak. Similar to the first NN shell, a bimodal distance distribution with two distinct second NN distances is necessary to satisfactorily represent the experimental data. Based on their average distances and relative weights, the authors assigned these two contributions to As-In-As and As-Ga-As pairs. As can be seen from Fig. 2.4a, the two distances remain close to their corresponding binary values and show only a small increase with increasing In content. Nevertheless, the two distributions are significantly broadened compared to the binaries accounting for the fact that each As atom has four In or Ga first NN that all affect its position [22]. Similarly, two scattering contributions are needed to represent the signal of the second NN shell around Ga atoms in $GaAs_xP_{1-x}$ corresponding to Ga-As-Ga and Ga-P-Ga pairs. The two distances are again closer to their binary values than to the VCA and exhibit only a moderate linear change with composition x (Fig. 2.4b). A bimodal second NN distance distribution for the common sublattice was also observed for other ternary alloys including $Cd_xZn_{1-x}Te$ [32, 33], $Zn_xMn_{1-x}Se$ [46], $Zn_xMn_{1-x}Te$ [47], and $ZnTe_xSe_{1-x}$ [51].

Regarding the mixed sublattice, the second NN shells around the In and Ga cations in $In_xGa_{1-x}As$ are again best fit with two scattering contributions each, corresponding to In and Ga second NN. The three cation distances In-As-In, In-As-Ga, and Ga-As-Ga are, however, much closer to each other and to the VCA than the two As second NN distances (Fig. 2.4a). Nevertheless, they consistently follow the relation In-As-In > In-As-Ga > Ga-As-Ga. Similar to the As sublattice, the widths of the three distributions are significantly increased compared to the binary alloys due to the increased number of atomic arrangements in the ternary alloy [22, 29]. For the mixed-anion alloy $GaAs_xP_{1-x}$, the As-Ga-As and As-Ga-P distances are also close to the VCA and follow the relation As-Ga-As > As-Ga-P (Fig. 2.4b). A similar behavior of the three different second NN distances of the mixed sublattice was further observed for $In_xGa_{1-x}P$ [29], $Cd_xZn_{1-x}Te$ [33], and $Zn_xMn_{1-x}Se$ [46]. The situation is somewhat less clear for the nitride alloys (see also Chap. 3). For $Ga_xAl_{1-x}N$, Miyano et al. [41] and Yu et al. [42] both report a Ga-N-Ga distance close to the VCA. However, while the former group also found the Ga-N-Al distance to be close to the VCA the latter authors observed a Ga-N-Al distance independent of composition and close to the second NN distance of AlN. Kachkanov et al. report the cation-cation distances of $In_xGa_{1-x}N$ to be again close to the VCA [44]. Differences between the In-N-Ga and Ga-N-In distances for low In contents, however, point to differences

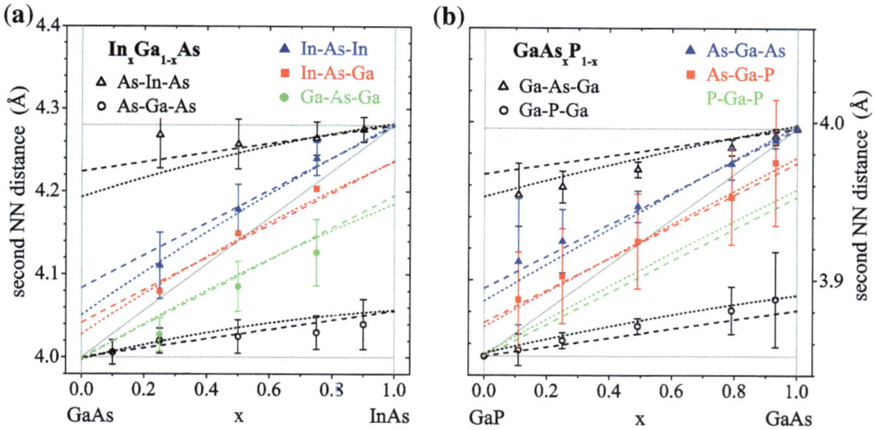

Fig. 2.4 Element-specific cation-cation and anion-anion second NN distances as a function of composition x for **a** the mixed-cation alloy $In_xGa_{1-x}As$ and **b** the mixed-anion alloy $GaAs_xP_{1-x}$. The experimental values by Mikkelsen and Boyce [22] and by Wu et al. [23] (symbols) are plotted together with the theoretical calculations by Cai and Thorpe [25] using $\epsilon \equiv a^{**} = 0.80$ determined from the first NN distances (*dashed lines*) and by Schabel and Martins [26] (*dotted lines*). The *thin grey lines* represent the binary values and the VCA

in the second NN environment for the two cations and were taken as an indication for phase segregation in these samples. Clustering effects in ternary semiconductor alloys are also discussed in Chap. 5.

While there exists an increased variation in the second NN distances of III–V and II–VI ternary alloys, the mixed sublattice already closely resembles the VCA. In contrast, the common sublattice is severely distorted with two distinct second NN distances close to the corresponding binary values. The atoms occupying the common sublattice are thus significantly more displaced from the ideal lattice sites than the atoms occupying the mixed sublattice.

Regarding the *third* NN distances, four cation-anion pairs can be distinguished that differ with respect to the atoms of the mixed sublattice. For $A_xB_{1-x}C$ alloys, they are A-C-A-C, A-C-B-C, B-C-A-C, and B-C-B-C. However, the third NN EXAFS signal is already significantly dampened and suffers from increasing overlap with other single or multiple scattering paths. This usually inhibits the determination of separate A-C-A-C and A-C-B-C distances. Likewise, the B-C-A-C and B-C-B-C distances cannot be distinguished experimentally. Nevertheless, the *average* In-P or Ga-P third NN distances have been determined for $In_xGa_{1-x}P$ from transmission EXAFS measurements at the In and Ga K-edge at 20 k [29]. The values agree well with each other and are identical to the VCA. The distance distributions are, however, strongly broadened compared to those of the binary compounds due to the averaging over different atomic arrangements in the ternary alloys.

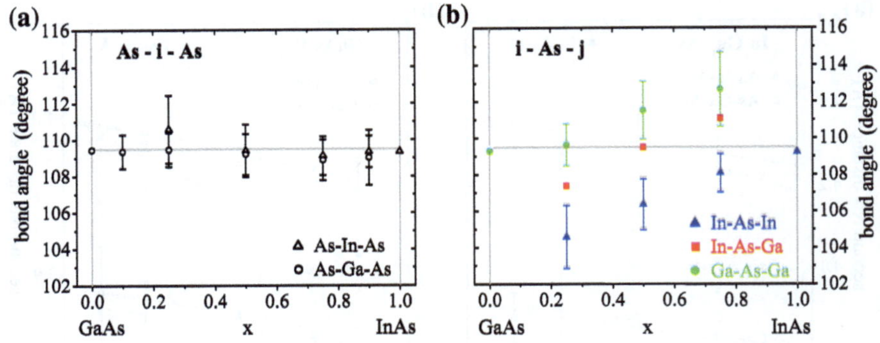

Fig. 2.5 **a** As-i-As and **b** i-As-j bond angles as a function of composition x calculated for $In_xGa_{1-x}As$ from the data reported by Mikkelsen and Boyce [22]. The ideal zincblende value of 109.5° is given as a *thin grey line*

2.3.3 Bond Angles

The tetrahedral bond angles can be evaluated from the first and second NN distances. Figure 2.5 plots the As-i-As and i-As-j angles calculated for $In_xGa_{1-x}As$ using the data given in [22]. Here i and j denote either In or Ga. The In-As and Ga-As first NN distances and the two As-i-As second NN distances remain close to their respective binary values. As a consequence, the As-i-As bond angle agrees well with the ideal zincblende value of 109.5° as shown in Fig. 2.5a. In contrast, the three cation-cation second NN distances are much closer to the VCA than to the binary values. The i-As-j bond angles thus deviate significantly from the ideal value and change linearly with composition x as can be seen from Fig. 2.5b. Similar findings were observed for $In_xGa_{1-x}P$ [29] and $GaAs_xp_{1-x}$ (using the data from [23]). This clearly demonstrates that bond bending is energetically favored over bond stretching and that the lattice mismatch is accommodated by bond angle relaxation rather than bond length relaxation.

2.4 First Shell Calculations

2.4.1 Models for the Dilute Limit

A number of different models have been proposed to describe the atomic-scale structure of binary and ternary semiconductor alloys. Many of the early works sought to predict the impurity bond length in the dilute limit and thus the relaxation parameter ϵ. The advantage of considering the dilute limit is the resulting symmetry around the impurity atom which simplifies the calculation. The most basic model is that of Shih et al. [52]. It considers an impurity atom B in an AC matrix and calculates the displacement of the surrounding first NN C atoms keeping all other atoms fixed. A harmonic potential, i.e. a simple spring, is assumed for the first NN interactions with a geometric approximation to correlate the B-C and A-C distances. A single force constant is used for both types of bonds. The result is independent of the nature of the A, B, and C atoms and depends only on the crystal geometry. For the zincblende structure, $\epsilon = 0.75$ is obtained.

More elaborate models use the valence force field (VFF) potentials by Keating [53] or Kirkwood [54] that describe the distortion energy in terms of contributions arising from changes of the bond lengths (bond stretching) and changes of the bond angles (bond bending). The corresponding bond stretching and bond bending force constants are derived for the binary compounds by fitting experimental macroscopic elastic constants and bulk moduli. The values reported by different groups typically differ by 10 % or less [25, 55, 56]. Alternatively, bond stretching and bond bending force constants can be determined directly from the microscopic atomic vibrations using temperature-dependent EXAFS measurements as presented in Chap. 6. It is then usually assumed that the binary force constants also apply for the ternary alloy system although recent theoretical studies suggest that this may not be strictly true [57].

Martins and Zunger [58] used the VFF potential by Keating applying different force constants for the A-C and B-C bonds and relaxing the first and second NN shells surrounding the impurity atom (Martins and Zunger I). Including the bond bending terms in the VFF potential and relaxing the second NN shell change the relaxation parameter ϵ in opposite directions [27, 58]. Not including both effects (Martins and Zunger II) thus yields better results compared to experimental values than including only one of the two. Both models further apply two additional approximations during the calculation. They include only the harmonic terms of the Keating potential and they approximate the geometric relation between A-C and B-C distances as done in the model by Shih et al. discussed above. The ϵ values obtained for $In_xGa_{1-x}P$ are listed in Table 2.2 together with the experimental results from [28, 29].

Chen and Sher considered not only the distortion energy due to bond stretching and bond bending but also 'chemical' effects such as differences in binding energy and chemically driven charge redistribution [56]. Their extensive work further calculates various VFF models that differ in whether or not bond bending terms are included in the potential and in how the relaxation of higher NN shells is treated. For the case of In impurities in GaP, for example, the ϵ values obtained with the different models vary by up to $\Delta\epsilon \sim 0.30$ illustrating the huge influence such choices have on the calculated results.

2.4.2 *Models for the Whole Compositional Range*

In ternary alloys, the atoms of the common sublattice are surrounded by one of five possible first NN configurations as shown schematically in Fig. 2.6. The first NN shell around the C anions in $A_xB_{1-x}C$ alloys, for example, is characterized by the number n of A cations varying between 0 and 4. The number of B first NN is

Table 2.2 Relaxation parameter ϵ for the In-P and Ga-P bonds of $In_xGa_{1-x}P$ obtained with different models proposed to describe the atomic-scale structure of ternary semiconductor alloys

Model	In-P	Ga-P	References
Shih et al.	0.75	0.75	[52]
Martins and Zunger I	0.63	0.73	[58]
Martins and Zunger II	0.73	0.77	[58]
Chen and Sher	0.70	0.80	[56]
Balzarotti et al.	0.78	0.73	[27]
Cai and Thorpe I	0.72	0.72	[25]
Cai and Thorpe II	0.66	0.77	[25]
Schabel and Martins	0.70	0.80	[26]
Silverman et al.	0.72	0.80	[59]
Srivastava et al.	0.82	0.78	[60]
Boyce and Mikkelsen	0.76 ± 0.05	0.80 ± 0.05	[28]
Schnohr et al.	0.80 ± 0.04	0.80 ± 0.04	[29]

The experimental values determined from EXAFS measurements are also listed for comparison

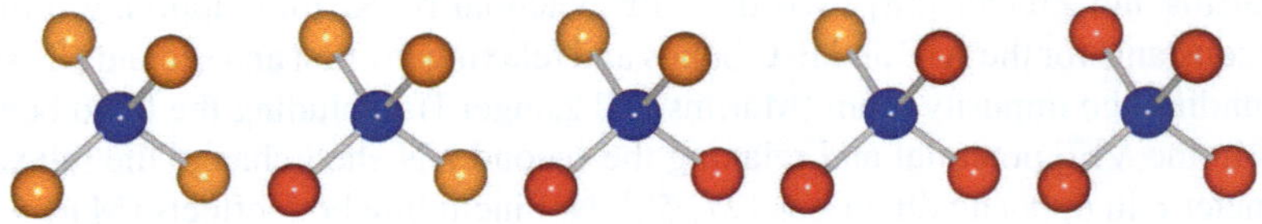

Fig. 2.6 Possible first NN configurations surrounding the atoms of the common sublattice (*blue*) in ternary III–V and II–VI semiconductor alloys with zincblende structure. The mixed sublattice, and thus the first NN shell, is populated with two different atomic species (*red* and *orange*). Images created with VESTA [7]

given by $4 - n$. Balzarotti et al. [27] have proposed the following model to calculate the *individual* composition-dependent A-C and B-C distances for each of the five configurations: (i) Based on Vegard's Law, the lattice constant $a(x)$ is derived as the weighted average of the two binary lattice constants. (ii) The four first NN cations are located on their ideal lattice sites, i.e. they occupy corners of a cube with the size $a(x)/2$. This assumption is justified by the second NN EXAFS results discussed in Sect. 2.3.2. They show that the atoms of the mixed sublattice are much closer to the ideal lattice sites than the atoms of the common sublattice. The same observation was made with X-ray diffraction studies and cluster simulations [61, 62]. (iii) The anion position and thus the first NN distances are obtained by minimizing the sum of the bond stretching terms of the Keating VFF potential. Neglecting the bond bending terms is reasonable as the bond bending force constants are significantly smaller than the bond stretching force constants [25, 55, 56, 63]. Furthermore, including neither bond bending nor second NN relaxation yields better results than including only one of the two effects as discussed above.

In a truly random alloy, the probability of finding a particular configuration at a certain composition x is given by the binomial Bernoulli distribution. The *average* composition-dependent A-C and B-C distances are thus obtained as the weighted average of the individual first NN distances of the five different configurations. Figure 2.3a plots the results obtained by Balzarotti et al. for $In_xGa_{1-x}As$ [27]. The average In-As and Ga-As distances change linearly with composition x and are in very good agreement with the experimental data. Similar results were also obtained for $In_xGa_{1-x}P$ [14], $Cd_xZn_{1-x}Te$ [32], and $Cd_xMn_{1-x}Te$ [27].

In a similar approach, Woicik has calculated the minimum-energy structural parameters of the five different first NN configurations in $In_xGa_{1-x}As$ relaxing not only the central As atom but also its first NN cations and its second NN As atoms while its third NN cations were fixed at their VCA positions [35]. Both, bond stretching and bond bending terms of the Keating VFF potential have been considered. The element-specific In-As and Ga-As bond lengths are then obtained as the weighted average over the five different 17-atom, 16-bond clusters assuming a binomial Bernoulli distribution. The results are in excellent agreement with the experimental values not only for $In_xGa_{1-x}As$ bulk alloys but also for strained thin films.

Thorpe and co-workers consider the alloy as a network of springs described by the Kirkwood VFF potential with one set of bond stretching and bond bending force constants [20, 25, 64]. Within this framework they calculate the element-specific first NN distance distributions using a Green's function formalism [64]. The average distance thus obtained for each atomic pair is a linear function of composition x with a slope determined by the topological rigidity parameter a^{**}. This property describes the competing effects of bond length and bond angle relaxation in the alloy structure and is identical to the relaxation parameter ϵ. Evaluating a^{**} with the Green's function formalism, Cai and Thorpe derived an interpolation formula that expresses a^{**} in terms of the ratio of bond bending to bond stretching force constants. For most semiconductor alloys, this ratio is between 0.1 and 0.2 resulting in a topological rigidity parameter of around 0.75 [64].

For $Si_{1-x}Ge_x$, a^{**} equals 0.707 and the model yields three parallel lines for the Si-Si, Si-Ge, and Ge-Ge first NN distances as shown in Fig. 2.2 [20]. While the relation between the bond lengths of the different atomic pairs is predicted correctly, the slope only matches the experimental values in case of Ge-Ge for which the best linear fit yields $a^{**} = 0.70 \pm 0.05$ [17]. In contrast, best fits to the Si-Ge and Si-Si data result in $a^{**} = 0.84 \pm 0.14$ and 0.94 ± 0.04, respectively [17]. As discussed in Sect. 2.2.1, this difference in slope is believed to be related to the different force constants of the various atomic pairs which are not considered in the model calculation.

For $A_xB_{1-x}C$ and AC_xD_{1-x} ternary alloys, a^{**} is taken as the mean value of the binary parent compounds (Cai and Thorpe I) and the model yields two parallel lines for the two different bond lengths as shown in Fig. 2.3 [25]. The calculation slightly overestimates the change of the first NN distances but is still in fairly good agreement with the experimental values. In an extension of their model, Cai and Thorpe take into account the different force constants of the binary compounds using an effective-medium approximation (Cai and Thorpe II) [25]. The resulting curves exhibit a slightly nonlinear behavior, however, the bowing is typically within the uncertainty of the experimental data and is thus difficult to verify.

2.4.3 Cluster and Supercell Calculations

Several groups have modeled the local structure of semiconductor alloys by simulating large atomic clusters. Weidmann and Newman, for example, relaxed the atomic positions of their $Si_{1-x}Ge_x$ supercells until the distortion energy taken as the Kirkwood VFF potential was minimal [65]. They found three distinct, almost parallel and slightly nonlinear curves for the first NN distances corresponding to Si-Si, Si-Ge, and Ge-Ge bonds. The distance distributions are nearly Gaussian in shape except for the dilute limit where a complex subpeak structure emerges that is determined by the configuration of the surrounding coordination shells.

In a similar way, Podgórny et al. simulated $In_xGa_{1-x}As$ clusters using the Keating VFF potential and obtained a bimodal first NN distance distribution corresponding to In-As and Ga-As bonds [66]. Schabel and Martins modeled the atomic-scale structure of eighteen III–V and eleven II–VI ternary alloys by minimizing the distortion energy also taken as the Keating VFF potential [26]. In all cases they observed two distinct bond length distributions corresponding to the two different types of first NN pairs. The average values exhibit a slightly nonlinear change with composition x and can be fitted with quadratic polynomials the parameters of which are reported for all 29 materials in [26]. The results for $In_xGa_{1-x}As$ and $GaAs_xP_{1-x}$ are plotted in Fig. 2.3. The values slightly overestimate the change of the first NN distances similar to the predictions by Cai and Thorpe.

The extensive work on $In_xGa_{1-x}P$ by Silverman et al. starts from a thermodynamic description of the alloy structure including energetic and entropic terms [59]. The distortion energy is again taken as a modified VFF potential while the configurational and vibrational degrees of freedom are treated using Monte Carlo calculations. The authors obtained distinct In-P and Ga-P bond lengths that change linearly with composition x and are close to the experimental values [29]. Srivastava et al. take a somewhat different approach using first-principles self-consistent local-density total-energy minimization to calculate the first NN distances of $In_xGa_{1-x}P$ [60]. The In-P and Ga-P bond lengths are again very different from each other, show a linear dependence on the composition x and yield a relaxation parameter $\epsilon \sim 0.80$ in excellent agreement with experiment [29]. For the *ordered* compounds $In_nGa_{4-n}P_4$ with $n = 1, 2, 3$, the calculated In-P and Ga-P first NN distances deviate even less from the binary values corresponding to $\epsilon \sim 0.95$.

2.4.4 Comparison of the Different Models

All models for ternary $A_xB_{1-x}C$ and AC_xD_{1-x} alloys discussed above yield a bimodal first NN distance distribution with bond lengths closer to the Pauling limit than to the VCA. The models differ, however, with regard to how much the first NN distances are predicted to change with composition x and whether they vary linearly or not. Table 2.2 lists the relaxation parameter ϵ obtained with the various models for the case of $In_xGa_{1-x}P$ together with the experimental results. Using the different values of bond stretching and bond bending force constants reported in the literature [25, 55, 56] for any given model yields only small differences in ϵ, typically less than 0.02. In contrast, the choice of approximations turns out to be a crucial factor in determining the resulting ϵ value. As discussed above, including neither the bond bending terms nor the relaxation of the second NN shell gives better results than including only one of the two effects. In the dilute limit, the model of Balzarotti et al. corresponds to Martins and Zunger II except that the two additional approximations used by the latter authors are not applied by the former group. A detailed analysis of these models reveals that using a harmonic approximation to the VFF potential does not influence the calculated ϵ values while approximating the geometric relationship between the A-C and B-C distances not only changes the absolute values of ϵ but also the ratio between them (see Table 2.2). Cai and Thorpe argue that the relaxation in semiconductors extends out to a long range [64]. Nevertheless, the simple models that only consider relaxation of the first NN shell yield ϵ values similar to those of the more complex calculations. It is therefore very difficult to judge the appropriateness of certain approximations without performing the complete calculation and comparing the results with experimental values such as those obtained from EXAFS measurements.

2.5 Modelling of Higher Shells

Some of the models discussed above also consider the structural parameters of the second coordination shell. Cai and Thorpe used their spring network model together with the Green's function formalism to develop expressions for the average distances of the five different second NN pairs in ternary $A_xB_{1-x}C$ and AC_xD_{1-x} alloys [25, 64]. For the common sublattice, they found two distinct second NN distances corresponding to the two possible bridging atoms on the mixed sublattice, for example C-A-C and C-B-C in the case of an $A_xB_{1-x}C$ alloy. These two distances remain close to the binary values over the whole compositional range. For $a^{**} \equiv \epsilon = 0.80$, the slope of the two parallel, linear curves amounts to 20 % of that of the VCA similar to what is observed for the first NN distances. For the mixed sublattice, three second NN distances can be distinguished corresponding to the three possible second NN pairs, for example A-C-A, A-C-B, and B-C-B for an $A_xB_{1-x}C$ alloy. These distances also vary linearly with composition x but they are much closer to the VCA than to the binary values. For $a^{**} \equiv \epsilon = 0.80$, the slope equals 70 % of that of the VCA. The curves for $In_xGa_{1-x}As$ and $GaAs_xP_{1-x}$ are shown in Fig. 2.4 using $a^{**} \equiv \epsilon = 0.80$ obtained from the best linear fit to the first NN distances (see Table 2.1). The model agrees well with the experimental data for all five second NN distances. Excellent agreement was also found for the cation-cation distances of $In_xGa_{1-x}P$ [29].

Podgórny et al. evaluated the second NN distances from their cluster calculations of $In_xGa_{1-x}As$ [66]. For the In-As-In and Ga-As-Ga pairs, they obtained broad Gaussian-shaped distance distributions the average values of which increase by approximately 0.2 Å when going from $x = 0$ to $x = 1$. Furthermore, the average In-As-In distance is about 0.08 Å larger than the average Ga-As-Ga distance for all compositions x studied. In contrast, the second NN As distance distribution is extremely broad and features multiple peaks. They can be grouped into an As-In-As and an As-Ga-As distribution with up to three peaks each. The average As-In-As distance is larger than the average As-Ga-As distance by approximately 0.2 Å while both values change by only 0.05 Å over the whole compositional range. These findings are in excellent agreement with the experimental data by Mikkelsen and Boyce (see Fig. 2.4a).

Similar second NN distance distributions were also obtained by Silverman et al. for $In_xGa_{1-x}P$ [59]. The authors further analysed the i-P-j, P-Ga-P, and P-In-P bond angles where i and j denote either In or Ga. In all three cases, they found broad distributions with angles as low as 100° and as high as 120° demonstrating that the lattice mismatch is accommodated to a large extent by bond angle relaxation. Nevertheless, all three distributions are mostly centered around the ideal value of 109.5° in good agreement with experimental findings. As can be seen in Fig. 2.5a, the average As-In-As and As-Ga-As angles of $In_xGa_{1-x}As$ remain close to the ideal value over the whole compositional range. Given the similarity of the materials, the same can be expected for $In_xGa_{1-x}P$. In contrast, the i-As-j and i-P-j angles strongly deviate from the ideal value depending on the kind of atoms involved (Fig. 2.5b and [29]). Nevertheless, the weighted average over all i-As-j or i-P-j angles still closely resembles the ideal value as predicted by the cluster calculations.

Podgórny et al. and Silverman et al. further found that both anions and cations are displaced from their ideal lattice sites, however, the displacement of the anions is much larger [59, 66]. The distortion of the common sublattice is thus significantly greater than the distortion of the mixed sublattice as already inferred from the experimental second NN distances discussed in Sect. 2.3.2. The same observation was also made by Schabel and Martins [26]. They show that each of the five second NN distance distributions consists of three different peaks determined by the configuration of the surrounding NN shells. For the atoms of the mixed sublattice, the spacing is rather small and the peaks mostly overlap leading to one broad distribution as observed by Podgórny et al. and Silverman et al. [59, 66]. In contrast, the spacing is large for the common sublattice resulting in a multimodal distribution as reported by the other two groups. The composition dependence of the average distances obtained by Schabel and Martins is again well represented by quadratic polynomials with a small nonlinear contribution. The parameters for all five second NN pairs are given in [26] for eighteen III–V and eleven II–VI ternary alloys. The results for $In_xGa_{1-x}As$ and $GaAs_xP_{1-x}$ are plotted in Fig. 2.4. Good agreement is observed with both the experimental data and the model by Cai and Thorpe. Similar findings are reported for $Cd_xZn_{1-x}Te$ [33].

The experimental $In_xGa_{1-x}As$ second NN distances are also well reproduced with the model by Woicik based on the structural parameters of the five different

17-atom, 16-bond $In_nGa_{4-n}As$ clusters with $n = 0, 1, 2, 3$, or 4 [35]. Similar to the first NN distances, good agreement between calculated and measured values is obtained for both bulk alloys and strained thin films. Regarding the *third* NN shell, the model predicts similar splittings and VCA behavior as for the cation second NN pairs. As discussed in Sect. 2.3.2, the various third NN distances cannot be distinguished experimentally. Nevertheless, the calculated *average* In-As and Ga-As third NN distances closely follow the VCA in perfect accord with the experimental results for $In_xGa_{1-x}P$ [29].

2.6 Conclusions

$Si_{1-x}Ge_x$ as a binary alloy and most III–V and II–VI ternary alloys are characterized by a size mismatch and hence by a difference in the bond lengths of the parent materials. Nevertheless, the alloys usually exhibit the same crystal structure as the parent compounds with a lattice constant that changes linearly, or close to linearly, with composition x. On the local atomic scale, however, significant distortions of the alloy structure are observed. Being chemically sensitive and a short-range probe, EXAFS provides an ideal tool for studying the element-specific atomic-scale structure of binary and ternary semiconductor alloys.

Distinct distance distributions were observed for the different first NN pairs in all semiconductor alloys studied. These element-specific bond lengths remain close to the values found in the parent compounds changing much less with composition than what would be expected from the lattice constant and the virtual crystal approximation. The lattice mismatch in tetrahedrally coordinated semiconductor alloys is thus accommodated mostly by bond angle relaxation and only to a small extent by bond length relaxation. By now, the bimodal character of the first NN distance distribution in III–V and II–VI ternary alloys was also deduced from other experimental techniques and is well described by numerous theoretical models. Nevertheless, the original discovery that semiconductor bond lengths are nearly preserved even in highly lattice-mismatched alloys is one of the great achievements of X-ray absorption spectroscopy. It is based on the unique capabilities of EXAFS that allow the direct, quantitative and unambiguous determination of element-specific first NN distances. Thanks to these investigations, a detailed picture of the atomic-scale structure of a large variety of semiconductor alloys has now been established.

In III–V and II–VI ternary alloys, the short-range structure of both sublattices deviates from the long-range crystallographic average albeit the distortion is significantly larger for the common sublattice than for the mixed one. Such atomic displacements have a remarkable influence on many important material properties, most prominently the band gap energy [12, 13] which is a crucial parameter for device applications. However, charge redistribution caused by the different properties of the atoms occupying the mixed sublattice also influences the band gap energy of semiconductor alloys [60, 67, 68]. Using experimental EXAFS data and the model by Balzarotti et al., these structural and electronic contributions to the band gap change could be successfully distinguished for $In_xGa_{1-x}P$ as a model system for

III–V ternary alloys [14]. Detailed knowledge of the atomic-scale structure of binary and ternary semiconductor alloys is therefore indispensable for a comprehensive understanding of these versatile and highly relevant group of materials.

References

1. R. Chen, T.T.D. Tran, K.W. Ng, W.S. Ko, L.C. Chuang, F.G. Sedgwick, C. Chang-Hasnain, Nat. Photonics **5**, 170 (2011)
2. S.I. Park, A.P. Le, J. Wu, Y. Huang, X. Li, J.A. Rogers, Adv. Mater. **22**, 3062 (2010)
3. S. Mokkapati, C. Jagadish, Mater. Today **12**, 22 (2009)
4. C. Ghezzi, R. Magnanini, A. Parisini, L. Tarricone, E. Gombia, M. Longo, Phys. Rev. B **77**, 125317 (2008)
5. M. Feng, J.N. Holonyak, G. Walter, R. Chan, Appl. Phys. Lett. **87**, 131103 (2005)
6. H. Cotal, C. Fetzer, J. Boisvert, G. Kinsey, R. King, P. Hebert, H. Yoon, N. Karam, Energ. Environ. Sci. **2**, 174 (2009)
7. K. Momma, F. Izumi, J. Appl. Crystallogr. **41**, 653 (2008)
8. S. Adachi, *Properties of Group-IV, III–V and II–VI Semiconductors* (Wiley, New York, 2005)
9. L. Vegard, Z. Phys. **5**, 17 (1921)
10. L. Nordheim, Ann. Phys. **401**, 607, 641 (1931)
11. L. Pauling, M.L. Huggins, Z. Kristallogr. **87**, 205 (1934)
12. A. Zunger, J.E. Jaffe, Phys. Rev. Lett. **51**, 662 (1983)
13. K.C. Hass, R.J. Lempert, H. Ehrenreich, Phys. Rev. Lett. **52**, 77 (1984)
14. C.S. Schnohr, J. Phys.: Condens. Matter **24**, 325802 (2012)
15. J.P. Dismukes, L. Ekstrom, R.J. Paff, J. Phys. Chem. **68**, 3021 (1964)
16. J.C. Woicik, K.E. Miyano, C.A. King, R.W. Johnson, J.G. Pellegrino, T.L. Lee, Z.H. Lu, Phys. Rev. B **57**, 14592 (1998)
17. J.C. Aubry, T. Tyliszczak, A.P. Hitchcock, J.M. Baribeau, T.E. Jackman, Phys. Rev. B **59**, 12872 (1999)
18. M.C. Ridgway, K.M. Yu, C.J. Glover, G.J. Foran, C. Clerc, J.L. Hansen, A. Nylandsted, Larsen. Phys. Rev. B **60**, 10831 (1999)
19. I. Yonenaga, M. Sakurai, Phys. Rev. B **64**, 113206 (2001)
20. N. Mousseau, M.F. Thorpe, Phys. Rev. B **48**, 5172 (1993)
21. Z.H. Sun, W.S. Yan, H. Oyanagi, Z.Y. Pan, S.Q. Wei, Phys. Rev. B **74**, 092101 (2006)
22. J.C. Mikkelsen, J.B. Boyce, Phys. Rev. B **28**, 7130 (1983)
23. Z. Wu, K. Lu, Y. Wang, J. Dong, H. Li, C. Li, Z. Fang, Phys. Rev. B **48**, 8694 (1993)
24. J.C. Mikkelsen, J.B. Boyce, Phys. Rev. Lett. **49**, 1412 (1982)
25. Y. Cai, M.F. Thorpe, Phys. Rev. B **46**, 15879 (1992)
26. M.C. Schabel, J.L. Martins, Phys. Rev. B **43**, 11873 (1991)
27. A. Balzarotti, N. Motta, A. Kisiel, M. Zimnal-Starnawska, M.T. Czyżyk, M. Podgórny, Phys. Rev. B **31**, 7526 (1985)
28. J.B. Boyce, J.C. Mikkelsen, *Ternary and Multinary Compounds* (Materials Research Society, Pittsburgh, 1987), p. 359
29. C.S. Schnohr, L.L. Araujo, P. Kluth, D.J. Sprouster, G.J. Foran, M.C. Ridgway, Phys. Rev. B **78**, 115201 (2008)
30. S. Hosokawa, T. Ozaki, N. Takata, N. Happo, H. Ikemoto, T. Shishido, K. Hayashi, J. Cryst. Growth **311**, 978 (2009)
31. C.J. Wu, Z.C. Feng, W.M. Chang, C.C. Yang, H.H. Lin, Appl. Phys. Lett. **101**, 091902 (2012)
32. P. Letardi, N. Motta, A. Balzarotti, J. Phys. C **20**, 2853 (1987)
33. V. Koteski, H. Haas, E. Holub-Krappe, N. Ivanovic, H.E. Mahnke, J. Alloys Comp. **371**, 138 (2004)

34. R.A. Mayanovic, W.F. Pong, B.A. Bunker, Phys. Rev. B **42**, 11174 (1990)
35. J.C. Woicik, J. Appl. Phys. **112**, 113515 (2012)
36. N. Liu, C.K. Shih, J. Geisz, A. Mascarenhas, J.M. Olson, Appl. Phys. Lett. **73**, 1979 (1998)
37. B. Grandidier, H. Chen, R.M. Feenstra, D.T. McInturff, P.W. Juodawlkis, S.E. Ralph, Appl. Phys. Lett. **74**, 1439 (1999)
38. R. Carles, G. Landa, J.B. Renucci, Solid State Commun. **53**, 179 (1985)
39. O. Pagès, A. Chafi, D. Fristot, A.V. Postnikov, Phys. Rev. B **73**, 165206 (2006)
40. V. Petkov, I.K. Jeong, J.S. Chung, M.F. Thorpe, S. Kycia, S.J.L. Billinge, Phys. Rev. Lett. **83**, 4089 (1999)
41. K.E. Miyano, J.C. Woicik, L.H. Robins, C.E. Bouldin, D.K. Wickenden, Appl. Phys. Lett. **70**, 2108 (1997)
42. K.M. Yu, W. Shan, C.J. Glover, M.C. Ridgway, W.S. Wong, W. Yang, Appl. Phys. Lett. **75**, 4097 (1999)
43. M. Katsikini, E.C. Paloura, F. Boscherini, F. D'Acapito, C.B. Lioutas, D. Doppalapudi, Nucl. Instrum. Methods B **200**, 114 (2003)
44. V. Kachkanov, K.P. O'Donnell, R.W. Martin, J.F.W. Mosselmans, S. Pereira, Appl. Phys. Lett. **89**, 101908 (2006)
45. J.C. Woicik, J.K.F. Ludwig, T.D. Moustakas, Appl. Phys. Lett. **100**, 162105 (2012)
46. W.F. Pong, R.A. Mayanovic, B.A Bunker, J.K. Furdyna, U. Debska, Phys. Rev. B **41**, 8440 (1990)
47. N. Happo, H. Sato, T. Mihara, K Mimura, S. Hosokawa, Y. Ueda, M. Taniguchi, J. Phys.: Condens. Matter **8**, 4315 (1996)
48. R.J. Iwanowski, K. Ławniczak-Jabłońska, I. Winter, J. Hormes, Solid State Commun. **97**, 879 (1996)
49. J.B. Boyce, J.C. Mikkelsen, Phys. Rev. B **31**, 6903 (1985)
50. A. Di Cicco, E. Principi, A. Filipponi, Phys. Rev. B **65**, 212106 (2002)
51. J.B. Boyce, J.C. Mikkelsen, J. Cryst. Growth **98**, 37 (1989)
52. C.K. Shih, W.E. Spicer, W.A. Harrison, A. Sher, Phys. Rev. B **31**, 1139 (1985)
53. P.N. Keating, Phys. Rev. **145**, 637 (1966)
54. J.G. Kirkwood, J. Chem. Phys. **7**, 506 (1939)
55. R.M. Martin, Phys. Rev. B **1**, 4005 (1970)
56. A.B. Chen, A. Sher, Phys. Rev. B **32**, 3695 (1985)
57. K. Biswas, A. Franceschetti, S. Lany, Phys. Rev. B **78**, 085212 (2008)
58. J.L. Martins, A. Zunger, Phys. Rev. B **30**, 6217 (1984)
59. A. Silverman, A. Zunger, R. Kalish, J. Adler, Phys. Rev. B **51**, 10795 (1995)
60. G.P. Srivastava, J.L. Martins, A. Zunger, Phys. Rev. B **31**, 2561 (1985)
61. V. Petkov, S.J.L. Billinge, Phys. B **305**, 83 (2001)
62. I.K. Jeong, F. Mohiuddin-Jacobs, V. Petkov, S.J.L. Billinge, S. Kycia, Phys. Rev. B **63**, 205202 (2001)
63. C.S. Schnohr, P. Kluth, L.L. Araujo, D.J. Sprouster, A.P. Byrne, G.J. Foran, M.C. Ridgway, Phys. Rev. B **79**, 195203 (2009)
64. Y. Cai, M.F. Thorpe, Phys. Rev. B **46**, 15872 (1992)
65. M.R. Weidmann, K.E. Newman, Phys. Rev. B **45**, 8388 (1992)
66. M. Podgórny, M.T. Czyżyk, A. Balzarotti, P. Letardi, N. Motta, A. Kisiel, M. Zimnal-Starnawska, Solid State Commun. **55**, 413 (1985)
67. J.E. Bernard, A. Zunger, Phys. Rev. B **36**, 3199 (1987)
68. R. Magri, S. Froyen, A. Zunger, Phys. Rev. B **44**, 7947 (1991)

Chapter 3
Wide Band Gap Materials

Maria Katsikini

Wide band gap semiconductors find applications in optoelectronics and in the fabrication of high-power and high-frequency microelectronic devices. This chapter reviews the main applications of XAFS for the characterization of the group-III nitrides, SiC, diamond and other II–VI and III–V wide band gap semiconductors. Some aspects, as for example the effect of ion implantation, temperature, pressure and alloying are discussed more extensively in other chapters of the book.

3.1 Introduction

Semiconductors with energy gap (E_g) larger than that of the commonly used Si (1.1 eV) and GaAs (1.43 eV) are termed as wide band gap (WBG) semiconductors. The main reason for their development was the necessity for extending the semiconductor technology towards light manipulation and high-power and high-frequency devices. Although the structure and some properties of most of the WBG semiconductors was basically known, the advance of various growth techniques led to the synthesis of materials with very good properties, low defect concentration and good control of their stoichiometry. The latter allowed tailoring of their energy gap by growing ternary and quaternary alloys. The most important candidate for high-power devices is SiC. Its WBG allows functioning at high temperatures (~900 °C) due to reduced thermal generation of carriers. It is characterized by high breakdown voltage and high drift velocity allowing operation at high switching frequencies (>20 kHz). Finally, it has high thermal conductivity and is attractive for the aerospace and nuclear industry due to radiation hardness. Diamond is also candidate for high power applications but necessitates higher processing temperatures. Another WBG semiconductor

M. Katsikini (✉)
School of Physics, Section of Solid State Physics, Aristotle University of Thessaloniki, 54124 Thessaloniki, Greece
e-mail: katsiki@auth.gr

C.S. Schnohr and M.C. Ridgway (eds), *X-Ray Absorption Spectroscopy of Semiconductors*, Springer Series in Optical Sciences 190,
DOI 10.1007/978-3-662-44362-0_3

that has potential for high-power devices (e.g. high-voltage Schottky diodes) and radio-frequency applications is GaN. Due to its direct band gap it is also attractive for cool lighting devices, namely light emitting diodes (LED) which are used for light production by non-incandescent sources. Although the first blue LED was fabricated with SiC, its efficiency was very low due to the indirect character of the energy gap. As a result, more research effort was directed towards GaN and its alloys with AlN and InN for the fabrication of high efficiency LEDs and laser diodes that cover the whole visible range up to UV. The advantage of using light with small wavelength in optical storage devices offers higher storage density compared to those using red light. Furthermore, proper combination of devices that emit red, green and blue light is utilized in panel displays and white-light emission. Another candidate for the fabrication of UV and blue-light emitting devices is ZnSe and its alloys. ZnO is also of current research interest due to its wide and direct gap and piezoelectric properties.

Some physical and structural properties of WBG semiconductors are listed in Table 3.1. The value of the E_g is mainly determined by the size of the atoms and the bond ionicity. Therefore, in the III–V and II–VI families the nitrides and the oxides, respectively, have wider gaps. As a general trend, as the bonds become more polar the gap tends to become direct. The bond ionicity is also related to the smaller electron effective mass which is a prerequisite for high-frequency microelectronic applications. The high values of Debye temperature are related to high thermal conductivity that allows easy transfer of the produced heat in high-power devices.

In most of the structures adopted by the WBG semiconductors, the atoms are tetrahedrally coordinated due to sp^3 hybridization except from the hexagonal BN where both the B and the N atoms are sp^2 bonded forming layers similarly to graphite. The main structures of the WBG semiconductors are depicted in Fig. 3.1. Diamond, which is the only elemental WBG semiconductor, crystallizes in the Fd3m space group. The compound WBG semiconductors of the form $M^{IV}N^{IV}$, $M^{III}N^{V}$ and $M^{II}N^{VI}$ crystallize mainly in the zincblende (ZB) and wurtzite (W) structures with space groups $F\bar{4}3m$ and $P6_3mc$, respectively. The former is comprised of two interpenetrated *fcc* while the latter by *hcp* lattices of M or N elements. In both structures each atom is tetrahedrally bonded with 4 atoms of the other kind. The ZB and W structures are similar when they are viewed along the (111) and (0001) axes, respectively. They differ only in the sequence of the atomic planes as indicated in Fig. 3.1. More specifically, the stacking sequence in the W structure is ABABA… while in the ZB is ABCABCA… As a result the ZB and W polytypes have similar nearest neighboring shell characteristics and therefore they can be hardly discriminated with extended X-ray absorption fine structure (EXAFS) analysis. Some of the WBG semiconductors exhibit polymorphism (or polytypism) and are stabilized in more than one structures depending on the growth conditions. The most stable polytype of BN is the hexagonal. It crystallizes also in the ZB phase whereas the W phase is achieved under high pressure conditions. The most characteristic case of polymorphism in WBG semiconductors is SiC that can be stabilized in more than 200 structures that differ mainly in the sequence of the atomic planes resulting in differences in the band structure as indicated in [3]. The only cubic polytype of SiC is the 3C and has an ABCABC stacking sequence. The most common hexagonal type is the 6H. The number 6

Table 3.1 Physical and structural properties of WBG semiconductors adopted from [1–6, 99]: E_g is the energy gap (d stands for direct and i for indirect), Θ_D is the Debye temperature at 300 K, f_i is the bond ionicity and a, c, u are the unit cell parameters

Semi-conductor	Type	E_g(eV)	Θ_D(K)	f_i	Space Group	a (Å)	c (Å)	u
Diamond	IV	5.5 (i)	1860	–	Fd3m	3.57	–	–
3C-SiC	IV	2.4 (i)	1200	0.177 (0.394)	$F\bar{4}3m$	4.36	–	–
4H-SiC	IV	3.2 (i)	1300		$P6_3mc$	3.07	10.05	–
6H-SiC	IV	3.0 (i)	1200		$P6_3mc$	3.08	15.12	–
c-BN	III–V	6.4 (i)	1700	0.256 (0.383)	$F\bar{4}3m$	3.62	–	–
h-BN	III–V	5.2 (d)	323		$P6_3/mmc$	2.50	6.66	–
AlN	III–V	6.2 (d)	970	0.449 (0.775)	$P6_3mc$ $F\bar{4}3m$	3.11 4.38	4.98 –	0.382 –
GaN	III–V	3.4 (d)	600	0.500 (0.770)	$P6_3mc$ $F\bar{4}3m$	3.19 4.5	5.18 –	0.377 –
InN	III–V	0.8 (d)	660	0.578 (0.859)	$P6_3mc$ $F\bar{4}3$ m	3.54 4.98	5.70 –	0.379 –
AlAs	III–V	2.2 (i)	446	0.274 (0.367)	$F\bar{4}3m$	5.66	–	–
AlP	III–V	2.5 (i)	588	0.307 (0.421)	$F\bar{4}3m$	5.46	–	–
GaP	III–V	2.3 (i)	445	0.374 (0.361)	$F\bar{4}3m$	5.45	–	–
ZnSe	II–VI	2.7 (d)	340	0.676 (0.740)	$F\bar{4}3m$ $P6_3mc$	5.67 4.00	– 6.54	–
ZnS	II–VI	3.7 (d) 3.9 (d)	352	0.623 (0.764)	$F\bar{4}3m$ $P6_3mc$	5.41 3.82	– 6.26	– 0.375
ZnTe	II–VI	2.3 (d)	260	0.546 (0.560)	$F\bar{4}3m$	6.10 4.27	– 6.99	– –
ZnO	II–VI	3.4 (d)	370	0.616	$P6_3mc$	3.25	5.21	0.382
CdSe	II–VI	1.9 (d) 1.8 (d)	230	0.699 (0.841)	$F\bar{4}3m$ $P6_3mc$	6.05 4.30	– 7.01	– 0.368
CdS	II–VI	2.6 (d) 2.5 (d)	210	0.685 (0.794)	$F\bar{4}3m$ $P6_3mc$	5.82 4.14	– 6.71	– 0.377

A number of properties vary between samples or films grown under different conditions. The ionicity values are according to Phillips [7] while those in parentheses correspond to local density approximation calculations by N. E. Christensen et al. [8, 9]

indicates the number of double (Si-C) atomic planes in the stacking sequence. Zig-zag lines drawn in the $(11\bar{2}0)$ plane as these shown in Fig. 3.1, are often used for the identification of the polytype. Controlling the stacking sequence of the SiC layers might be proved a useful method for band-gap engineering taking into account that the E_g varies with the polytype (see Table 3.1). The fractional positions of the M and N type atoms in the unit cell for the various structures adopted by the WBG semiconductors are:

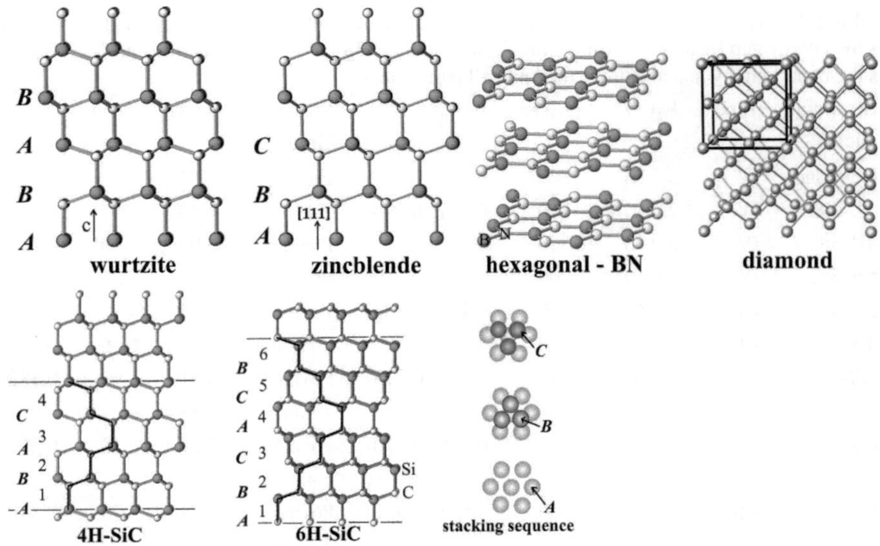

Fig. 3.1 Structures adopted by the WBG semiconductors. A, B and C denote the close-packing planes of the atomic basis in the unit cell (or equivalently the Si-C diatomic layers). The black zig-zag lines are used for the determination of the SiC polytype where the number indicates the double layers in the stacking sequence in the $(11\bar{2}0)$ plane

F$\bar{4}$3m (ZB)

	x	y	z
M:	0	0	0
N:	$\frac{1}{4}$	$\frac{1}{4}$	$\frac{1}{4}$

P6$_3$mc (W)

	x	y	z
M:	$\frac{1}{3}$	$\frac{2}{3}$	0
N:	$\frac{1}{3}$	$\frac{2}{3}$	u

P6$_3$/mmc (hex-BN)

	x	y	z
M:	$\frac{1}{3}$	$\frac{2}{3}$	$\frac{1}{4}$
N:	$\frac{1}{3}$	$\frac{2}{3}$	$\frac{3}{4}$

Fd3m (diamond)

	x	y	z
M:	$\frac{1}{8}$	$\frac{1}{8}$	$\frac{1}{8}$

4H-SiC [Ref. 10]

Si	0	0	0
Si	$\frac{1}{3}$	$\frac{2}{3}$	$\frac{4}{16}$
C	0	0	$\frac{3}{16}$
C	$\frac{1}{3}$	$\frac{2}{3}$	$\frac{7}{16}$

6H-SiC [Ref. 11]

Si	0	0	0
Si	$\frac{1}{3}$	$\frac{2}{3}$	$\frac{4}{24}$
Si	$\frac{2}{3}$	$\frac{1}{3}$	$\frac{8}{24}$
C	0	0	$\frac{9}{24}$
C	$\frac{1}{3}$	$\frac{2}{3}$	$\frac{1}{24}$
C	$\frac{2}{3}$	$\frac{1}{3}$	$\frac{5}{24}$

In the case of the W structure the u-parameter, otherwise called internal structural parameter, corresponds to the displacement of the N atom sublattice relative to the M atom sublattice along the c-axis. The ideal values for the c/a ratio and the u-parameter are $\sqrt{8/3} = 1.633$ and $3/8 = 0.375$, respectively. In this case the nearest neighbor (*nn*) distances within the 1st and the 2nd coordination shells are equal. However, deviation of the c/a ratio and u-parameter from the ideal values results in slight distortion of the tetrahedra.

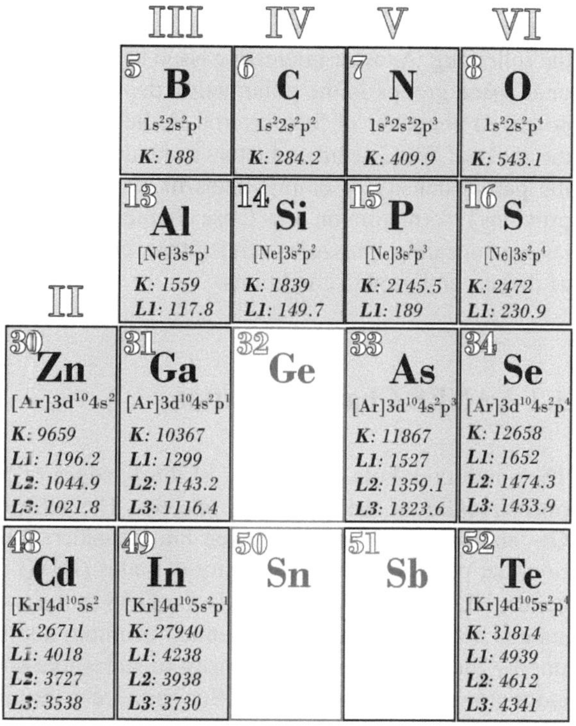

Fig. 3.2 Part of the periodic table that contains the elements related to the formation of WBG semiconductors. The electronic configuration and the energy of the absorption edges are also included for each element

The part of the periodic table that contains the elements that participate in the formation of WBG semiconductors is shown in Fig. 3.2. The electronic configuration and the core electron binding energies are also indicated for each atom. Both hard and soft X-rays are utilized for the study of semiconductors by means of XAFS (for the experimental aspects see Sect. 1.3.1). Application of the fluorescence yield mode is advantageous compared to the transmission mode in the case of thin layers grown epitaxially on substrates. It is also useful for the electronic discrimination of edges which are very close each other as for example for the rejection of the O parasitic signal and the Ga-L-edge due to 2nd order monochromator diffraction from the N-K-edge spectra of GaN. Electron yield detection is also used but charging effects due to the WBG character might result in spectra distortion as shown in [12, 13].

The core-hole lifetime, τ, that is the period of time during which the hole formed in a core shell after the absorption process remains empty, is an important parameter that affects the quality of the near edge XAFS spectra—traditionally the terms near-edge X-ray absorption fine structure (NEXAFS) and X-ray absorption near edge structure (XANES) are used when soft or hard X-rays, respectively, are utilized. Thus, the primary cause of natural broadening of the NEXAFS peaks is the natural width $\Gamma = \hbar/\tau$ which increases with Z. For the first row elements, Γ is in the order of 0.1 eV resulting in NEXAFS spectra with very fine structure allowing the observation of core-exciton peaks and vibronic transitions as it will be discussed in the following. Another interesting issue in semiconductors that crystallize in hexagonal space groups is the polarization dependence of the spectra due to the linear polarized character of Synchrotron Radiation. The main application of XAFS for the study of WBG semiconductors is the determination of structural parameters and the partial density of empty states in the conduction band. In addition to that it provides information on how these characteristics are affected by alloying, implantation, annealing, pressure, temperature, crystal size, polarization, strain, presence of defects and growth conditions.

3.2 XANES Characterization

The small core-hole broadening of the K-edges of N, O, B and C results in very fine structure of the near edge spectra. The NEXAFS spectra at the Al-K and Ga- and In-L-edge are also taken into consideration. Contrary to optical absorption that provides the joint density of states (DOS) in the valence and conduction bands, NEXAFS provides directly the DOS in the conduction band since the initial state is a core state with well defined quantum numbers. However, selection rules impose the projection of the density of states to a specific orbital of the absorbing atom providing the partial DOS. Furthermore, core-hole relaxation effects should be also considered since they affect the energy of the final state (see Sect. 1.2).

In order to compare spectra recorded at the same edge of different samples, from the same sample after processing (irradiation, annealing etc.) or recorded at different angles of incidence, the spectra should be normalized after subtraction of a proper, usually linear, background. The normalization can be done to the edge jump (see Sect.1.4.1) or equivalently to the atomic limit that is the energy where neither NEXAFS transitions nor EXAFS oscillations occur. The latter is helpful when good quality EXAFS spectra can not be recorded. According to Bianconi [14], the atomic limit is roughly defined as the energy that corresponds to the photoelectron wavenumber, k, with wavelength, λ, equal to the distance of the absorbing atom to its first neighbors i.e. $kR \sim 2\pi$.

3.2.1 Polarization Dependent Measurements

The linear polarization of Synchrotron Radiation is important for the study of semiconductors that crystallize in the W or hexagonal structure, since angular dependent measurements provide information on the type of the final state. The spectra of (0001) wurtzite AlN, GaN and InN recorded at the N-K-edge exhibit strong polarization dependence as discussed in [15]. Representative spectra recorded at near-normal and near-grazing incidence as well as at the magic angle, are shown in Fig. 3.3. The magic angle is approximately 55° with respect to the sample surface and its meaning

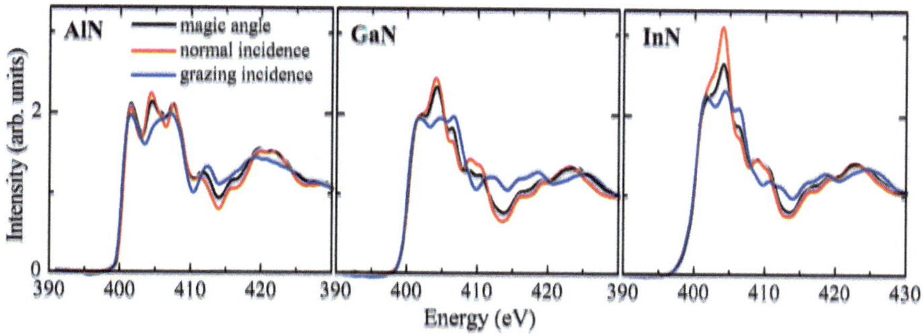

Fig. 3.3 Angular dependent N-K-edge NEXAFS spectra of (0001) wurtzite group-III nitrides

will be discussed later in this Section. The sample was positioned vertically and thus the c-axis was lying on the horizontal plane. The sample was rotated around a vertical axis so that the angle between the impinging beam and the c-axis was varying. The spectra shown in Fig. 3.3 exhibit a strong angular dependence, which is also referred as polarization dependence or linear dichroism. Such an angular dependence is not observed in samples with cubic symmetry, in powder or amorphous samples as well as in polycrystalline samples with random orientation of crystallites. Some of the NEXAFS peaks in the spectra shown in Fig. 3.3 appear more intense at near-grazing while others at near-normal incidence. The peaks correspond to maxima of the partial DOS in the conduction band. In a different approach, due to the small mean free path of the photoelectron that limits the sensitivity of XAFS to regions of about a nm or less, the bonding environment of the absorbing atom can be considered in a molecular orbital approach, taking into account the point symmetry around the absorbing atom. However, the electronic states for such a molecule, which is the tetrahedron around the absorbing atom, are modified by the potential of the crystal. Thus, it can be considered that the NEXAFS peaks correspond to transitions from the 1s state of the N atom to final states of antibonding molecular orbitals with p character of the tetrahedron formed due to sp^3 hybridization.

The position of the NEXAFS peaks can be determined after proper fitting. A step function can be used to take into account the transition probability to final states in the continuum whereas a number of Gaussians is used to account for the transitions to final bound states of antibonding molecular orbitals as it is proposed by Stöhr [16]. Representative fitting of a NEXAFS spectrum of W-GaN recorded at 80° is shown in Fig. 3.4. The type of the final p-state of the NEXAFS peaks can be deduced from the angular dependence of the area under the peaks. More specifically, as described in Chap. 1 (see (1.3)), the absorption coefficient, μ, is proportional to the transition rate between the core level $|i\rangle$ and the final $|f\rangle$ states given by the equation, assuming the validity of the dipole approximation:

$$\mu \propto \left|\langle f|\,\hat{\varepsilon}\cdot\vec{r}\,|i\rangle\right|^2 \rho \qquad (3.1)$$

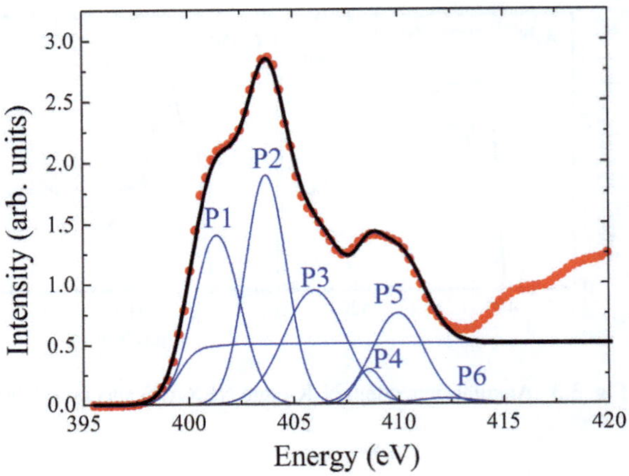

Fig. 3.4 Fitting of N-K-edge NEXAFS spectrum of GaN recorded at near-normal incidence

Table 3.2 Character table of the C_{6v} point group symmetry as defined in [17]

C_{6v}	E	$2C_6$	$2C_3$	C_2	$3\sigma_v$	$3\sigma_d$		
A_1	1	1	1	1	1	1	z	$x^2 + y^2$, z^2
A_2	1	1	1	1	−1	−1	R_z	
B_1	1	−1	1	−1	1	−1		
B_2	1	−1	1	−1	−1	1		
E_1	2	1	−1	−2	0	0	(x, y)(R_x, R_y)	(xz, yz)
E_2	2	−1	−1	2	0	0		(x^2-y^2, xy)

where ρ is the density of empty final states and $\hat{\varepsilon}$ is the polarization vector of the X-ray beam. The operator $\hat{\varepsilon} \cdot \vec{r}$ projects the orbitals along the direction of polarization of the beam. Therefore, in crystalline solids, the dipole matrix element, and consequently the intensity of the NEXAFS peaks, depends on the angle of incidence of the beam. The angular dependence of the NEXAFS peaks can be explained using group theory as indicated in [16, 17]. Contrary to X-ray diffraction that due to translational symmetry depends mainly on space group symmetry, spectroscopy is interpreted in terms of point group symmetry operations. The point symmetry of the W structure is C_{6v} and Table 3.2 is its character table. Since x^2+y^2, z^2 that represents a spherically symmetric distribution, forms a basis of the A_1 irreducible representation, the 1s initial state is an a_1 state. Due to dipole selection rules, the final state should be a p state and thus it should transform either like the A_1 (in the case of a p_z orbital) or like the irreducible representation E_1 in the case of p_x, p_y orbitals since x and y are basis functions of E_1. Mixing of orbitals occurs when they belong to the same irreducible representation. An electric dipole transition is allowed if the direct product of the initial and final states is or contains the irreducible representations to which x, y, z belong. Since $A_1 \times A_1 = A_1$ and $A_1 \times E_1 = E_1$ the allowed transitions are the $1a_1 \rightarrow a_1^*$ and $1a_1 \rightarrow e_1^*$. The $a_1{}^*$ state, which results from mixing of s and p_z orbitals, can be considered as a vector orbital (electronic charge distribution along

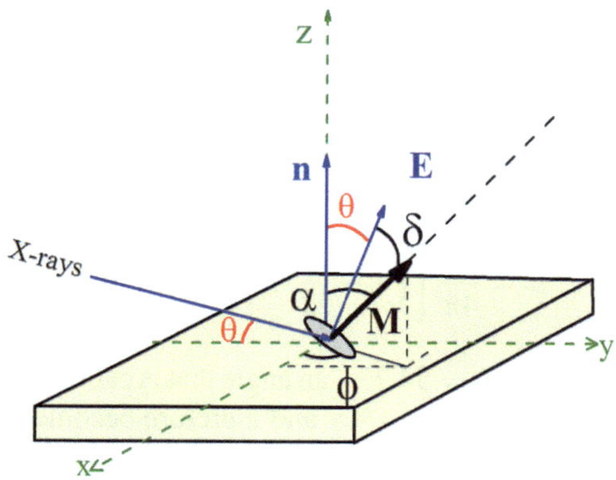

Fig. 3.5 Angles defining the orientation of a molecular orbital relative to the sample and the polarization vector of the impinging X-ray beam. For vector orbitals, **M** points towards the direction of maximum electron charge density along the orbital's axis. In plane orbitals, **M** is normal to the electron charge distribution

the z axis) and will be strongest when the electron field vector is parallel to z, i.e. for grazing incidence in c-plane GaN. The $e_1{}^*$ final state, which results from mixing of p_x and p_y orbitals, can be considered as a plane orbital and will be strongest when the electron field vector is parallel to the (x, y) plane, i.e. for normal incidence. The peaks P1, P3 and P6 shown in Fig. 3.4 correspond to transitions to final vector (s, p_z) orbitals while P2, P4, P5 to (p_x, p_y) plane orbitals. Similar angular dependence has been observed in AlGaN and InGaN alloys in [15, 18, 19]. The angular dependence of the N-K-edge NEXAFS spectra has been applied for the determination of the orientation of GaN crystallites formed by nitridation of GaAs layers by Lübbe et al. [20] and for the study of the effect of thickness of InN layers on the orientation of the microcrystallites [21].

The intensity of the NEXAFS peaks depends on the angle between the polarization vector of the beam and the electron charge distribution of the orbital. The angles that define the orientation of a vector orbital, **M**, or a plane orbital with the normal to the planar electron charge distribution represented by **M**, are shown in Fig. 3.5. The matrix element of a NEXAFS resonance in 3.1 in the case of a (s, p_z) vector or a (p_x, p_y) plane orbital depends on the angle δ between the electric field vector **E** and **M**. More specifically, the absorption cross section is given by the equation:

$$\mu(\delta) = \frac{\mu_0}{4\pi}\left[1 + \frac{1}{2}\beta_m\left(3\cos^2\delta - 1\right)\right] \tag{3.2}$$

where μ_0 is the photoabsorption cross section for random bond orientation and β_m takes the value -1 for transitions from a 1s initial state to final π states and the value 2 for transitions to final σ states, as indicated in [16, 22]. According to Stöhr et al. [22], in order to take into account the orientation of the orbital relative to the sample, the term $\cos^2\delta$ can be written as

$$\cos^2\delta = \cos^2\theta\cos^2\alpha + \sin^2\theta\sin^2\alpha\cos^2\phi + 2\sin\alpha\cos\alpha\sin\theta\cos\theta\cos\phi \tag{3.3}$$

where ϕ is the azimuthial angle, as shown in Fig. 3.5. In bulk samples the angle ϕ can take any value. Moreover, in samples with the c-axis normal to the surface, the normal to the sample surface is parallel to a rotation axis of sixth order. The mean values of $\cos\phi$ and $\cos^2\phi$ in the range 0–2π are 0 and 0.5, respectively. Therefore, 3.2 transforms to:

$$\mu = \frac{\mu_0}{4\pi}\left[\left(1 - \frac{\beta_m}{2} + \frac{3}{4}\beta_m \sin^2\alpha\right) + \frac{3}{2}\beta_m\left(1 - \frac{3}{2}\sin^2\alpha\right)\cos^2\theta\right] \qquad (3.4)$$

In 3.4 for $\theta = 54.73°$, an angle that is called magic angle, the absorption cross section takes the value $\mu_0/4\pi$ and therefore becomes independent of the orbital orientation i.e. angle α. Thus, the directions of maximum electron charge density can be deduced by fitting the angular dependence of the intensities of the NEXAFS peaks using 3.4, after normalization with corresponding values from the spectra recorded at the magic angle.

The hexagonal polytype of BN exhibits a more prominent angular dependence compared to nitrides that crystallize in the W structure due to the characteristic sp^2 bonding that results in a layered structure as described in [23, 24]. According to Brown et al. [25] the point symmetry of hexagonal BN is D_{3h} and is expected to exhibit dichroic absorption in a similar way as W crystals with C_{6v} point symmetry, as it is observed by Brouder [26]. Transitions from the initial 1s state are to final states with (p_x, p_y) contribution (σ^* states) or to states with p_z character (π^* states). The vector orbitals π^* can be described as vectors parallel to the c-axis and normal to the hexagonal rings (see Fig. 3.1). The N- and B-K-edge spectra of hexagonal BN recorded at different angle of incidence, θ, are shown in Fig. 3.6. The c-axis was normal to the sample surface. In the B-K-edge spectra a sharp peak is observed at 192 eV whereas broader peaks appear above the energy of 197 eV. The intensity of the former decreases drastically with the angle of incidence. Similarly, in the N-K-edge spectra the peaks that appear just above the absorption threshold, at ~401 eV, lose intensity as the angle of incidence increases. Muramatsu et al. [23] used discrete variational—Xα molecular orbital calculations for the density of unoccupied states and identified the π character of the final states for the transitions around 192 and 401 eV at the B- and N-K-edge, respectively. The complete disappearance of these peaks at normal incidence has been observed by Moscovici et al. [24] in the case of highly oriented pyrolytic BN with the c-axis normal to the surface, verifying the p_z character of the final states.

The angular dependence of the NEXAFS spectra can be also approached by taking into account the orientation dependence of the absorption tensor M in $\mu \propto \hat{\varepsilon} \cdot M \cdot \hat{\varepsilon} = \sum_{jk} M_{jk}\varepsilon_j\varepsilon_k$ where the vector $\hat{\varepsilon} = [\sin\theta\cos\varphi, \sin\theta\sin\varphi, \cos\theta]$ describes the polarization of the X-ray beam. $M_{jk} \propto \sum_f \langle\psi_i|r_j\,|\psi_f\rangle\langle\psi_f|\,r_k\,|\psi_i\rangle$ are the tensor elements of the absorption tensor M that sums over the dipole transitions to all final states. Another method for the study of the angular dependence that utilizes spherical tensors can be found in the paper by C. Brouder [26].

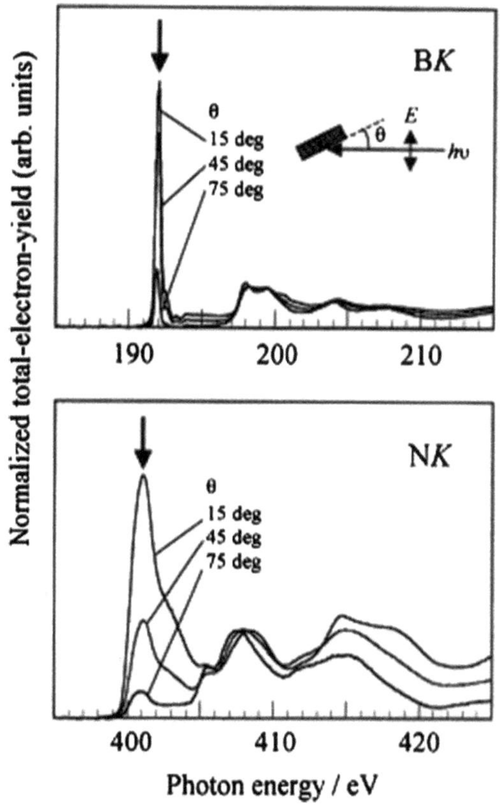

Fig. 3.6 B-K- (*top*) and N-K- (*bottom*) edge angular resolved NEXAFS spectra of hexagonal BN. Reprinted from Spectrochimica Acta A, Vol. 59, Y. Muramatsu, T. Kaneyoshi, E. M. Gullikson, R. C. C. Perera, Angle-resolved soft X-ray emission and absorption spectroscopy of hexagonal boron nitride, pages 1951–1957, © (2003), with permission from Elsevier

3.2.2 Polymorphism and Multiphase Materials

The N-K-edge NEXAFS spectra of the W polytypes of the binary compounds AlN, GaN and InN shown in Fig. 3.3, reveal that the NEXAFS spectra are fingerprints of the cation bonded to the absorbing atom. A closer inspection of the part of the spectrum around the absorption edge, reveals that as the atomic number of the cation, Z, increases (from AlN to InN), the absorption edge shifts towards lower energies indicating a red shift of the conduction band minimum associated with the smaller E_g (see Table 3.1). Furthermore, the absorption edge becomes steeper as the Z of the cation decreases. This variation could be related to the smaller curvature of the conduction band minimum resulting in larger effective mass, m_e^*, of the electron around the band minimum. Values of m_e^*/m_0 equal to 0.3, 0.2 and 0.1 for AlN, GaN and InN, respectively, were reported in [27–29].

The NEXAFS spectrum is not only fingerprint of the group-III element bonded to N but also of the symmetry. Although the nearest neighboring coordination of the W and ZB GaN polytypes is quite similar (see Sect. 3.3.1), the N-K-edge spectra exhibit differences. The spectra of the W and ZB polymorphs of GaN are compared

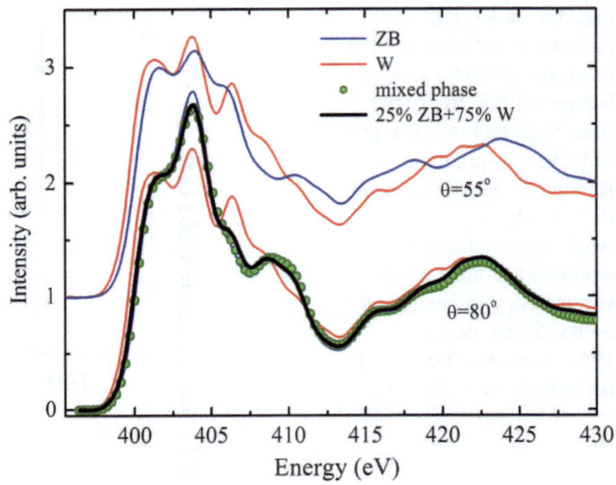

Fig. 3.7 N-K-edge NEXAFS spectra of W and ZB polytypes of GaN recorded at the magic angle and at near normal incidence. The spectrum of a mixed phase sample can be simulated as a weighted average of the spectra of the W and ZB phases

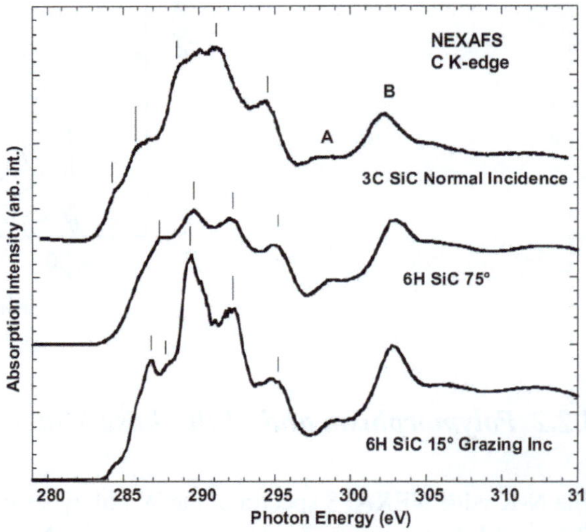

Fig. 3.8 C-K-edge NEXAFS spectra of SiC polytypes. The spectra of the cubic (3C) polytype do not depend on the angle of incidence contrary to the spectra of the hexagonal (6H) that exhibit characteristic angular dependence. The spectra were kindly provided by Dr. Maddalena Pedio, Instituto Officina dei Materiali IOM-CNR Laboratorio TASC, Italy

in Fig. 3.7 for near normal incidence ($\theta = 80°$) and incidence at the magic angle in order to avoid polarization effects. The C-K-edge NEXAFS spectra of the 3C and the 6H polytype of SiC are shown in Fig. 3.8 and their differences are discussed by Pedio et al. [30]. In a similar way to GaN, the 6H polytype exhibits prominent angular dependence. Furthermore, although the *nn* bonding configuration of the atoms is the same in all the SiC polytypes, the different long range order affects significantly the energy gap (see Table 3.1) as well as the electronic structure that is evidenced by the different energy positions of the maxima of the spectra shown in Fig. 3.8. However, the spectra exhibit strong similarities in the region above 295 eV (~15 eV above the absorption edge). Comparison of the NEXAFS spectra recorded at the Si-L_3 and C-K-edges of the 3C and 4H polytypes can be found in [31]. The authors discuss the differences between the spectra of the two polytypes in relation to their band structure characteristics. Particularly the position of the absorption edge in SiC was found to be sensitive to the variations of the energy gap of the various polytypes because the valence band maximum remains unaffected, as it is verified by combined X-ray emission and absorption measurements by Lüning et al. [32].

The observed differences in the NEXAFS spectra of the various semiconductor polymorphs can be used for the estimation of their percentage in mixed-phase samples using the method of linear combination fitting of the spectra (see Sect. 1.4). For example, growth of GaN on (0001) Al_2O_3 substrates results in the stabilization of the W polytype. ZB GaN has been grown on (100) Si whereas an attempt to grow hexagonal GaN on (111) Si resulted in samples with large concentration of stacking faults, i.e. introduction of ABC stacking regions (ZB phase) in the ABAB (W phase) sequence. The NEXAFS spectrum of the mixed phase sample is fitted as a weighted average (linear combination) of the spectra of the ZB and W polymorphs in Fig. 3.7. The linear combination of spectra approach is applicable due to the localized character of the NEXAFS technique that stems from the small mean free path of the photoelectron compared to the excitation volume of the beam. The latter can be estimated by the information depth, which for measurements in the fluorescence yield mode can be calculated using the methodology described by Tröger et al. [33] and the cross section of the beam projected on the sample surface. Determination of the component percentage in mixed phase samples by means of the linear combination method using spectra recorded at different angles of incidence reduces the uncertainty. A similar procedure has been followed for the determination of the hexagonal and ZB phases in BN films using both spectra recorded at the N- and B-K-edges by Zhou et al. [34].

Another example on the application of the linear combination of spectra is the determination of the percentage of sp^2 bonded C in diamond-like films. The amount of the sp^2 phase in the sample is usually determined by the growth conditions. The ratio of diamond (sp^3) to non-diamond (sp^2) carbon affects the film properties like energy gap, electrical conductivity, optical transparency, hardness etc. as discussed in [35, 36]. The NEXAFS spectra of diamond, highly oriented pyrolytic graphite (HOPG) and amorphous carbon are illustrated in Fig. 3.9. The different symmetry and bonding configuration of C results in significantly different NEXAFS spectra. The peak that appears at $\sim$285 eV is assigned to $1s \rightarrow \pi^*$ transitions to antibonding states of carbon and is characteristic of the sp^2 configuration. The intensity of this peak in the HOPG exhibits strong polarization dependence, similarly to the corresponding peak in BN (see Fig. 3.6). The $1s \rightarrow \pi^*$ peak is also present in the spectra of amorphous carbon, although the rest of the spectrum is characterized by the absence of fine structure at energies higher than 290 eV. The same behavior is also observed in samples with no preferential orientation. Therefore, the presence of a peak at $\sim$285 eV in the spectra of diamond-like materials is a fingerprint of the existence of sp^2 bonded C. Fayette et al. [36] reconstructed the NEXAFS spectra of diamond-like materials prepared by chemical vapor deposition as a linear combination of three reference samples: diamond, amorphous carbon and HOPG. As in the case of W nitrides, in order to avoid the polarization dependence of the HOPG

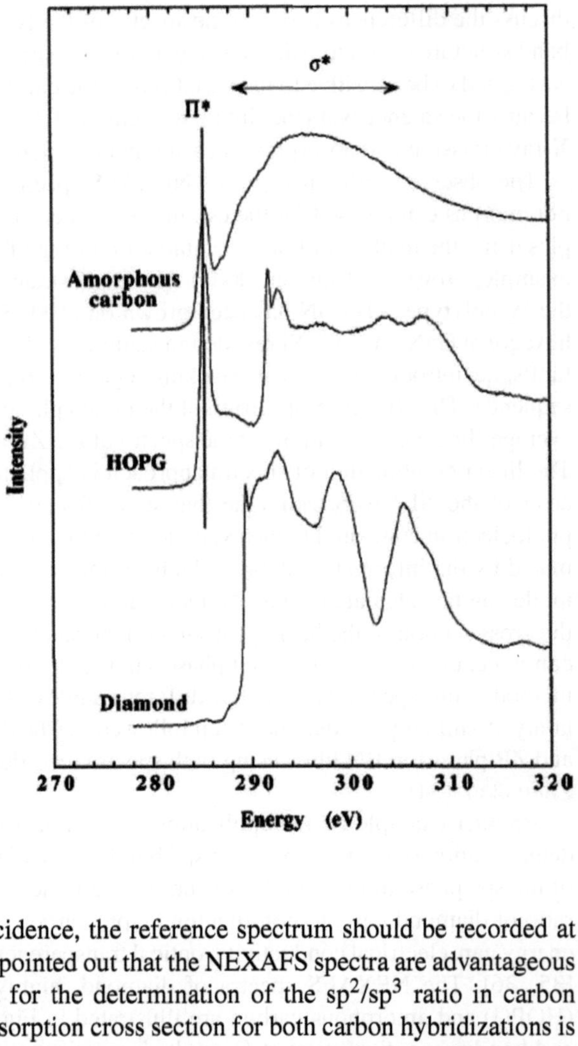

Fig. 3.9 C-K-edge NEXAFS spectra of natural IIA diamond, highly oriented pyrolytic graphite (HOPG) and amorphous carbon films prepared by ion-beam sputtering. Reprinted figure with permission from L. Fayette, B. Marcus, M. Mermoux, G. Tourillon, K. Laffon, P. Parent, F. Le Normand, Phys. Rev. B, **57**, 14123 (1998). © (1998) by the American Physical Society

reference on the angle of incidence, the reference spectrum should be recorded at the magic angle. It should be pointed out that the NEXAFS spectra are advantageous compared to Raman spectra for the determination of the sp^2/sp^3 ratio in carbon thin films, since the X-ray absorption cross section for both carbon hybridizations is the same and NEXAFS is not very sensitive to particle size effects as described in [37, 38].

3.2.3 Core Exciton in Diamond

The excitation of an atom after the absorption process results in the formation of excitons that can be considered as the interaction of photoelectron with the 1s core-hole. In semiconductors with high dielectric constant, the electron-hole interactions are strongly screened. According to Bassani [39] the resulting excitons are of Wannier type and they can be described by a hydrogen-like model. Absorption that involves excitonic states can be observed in the near edge spectra of diamond, since the C-K-edge core-hole width is smaller than 0.1 eV, as reported in [16] and thus smaller than the binding energy of the exciton. The observation of excitonic absorption is allowed provided that good experimental resolution has been achieved and the materials are of good quality. Such a C 1s excitonic absorption appears as a very sharp peak at about 289 eV in the spectrum of diamond shown in Fig. 3.9. Morar et al. [40] fitted the near edge region of the XAFS spectrum of diamond in order to determine the energy of the excitonic transition using the theoretical absorption probability $\Phi(h\nu)$ convoluted with a Gaussian lineshape, so as to take into account the experimental broadening:

$$\Phi(h\nu) \propto \left\{ \sum_{n=1}^{\infty} \frac{4\Delta E_{ex}}{n^3} \frac{\frac{\Gamma}{2\pi}}{\left[h\nu - \left(E_{CBM} - \frac{\Delta E_{ex}}{n^2}\right)\right]^2 + \left(\frac{\Gamma}{2}\right)^2} \right\} + \left[\int_{E'>E_{CBM}}^{\infty} \frac{e^{\pi\gamma}}{\sinh(\pi\gamma)} \frac{\frac{\Gamma}{2\pi}}{(h\nu - E')^2 + \left(\frac{\Gamma}{2}\right)^2} dE' \right] \quad (3.5)$$

where $h\nu$ is the photon energy, ΔE_{ex} is the ground state excitonic binding energy, Γ is the Lorentzian width, E_{CBM} is the conduction band minimum energy and $\gamma = \Delta E_{ex}/\sqrt{E' - E_{CBM}}$. The first term represents a series of discrete excitonic states of quantum number n whereas the second, transitions to the continuum assuming a parabolic band shape around the conduction band minimum. In both terms convolution with Lorentzian lineshapes is applied in order to account for the core-hole broadening. The modeling revealed the value of $\Delta E_{ex} = 189$ meV which is in very good agreement with the value theoretically predicted assuming a hydrogen-like model for the core exciton in diamond. As reported in [41, 42], the excitonic peak in diamond is very sensitive to the presence of defects and lattice disorder and disappears even after implantation with low-energy ions at low fluences where the NEXAFS peaks assigned to the density of empty states remain unaffected. Annealing of the implanted diamond was not proved adequate to recover the excitonic states indicating that core excitons are very sensitive to the crystallinity and the absence of defects. Therefore, as proposed by Coffman et al. [37], the intensity of the exciton peak can be used as an indicator for the assessment of the quality of the diamond films.

3.2.4 Ion Implantation and Defects

Ion implantation is used for semiconductor doping and synthesis as well as for electrical isolation. It is also used for the study of damage accumulation in crystalline materials. Extended discussion on disordered semiconductors studied by XAS can

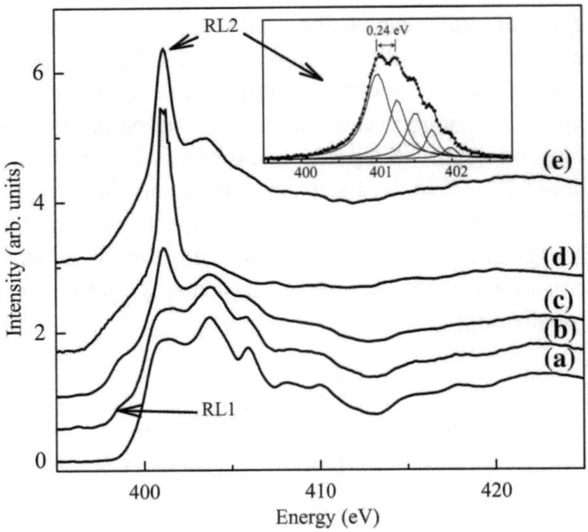

Fig. 3.10 N-K-edge NEXAFS spectra of implanted GaN recorded at the magic angle: **a** as-grown sample and implanted with fluences **b** 5×10^{13} **c** 1×10^{15} **d** 5×10^{15} and **e** $1 \times 10^{17}\,cm^{-2}$. (**b**–**d**) Indium implants at 700 keV, 77 K and **e** O implant, 70 keV, room temperature. A magnification of the spectrum around the resonance RL2, recorded with better energy resolution, is shown in the inset

be found in Chaps. 7 and 8 while XAS studies of dopants in semiconductors are discussed in Chap. 4.

The N-K-edge NEXAFS spectra of GaN subjected to implantation with O and In ions are shown in Fig. 3.10. The main effects of ion implantation are:

(a) Broadening of the NEXAFS peaks as the implantation fluence increases due to the increased lattice damage that results in locally different bonding modifications of the N atom (spectra a,b,c in Fig. 3.10). In this case the spectrum represents a statistic average of the spectra of the various bonding environments. Above a fluence threshold the NEXAFS peaks smear out (spectrum d). From this point of view, the NEXAFS spectra offer an experimental tool for the determination of the amorphization fluence threshold.
(b) Emergence of resonance lines at energy positions that do not depend on the type of the implant and can be attributed to transitions to empty defect-related states. RL1 appears about 1.5 eV below the absorption edge while RL2 at approximately 1 eV above the absorption edge. RL1 has been attributed in [43] to the N split-interstitial defect which is formed by two N atoms that occupy the same lattice site. According to theoretical calculations by Neugebauer et al. [44], the interaction between the two N atoms results in the formation of empty states with p component exactly in the middle of the gap. RL2 has been observed in the spectra of GaN implanted with various ions in [45–47] and has been assigned to $1s \rightarrow \pi_g^*$ transitions of molecular nitrogen trapped in the GaN matrix in [45–47]. It is asymmetric and exhibits a fine structure that can be simulated using a number of Lorentzians which are equidistant. Petravic et al. [48] attributed the fine structure of the peak to the vibronic character of the final states because they combine changes in both the electronic and the vibrational state of the N_2 molecule. The formation of molecular nitrogen in the form of bubbles after

implantation is also substantiated by transmission electron microscopy characterization results reported in [49]. The area under the resonance lines can be in principle related to the concentration of the corresponding defects in the sample. N split-interstitials are formed at much lower implantation fluences compared to N_2 while the concentration of both increases with the fluence and is affected by the type of the implant and the implantation energy and temperature [50]. Annealing at 800 °C for 15 s can effectively remove both implantation-induced defects in In implanted layers. The N-K-edge NEXAFS spectra of disordered GaN:O films were used in a similar manner by Ruck et al. [46] for the determination of the N_2 content in the films. The broadening of the N_2 1s $\rightarrow \pi_g^*$ resonance in various compound semiconductors subjected to low-energy N ion bombardment has been studied by Petravic et al. [48]. They found a direct relation of the width of the peaks that comprise the resonance line with the lattice constant indicating that the broadening of the lines is related to the probability of the electron to escape from the π^* orbital to the semiconductor matrix.

The sensitivity of the NEXAFS spectra for the identification of defects has been also demonstrated in BN by Jimenez et al. [51] and Caretti et al. [52]. The B-K-edge NEXAFS spectra of hexagonal BN prepared in the form of powder and thin film are shown in Fig. 3.11 along with the spectrum of an Ar^+ sputtered powder sample. The spectrum of the powder sample shows the characteristic 1s $\rightarrow \pi^*$ resonance (indicated with W) at 192 eV of sp^2 bonded B. However, in a thin BN film grown by N_2^+ assisted pulsed laser deposition (PLD), three additional peaks (denoted as X, Y, Z in Fig. 3.11) appear at 192.6, 193.2 and 193.8 eV, respectively, indicating that this growth procedure results in defective layers. The peak W evolves in four peaks (W-Z) in a similar way also in the Ar^+ sputtered BN power. Furthermore, implantation of BN films with N_2^+ions was found to result in enhancement of the

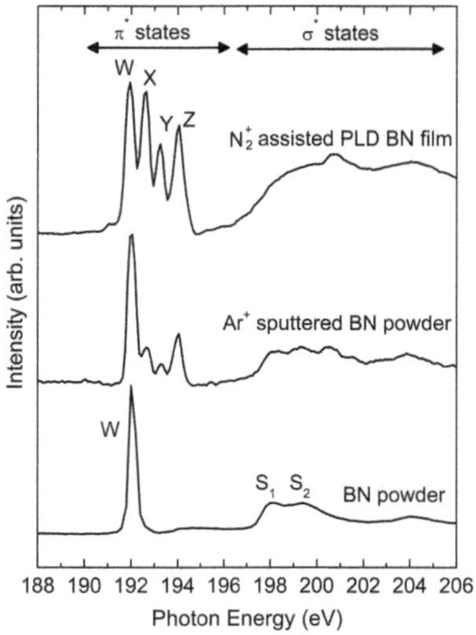

Fig. 3.11 B-K-edge NEXAFS spectra of hexagonal BN in the form of powder (as prepared and Ar^+ sputtered) and a thin BN film prepared by N_2^+ assisted pulsed laser deposition (PLD). The figure is kindly provided by Dr. Ignacio Jiménez, Instituto de Ciencia de Materiales de Madrid, Spain

intensity of the X-Z peaks. These peaks were attributed by Jiménez et al. [51] to the formation of N void defects. In another interpretation, the decoration of these voids with oxygen has been considered and the X-Z peaks were assigned to BN_2O, BNO_2, BO_3 bonding modifications, respectively. The proposed assignment is based on the charge transfer from the B atom to its ligands due to the difference in their electronegativities. According to Caretti and Jiménez [52], the substitution of a N atom with an O results in a blue-shift of the $1s \rightarrow \pi^*$ resonance by about 0.6 eV.

3.2.5 Near Edge Spectra Simulations

As mentioned in Chap. 1, the near edge part of the XAFS spectrum is very sensitive to multiple scattering and thus the path-by-path fitting of the spectra is not applicable because it necessitates a lot of computing power and time. Therefore, simulations are performed using various theoretical methods based on hypothetical structures. Despite this drawback, the significance of the multiple scattering contributions renders XANES very sensitive in the bonding geometry even in the case of polymorphs with similar nearest neighbor shell characteristics which are hardly distinguishable by EXAFS.

In order to predict theoretically the NEXAFS spectra it is necessary to describe the initial state, the final state and the transition probability based on 3.1. According to selection rules imposed by the dipole approximation, in the spectra recorded at the K edge, the final state is a p state and thus the projected partial density of empty states should be considered. In the case of L edges, mainly d final states are probed since the probability to probe states with angular quantum number $\ell - 1$, is considerably smaller. However, according to de Groot [53], the presence of the core-hole distorts slightly the empty DOS since the atom is in an excited state, is left with Z-1 electrons and the core potential is slightly different. Therefore, quasi-particle models should be applied (see Sect. 1.2.2). In crystalline semiconductors band structure calculation methods can be employed. Furthermore, characteristics of the XANES spectra, e.g. maxima, are often related to characteristics of band structure. For example the dip at approximately 303 eV in the XANES spectrum of diamond shown in Fig. 3.9, is attributed in [54] to a secondary gap in its conduction band structure. On the other hand, the use of multiple scattering calculations for a cluster of atoms around the absorbing atom are less time consuming, there is no restriction to crystalline phases and can predict well also the absorption cross section for transitions to the continuum. A description of the theoretical methods used for the calculations of XANES can be found in [55].

The ground state density of empty states, namely without taking into account the core-hole effects, of BN, InN, GaN, AlN, ZnO, SiC, ZnS, CdS, GaN, CdSe has been calculated using the self consistent orthogonalized linear combination of atomic orbitals (OLCAO) in the local density approximation (LDA) by Xu and Ching [56]. The same authors have reported the partial density of states of the W, ZB and hexagonal polytypes of BN in [57]. Mizoguchi et al. [58] used the OLCAO method for the simulation of the K- and L_3-XANES spectra of AlN, GaN, InN and ZnO. In this case, the core-hole effects in the final state have been taken into account and the matrix elements for the dipole transition were also calculated. Relatively large supercells were used in order to minimize the interaction between the core-holes and the polarization dependence of the spectra was also predicted. The K- and L- edge spectra of group II-tellurides are compared with self-consistent linear muffin-tin-orbital (LMTO) calculations by Kisiel et al. up to about 15 eV above the absorption edge [59] without the inclusion of core-hole corrections. Since this method fails to determine accurately the position of the absorption edge, the energy scales of the experimental and theoretical spectra should be aligned. Lambrecht et al. [60] have used the LMTO method for the simulation of the Ga $M_{2,3}$ and N-K-edge NEXAFS spectra of GaN without taking into account core-hole effects. They also reproduced the NEXAFS spectra from the glancing angle X-ray reflectivity spectra using the Kramers-Kronig formulation. The WIEN2k code that employs the density functional (DFT) augmented plane wave (APW) scheme has been also used for the reproduction of the main features of the spectra of AlN and GaN taking also into account core-hole and polarization effects by Magnuson et al. [61, 62]. The same code has been used for the simulation of the C- and Si-K-edge spectra of the 6H and 3C polytypes of SiC by Liu et al. [63]. Although the simulations agree with the experimental spectra recorded at the Si-K-edge, at the C-K-edge the theory fails to reproduce the spectra as far as the positions of the maxima and the relative intensity of the peaks is concerned. The influence of various treatments for the inclusion of core-hole effects in the Ga-K-edge spectra of GaP calculated using the WIEN2k code were investigated by Li et al. [64].

Multiple scattering calculations were performed in real-space clusters of a number of atoms around the absorbing atom. Such calculations with clusters of about 100 atoms for the W and ZB polytypes of CdS were published by Levelut et al. [65]. The differences in the spectra of the W and ZB polytypes were explained after taking into account the contribution of distant neighboring shells in the absorption cross section and their collinearity with the second shell (strong focusing effect). The angular dependence of the spectra of the W polytype is also discussed [26]. Multiple scattering simulations using the FEFF code [66] have been performed in the K- and L- edge spectra of both Zn and S of the ZB and W polytypes of ZnS by Gilbert et al. [67]. FEFF simulations that yielded very good agreement with the experiment have been also performed in [68] for the W and the rocksalt phase of ZnO, the latter being stabilized under high pressure. Clusters containing 27 and 93 atoms were used for the self consistent field (SCF) and the full multiple scattering (FMS) calculations, respectively. The effect of the value of the u-parameter on the Zn-K-edge spectra of ZnO was investigated by means of FEFF by Haug et al. [69]. They reported that variation of the u-parameter from 0.37 to 0.39 did not affect significantly the Zn-K-edge XANES spectra. FEFF simulations of the N-K-edge spectra of group III nitrides have been also reported by Guo et al. [70] and Chassé et al. [71].

3.3 EXAFS Characterization

3.3.1 Binary Compounds

The similarity of the W and ZB structures when viewed along the (0001) and (111) axes, respectively, results in comparable *nn* shell parameters up to the fourth shell. To be more precise, in the ZB phase the 4 and 12 atoms that comprise the 1st and 2nd shells, respectively, are equidistant to the central atom. Contrary to that, in a binary compound MN that crystallizes in the W structure, the tetrahedron around M (and N) is distorted with the N (M) atom that is located on the c-axis being at the distance $R^{\rm in}_{\rm M-N} = u \cdot c$ and the other three, which are positioned in the base of the tetrahedron and out of the c-axis, at a distance $R^{\rm out}_{\rm M-N} = \sqrt{\frac{a^2}{3} + c^2\left(\frac{1}{2} - u\right)^2}$. In the second *nn* shell, 6 atoms are located in the same plane (the basal, c-plane) at a distance equal to $R^{\rm in}_{\rm M-M} = a$ and three above and below that plane at distance $R^{\rm out}_{\rm M-M} = \sqrt{\frac{a^2}{3} + \frac{c^2}{4}}$. The slight distortion of the tetrahedron and the splitting in the distances in the 2nd *nn* shell for the W structure due to the deviation of the c/a and u from the normal values can not be easily identified by EXAFS. However, acquisition of EXAFS spectra at normal and grazing incidence results in enhancement of the different shell contributions, depending on the angle between the polarization vector of the X-ray beam and the bond direction, offering a method for the determination of the two distances within the shell. Furthermore, in the case of thin films which are often subjected to usually biaxial strain, the c/a ratio deviates further from the normal value resulting in larger variation of the *nn* distances within one shell as reported in [72, 73].

Due to the tetrahedral coordination of most of the WBG semiconductors, the contribution of multiple scattering paths, mainly triangular paths, is generally small. Especially when the 1st *nn* shell consists of light elements (e.g. N, P, O) the multiple scattering contribution in the EXAFS region is even less significant as it has been shown by Wei et al. [74]. The triangular paths DS1 and DS2, shown in Fig. 3.12, in tetrahedrally coordinated semiconductors are expected to contribute in the Fourier

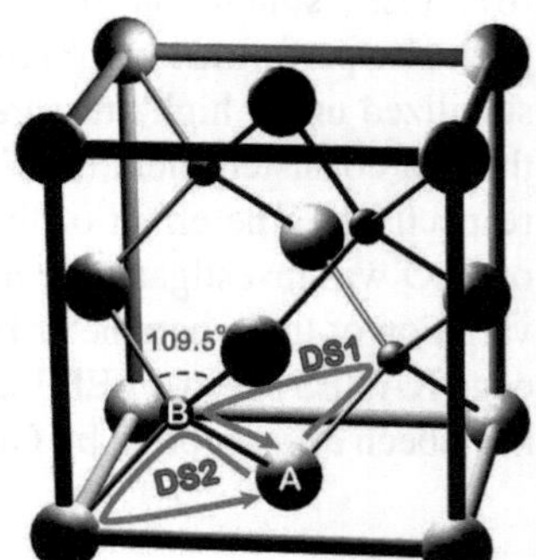

Fig. 3.12 Triangular scattering paths in the ZB structure (central atom: A)

transform amplitude between the 2nd and 3rd *nn* shells. The situation is similar in semiconductors that crystallize in the W structure although the slight distortion of the tetrahedron results in a larger number of DS paths with smaller degeneracy. Compared to GaP, where the DS contribution in the magnitude of the Fourier transform of the spectrum recorded at the Ga-K-edge is small, in GaSb significant damping of the second shell peak has been observed by Wei et al. [74]. Similarly, for the fitting of the ZnSe EXAFS spectra reported in [75], only single scattering contributions were necessary in both the Se and Zn-K-edges for $k \geq 3\text{Å}^{-1}$. In case of negligible multiple scattering contributions, provided that the *nn* shell contributions in the Fourier transform magnitude are well discriminated and can be filtered without overlap, the amplitude-ratio method can be implemented for the determination of the cumulants.

3.3.2 Effect of Temperature

As discussed in Chap. 1, the Debye-Waller factors can be considered as having a static and a thermal component. The contribution due to static disorder is relatively small in crystals with low defect concentration. The thermal component increases with temperature and the rate depends on the strength of the bonds. A simple model that relates the frequency of the thermal atomic vibrations with the Debye-Waller factors is the Einstein model that assumes a unique frequency, ω_E, for all the vibrational modes in a lattice. The Einstein temperature is defined as $\Theta_E \equiv \hbar\omega_E/k_B$ where $\hbar$ and k_B are the Planck and Boltzmann constants. According to the Einstein model the dependence of the thermal component of the Debye-Waller factor, σ^2, can be fitted using the equation:

$$\sigma^2 = \frac{\hbar^2}{2\Theta_E k_B \mu} \coth\left(\frac{\Theta_E}{2T}\right) \tag{3.6}$$

where μ is the reduced mass of the corresponding atom pair. The values of σ^2 versus temperature for the case of GaN, InN and ZnSe are fitted in Fig. 3.13 using 3.6 as it also described in [76–78]. It is evident that for $T \to 0$, $\sigma^2 \to \frac{\hbar}{2\omega_E \mu}$, i.e. it is inversely proportional to the effective mass of the atom pair. The value of Θ_E for the Ga-Ga interaction in GaN is found equal to 300 K and for the In-In interaction in InN equal to 190 K. In ZnSe, Θ_E is found equal to 295 K for the Zn-Se interaction (1st *nn* shell) and 165 K for the Se-Se interaction in the 2nd *nn* shell as reported in [78]. Dalba et al. [79] determined an Einstein temperature equal to 228 K for the Se-Cd pair in wurtzite CdSe. Corrections due to anharmonicity of the atomic vibrations are also discussed in the same paper. Anharmonicity effects become important as the temperature increases and the inclusion of the fourth cumulant in the EXAFS analysis becomes necessary, as it is discussed in Sect. 1.4.2. Further discussion of thermal effects on the EXAFS spectra of semiconductors can be found in Chap. 6.

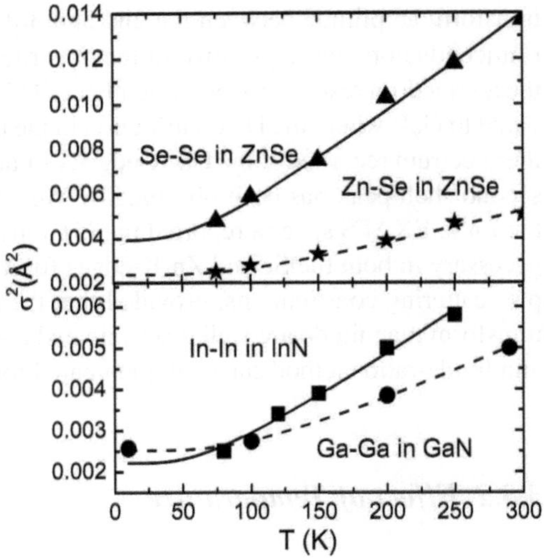

Fig. 3.13 Temperature dependence of 1st and 2nd nearest neighboring Debye-Waller factors in GaN and InN estimated by Katsikini et al. [76, 77] and in ZnSe determined by Diop et al. [78]. The lines correspond to fitting using the Einstein model

3.3.3 Alloying

Alloying in group-III nitrides is an important issue since many applications in micro- and opto-electronics necessitate tailoring of the materials properties, e.g. band gap, unit cell parameters etc. EXAFS spectroscopy is well suited for the study of the local bonding in alloys as it can detect local distortions and deviations of the nearest neighbor distances from the average values predicted using the unit cell parameters determined by X-ray diffraction (XRD). Application of XAFS for the study of alloying in semiconductors is more extensively discussed in Chap. 2. In random nitride alloys, where the different type of group-III elements are randomly distributed in the cation sublattice, the main issue is the amount of deviation of the 1st and 2nd *nn* distances from the values predicted by the virtual crystal approximation (VCA). According to VCA, an alloy can be represented by a lattice that has the average properties of the constituent binary pure semiconductors. Complications in the EXAFS analysis might arise due to ordering, phase separation and strain in thin films. A summary of the dependence of the nearest neighbor distances on the In content in $In_xGa_{1-x}N$ layers is shown in Fig. 3.14. The symbols indicate values that were obtained by means of EXAFS analysis and are reported in [80–83] while the solid lines correspond to theoretical calculations by Ferhat and Bechstedt [84] and Saito and Arakawa [85]. The dashed line corresponds to the VCA values assuming the validity of the law of Vegard.

The dispersion of the measured *nn* distances in Fig. 3.14 could be attributed to uncertainties in the fitting of the EXAFS spectra as well as to the different growth conditions and film thicknesses that might result in deviations from the stoichiometry, biaxial strain, high concentration of extended defects, ordering or phase separation. The latter is considered to be important in $In_xGa_{1-x}N$ alloys with $x > 0.4$. According

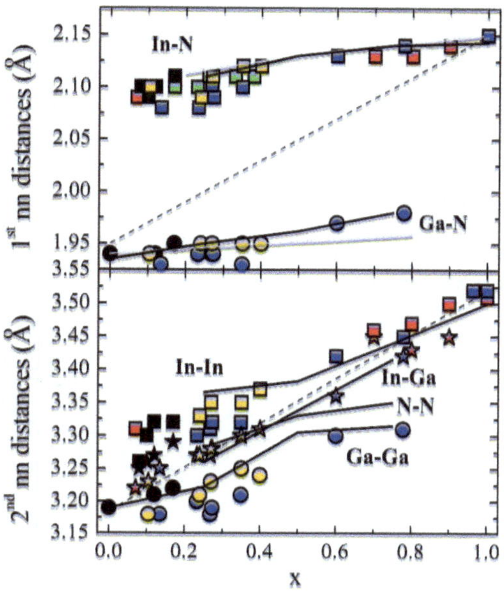

Fig. 3.14 1st and 2nd *nn* distances of $In_xGa_{1-x}N$ alloys. *Experiment*: Squares: In-N (*top*) and In-In (*bottom*), circles: Ga-N (*top*) and Ga-Ga (*bottom*), asterisks: In-Ga. O' Donnel et al. [80] (*green*), Katsikini et al. [81] (*black*), [83] (*red*), Kachkanov et al. [82] MOVPE (*orange*) and MBE (*blue*) samples. *Theory:* Ferhat & Behstedt [84] (*black line*) and Saito & Arakawa [85] (*grey line*). *Dashed lines* correspond to the expected values according to the VCA

to Fritsch et al. [86], the local ordering is also expected to result in slight deviations from the expected values in the 1st *nn* shell distances. The results summarized in Fig. 3.14 reveal that there is agreement between experimental and theoretical results. The 1st *nn* distances deviate significantly from the values predicted by the VCA and exhibit a bimodal behavior. Although the 2nd *nn* distances also appear to split according to the type of the neighboring atom, the deviation from the VCA is smaller. This behavior indicates that the change in the average unit cell parameter due to alloying is locally accommodated by bond angle deformation rather than bond length variation and can be attributed to the polar character of the bonds in group-III nitrides. AlGaN alloys exhibit a similar behavior as it is theoretically predicted by Ferhat and Bechstedt [84] and experimentally verified by Yu et al. [87]. XAS studies of nitrogen—hydrogen complexes in hydrogenated GaAsN are reviewed in Chap. 5 while studies of GaN and AlGaN nanostructures are discussed in Chaps. 12 and 13.

3.3.4 Ion Implantation

The lattice damage induced by ion implantation is not only evident in the NEXAFS spectra, as discussed in Sect. 3.2.4, but also in the magnitude of the Fourier transform of the EXAFS spectra. An example is shown in Fig. 3.15 for 6H-SiC implanted at 77 K with Na ions at fluences ranging two orders of magnitude. The spectra were recorded at the Si-K-edge by Bolse [88]. In SiC both the Si and C atoms are tetrahedrally coordinated. In the as-grown sample shown on the top, bonding

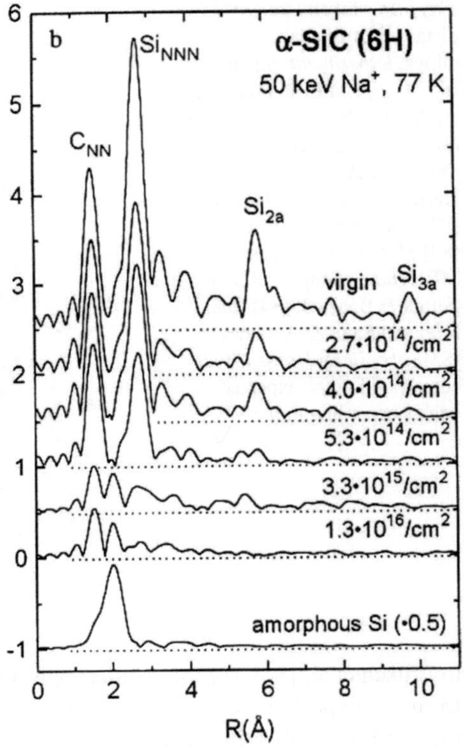

Fig. 3.15 Si-K-edge radial distribution functions of 6H-SiC implanted with Na ions at 77 K. Reprinted from Nuclear Instruments and Methods B, Vol. 148, W. Bolse, Amorphization and recrystallization of covalent tetrahedral networks, pp. 83–92, © (1999), with permission from Elsevier

contributions in the Fourier transform can be clearly discriminated up to the distance of 9 Å. As the implantation fluence increases the intensity of the peaks in the Fourier transform decreases. This effect is more pronounced at more distant shells. When the fluence exceeds the value of $3.3 \times 10^{15}\,cm^{-2}$, the formation of Si-Si bonds is observed (an additional peak appears in the Fourier transform after the peak denoted with C_{NN}). Chemical effects have been also observed in GaN heavily implanted with Si, where the formation of Si-N bonds has been detected in [89]. The use of EXAFS for the study of disordered semiconductors is discussed in more detail in Chaps. 7–9.

3.3.5 Effect of Pressure

EXAFS studies of semiconductors under high pressure provide information on bond compressibility and transition to other phases along with their identification (see Chap. 9). According to Aquilanti et al. [90], binary semiconductors of the form $A^N B^{8-N}$, i.e. those with atoms that have their outer shell filled with 8 electrons, when subjected to compression are transformed from the four-fold coordinated structures (diamond, W, ZB) to six-fold (rocksalt, β-tin) and to eight-fold (CsCl) coordinated. A systematic absence of the rocksalt or the β-tin phase is observed in the more covalent or more ionic semiconductors, respectively. The *nn* distance, which is directly related to the unit cell parameters, decreases monotonically with the pressure until an abrupt change that signifies that a phase transition occurs. In the ZB structure, the 1st *nn* distance is directly related to the a lattice constant since it is equal to $\sqrt{\frac{3}{4}}a$. In the W phase, the 2nd *nn* shell distance is the average of two distances as discussed in Sect. 3.3.1 whereas fitting of the 1st *nn* shell as well, provides information on the u-parameter. The distance *versus* pressure dependence can be fitted using the Murnaghan's equation:

$$d = d_0 \left(1 + \frac{B_0' P}{B_0}\right)^{-\frac{1}{3B_0'}} \tag{3.7}$$

where B_0 and d_0 are the bulk modulus and the interatomic distance at ambient conditions and $B_0' = \frac{dB}{dP}\big|_{P=0}$. Table 3.3 summarizes EXAFS results on the effect of pressure on WBG semiconductors.

Decremps et al. [68] fitted both the 1st and 2nd *nn* neighboring shells of the W-ZnO and found that the u-parameter is not significantly affected by the hydrostatic pressure. De Campos et al. [92] determined the pressure transition from the ZB to the rocksalt structure in nano-crystalline ZnSe at 15 GPa that is 1.5 GPa higher than the transition pressure in the crystalline phase. The XANES spectra are also sensitive in the application of pressure. They can be simulated using atom clusters with coordinates modified according to the changes in the lattice parameters upon pressure application. The linear combination of spectra can be also used for the determination of the percentage of the low- and high-pressure phases in the pressure region where the phase transition takes place as shown by Besson et al. [98].

Table 3.3 Phase transitions of WBG semiconductors identified by EXAFS and bulk modulus at ambient conditions (B_0) for the low-pressure phase

Material	Phase at ambient	Phase transition pressure (GPa)	B_0 (GPa)	Structure after depressurization	References
ZnTe	ZB	9.5 (cinnabar) 12 (rocksalt)	56 ± 5	ZB	[91]
nc-ZnSe	ZB	15 (rocksalt)	104.5		[92]
ZnS	ZB	15 (rocksalt)	82	ZB	[93, 94]
ZnO	W	9 (rocksalt)	181	W	[68]
GaN	W	50 (rocksalt)	245	W	[95]
GaP	ZB	26 (β-tin or Cmcm)	87.4	Partially amorphous	[96, 97]

3.4 Summary

In this chapter, applications of XAFS spectroscopy for the characterization of wide band gap semiconductors are presented. The effect of growth conditions, symmetry, composition, polymorphism, ion implantation, temperature and pressure are discussed. Examples on theoretical modeling of the near edge part of the spectra are also included.

References

1. L. Börnstein, *Group III Crystal and Solid State Physics*, vol 17a, b (Springer, Berlin, 1982)
2. S. Adachi, *Handbook on Physical Properties of Semiconductors*, vol. 3 (II–VI semiconductors), Springer, US (2004)/vol. 2 (III–V semiconductors), (Kluwer Academic Publisher, New York, 2004)
3. G.L. Harris (ed.), *Properties of Silicon Carbide* (Inspec, London, 1995)
4. S. Kasap, P. Capper (eds.), *Handbook of Electronic and Photonic Materials* (Springer, New York, 2007)
5. H.T. Grahn, *Introduction to Semiconductor Physics* (World Scientific Publishing, Singapore, 1999)
6. Ioffe Physico-technical Institute Electronic Archive, http://www.ioffe.ru/SVA/NSM/
7. J.C. Phillips, Rev. Mod. Phys. **42**, 317 (1970)
8. N.E. Christensen, S. Satpathy, Z. Pawlowska, Phys. Rev. B **36**, 1032 (1987)
9. N.E. Christensen, I. Gorczyca, Phys. Rev. B **50**, 4397 (1994)
10. A. Bauer, Ph. Reischauer, J. Kräusslich, N. Schell, W. Matz, K. Goetz, Acta Crystalogr. A **57**, 60 (2001)
11. G.C. Capitani, S. Di Perro, G. Tempesta, Am. Mineral. **92**, 403 (2007)
12. M. Katsikini, E.C. Paloura, T.D. Moustakas, J. Appl. Phys. **83**, 1437 (1998)
13. M. Katsikini, E.C. Paloura, M. Fieber-Erdmann, J. Kalomiros, T.D. Moustakas, H. Amano, I. Akasaki, Phys. Rev. B. **56**, 13380 (1997)
14. A. Bianconi, in *X-ray Absorption: Principles, Applications Techniques of EXAFS, SEXAFS and XANES*, ed. by D.C. Koningsberger, R. Prins (Wiley, New York, 1988), pp. 573–662
15. M. Katsikini, E.C. Paloura, M. Fieber-Erdmann, E. Holub-Krappe, D. Korakakis, T.D. Moustakas, J. Electron Spectrosc. **101–103**, 695 (1999)
16. J. Stöhr, *NEXAFS Spectroscopy* (Springer, Berlin, 1992)
17. F.A. Cotton, *Chemical Applications of Group Theory* (Willey, New York, 1971)
18. L.-C. Duda, C.B. Stagarescu, J. Downes, K.E. Smith, D. Korakakis, T.D. Moustakas, J. Guo, J. Nordgren, Phys. Rev. B **58**, 1928 (1998)
19. P. Ryan, C. McGuinness, J.E. Downes, K.E. Smith, D. Doppalapudi, T.D. Moustakas, Phys. Rev. B **65**, 205201 (2002)
20. M. Lübbe, P.R. Bresseler, W. Braun, T.U. Kampen, D.R.T. Zhan, J. Appl. Phys. **86**, 209 (1999)
21. I.J. Lee, H.-J. Shin, M.K. Lee, H.-K. Kim, Appl. Phys. Lett. **82**, 2981 (2003)
22. J. Stöhr, R. Jaeger, Phys. Rev. B. **26**, 4111 (1982)
23. Y. Muramatsu, T. Kaneyoshi, E.M. Gullikson, R.C.C. Perera, Spectrochim. Acta A **59**, 1951 (2003)
24. J. Moscovici, C. Loupias, Ph Parent, G. Tourillon, J. Phys. Chem. Solids **57**, 1159 (1996)
25. F.C. Brown, R.Z. Bachrach, M. Skibowski, Phys. Rev. B **13**, 2633 (1976)
26. C. Brouder, J. Phys.: Condens. Matter **2**, 701 (1990)
27. K. Miwa, A. Fukumoto, Phys. Rev. B **48**, 7897 (1993)
28. S. Elhamri, R.S. Newrock, D.B. Mast, M. Ahoujja, W.C. Mitchel, J.M. Redwing, M.A. Tischler, S.J.W. Flynn, Phys. Rev. B **57**, 1374 (1998)

29. Y.C. Yeo, T.C. Chong, M.F. Li, J. Appl. Phys. **83**, 1429 (1998)
30. M. Pedio, A. Giglia, N. Mahne, S. Nannarone, S. Giovannini, C. Cepek, F. Boscherini, R. Carboni, M. Benfatto, S. Della, Longa. Phys. Scripta **T115**, 308 (2005)
31. M. Tallarida, D. Schmeisser, F. Zheng, F.J. Himpsel, Surf. Sci. **600**, 3879 (2006)
32. J. Lüning, S. Eisebitt, J.-E. Rubensson, C. Ellmers, W. Eberhardt, Phys. Rev. B **59**, 10573 (1999)
33. L. Tröger, D. Arvanitis, K. Baberschke, H. Michaelis, U. Grimm, E. Zschech, Phys. Rev. B **46**, 3283 (1992)
34. X. Zhou, T.-K. Sham, W. Zhang, C.-Y. Chan, I. Bello, S.-T. Lee, Ha. Hofsäss. Anal. Chem **78**, 6314 (2006)
35. A. Gutiérrez, M.F. López, I. García, A. Vázquez, J. Vac. Sci. Technol. A **15**, 294 (1997)
36. L. Fayette, B. Marcus, M. Mermoux, G. Tourillon, K. Laffon, P. Parent, F. Le Normand, Phys. Rev. B **57**, 14123 (1998)
37. F.L. Coffman, R. Cao, P.A. Pianetta, S. Kapoor, M. Kelly, L.J. Terminello, Appl. Phys. Lett. **69**, 568 (1996)
38. D.M. Gruen, A.R. Krauss, C.D. Zuiker, R. Csencsits, L.J. Terminello, J.A. Carlisle, I. Jimenez, D.G.J. Sutherland, D.K. Shuh, W. Tong, Appl. Phys. Lett. **68**, 1640 (1996)
39. F. Bassani, Appl. Opt. **19**, 4093 (1980)
40. J.F. Morar, F.J. Himpsel, G. Hillinger, G. Hughesn, J.L. Jordan, Phys. Rev. Lett. **54**, 1960 (1985)
41. L.J. Huang, I. Bello, W.M. Laum S. -T. Lee, P. A. Stevens, B. D. DeVries, J. Appl. Phys. **76**, 7483 (1994)
42. A. Laikhtman, I. Gouzman, A. Hoffman, G. Comtet, L. Hellner, G. Dujardin, J. Appl. Phys. **86**, 4192 (1999)
43. M. Katsikini, F. Pinakidou, E.C. Paloura, W. Wesch, Appl. Phys. Lett. **82**, 1556 (2003)
44. J. Neugebauer, C.G. Van de Walle, Phys. Rev. B **50**, 8067 (1994)
45. M. Petravic, P.N.K. Deenapanray, V.A. Coleman, K.-J. Kim, B. Kim, G. Li, J. Appl. Phys. **95**, 5487 (2004)
46. B.J. Ruck, A. Koo, U.D. Lanke, F. Budde, S. Graville, H.J. Trodahl, A. Bittar, J.B. Metson, V.J. Kennedy, A. Markwitz, Phys. Rev. B **70**, 235202 (2004)
47. M. Katsikini, F. Pinakidou, E.C. Paloura, F. Boscherini, E. Wendler, W. Wesch, Mat. Sci. Eng. B **152**, 132 (2008)
48. M. Petravic, Q. Gao, D. Llewellyn P.N.K. Deenapanray, D. Macdonald, C. Crotti, Chem. Phys. Lett. **425**, 262 (2006)
49. S.O. Kucheyev, J.S. Williams, S.J. Pearton, Mat. Sci. Eng. R **33**, 51 (2001)
50. M. Katsikini, F. Pinakidou, E.C. Paloura, E. Wendler, W Wesch, R. Manzke. J. Phys. Conf. Ser. **190**, 012065 (2009)
51. I. Jiménez, A. Jankowski, L.J. Terminello, J.A. Carlisle, D.G.J. Sutherland, G.L. Doll, J.V. Mantese, W.M. Tong, D.K. Shuh, F.J. Himpsel, Appl. Phys. Lett. **68**, 2816 (1996)
52. I. Caretti, I. Jiménez, J. Appl. Phys. **110**, 23511 (2011)
53. F.M.F. de Groot, J. Electron Spectrosc. **67**, 529 (1994)
54. L.S. Pan (ed.) *Diamond: Electronic Properties and Applications* (Kluwer, Massachusetts, 1995)
55. D.D. Vvedensky, in *Unoccupied Electronic States: Fundamentals for XANES, EELS, IPS and BIS*, ed. by J.C. Fuggle, J.E. Inglesfield (Springer, Berlin, 1992), pp. 139–176
56. Xu Yong-Nian, W. Y. Ching. Phys. Rev. B **48**, 4335 (1993)
57. Xu Yong-Nian, W. Y. Ching. Phys. Rev. B **44**, 7787 (1991)
58. T. Mizoguchi, I. Tanaka, S. Yoshioka, M. Kunisu, T. Yamamoto, W.Y. Ching, Phys. Rev. B **70**, 045103 (2004)
59. A. Kisiel, G. Dalba, P. Fornasini, M. Podgórny, J. Oleszkiewicz, F. Rocca, E. Burattini, Phys. Rev. B **39**, 7896 (1989)
60. W.R.L. Lambrecht, S.N. Rashkeev, B. Segall, K. Lawniczak-Jablonska, T. Suski, E.M. Gullikson, J.H. Underwood, R.C.C. Ferera, J.C. Rife, I. Grzegory, S. Porowski, D.K. Wickenden, Phys. Rev. B **55**, 2612 (1997)
61. M. Magnuson, M. Mattesini, C. Höglund, J. Birch, L. Hultman, Phys. Rev. B **80**, 155105 (2009)
62. M. Magnuson, M. Mattesini, C. Höglund, J. Birch, L. Hultman, Phys. Rev. B **81**, 085125 (2010)

63. L. Liu, Y.M. Yiu, T.K. Sham, L. Zhang, Y. Zhang, J. Phys. Chem. C **114**, 6966 (2010)
64. Q. Li, H. Wang, Y. Xia, T. Cui, Y. Ma, G. Zou, Phys. B **404**, 643 (2009)
65. C. Levelut, Ph Sainctavit, A. Ramos, J. Petiau, J. Phys.: Condens. Matter **7**, 2353 (1995)
66. A.L. Ankudinov, B. Ravel, J.J. Rehr, S.D. Conradson, Phys. Rev. B **58**, 7565 (1998)
67. B. Gilbert, B.H. Frazer, H. Zhang, F. Huang, J.F. Banfield, D. Haskel, J.C. Lang, G. Srajer, G. De Stasio, Phys. Rev. B **66**, 245405 (2002)
68. F. Decremps, F. Datchi, A.M. Saitta, A. Polian, S. Pascarelli, A. Di Cicco, J.P. Itié, F. Baudelet, Phys. Rev. B. **68**, 104101 (2003)
69. J. Haug, A. Chassé, M. Dubiel, 1 Ch. Eisenschmidt, M. Khalid, P. Esquinaz. J. Appl. Phys. **110**, 063507 (2011)
70. Q. Guo, J. Ding, T. Tanaka, M. Nishio, H. Ogawa, Appl. Phys. Lett. **86**, 111911 (2005)
71. T. Chassé, K.H. Hallmeier, J.-D. Hecht, F. Frost, Surf. Rev. Lett. **9**, 381 (2002)
72. F. Boscherini, R. Lantier, A. Rizzi, F. D'Acapito, S. Mobilio, Mat. Sci. Eng. B **86**, 225 (2001)
73. Y.L. Wu, Z.C. Feng, J.-F. Lee, W. Tong, B.K. Wagner, I. Ferguson, W. Lu, Thin Solid Films **518**, 7475 (2010)
74. Shiqiang W, Zhihu S, J. Phys.: Condens. Matter **17**, 8017 (2005)
75. D. Diop, R. Grisenti, Phys. B **208&209**, 89 (1995)
76. M. Katsikini, H. Rossner, M. Fieber-Erdmann, E. Holub-Krappe, T.D. Moustakas, E.C. Paloura, J. Synchrotron Radiat. **6**, 561 (1999)
77. M. Katsikini, F. Pinakidou, E.C. Paloura, Ph Komninou, A. Georgakilas, E. Welter, Phys. Status Solidi A **205**, 2611 (2008)
78. D. Diop, R. Grisenti, Phys. B **208&209**, 164 (1995)
79. G. Dalba, P. Fornasini, R. Grisenti, D. Pasqualini, D. Diop, F. Monti, Phys. Rev. B **58**, 4793 (1998)
80. K.P. O'Donnell, J.F. W. Mosselmans, R.W. Martin, S. Pereire, M.E. White, J. Phys.: Condens. Matter **13**, 6977 (2001)
81. M. Katsikini, E.C. Paloura, F. Boscherini, F.D' Acapito, C.D. Lioutas, D. Doppalapudi, Nucl. Instrum. Methods B **200**, 114 (2003)
82. V. Kachkanov, K.P. O'Donnel, S. Pereire, R.W. Martin, Philos. Mag. **87**, 1999 (2007)
83. M. Katsikini, F. Pinakidou, E.C. Paloura, Ph. Komninou, E. Iliopoulos, A. Adikimenakis, A. Georgakilas, E. Welter, Phys. Status Solidi A **205**, 2593 (2008)
84. M. Ferhat, F. Bechstedt, Phys. Rev. B **65**, 07523 (2002)
85. T. Saito, Y. Arakawa, Phys. Rev. B **60**, 1701 (1999)
86. J. Fritsch, O.F. Sankey, K.E. Schmidt, J.B. Page, J. Phys.: Condens. Matter **11**, 2351 (1999)
87. K.M. Yu, W. Shan, C.J. Glover, M.C. Ridgway, W.S. Wong, W. Yang, Appl. Phys. Lett **75**, 4097 (1999)
88. W. Bolse, Nucl. Instrum. Methods B **148**, 83 (1999)
89. M. Katsikini, F. Pinakidou, E.C. Paloura, F. Boscherini, J. Appl. Phys. **101**, 083510 (2007)
90. G. Aquilanti, S. Pascarelli, J. Phys.: Condens. Matter **17**, 1811 (2005)
91. A. San-Miguel, A. Polian, M. Gauthier, J.P. Itié, Phys. Rev. B **48**, 8683 (1993)
92. C.E.M. Campos, J.C. de Lima, T.A. Grandi, J.P. Itié, A Polian, A. Michalowicz, J. Phys.: Condens. Matter **17**, 5187 (2005)
93. J. Itiéf Baudelet, E. Dartyge, A. Fontaine, H. Tolentino, A. San-Miguel, High Pressure Res. **8**, 697 (1992)
94. J.P. Itié, High Pressure Res. **7**, 354 (1991)
95. P. Perlin, C. Jauberthie, Carillon, J. P. Itie, A. San Miguel, I. Grzegory, A. Polian. Phys. Rev. B **45**, 83 (1992)
96. J.P. Itie, A. Polian, C. Jauberthie-Carillon, E. Dartyge, A. Fontaine, H. Tolentino, G. Tourillon, Phys. Rev. B **40**, 9709 (1989)
97. G. Aquilanti, H. Libotte, W.A. Crichton, S. Pascarelli, A, Trapanati, J.-P. Itié. Phys. Rev. B **76**, 064103 (2007)
98. J.M. Besson, J.P. Itié, A. Polian, G. Weill, J.L. Mansot, J. Gonzalez, Phys. Rev. B **44**, 4214 (1991)
99. L. Börnstein, *Group III Crystal and Solid State Physics*, vol 22 (Springer, Berlin, 1987)

Chapter 4
Dopants

Federico Boscherini

Abstract The applicability of XAFS to the structural study of dilute atoms is one of its main characteristics. It has been used extensively to study dopants in semiconductors, providing in many cases a structural basis for the understanding of physical properties. After a brief introduction to the use of XAFS in this field a review of the literature will be presented, ranging from the pioneering studies of As in amorphous Si to recent investigations on semiconductor systems of current interest.

4.1 Introduction to X-Ray Absorption Fine Structure Investigations of Dopants

4.1.1 General Aspects

The study of dopants in semiconductors using X-ray absorption fine structure (XAFS) is one of the fields which has best exploited the characteristics of the technique and has contributed to its quantitative and qualitative development over the past 20 years. The application of XAFS to this field of investigation is essentially made possible by the selectivity offered by appropriately choosing photon-in (absorption edge) and photon-out (fluorescence line) energies.

In fact, one of the main characteristics of XAFS is its chemical sensitivity: by tuning the impinging (photon-in) photon energy to the absorption edge of a particular atom its local structure can be probed. By exploiting the broad energy interval of synchrotron radiation sources in the vacuum ultraviolet, soft and hard X-ray ranges, the K and/or L edges of most elements in the periodic table can be excited and their local environment studied. Chemical sensitivity is not unique to XAFS (nuclear magnetic resonance and Mössbauer spectroscopy also exhibit it) but the applicability to most elements is certainly notable.

F. Boscherini (✉)
Department of Physics and Astronomy, University of Bologna,
Viale C. Berti Pichat 6/2, 40127 Bologna, Italy
e-mail: federico.boscherini@unibo.it

C.S. Schnohr and M.C. Ridgway (eds.), *X-Ray Absorption Spectroscopy of Semiconductors*, Springer Series in Optical Sciences 190,
DOI 10.1007/978-3-662-44362-0_4

Appropriate choice of the type of decay channel of the photoexcited atoms (fluorescence X-rays or Auger electrons) allows to isolate their contribution to the total cross section, thus drastically reducing background and permitting to probe the local structure of dilute elements. In the study of dopants in the bulk, thin films or heterostructures fluorescence yield (FLY) detection is mostly used. In the FLY detection mode the intensity of a certain energy region of the fluorescence radiation (the "region of interest") is measured as a function of the exciting photon energy; as we will show in the next section, for dilute elements this measurement is directly proportional to the X-ray absorption cross section of the excited atom. Electron yield detection is also used when the dopant atoms are in the near surface region. Various types of electron detection have been developed, among which are total electron yield (TEY), partial electron yield (PEY) and conversion electron yield (CEY). A grazing incidence geometry can be used to enhance surface sensitivity.

The dopant concentration which can be detected with XAFS depends on the atomic number of the dopant and the matrix, brilliance of the photon source, experimental set-up and detector; as a rough guide, EXAFS data with reasonable signal-to-noise ratio can be obtained on volume or areal concentrations of the order of 10^{18} atoms/cm^3 or 10^{15} atoms/cm^2, respectively, while XANES data can be recorded on one order of magnitude lower concentrations. This low concentration is a notable advantage of XAFS compared to other techniques with structural sensitivity (for example to X-ray diffraction, which is essentially inapplicable to dopants) but is considerably higher than that which can be probed by some defect-sensitive but structurally insensitive techniques such as deep level transient spectroscopy (DLTS). The main limitation of XAFS to study dopants is the fact that it intrinsically measures the average structure of all atoms of the photoexcited element; if these atoms are present in different local atomic environments corresponding to different incorporation sites in the host lattice, disentangling their contribution can be difficult. However, we note that use of ab initio structural simulations in data analysis has allowed great advances in tackling this problem (see especially work on As ultra shallow junctions reported in Sect. 4.2.3).

4.1.2 Experimental Methods

4.1.2.1 Fluorescence Detection

As illustrated in Chap. 1, transmission measurement of XAFS is the most commonly used experimental configuration. It relies on the measurement of the intensity of monochromatic photons incident (I_0) and transmitted (I_T) by a sample of thickness x. The absorption coefficient $\mu(\hbar\omega)$ is related to the measured quantities by

$$\mu(\hbar\omega)\, x = \ln\left[\frac{I_0(\hbar\omega)}{I_T(\hbar\omega)}\right]. \tag{4.1}$$

The experiment provides a quantity which is linearly dependent on the expression on the right hand side of (4.1) and thus on the absorption coefficient, which is what one must measure in an XAFS experiment.

In FLY detection [1], a photon detector is used to measure the fluorescence intensity in a certain "region of interest" as a function of energy. In the most common experimental geometry using horizontally polarized X-rays, the detector is placed in the horizontal plane at right angles to the incident beam to minimize the intensity of elastic scattering of the impinging X-ray beam; in this geometry, the intensity of fluorescence photons, $I_f(\hbar\omega)$ of atom A embedded in a matrix M for a sample thickness d, can be easily shown to be [2]:

$$I_f(\hbar\omega) = I_0(\hbar\omega)\frac{\Omega}{4\pi}\varepsilon_f\frac{\mu_A(\hbar\omega)}{\mu_{tot}(\hbar\omega)+\mu_{tot}(\hbar\omega_f)\tan\theta} \times \left[1-\exp\left\{-\left(\mu_{tot}(\hbar\omega)+\mu_{tot}(\hbar\omega_f)\tan\theta\right)d\right\}\right] \quad (4.2)$$

where ε_f is the fluorescence yield, Ω is the solid angle subtended by the detector, $\hbar\omega_f$ is the energy of the fluorescence photons, θ is the angle between the impinging beam and the sample surface and the total absorption coefficient is the sum of contributions due to atomic species A and that due to the matrix M: $\mu_{tot} = \mu_A + \mu_M$. Since μ_{tot} depends on μ_A, (4.2) is not in general proportional to the quantity one wants to measure, that is $\mu_A(\hbar\omega)$. This proportionality is recovered in two limiting cases: thin samples (for which the series expansion of the exponential in (4.2) can be truncated at the term linear in d, typically below 100 nm) and for thick samples in which atom A is dilute ($\mu_{tot} \cong \mu_M$, a condition which depends on the angle θ but usually occurs below a few atomic %). The latter condition is met by dopants in semiconductors and is thus of interest for this chapter.

Different types of detectors for FLY mode have been used, but the most common are based on hyper-pure Ge diodes or Si drift diodes (SDD) [3–5]. These detectors often are in the form of multi-element arrays and have an energy resolution in the range of 100–200 eV at 6 keV, a reasonable trade-off allowing a good counting rate and efficient background rejection. Since pulse counting electronics is used, care must be taken to avoid, or correct, dead-time effects [6].

The lowest concentration which can be measured in a FLY-XAFS experiment ultimately depends on the magnitude of the inelastic scattering background in the specific experimental geometry. Takeda et al. [7] have studied in detail the factors which determine the lowest concentration which can be detected; they have come to the conclusion that, for the specific case of the L_{III} edge of Er in GaAs or InP, the lowest concentration measurable is $1\text{–}5 \times 10^{14}$ atoms/cm^2.

In the above description, core hole creation and emission of fluorescence radiation are viewed as separate processes. This is an adequate approach for all cases we will describe in this chapter but may fail in highly correlated materials in the near edge region; in these cases it is necessary to describe the process as (resonant) inelastic scattering (RIXS) [8]. In this context high resolution crystal spectrometers have been developed which allow high energy resolution analysis of the spectrum of emitted photons. New variations of fluorescence detected XAFS have been developed, such as use of K_β detection [9] and use of only a portion of the fluorescence spectrum—so called high energy resolution fluorescence detection (HERFD) [10]; these techniques are useful when multiple oxidation or spin states of the excited atom are present and might be of future interest in the study of dopants in semiconductors.

4.1.2.2 Electron Yield Detection

Following photoexcitation the core hole may decay with the emission of an Auger electron; fluorescence and Auger electron emission are competing processes, the probability of each summing to one. The Auger electron creates a cascade of lower energy ("secondary") electrons due to multiple ionization and inelastic scattering processes. Various electron yield detection methods have been developed, depending on which electrons are detected and the detection method. The experimentally simplest method, often used in the soft X-ray region and termed Total Electron Yield (TEY), consists in measuring the drain current using an electrometer which connects the sample to ground. A variation on TEY is Partial Electron Yield (PEY) in which only electrons of a given energy range are detected, with the use of retarding grids and electron detectors. In the hard X-ray range, cross-sections are smaller than at lower energies and the most common method consists in recording the signal due to radiationless decay using a He-filled ionization chamber in which the sample acts as the anode [11–14]: He gas acts as an electron multiplier and an amplified signal is recorded from the cathode; this method is also termed Conversion Electron Yield (CEY).

The main peculiarity of all electron detection modes is the surface sensitivity, which is guaranteed by the limited electron mean free path. Hence, these modes are particularly useful for the investigation of the near surface region or of interfaces. Erbil et al. [15] have reported a detailed study of the depth sensitivity of TEY mode, concluding that it depends on the range of the original Auger electrons and that it generally is in the range of a few 100 to a few 1,000 Å. Schroeder et al. [16] have compared TEY and CEY in detail, concluding that the former is more surface sensitive than the latter.

4.1.2.3 Electrical Detection

Some "non-conventional" methods to detect XAFS have been proposed, each with its own advantages, limitations and pitfalls. We briefly review electrical detection since it is the most relevant to dopants in semicondutors.

The electrical detection of XAFS has been investigated by some groups with the objective of determining whether the local structure of only those atoms contributing to the electrical signal due to defects could be determined, leading to an attractive site-selective version of XAFS.

Photocurrent detected XAFS spectra have been reported by Boehme et al. [17] and Hu et al. [18] for an Al–GaAs diode and for bulk GaAs, respectively. Their common conclusion is that these spectra contain, for these concentrated samples, the same structural information as the more traditional transmission mode ones, once corrections for sample thickness are taken into account.

Ishii et al. [19–21] have proposed a capacitance-detected version of XAFS, the specific system studied being a Se-doped AlGaAs–Al diode in which the semiconductor exhibits the so-called DX deep level electron trap. It is argued by these authors that, since changes in capacitance are due to electrons localized in the traps, it is possible to determine the local structure of only those atoms at a short distance from the trap itself. In fact, their spectra at the Ga edge recorded in the fluorescence and in the capacitance mode exhibit significant differences. Ishii has also proposed variations of the capacitance method; by varying the diode bias potential a depth selectivity is proposed [22], while use of scanning capacitance probes is proposed to provide lateral resolution on the length scale of the AFM [23–26].

The site-selectivity of capacitance-detected XAFS has been challenged in detail by Bollmann et al. [27]. These authors have examined samples exhibiting both the DX deep level and antisite defects; spectra measured in the fluorescence and capacitance mode *do not* show any significant differences. Also by modeling the X-ray induced current and capacitance changes, Bollmann et al. conclude that, since the defect centers can be ionized also by electrons (directly or indirectly) generated by core ionization of atoms distant from the defect center themselves, no site-selectivity results. It is clear that in order for this detection mode to be accepted by the scientific community more work is necessary to resolve the issue of selectivity.

4.1.2.4 Grazing Incidence

In the X-ray range the index of refraction can be written as

$$n(\hbar\omega) = 1 - \delta(\hbar\omega) + i\beta(\hbar\omega) \tag{4.3}$$

where both δ and β are positive and $\ll 1$; δ and β describe dispersion and absorption processes, respectively and are related by the Kramers–Kronig relations. For Si at $\lambda \sim 1.2\,\text{Å}$ ($\hbar\omega = 10\,\text{keV}$) $\delta \cong 9.8 \times 10^{-6}$ and $\beta \cong 1.5 \times 10^{-7}$. As a consequence of the very small value of δ, an X-ray beam impinging on the flat surface of a sample may undergo *total external reflection* if the angle of incidence (α, taken to be the angle between the beam's direction and the surface) is smaller than the critical angle,

$$\alpha_c = \sqrt{2\delta}. \tag{4.4}$$

For Si at $\lambda \sim 1.2\,\text{Å}$, $\alpha_c \cong 4.5\,\text{mrad}$. For $\alpha < \alpha_c$ the reflectivity approaches 1 and inside the sample the beam is confined to the near surface region; compact expressions for the penetration depth ℓ in the direction normal to the surface can be derived [28] for three cases (the values listed are for Si at $\lambda \sim 1.2\,\text{Å}$):

$$\begin{aligned} \alpha \ll \alpha_c: \quad & \ell = \tfrac{1}{2k\sqrt{2\delta}} \approx 2.2\,\text{nm} \\ \alpha = \alpha_c: \quad & \ell = \tfrac{1}{2k\sqrt{\beta}} \approx 26\,\text{nm} \\ \alpha = 10\alpha_c: \quad & \ell = \tfrac{\alpha}{2k\beta} \approx 29\,\mu\text{m} \end{aligned} \tag{4.5}$$

The last expression in (4.5) is just the usual penetration length determined by the absorption coefficient ($\mu = 2k\beta$) projected on the surface normal. Moreover, it can be shown that for $\alpha = \alpha_c$ the beam intensity at the surface is four times the intensity of impinging beam. The reflectivity is a complicated function of α, δ and β, which approaches 1 for $\alpha \ll \alpha_c$ and 0 $\alpha \gg \alpha_c$. It is clear that a strong confinement of the X-ray beam in the near surface region can be obtained by using grazing incidence.

The possibility of using grazing incidence to probe the surface region of materials was illustrated by Parratt [29]. As far as XAFS is concerned, Barchewitz et al. [30] were the first to record fine structure via the reflectivity, $R\,(\hbar\omega)$, and to suggest the relation with the absorption spectrum. Subsequently, Martens and Rabe [31] provided an in-depth discussion, opening the way to what has been called the "RefleXAFS" technique. The fine structure in $R\,(\hbar\omega)$ is related to the "usual" XAFS spectrum $\chi(k)$. However, it is not straightforward, in the general case, to obtain $\chi(k)$ from $R\,(\hbar\omega)$ since the latter quantity depends in a complicated way on both $\delta\,(\hbar\omega)$ and $\beta\,(\hbar\omega)$; of course, the "usual" XAFS spectrum is contained in $\beta(\hbar\omega)$. A simplification arises sufficiently far from the edge and for $\alpha < 0.8\alpha_c$, since it can be shown [32] that

$$\mu\,(\hbar\omega) = \frac{1 - R\,(\hbar\omega)}{1 + R\,(\hbar\omega)} \tag{4.6}$$

A number of methods have been proposed to extract an XAFS spectrum from the reflectivity, each with its advantages and limitations [31–34]. Instead of measuring the reflectivity, FLY or TEY may be used, of course [35]; in the case of a dilute or thin sample, a spectrum recorded in this way is directly proportional to $\mu(\hbar\omega)$ and does not require corrections [36]. Grazing incidence XAFS is a powerful tool to study the near surface region of samples and has been used also to study dopants in semiconductors.

4.1.2.5 Micro-XAFS

The great improvements in the brightness of synchrotron sources has been the possibility of obtaining smaller focal spot sizes and to design instruments capable of increasingly higher spatial resolution. In fact, XAFS spectra with spatial resolutions near or below 1 μm can be recorded at a number of synchrotron radiation laboratories (see for example [37]).

There is a great interest in performing XAFS with high lateral resolution (micro-XAFS) [38] since it allows a new level of description of heterogeneous samples, combining microscopy with the atomic-scale structural information obtainable from XAFS; with sufficiently small focal spots the local atomic environment of single nanostructures might be determined. Applications of micro-XAFS to semiconductors have been relatively limited so far, but it is expected they will be more numerous in the future. Micro-XAFS is particularly useful when the samples investigated exhibit lateral inhomogeneities and it is not surprising that many applications in the field of semiconductors have been on samples the growth of which has yet to be optimized, for example dilute magnetic semiconductors based on GaN, While photon-based techniques might maybe never reach the sub-Å spatial resolution obtainable in electron microscopy, the advances in recent years have been really dramatic since they involve many orders of magnitude improvement in spatial resolution. In the future, micro-XAFS will certainly become a powerful tool in materials science, complementing spatially averaged techniques which have been the key to the success and widespread use of X-rays for structural studies.

4.2 A Review of XAFS Investigations of Dopants

The introductory textbook description of doping in semiconductors states that a foreign atom (e.g. As) with one electron more (or less) in the valence shell than the matrix atoms (e.g. Si) occupies a substitutional lattice position; it is thus "forced" in a bonding configuration which releases a charge carrier (an electron or a hole). The reality is of course much more complex: foreign atoms often occupy non substitutional positions and/or form complexes or precipitates which do not release charge carriers; this occurs especially at high concentration, leading to charge carrier saturation. XAFS has played an important role in elucidating, in selected cases, the real local structure of dopants in semiconductors. In this section these studies will be briefly reviewed; the selection of papers cited highlights the cases in which a good understanding of physical properties has been obtained on the basis of XAFS (thus going beyond a mere descriptive approach) and is in part the result of a personal choice. We note that the application of XAFS to semiconductor heterostructures and nanostructures [39] and the application of advanced XAFS methods in semiconductor science [40] have been recently reviewed.

4.2.1 Amorphous Semiconductors

Amorphous silicon (a-Si) and hydrogenated a-Si (a-Si:H) exhibit semiconducting properties with lower carrier mobility with respect to crystalline Si. However, they can be easily and cheaply deposited as thin films at low temperatures on a variety of substrates, even on large areas. They are used in solar cells and thin film transistors for liquid crystal displays.

Amorphous semiconductors lack long range order and in the case of group IV materials their structure can generally be described as a random arrangement of locally ordered tetrahedral units (see also Chap. 7). Since in conventional wisdom it is long range order which imposes fourfold coordination to an atom (e.g. As or P) which does not have four valence electrons, before 1975 it was believed that amorphous semiconductors could not be doped; a group III or group V impurity would thus always have three nearest neighbors, following elementary valence chemistry. The discovery that limited doping of hydrogenated amorphous silicon (a-Si:H) was, in fact, possible opened the way to applications of this material. The pioneering application of XAFS to this issue was extremely important to explain the structural origin of doping and also to illustrate the capabilities of the technique (which at the time was still in its infancy).

Knights et al. [41] studied As impurities in a-Si:H in the concentration range from 1 to 12 %, with measurements actually performed in the transmission mode (fluorescence detection was being developed at the same time); thin films were deposited by chemical vapor deposition from silane/arsine mixtures. These authors showed that while at the high concentration range As is always threefold coordinated with decreasing As content some fourfold coordinated sites are found (this was deduced from an increase in the average coordination number); the former sites are electrically inactive while the latter are electrically active and are thus responsible for doping in a-Si:H. Subsequent modeling studies [42, 43] showed that the number of electrically active sites scale as the square root of the total impurity concentration, due to an equilibrium between active impurities on one hand and inactive impurities plus Si dangling bonds on the other.

As and Ga dopants ion implanted in a-Si:H in the concentration range 5×10^{19}–10^{21} atoms/cm^3 were studied by Greaves et al. [44] using grazing incidence and fluorescence detection. From an analysis of the envelope of the EXAFS oscillations these authors were able to detect the presence, along with the threefold site, of the structurally ordered, electrically active, fourfold site and to confirm the square root dependence of their number on the impurity concentration.

The case of hydrogenated amorphous Germanium appears to be different. In fact, Dalba et al. studied Ga, Sb and Bi dopants in a-Ge:H films deposited by radio frequency sputtering in the concentration range 1.5×10^{18}–4.5×10^{20} atoms/cm^3 [45, 46]. They were able to explicitly show the change from threefold to fourfold coordination but their data is not compatible with the equilibrium model described for a-Si:H.

4.2.2 Crystalline Silicon: Bulk

The prototypical dopant system—arsenic in crystalline Si—has (not surprisingly) been the object of numerous investigations. Pioneering studies were published by an IBM group in 1986 [47] and 1988 [48]. Si (001) wafers were ion implanted at 100 keV with doses 6×10^{15}–6×10^{16} atoms/cm^2 and were subsequently furnace or laser annealed. The first study focused on the issue of structural relaxation: the authors found that the first shell As–Si interatomic distance was 0.06 Å greater than the Si–Si bond length in the crystal, while the second shell distance around As was very close to the value in Si bulk. This fact is not surprising since it has been well established that semiconductor bonds are rather "stiff" (see also Chap. 2). The second study was among the first to combine ab initio structural calculations with XAFS. In the higher concentration samples the authors detected a weak contribution due to As atoms in the second coordination shell around As itself and related this to the formation of dopant—vacancy complexes, specifically a Si vacancy surrounded by four As atoms: (v-As_4). A more complex picture emerged from a subsequent study by Allain et al. [49] performed on similar samples. For annealing temperatures lower that 750 °C they found the formation of inactive clusters involving about seven As atoms around one vacancy while at higher annealing temperatures they suggest the coexistence of these clusters with SiAs precipitates. These studies were important in showing the potential of XAFS to study dopants; however, the conclusion would nowadays be judged as merely indicative and a state of the art analysis would now attempt at reproducing the full XAFS spectrum on the basis of a structural simulation (this has recently been done for As dopants in Si ultra shallow junctions, which will be described in the next section). An intrinsic difficulty in analyzing data from this system is that vacancies affect the XAFS spectrum by reducing the apparent coordination number, a parameter which is notoriously affected by a high uncertainty.

More recently, Koteski et al. [50] have focused on the issue of structural relaxation around substitutional As in Si (concentration 4×10^{18} atoms/cm^3), complementing the experiment with ab initio density functional theory (DFT) calculations. Data analyzed up to the fourth coordination shell confirmed an expansion of the first coordination shell distance around As, in agreement with simulations.

An alternative explanation for charge carrier saturation in Si has been proposed by a group from Bell laboratories [51]. They studied a set of Sb doped Si samples grown by molecular beam epitaxy in the concentration range 7×10^{19}–1.5×10^{21} atoms/cm^3; data analysis was aided by total energy calculations of defect structures. These authors propose that donor pair complexes which involve two separated but interacting dopants with no Si vacancies are responsible for charge carrier saturation. Their data analysis was based on qualitative comparison of the experimental spectrum with simulations of the XAFS signal based on the geometry of the proposed donor pair complexes on the basis of ab initio simulations; this approach was in its infancy at the time and would then be much extended and used in a more quantitative fashion. The same group subsequently extended their study to "δ-doped" Si [52] in which Sb dopants are confined to a quasi two dimensional layer. This technique allows to obtain very high electrical activation. The electrical data correlates with the Sb L_3 edge XAFS measurements which indicates reduced formation of donor pair complexes in this case; in Fig. 4.1 we report free carrier densities versus Sb concentration for various samples and in Fig. 4.2 the Fourier Transforms of some samples; the presence of donor pairs is indicated by the low amplitude of the second and third shell peak between 3 and 4 Å. XAS studies of complexes in Si and other semiconductors are also discussed in Chap. 5.

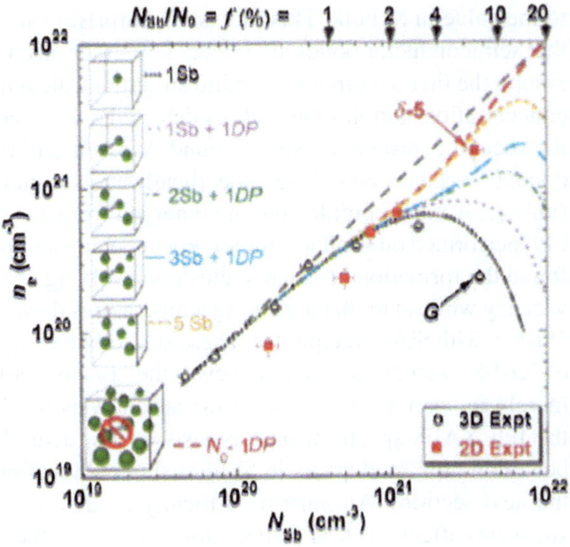

Fig. 4.1 Free carrier densities in Sb doped Si: 3D refers to usual doping in the bulk while 2D refers to "δ-doping". Reprinted with permission from [52]. © The American Physical Society

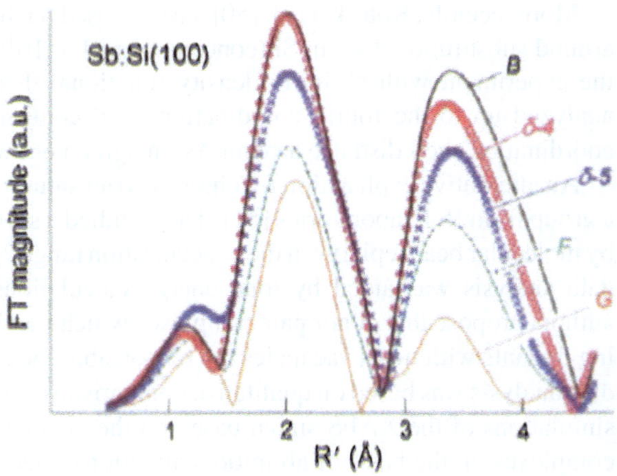

Fig. 4.2 Corresponding Fourier transforms of Sb L_3 edge XAFS data; the reduced intensity of the peak at ∼3.5 Å is due to formation of donor pair complexes. Samples B and G are bulk doped at low and high concentrations, respectively, while δ-4 and δ-5 are δ doped samples. Reprinted with permission from [52]. © The American Physical Society

4.2.3 *Crystalline Silicon: Ultra Shallow Junctions*

In the continuing trend toward miniaturization of microelectronic components a crucial recent issue has been to determine the best processing method for the fabrication of ultra shallow junctions (USJs) in Si; USJs are regions with greater than equilibrium dopant concentrations confined to the top 10–20 nm of the Si wafer and which are required to have the sharpest possible concentration profile. USJ are fabricated by very low (a few keV) ion implantation coupled to various annealing treatments. XAFS has been able to elucidate the connection between the local structure of As and B dopants and their physical properties.

Arsenic USJs have been studied by fluorescence detected grazing incidence XAFS in similar studies performed in the same period of time (on the same beamline) by d'Acapito et al. [53] and by Giubertoni et al. [54]. The two studies will be shortly described separately and a brief common conclusion presented.

- The first group [53] studied a set of samples deposited at 5 keV (estimated projected range 8 nm) at a fluence of 1×10^{15} atoms/cm^2 and subjected to various rapid thermal annealing processes. Variation of the grazing angle allowed varying the probe depth from 6 to 300 nm. Interpretation of XAFS data was aided by ab initio structural simulations of the equilibrium structure of substitutional As (As_{Si}) and vacancy—As complexes involving from one to four As atoms (v-As_n). They detected the presence of mostly As_{Si} accompanied to various amounts of v-As_n, the relative fractions changing with distance from the surface; a considerable degree of structural disorder was found in the top surface regions.
- The second group of investigators [54] combined XAFS and ab initio structural simulations of defect structures with secondary mass spectrometry SIMS to measure the As concentration, sheet resistance measurements and Hall effect to measure the carrier concentration and thus the fraction of active As dopants. Samples were obtained by implantation at 2 keV at a dose of 1 and 3×10^{15} atoms/cm^2 followed by laser submelt annealing or rapid thermal annealing; the latter process was followed by etching to remove the disordered topmost layers. The authors were able to fit the EXAFS data with a linear combination of the signal expected for As_{Si} and ν-As_n defects and thus obtain relative fractions of these different local As environments. They found a good correspondence between the fraction of active As sites determined from transport measurements and the relative amount of As_{Si} sites determined from EXAFS. This is illustrated by Fig. 4.3 in which the Fourier Transforms of the As K-edge EXAFS for various implanted samples are compared to a reference sample of substitutional As (topmost trace). The trace at the bottom corresponds to an as—implanted and amorphized sample and the intermediate ones to higher activation levels (3.4, 29.5 and 27.3 % for samples 4, 2 and 3); it can be noticed that the lineshapes clearly become increasingly similar to the reference as the active fraction increases from bottom to top. The same group subsequently reported on the deactivation of the dopants following thermal treatment [55].

Both studies just reported clearly show that As is generally present in a range of local structures: As_{Si}, ν-As_n ($n = 1$–4) defects and SiAs precipitates, the first correlating well with the fraction of electrically active As (reasonable on the basis of textbook physics!). It can be commented that these results could not have been obtained without a significant contribution of ab initio structural simulations; a simple analysis based only on the determination of the average coordination numbers in the first few coordination shells would certainly have failed to catch the details inherent in the distribution of local environments in this rather complex system.

In the case of boron, USJs have recently been realized using an approach known as solid state epitaxy regrowth; the dopant is implanted into a preamorphized substrate and then activated during the recrystallization of the amorphous phase.

Fig. 4.3 Magnitudes of the Fourier transforms of As K-edge EXAFS for As USJ in Si. The various traces are described in the text. Reprinted with permission from [54]. © The American Institute of Physics

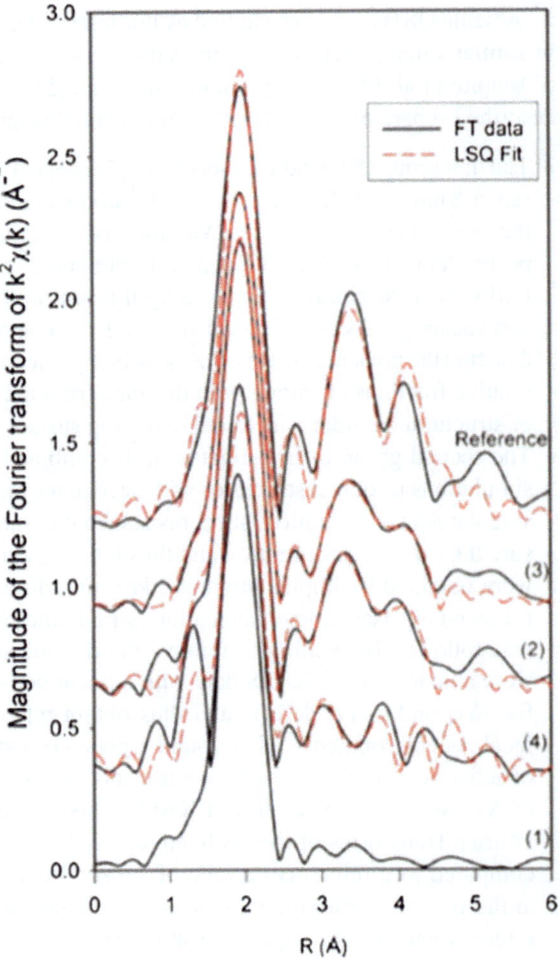

An active level of substitutional boron of about 2–4 × 10^{20} atoms/cm^3 can be reached, well above the solubility limit in the crystalline phase. The structural origin of the significant fraction of electrically inactive B atoms above this activation level and the determination of the processing step at which inactive B forms have been addressed using B K-edge XANES by De Salvador et al. [56]. The B K edge is found at a very low energy (190 eV) at which XAFS investigations are quite few, due to a number of technical problems which make such measurements rather challenging. Samples were preamorphized with a Ge 20 keV beam, implanted with B at 500 eV (1 × 10^{15} atoms/cm^2) and subsequently thermally annealed. The XANES experiment was combined with SIMS to determine the B concentration profile in the sample and Rutherford Basckscattering Spectrometry (RBS) to probe the thickness of the amorphous layer. By correlating the spectral signature of B sp^2 hybridized orbitals with the position of the amorphous–crystal interface it was possible to demonstrate that B dimers [57] are responsible for charge carrier saturation, that they form in the amorphous matrix and then are transferred to the crystalline one. This study shows how a rather simple spectral interpretation based on molecular orbital description can be applied in the field of semiconductors. We close this section by noting that Sahiner et al. [58] have detected Ge–B bonding in this system.

4.2.4 Solar Grade Silicon

In recent years the demand for high purity Si has increased to the point of exceeding supply; in the field of photovoltaics, alternative solutions have thus been sought, one being the use of "solar grade silicon" which contains much higher concentrations of transition metal impurities. Since these impurities decrease charge carrier diffusion lengths and thus the efficiency of solar cells, ways to reduce these detrimental effects have been sought. In this context, micro XAFS investigations performed by Buonassisi et al. [59] have been very important in suggesting a processing recipe

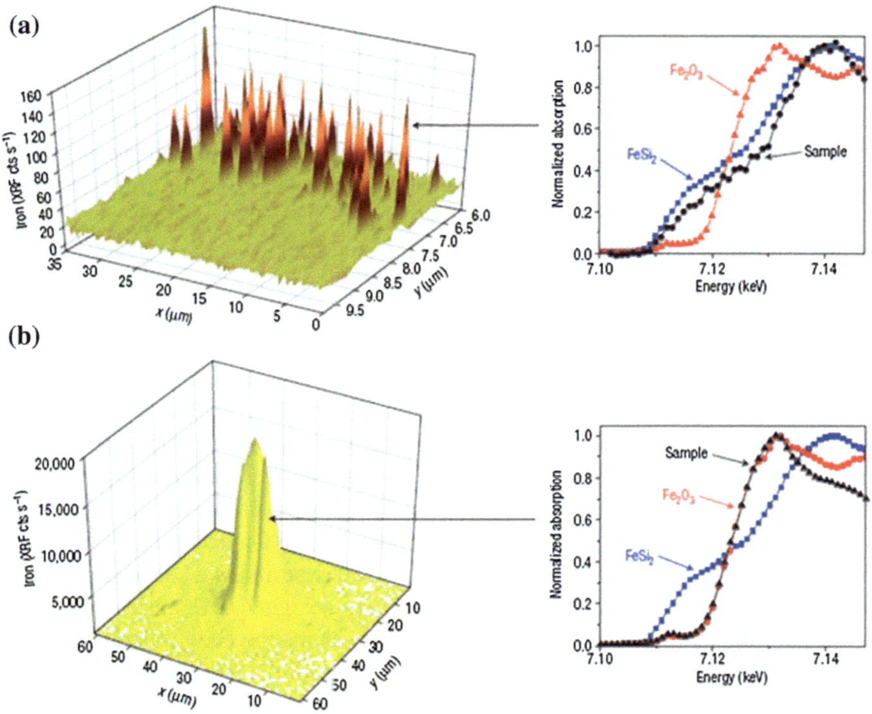

Fig. 4.4 Micro fluorescence maps and micro-XANES of two types of Fe precipitates in solar grade silicon. Reprinted with permission from [59]. © Nature publishing group

which minimizes the adverse effects of metal impurities. This paper shows how XAFS measurements can provide an explanation of a processing technique.

The authors use a highly focussed X-ray beam to perform in combination fluorescence mapping, X-ray beam induced current (XBIC) and XANES spectroscopy with a high spatial resolution. Their main result is illustrated in Fig. 4.4, which reports the spatial distribution and XANES spectra of two types of metal defect in commercial solar grade Si: rather small (20–30 nm) and densely packed Fe silicide nanoprecipitates and bigger (several μm) and sparsely distributed Fe oxide precipitates. The relative abundance of these two kinds of defect can be controlled by the cooling rate of the crystal from the melt, a slower cooling rate producing the bigger and sparsely distributed oxide precipitates. Electrical measurements showed that in this case carrier diffusion lengths are a factor of four higher than for a sample with densely packed, small precipitates, thus indicating the best processing route. These micro-XAFS studies are important as they provide a structural and chemical foundation for the processing recipe which leads to the optimization of transport properties. Also, they have illustrated the power of microbeam X-ray techniques in the characterization of inhomogeneous semiconductor materials. There has been much progress in the field recently, as testified by the construction of dedicated experimental stations, for example at the ESRF [60].

4.2.5 Gallium Arsenide

The site of Si dopants in GaAs has been the subject of thorough investigations published in 1993 and 1995 by Schuppler et al. [61, 62]. At the time, electrical deactivation in GaAs was thought to be due to an "autocompensation" mechanism which originates in the amphoteric nature of the Si dopant in GaAs; a substitutional Si on a Ga site acts as a donor while a substitutional Si on an As site acts as an acceptor and both sites are possible. In the conventional wisdom at the time, deactivation was thought to be due to the presence, at high concentrations, of both sites for Si. Schuppler et al. were able to convincingly demonstrate that Si dimers and small clusters are also present and contribute significantly to deactivation.

The papers by Schuppler et al. are an excellent example of application of XAFS to a complex materials science problem. The samples were specifically deposited to solve the issue at hand, the challenging XAFS measurements (a weak Si K_α signal nearly superposed to a strong background due to Ga and As L_α lines) were made possible by the use of advanced detectors and experimental set-up and the data analysis relied on the comparison with specially designed reference compounds. Nowadays this study would have been aided by (or maybe even relied on) refined structural and spectral simulations; nevertheless, the direct and empirical approach of these papers published nearly 20 years ago is very convincing.

In the first paper [61] the authors studied a set of eight samples deposited by MBE, half of them deposited on GaAs(001) substrates and the others on GaAs(311)A substrates; at low concentrations the incorporation site is known to be a substitutional

Ga site in the former case and a substitutional As site in the latter. For each substrate, four concentrations were investigated: below, at, above and well above the critical concentration of 1×10^{18} atoms/cm^3 (labeled L, M, H and VH in the figures). The Si K-edge XANES spectra (reproduced in Fig. 4.5) clearly illustrate the changing local environment of Si: for both sets of samples a single site at low concentrations is found while at higher concentrations the spectra suggest a more complex structure. While the spectral evolution is compatible with the presence of both As and Ga substitutional sites the authors found that it was impossible to fit the highest concentration samples with only these two contributions. The issue was solved in the 1995 paper [62] with the use of EXAFS spectra of the samples and of a set of SiGe random alloy reference compounds; SiGe alloys (see also Chap. 2) can serve as a reference for the local coordination of Si in GaAs in view of the similarity of backscattering function of Ga, Ge and As. In Fig. 4.6 the Fourier Transforms of selected data is reported; it is apparent that the very high concentration sample has a signal very similar to the $Si_{0.20}Ge_{0.8}$ sample, demonstrating that significant Si–Si bonding occurs at very high concentrations.

Formation of Si dimers and clusters was also found in GaAs δ-doped layers and Si/GaAs superlattices by Boscherini et al. [63]. The samples consisted in multiple periods of Si layers separated by GaAs spacer layers deposited by MBE, specifically

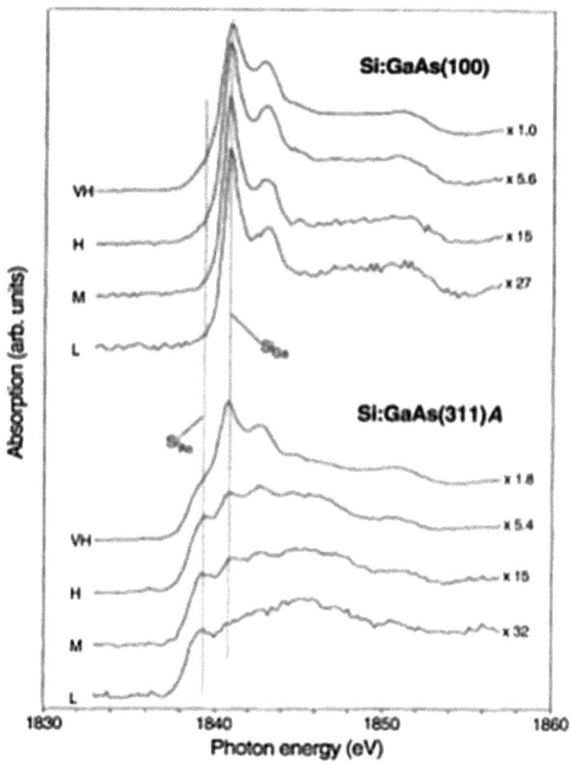

Fig. 4.5 Si K-edge XANES for Si doped GaAs, deposited on GaAs(001) and GaAs(311)A substrates as a function of Si concentration. Reprinted with permission from [62]. © The American Physical Society

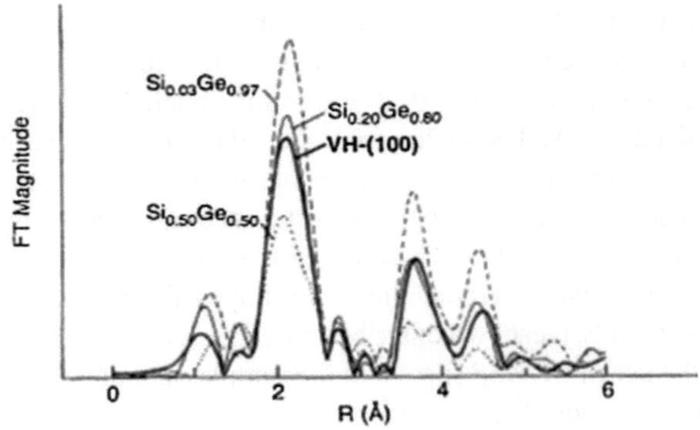

Fig. 4.6 Fourier transforms of Si K-edge EXAFS for the very high concentration sample deposited on GaAs(001) and selected SiGe alloys. Reprinted with permission from [62]. © The American Physical Society

deposited to perform the XAFS experiment; the equivalent thickness of the Si layers ranged from 0.02 to 4 monolayers and the number of periods from 500 to 40. Also in this case the presence of a considerable number of Si–Si bonds (never less than two bonds out of the four nearest neighbors) was unambiguously detected by analysis of the EXAFS data.

The related system consisting of Mn δ-doped GaAs has been recently studied by d'Acapito et al. [64]. This system is of interest for the development of ternary magnetic semiconductors. In samples grown at 300 °C Mn was found to always occupy a substitutional Ga site. Of interest in this study is that the authors were able to overcome the difficulty of distinguishing between Ga and As nearest neighbors from EXAFS (the backscattering functions are very similar) by analyzing the XANES spectrum with the aid of ab initio spectral simulations in the real space multiple scattering approach [65, 66]; the simulated XANES is very different for Ga and As substitutional sites—a consequence of the difference charge densities which strongly affect scattering of low energy near edge electrons. In samples grown at a 450 °C the authors detected the initial formation of MnAs precipitates. Interestingly, co-doping with Be causes the formation of a new defect site—Mn interstitials; this is due to the shift of the Fermi level and to the reduction of the formation energy for such defects. XAS studies of GaMnAs are discussed in detail in Chap. 15.

We close this section by citing two papers [67, 68] which report studies of the incorporation mechanism of N in GaAs using high resolution N K-edge XANES, a method which will be discussed in more detail in the next section.

4.2.6 Zinc Oxide

ZnO is a direct and wide band gap (3.3 eV) semiconductor which has recently attracted considerable interest as a possible optoelectronic material. ZnO is inexpensive, mechanically hard, and bulk substrates are available. One major limitation in the the wide-scale use of ZnO for device applications is the difficulty to dope it

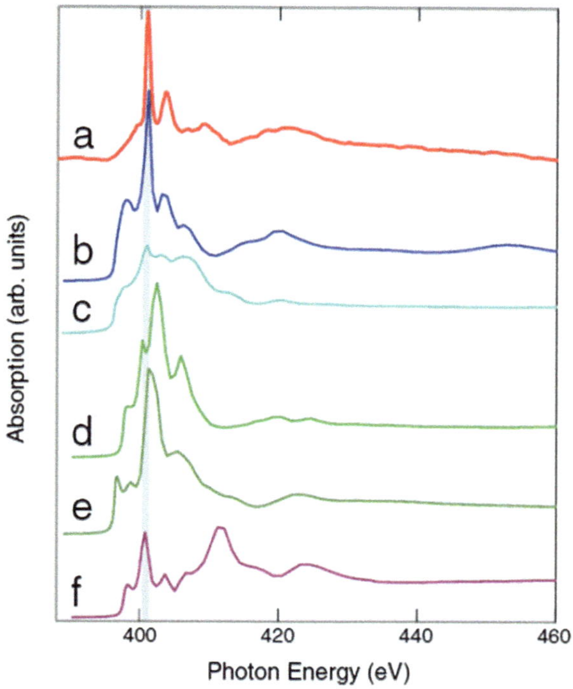

Fig. 4.7 Experimental (**a**) and simulated XANES spectra for (**b**) N on an O site, (**c**) N_2 on an O site, (**d**) N in a tetrahedral interstitial, (**e**) N in a Zn site, and (**f**) N in an octahedral interstitial. The gray band around 400 eV shows the expanded scale of Fig. 4.5b for reference. Reprinted with permission from [69]. © The American Physical Society

p type. Nitrogen is widely considered to be the most promising *p*-type dopant due to its similar size to oxygen but while there have been reports of *p*-type doping in the literature, they have been plagued by lack of reproducibility. N K-edge XANES experiments by Fons et al. [69] have contributed to explaining why this is so.

The authors studied samples grown by MBE on ZnO substrates with a N concentration of 3×10^{20} atoms/cm^3. Half of the samples were subjected to rapid thermal annealing. The N K-edge (~400 eV) XANES spectra, recorded with a windowless Ge detector, were analyzed with the help of ab initio DFT simulations of the equilibrium structure of various incorporation sites of N in ZnO and subsequent spectral simulations in the full multiple scattering approach. The main results are reported in Figs. 4.7 and 4.8. In Fig. 4.7 the experimental spectrum for an as—deposited sample (trace a) is compared to simulations for N in various sites, as detailed in the figure caption; it is clear that the best comparison is for N in a substitutional O site. In Fig. 4.8 an expanded view of the spectrum is shown for an annealed sample and compared to that of a molecular N_2 standard; the fine structure in the spectrum is due to vibrational splitting of the electronic transitions. It is clear that annealing causes formation of N_2 bubbles. Overall this study showed that N does in fact occupy a substitutional O site where it acts as an acceptor but that it is rather unstable. This result has been confirmed in its main conclusions by another experiment [70] performed on samples grown by thermal oxidation of Zn_3N_2. From the methodological point of view this study highlights two important aspects. Firstly, despite very low fluores-

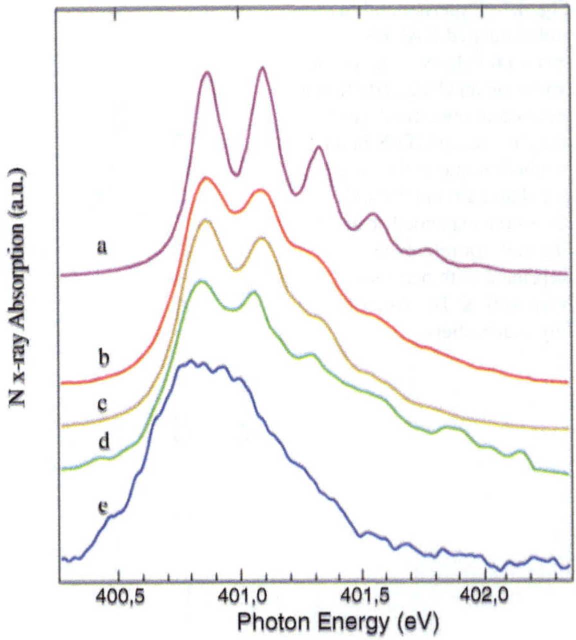

Fig. 4.8 N XAFS data for (**a**) a molecular nitrogen standard and an RTA ZnO:N film with the electric vector directed along the (**b**) *c* axis, (**c**) 45 °, in the (**d**) *ab* plane, and (**e**) as grown at 45 °. Reprinted with permission from [62]. © The American Physical Society

cence yields, soft X-ray XAFS is feasible even at low concentrations with advanced instrumentation; secondly, ab initio structural and spectral simulations of XANES, which are sensitive to the three dimensional structure around the photoexcited atom, are a powerful tool to determine the incorporation site of dopants in semiconductors. In the specific case this study shows the presence of molecular nitrogen can be detected using high resolution K-edge XANES due to the vibrational splitting. This feature has been used in a more recent study of N ion implanted in ZnO [71].

Recently there has been a great deal of interest in doping ZnO with magnetic ions in order to obtain dilute magnetic semiconductors. These studies are reviewed in Chaps. 5, 13 and 16.

4.2.7 Other Semiconductors

We close this chapter by briefly reporting work on semiconductors which have been less investigated.

Cadmium telluride has the zincblende structure, a direct band gap of ~1.5 eV and is used in thin film solar cells, IR and room temperature X and gamma ray detector. Mahnke et al. [72] have comparatively studied the local relaxations around As (acceptor), Se (isoelectronic) and Br (donor) dopants in CdTe in the dilute limit in which no clustering or compensation mechanisms occur. They find that the first shell bond length between the dopant and the nearest neighbors varies significantly with respect to the Cd–Te bond length, in good agreement with first principles calculations. For the case of Br they present preliminary evidence for the formation of the so-called A center (a Cd vacancy associated to the donor atom). More recently, Fraboni et al. [73] have coupled electrical measurements with Ge K-edge XANES to study the variation of the Ge local environment with concentration and annealing treatments. These authors were able to detect Ge clustering from the XANES lineshape at concentrations of the order of 1×10^{16} atoms/cm^3 and correlate it to the formation of electrically active deep donor traps, one located 0.31 eV below the conduction band edge and the other one at midgap, with an activation energy of 0.82 eV. This is one of the few examples in which the local structure determined from XAFS has been directly correlated with the electrical signature of a particular defect. Doping of II–VI semiconductors is also discussed in Chaps. 12, 13 and 16.

Gallium nitride is a mechanically hard and stable, wide gap (~3.4 eV) semiconductor which crystallizes in the wurtzite structure. It is used mainly in optoelectronics and for high power transistors. A preliminary report on Si doping has been reported by Li et al. [74], who suggested the formation of Si_3N_4 precipitates for concentrations of the order of 4×10^{19} atoms/cm^3. A quite extensive study of O dopants in GaN has been published by Katsikini et al. [75]. O is a common contaminant in GaN and in a N substitutional site acts as a donor. These authors studied samples grown by MBE implanted with 70 keV O ions at fluences from 1×10^{15} to 1×10^{17} atoms/cm^2; the higher fluence implantation was found to completely amorphise the sample. The spectra were interpreted on the basis of simulations performed in the multiple scattering approach assuming various models: O interstitial, O substitutional in N sites, Ga and N vacancies, and various polymorphs of Ga_2O_3. The lattice disorder was modelled by displacing atoms from their equilibrium positions by adding to their Cartesian coordinates random numbers that belong to normal distributions. The authors concluded that in the low fluence limit, the O atoms occupy interstitial sites preferentially in the empty channels aligned to the c-axis and in the atomic planes containing the Ga atoms, at intermediate fluence more N atoms are displaced from their positions which are then occupied by substitutional O and finally at the highest fluence a GaOxNy phases is formed. XAS studies of GaN and other wide band gap semiconductors are also presented in Chap. 3 while Eu and Mn doping of GaN nanostructures is discussed in Chaps. 12 and 13, respectively.

There have been numerous investigations of transition metal dopants in GaN since this is possible route to obtaining a dilute magnetic semiconductor. These studies are review in Chap. 15.

References

1. J. Jaklevic, J.A. Kirby, M.P. Klein, A.S. Robertson, G.S. Brown, P. Eisenberger, Solid State Commun. **23**, 679 (1977)
2. P.A. Lee, P.H. Citrin, P. Eisenberger, B.M. Kincaid, Rev. Mod. Phys. **53**, 769 (1981)
3. H. Oyanagi, M. Martini, M. Saito, Nucl. Instrum. Methods A **403**, 58 (1998)

4. G. Derbyshire, K.C. Cheung, P. Sangsingkeow, S.S. Hasnain, J. Synchrotron Radiat. **6**, 62 (1999)
5. J. Goulon, A. Rogalev, G. Goujon, C. Gauthier, E. Moguline, A. Sole, S. Feite, F. Wihelm, N. Jaouen, C. Goulon-Ginet, P. Dressler, P. Rohr, M.O. Lampert, R. Henck, J. Synchrotron Radiat. **12**, 57 (2005)
6. G. Ciatto, F. d'Acapito, F. Boscherini, S. Mobilio, J. Synchrotron Radiat. **11**, 278 (2004)
7. Y. Takeda, H. Ofuchi, H. Kyouzu, R. Takahashi, M. Tabuchi, J. Synchrotron Radiat. **12**, 494 (2005)
8. F. de Groot, A. Kotani, *Core Level Spectroscopy of Solids* (CRC Press, Boca Raton, 2008)
9. P. Glatzel, L. Jacquamet, U. Bergmann, F. de Groot, S.P. Cramer, Inorg. Chem. **41**, 3121 (2002)
10. P. Glatzel, U. Bergmann, Coord. Chem. Rev. **249**, 65 (2005)
11. M.E. Kordesh, R.W. Hoffman, Phys. Rev. B **29**, 491 (1984)
12. T. Guo, M.L. den Boer, Phys. Rev. B **31**, 6233 (1985)
13. G. Tourillon, E. Dartyge, A. Fontaine, M. LeMonnier, F. Bartol, Phys. Lett. A **121**, 251 (1987)
14. C.E. Bouldin, R.A. Forman, M.I. Bell, Phys. Rev. B **35**, 1429 (1987)
15. A. Erbil, G.S. Cargil III, R. Frahm, R.F. Boehme, Phys. Rev. B **37**, 2450 (1988)
16. S.L.M. Schroeder, G.D. Moggridge, R.M. Ormerod, T. Rayment, R.M. Lambert, Surf. Sci. **324**, L371 (1995)
17. R.F. Boheme, G.S. Cargill III, W. Weber, T. Jackson, J. Appl. Phys. **58**, 811 (1985)
18. T.D. Hu, Y.N. Xie, S. Qiao, Y. Hai, Y.L. Jin, D.C. Xian, Phys. Rev. B **50**, 2216 (1994)
19. M. Ishii, Y. Yoshino, K. Takarabe, O. Shimomura, Appl. Phys. Lett. **74**, 2672 (1999)
20. M. Ishii, Y. Yoshino, K. Takarabe, O. Shimomura, J. Appl. Phys. **88**, 3962 (2000)
21. M. Ishii, Phys. Rev. B **65**, 085310 (2002)
22. M. Ishii, Jpn. J. Appl. Phys. **40**, 7129 (2001)
23. M. Ishii, Phys. B **308–310**, 1153 (2001)
24. M. Ishii, Jpn. J. Appl. Phys. **41**, 4415 (2002)
25. M. Ishii, Phys. B **340–342**, 1142 (2003)
26. M. Ishii, N. Rigopoulos, N. Poolton, B. Hamilton, Phys. B **376–377**, 950 (2006)
27. J. Bollmann, S. Knack, J. Weber, V. Koteski, H.-E. Mahnke, S. Welter, Phys. Rev. B **68**, 125206 (2003)
28. J. Als-Nielsen, D. McMorrow, *Elements of Modern X-ray Physics* (Wiley, New York, 2001)
29. L.G. Parratt, Phys. Rev. **95**, 359 (1954)
30. R. Barchewitz, M. Cremonese-Visicato, G. Onori, J. Phys. C: Solid State Phys. **11**, 4439 (1978)
31. G. Martens, P. Rabe, Phys. Status Solidi (a) **58**, 415 (1980)
32. S. Pizzini, K.J. Roberts, G.N. Greaves, N. Harris, P. Moore, E. Pantos, R.J. Oldman, Rev. Sci. Instrum. **60**, 2525 (1989)
33. B. Poumellec, R. Cortes, F. Lagnel, G. Tourillon, Phys. B **158**, 282 (1989)
34. S.M. Heald, H. Chen, J.M. Tranquada, Phys. Rev. B **38**, 1016 (1988)
35. F. D'Acapito, I. Davoli, P. Ghigna, S. Mobilio, J. Synchrotron Radiat. **10**, 260 (2003)
36. D.T. Jiang, E.D. Crozier, B. Heinrich, Phys. Rev. B **44**, 6401 (1991)
37. G. Martinez-Criado, R. Tucoulou, P. Cloetens, P. Bleuet, S.B.J. Cauzid, I. Kieffer, E. Kosior, S. Labouré, S. Petitgirard, A. Rack, J.A. Sans, J. Segura-Ruiz, H. Suhonen, J. Susini, J. Villanova, J. Synchrotron Radiat. **19**, 10 (2012)
38. G. Martinez, in *Criado in Characterization of Semiconductor Heterostructures and Nanostructures*, 2nd edn., ed. by C. Lamberti, G. Agostini (Elsevier, Waltham, 2012).
39. F. Boscherini, in *Characterization of Semiconductor Heterostructures and Nanostructures*, 2nd edn., ed. by C. Lamberti, G. Agostini (Elsevier, Waltham, 2012)
40. F. d'Acapito, Semicond. Sci. Technol. **26**, 064004 (2011)
41. J.C. Knights, T.M. Hayes, J.C. Mikkelsen Jr., Phys. Rev. Lett. **39**, 712 (1977)
42. R.A. Street, Phys. Rev. Lett. **49**, 1187 (1982)
43. J. Robertson, J. Phys. C **17**, L349 (1984)
44. G.N. Greaves, A.J. Dent, B.R. Dobson, S. Kalbitzer, S. Pizzini, G. Müller, Phys. Rev. B **45**, 6517 (1992)

45. G. Dalba, P. Fornasini, R. Grisenti, F. Rocca, D. Comedi, I. Chambouleyron, Appl. Phys. Lett. **74**, 281 (1999)
46. G. Dalba, P. Fornasini, R. Grisenti, F. Rocca, D. Comedi, I. Chambouleyron, Appl. Phys. Lett. **81**, 625 (2002)
47. A. Erbil, W. Weber, G.S. Cargill III, R.F. Boehme, Phys. Rev. **34**, 1392 (1986)
48. K.C. Pandey, A. Erbil, W. Weber, G.S. Cargill III, R.F. Boehme, D. Vanderbilt, Phys. Rev. Lett. **61**, 1282 (1988)
49. J.L. Allain, J.R. Regnard, A. Bourret, A. Parisini, A. Armigliato, G. Tourillon, S. Pizzini, Phys. Rev. B **46**, 9434 (1992)
50. V. Koteski, N. Ivanovic, H. Haas, E. Holub-Krappe, H.E. Mahnke, Nucl. Instrum. Methods Phys. Res. B **200**, 60 (2003)
51. D.J. Chadi, P.H. Citrin, C.H. Park, D.L. Adler, M.A. Marcus, H.J. Gossman, Phys. Rev. Lett. **79**, 4834 (1997)
52. P.H. Citrin, D.A. Muller, H.J. Gossman, R. Vanfleet, P.A. Northrup, Phys. Rev. Lett. **83**, 3234 (1999)
53. F. d'Acapito, S. Milita, A. Satta, A. Colombo, J. Appl. Phys. **102**, 043524 (2007)
54. D. Giubertoni, G. Pepponi, S. Gennaro, M. Bersani, M.A. Sahiner, S.P. Kelty, R. Doherty, M.A. Foad, M. Kah, K.J. Kirby, J.C. Woicik, P. Pianetta, J. Appl. Phys. **104**, 103716 (2008)
55. D. Giubertoni, G. Pepponi, M.A. Sahiner, S.P. Kelty, S. Gennaro, M. Bersani, R. Doherty, M.A. Foad, M. Kah, K.J. Kirby, J.C. Woicik, P. Pianetta, J. Vac. Sci. Technol. B **28**, C1B1 (2010)
56. D. De Salvador, G. Bisognin, M. Di Marino, E. Napolitani, A. Carnera, H. Graoui, M.A. Foad, F. Boscherini, S. Mirabella, Appl. Phys. Lett. **89**, 241901 (2006)
57. A. Vailionis, G. Glass, P. Desjardins, D.G. Cahill, J.E. Greene, Phys. Rev. Lett. **82**, 4464 (1999)
58. M.A. Sahiner, D.F. Downey, S.W. Novak, J.C. Woicik, D.A. Arena, Microelectron. J. **36**, 522 (2005)
59. T. Buonassisi, A.A. Istratov, M.A. Marcus, B. Lai, Z. Cai, S. Heald, E.R. Weber, Nat. Mater. **4**, 676 (2005)
60. J. Villanova, J. Segura-Ruiz, T. Lafford, G. Martinez-Criado, J. Synchrotron Radiat. **19**, 521 (2012)
61. S. Schuppler, D.L. Adler, L.N. Pfeiffer, K.W. West, E.E. Chaban, P.H. Citrin, Appl. Phys. Lett. **63**, 2357 (1993)
62. S. Schuppler, D.L. Adler, L.N. Pfeiffer, K.W. West, E.E. Chaban, P.H. Citrin, Phys. Rev. B **51**, 10527 (1995)
63. F. Boscherini, N. Ferretti, B. Bonanni, D. Orani, S. Rubini, M. Piccin, A. Franciosi, Appl. Phys. Lett. **81**, 1639 (2002)
64. F. d'Acapito, G. Smolentsev, F. Boscherini, M. Piccin, G. Bais, S. Rubini, F. Martelli, A. Franciosi, Phys. Rev. B **73**, 035314 (2006)
65. A.L. Ankudinov, B. Ravel, J.J. Rehr, S.D. Conradson, Phys. Rev. B **58**, 7565 (1998)
66. Y. Joly, Phys. Rev. B **63**, 125120 (2001)
67. A. Bozanic, Z. Majlinger, M. Petravic, Q. Gao, D. Llewellyn, C. Crotti, Y.W. Yang, J. Vac. Sci. Technol. A **26**, 592 (2008)
68. Z. Majlinger, A. Bozanic, M. Petravic, K.J. Kim, B. Kim, Y.W. Yang, J. Appl. Phys. **104**, 063527 (2008)
69. P. Fons, H. Tampo, A.V. Kolobov, M. Ohkubo, S. Niki, J. Tominaga, R. Carboni, F. Boscherini, S. Friedrich, Phys. Rev. Lett. **96**, 045504 (2006)
70. C.W. Zou, X.D. Yan, J. Han, R.Q. Chen, W. Gao, J. Metson, Appl. Phys. Lett. **94**, 171903 (2009)
71. P. Hofmfmann, C. Pettenkofer, Phys. Status Solidi B **248**, 327 (2011)
72. H.-E. Mahnke, H. Haas, E. Holub-Krappe, V. Koteski, N. Novakovic, P. Fochuk, O. Panchuk, Thin Solid Films **480–481**, 279 (2005)
73. B. Fraboni, F. Boscherini, P. Fochuk, A. Cavallini, J. Appl. Phys. **110**, 053706 (2011)
74. Z.H. Lu, T. Tyliszczack, B. Broderson, A. Hitchcock, J.B. Webb, H. Tang, J. Bardwell, Appl. Phys. Lett. **75**, 534 (1999)
75. M. Katsikini, F. Boscherini, E.C. Paloura, Local bonding geometry of oxygen implanted in GaN: a depth-dependent study. J. Nanosci. Nanotechnol. **10**, 6260 (2010)

Chapter 5
Complexes and Clusters

Gianluca Ciatto

Crystalline semiconductors often contain more complex defective structures than the simple point defects which may be generated, for example, by intentional doping. These structures can be homogenously distributed in the semiconductor matrix and, in several cases, can dramatically modify the material properties with important implications in technological applications. In this chapter, we illustrate several examples in which XAS, coupled with state-of-art simulation tools, was able to gather important information on complexes and clusters and, in some case, to directly resolve their three-dimensional structure.

5.1 Definition of Complexes and Clusters

Before starting any discussion on this subject, it is worth defining well what we are talking about. In this chapter, as "complex" we define a configuration in which two or more atomic "entities" are correlated one with the other in a particular geometry. An "entity" can be an impurity or dopant atom intentionally or unintentionally incorporated in the lattice, a vacancy, or also an atom of the semiconductor matrix. However, at least one of the two entities constituting the complex is different from a matrix atom located in its crystallographic site. Therefore, a single dopant, or interstial, or vacancy, does not constitute a complex since a correlation between at least two entities is needed; co-doping with two atomic species does not automatically generate complexes unless they are related to each other in a definite short-range configuration. In the following sections, we will show several examples in which the formation of complexes dramatically affects the physical properties of the materials (optical band gap, lattice parameter, magnetism, etc.) and illustrate the potential of

G. Ciatto (✉)
Synchrotron SOLEIL, L'Orme des Merisiers, BP 48,
Saint-Aubin 91192, Gif sur Yvette CEDEX, France
e-mail: gianluca.ciatto@synchrotron-soleil.fr

C.S. Schnohr and M.C. Ridgway (eds.), *X-Ray Absorption Spectroscopy of Semiconductors*, Springer Series in Optical Sciences 190,
DOI 10.1007/978-3-662-44362-0_5

X-ray absorption spectroscopy (XAS) for solving defect-complexes structures. Of course, a complex is characterized by the relative orientation of its constituent elements in the real space (interatomic distances and angles), therefore there is a strong interest in applying X-ray absorption near edge structure (XANES) since it is more sensitive to the three-dimensional structure of defects than extended X-ray absorption fine structure (EXAFS), which to a large extent performs a one-dimensional projection of the structure on the interatomic distance axis.

The definition of "cluster" is probably different from what the reader expects: in fact, we do not mean a situation in which a new separate phase, different from the matrix one, precipitates in the sample. This is the case, for example, for the formation of metallic clusters in magnetic semiconductors, which will be addressed in Chaps. 14–16. In contrast, in this chapter, the term "clustering" ("anticlustering") identifies the situation in which like-atoms tend to get closer (farer) from each other conserving the same crystallographic structure. In a ternary semiconductor alloy with a mixed sublattice of the kind $A_xB_{1-x}C$ (see also Chap. 2), these phenomena can be quantified by the Cowley's short range order parameters [1]:

$$\alpha_j = 1 - \frac{P_j}{x}$$

where P_j is the probability of finding an A atom being *j*th nearest-neighbor of a B one in the mixed sublattice. $\alpha_j > 0$ indicates an association of like atoms (clustering), while $\alpha_j < 0$ indicates an association of unlike atoms (anticlustering). For completely random alloys, $\alpha_j = 0$ for any value of the j index (i.e. for any atomic shell in the mixed sublattice). Deviations from random atomic distribution, driven by strain energy modifications, have been predicted for ternary semiconductor alloys such as GaAsN and GaInN via large supercell calculations and are expected to strongly affect the emission properties of the materials [2]. We will show that XAS, and in particular EXAFS, is a powerful tool for seeking and quantifying such deviations.

5.2 Modeling and Data Analysis Approaches

5.2.1 Conventional XAS Analysis of Complexes/Clusters

Most of the EXAFS works on crystalline semiconductors available in literature employ a data analysis based on the path fitting approach described in Chap. 1. The starting model structure, used to calculate the effective scattering amplitude and phase shifts of single and multiple scattering paths, is often taken from crystallography and, when dealing with ternary or more complex alloys, the structures of two or more binary (or ternary) constituent compounds may be used as a model for scattering path generation. Therefore, since in ternary/quaternary alloys local interatomic distances neither are identical to values they have in the constituent compounds nor do they follow the Vegard's law (Chap. 2), the model is usually not effective in reproducing the experimental distances. Therefore, interatomic distances have to be varied freely in the least-square fit, along with the Debye-Waller (DW) factors, in order to obtain a satisfactory agreement with the experimental data.

Although it is a priori possible to use this strategy in the analysis of complexes and clusters, this has some limitations related to the total number of variables of the fit. In fact, complexes or clusters models usually need additional variables to be described and in some case combinations of different local environments are mandatory in order to reproduce well the experimental data. If too many variables are used in the fit, strong correlations among them can dramatically affect the accuracy of structural parameters evaluation, increasing the error bars. This imposes to fix or restrain the maximum number of cumulants for each scattering path, which can be done using more realistic models (based on valence force field or density functional theory calculations), as explained in the following.

Also in the simulation of XANES spectra, often non-relaxed atomic clusters are used as input file for calculations performed by using, for example, the FEFF or FDMNES [3] programs. However, in the case of complex or cluster formation, the local environment is often strongly distorted and the different three-dimensional arrangement of the atoms with respect to the defect-free material affects the multiple scattering paths, which are particularly important in this spectral region. Hence, even for XANES, a better agreement between calculated and measured spectra is normally obtained by employing properly relaxed input clusters.

In the examples presented in this chapter, conventional analysis has been employed in the EXAFS (but not XANES) analysis of Mn–H complexes in GaMnAs (Sect. 5.3.2), in the study of defect complexes in (Er,O)-doped Si (Sect. 5.3.4), and to look for S clustering in ZnSSe (Sect. 5.4.2).

5.2.2 Valence Force Field-Based XAS Analysis of Complexes/Clusters

Valence force field (VFF) calculations are an ideal and relatively simple tool to evaluate the local distortions of the bond lengths around a substitutional impurity in zincblende semiconductors. The VFF approach has been already introduced in Chap. 2, thus we remind to that chapter for an introduction to the method. In this chapter, VFF calculations have been applied to the quantitative analysis of clustering and anticlustering phenomena. In the examples reported in Sect. 5.4, the static displacement of the atoms from the average crystal sites was determined by minimizing the Keating's potential within the strategy formerly proposed by Glas for the analysis of transmission electron microscopy experiment [4]. In this strategy, the positions of all atoms inside a large cluster are relaxed via numerical minimization of the many-variables potential function. Clusters of 20 atomic shells around the absorber are used in order to generate the input files for calculation of the theoretical EXAFS signals (which is performed by FEFF). Nonrandom input clusters corresponding to "clustering" situations are obtained by placing like-atoms around the absorber one to realize the particular short-range order (SRO) of interest. In order to establish the boundary conditions necessary for VFF minimization, these clusters are embedded into much larger crystals consisting of 30 cubic unit cells in each direction. In the embedding part of the crystal, atoms are placed at random. The equilibrium atomic positions of all the atoms in the large crystals are found by minimizing the VFF energy.

This VFF approach has considerable advantages with respect to a conventional analysis since it not only permits a more precise calculation of the effective scattering phase shifts, but can also reproduce the broadening of the distance distribution caused by structural disorder when some SRO configuration breaks the symmetry of the perfect random crystal. In the analysis of a sample series characterized by clustering (anticlustering), if one possesses a reference sample with similar structure in which SRO does not exist or is negligible, it is possible to estimate the systematic error done by the multiple scattering code in the calculation of the phase shifts along with the thermal part of DW factors from the reference sample and then fix it in the analysis of the series. Since the determination of bond lengths and distances distribution done by VFF is good enough, no more variable for the interatomic distances and DW factors is a priori necessary in the fit, which results in negligible correlation between the few remaining variables (weight of different possible SRO configurations and energy edge shift) and in smaller error bars on the extracted SRO parameters. Even in the calculations of XANES spectra in the full multiple scattering (FMS) approach, VFF-relaxed input clusters allow one to obtain a convincingly improved agreement with the experimental spectra of alloys with mixed atomic shells.

5.2.3 Density Functional Theory-Based Analysis of Complexes/Clusters

Even if the VFF approach described above is rather powerful in some case, its application range remains quite limited. In fact, it works only for substitutional impurities and requires knowledge of the force constants and/or elastic constants of the binary constituent alloys. If on the one hand VFF is useful in analyzing clusters, on the other hand the formation of complexes is normally more difficult to deal with. A complex usually features the presence of one or more atoms in non-substitutional position and, sometimes, binary constituent alloys do not exist. This is the case, for example, of the hydrogen complexes in GaAsN or GaMnAs discussed in Sects. 5.3.1 and 5.3.2, where force constants relative to the N–H or Mn–H bond stretching are not available. In other cases, vacancies are involved in the complex structure, which induce important modifications of the local structure and orbitals distribution with respect to a simple substitutional defect (Sect. 5.3.3).

In these situations, ab initio calculations are necessary in order to generate a model cluster with a realistic relaxation of the atomic positions. At this scope, in several examples reported in Sect. 5.3, density functional theory (DFT) methods were used under different approximations: local density approximation (LDA) was used in case of *p* valence orbitals, while spin-polarized approaches based on the Hubbard U (GGA+U) [5] approximation were used when dealing with *3d* orbitals. A supercell approach using separable ab initio pseudopotentials [6, 7] was employed, geometry optimization was performed by fully relaxing the atomic positions in 64-atom supercells with the PWSCF or VASP code. When the DFT-generated clusters were used as an input for XANES simulations in the FMS approach, supercells were repeated eight times in order to obtain large clusters of about 500 atoms. The use of large clusters is in fact mandatory to reproduce the region of the absorption spectrum close to the edge, where the photoelectron mean-free path can reach a few tens of angstroms.

As for the VFF approach, the relaxed interatomic distances calculated by DFT are given as input to the FEFF and FDMNES codes to calculate the EXAFS backscattering amplitudes and phase shifts in multiple scattering mode, or the XANES spectra in the FMS mode. The DFT approach has all the advantages of the VFF in terms of variables minimization in the EXAFS fit and it is more flexible, at the cost of more sophisticated and time-consuming theoretical calculations.

5.3 Complexes

In this section, we present some examples in which XAS demonstrated to be a very useful tool for addressing the structure of defect complexes in crystalline semiconductors. The chosen examples cover complex formation in III–V epilayers (N–H complexes in GaAsN and M–H complexes in GaMnAs), in II–VI (Co–O vacancy complexes in ZnCoO) and group-IV materials (Er at O-decorated multi-vacancies in Si). The selection is, however, not exhaustive and XAS has been useful in other cases, for example in the study of As-decorated vacancy defects in As-doped Si [8] (discussed in Chap. 4) or Cl–Zn vacancy complexes in Cl-doped ZnSe [9]. Most of the examples described in the present chapter, which are the outcome of rather recent research, have in common the coupling between XAS and state of art theoretical calculations aimed at extracting more information on the complex structure.

5.3.1 Nitrogen–Hydrogen Complexes in Dilute Nitrides

Dilute nitride alloys, such as GaAsN and InGaAsN, are semiconductors characterized by strong and counterintuitive modifications induced by nitrogen on the physical properties of the host lattice: a large reduction of the optical band gap [10], and an atypical behavior of the electron effective mass [11] as a function of nitrogen

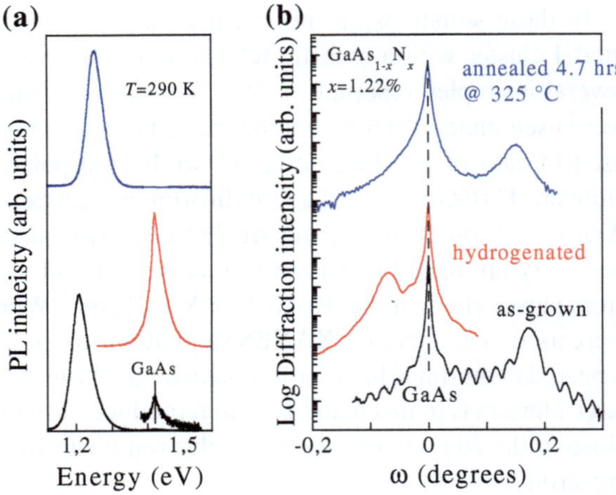

Fig. 5.1 **a** Photoluminescence (PL) spectrum of untreated (*black*), hydrogenated (*red*) and annealed (*blue*) $GaAs_{0.9878}N_{0.0122}$ samples; **b** XRD rocking *curves* taken around the (004) crystal plane reflection on the same epilayers

concentration. The band gap reduction renders these materials particularly attractive both for the manufactory of vertical-cavity laser emitting at the minima of absorption in silica optical fibers [12] and as 1-eV gap component in high-efficiency multi-junction solar cells [13]. Even more surprisingly, the dramatic effects of nitrogen incorporation can be fully reversed via hydrogenation of the samples by a low energy ion gun [14, 15].

In Fig. 5.1, the most important effects on the optical and structural properties of hydrogenation on a 300 nm-thick $GaAs_{1-y}N_y$ ($y \sim 1\,\%$) epilayer grown by solid source molecular-beam epitaxy (MBE) on GaAs(001) substrate are shown [15, 16]. Figure 5.1a shows photoluminescence (PL) spectra taken on the epilayer before and after hydrogenation. The intense band that peaked at 1.21 eV in the bottommost spectrum, recorded on the untreated sample, is due to carriers recombining in the $GaAs_{1-y}N_y$ layer. The much less intense band at 1.42 eV is caused by carrier recombination in the GaAs buffer layer. Hydrogen irradiation changes dramatically the PL (middle spectrum). No carrier recombination can be observed from the $GaAs_{1-y}N_y$ band-gap energy and the wavelength of all emitted photons is blueshifted at the band-gap energy of GaAs. In Fig. 5.1b the (004) X-ray diffraction (XRD) rocking curves (RCs) of the as grown epilayer (bottommost curve), and of the same sample after hydrogenation (middle curve) are shown. By comparing the RCs of the as-grown and of the hydrogenated samples, it is clear that the hydrogenation process is the origin of the $GaAs_{1-y}N_y$ epilayer (004) Bragg-peak shift from larger to smaller angles with respect to the substrate peak. The original epilayer tensile strain, due to the insertion of substitutional N in GaAs, is indeed reversed into a compressive one.

In summary, hydrogenation brings about a reopening of the optical band gap, passivating the effects of nitrogen and an enlargement of the alloy lattice parameter. It also turns out (topmost curves) that both effects can be reversed by a long thermal annealing at 325 °C. Such effects are technologically relevant since hydrogen is present in most growth processes and mass-production steps of semiconductor

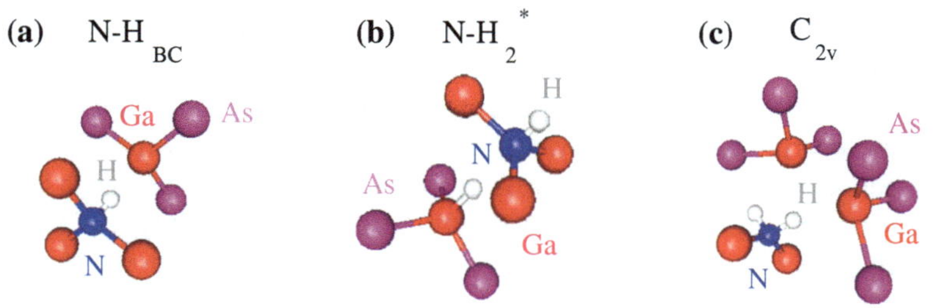

Fig. 5.2 Sketch of the most stable N–H complexes.

devices (for example, in MOCVD growth). Moreover, the possibility of obtaining structures with lateral band-gap variation on the micrometer scale by hydrogenating dilute nitrides using patterned masks has been recently demonstrated [17] and, with an improvement of miniaturization, hydrogenation will have the possibility to produce nanostructures with in-plane confinement, with potential impact in the engineering of optical circuits.

The effects of H irradiation in GaAsN, InGaAsN, and GaPN have inspired many theoretical studies aimed at discovering the origin of the experimental observations [18]. Considering the reactivity of H and N atoms and the reversibility of the phenomena, the most intuitive explication is provided by the formation of microscopic hydrogen–nitrogen complexes which sufficiently low dissociation energy to be released by a moderate thermal annealing. Different mono hydrogen–nitrogen, dihydrogen–nitrogen, and multi-hydrogen complexes have been proposed by several groups over the years and the most relevant examples are sketched in Fig. 5.2. For a long time it has been believed that an in-line dihydrogen–nitrogen complex, named N–H_2^*, was responsible for hydrogen-related effects in dilute nitride. In this complex, two strong Ga-H_{BC} and N–H_{AB} bonds are formed, which involve, respectively, a H_{BC} atom in bond center (BC) position between the Ga and N atoms and a H_{AB} atom in the opposite position with respect to the same nitrogen atom (antibonding position, AB). In fact N–H_2^* is supposed to cause a blue shift of the optical band gap and an enlargement of the alloy lattice parameter.

The hypothesis of the N–H_2^* complex was defied by infrared absorption (IR) measurements [19] which showed that, if on the one hand the local vibrational mode related to the Ga-N bond disappears in GaAsN after H irradiation, on the other hand no experimental evidence exists for the Ga-H_{BC} mode, contrary to the expectations of the N–H_2^* model. Moreover, IR detected instead two N–H stretching modes due to a *same* N–H complex. At the same time, theoretical studies based on computation of the defect energetics proposed that a new dihydrogen complex may form in hydrogenated GaAsN [18]. This complex consists of *two* H_{BC} atoms bound to a *single* N atom. The complex (indicated as C_{2v} in Fig. 5.2), has a quasi C_{2v}-symmetry since one of the H_{BC} ions is located off the N-Ga axes. The C_{2v} complex can lead to a recovery of the band-gap energy, while the recovery of the lattice parameter is more complex and depends on the charge state of the complex.

In order to understand which N–H complex actually forms in the alloy, it was evident the need of a technique able to provide more direct and quantitative information on its local structure than IR spectroscopy: the following shows that N K-edge soft X-ray absorption spectroscopy was the eligible tool to address this scientific case. In this section we present XANES measurements performed on the branch line of the ALOISA beamline of the ELETTRA synchrotron radiation facility in Trieste, Italy. The energy dependence of the N absorption cross section was monitored by recording the intensity of N Kα fluorescence using a windowless hyperpure Ge detector. The XANES spectra were simulated in a real-space FMS approach using the FEFF code. As described in Sect. 5.2.3, atomic coordinates for each N–H complex model were taken from DFT calculations [18], repeating eight times a 64-atoms supercell. The method was first tested on H-free $GaAs_{1-y}N_y$: Fig. 5.3a shows a comparison of the nitrogen K-edge XANES experimental spectrum for a $GaAs_{0.986}N_{0.014}$ sample with a simulation based on the atomic coordinates obtained by calculations. The correspondence between the "well known" experimental spectrum and the simulated one is excellent, which demonstrates the soundness of the approach to investigate "unknown" structures, as that of hydrogenated $GaAs_{1-y}N_y$, testing different proposed models and comparing the relative calculated XANES with the experimental one.

Figure 5.3b shows the experimental N K-edge XANES for a hydrogenated $GaAs_{0.986}N_{0.014}$ sample along with simulations for the N–H complexes introduced

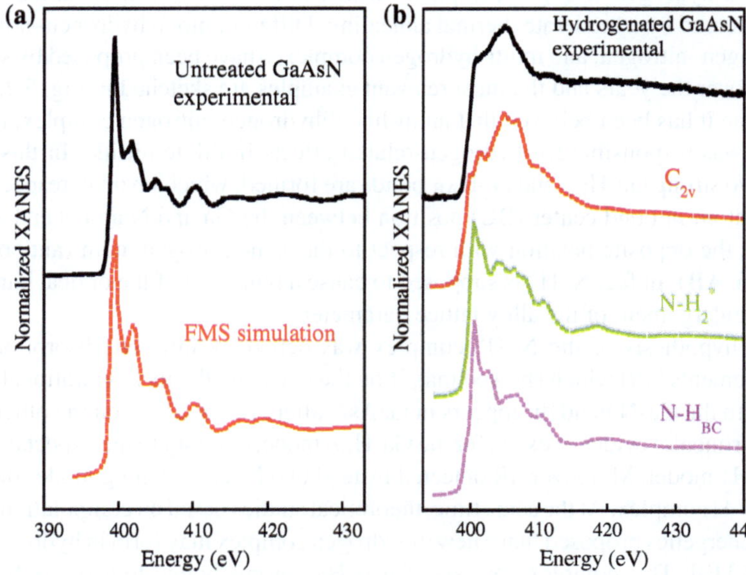

Fig. 5.3 **a** Experimental and simulated N K-edge XANES for as grown $GaAs_{0.986}N_{0.014}$ before exposure to hydrogen. **b** Experimental N K-edge XANES for hydrogenated $GaAs_{0.986}N_{0.014}$ (*top*) and corresponding FMS simulations based on the atomic coordinates calculated via DFT in presence of the most stable monohydrogen– and dihydrogen–nitrogen complexes sketched in Fig. 5.2. Modified from [20]

above. The variations of the XANES lineshape when simulating the different N–H complexes are evident and testify to the sensitivity of the method to H-driven changes in the local environment of nitrogen. It is straightforward that only the C_{2v} complex reproduces the main features of the experimental spectrum; in particular the fact that the absorption coefficient does not decrease suddenly after the white line but rather exhibits an extra peak between 405 and 408 eV.

Quite the reverse, the simulations for the monohydrogen–nitrogen complex (N–H_{BC}) and the N–H_2^* dihydrogen–nitrogen complex show large discrepancies with the experimental XANES spectrum: this means that these structural models are incorrect and can be discarded. These results gave firm evidence that the N–H complex forming in hydrogenated GaAsN has C_{2v} symmetry [20] and accounted for the existence of two N–H stretching modes related to a same N–H complex, as suggested by IR measurements and some theoretical work.

Details on the XAS investigation of the effects of epitaxial strain, the charge state, the presence of H satellites and bond lengths in these hydrogenated samples can be found in other literature papers [21, 22]. Here, we want to point out that the DFT-based XANES method allows a much more direct comparison between experiment and simulation than IR spectroscopy: in fact a full comparison of the experimental spectrum (energy position of the spectral features, relative intensities and width) with the simulations is possible, which renders an accidental agreement with the experimental data extremely improbable. Moreover, XANES inherently probes all the atomic geometries present, weighing the contribution of each one to the average spectrum with its relative abundance, thus avoiding the interpretative problems of IR due to the varying cross sections of vibrational modes.

5.3.2 Manganese–Hydrogen Complexes in GaMnAs

$Ga_{1-x}Mn_xAs$ (see also Chap. 15) is a ferromagnetic semiconductor of interest for applications in spintronics devices [23], in which the spin degree of freedom of the charge carriers can be used to enhance the capabilities of conventional electronics, combining semiconducting and magnetic properties. In this alloy, which follows the Dietl's dilute magnetic semiconductor model [24] for relatively low x, both the localized magnetic moments as well as the itinerant holes mediating the ferromagnetic coupling between them are introduced by manganese acceptors [25]. Post-growth hydrogenation has been proposed as a way of switching off or influencing the magnetism in this material [26]: in fact the conductivity in $Ga_{1-x}Mn_xAs$ films can be reduced by several orders of magnitude upon exposure to hydrogen plasma [27, 28]. The decrease in conductivity goes with a loss of the ferromagnetic behavior [26, 29], which is not surprising since ferromagnetism is carrier-mediated in this alloy. Since Mn is naturally an acceptor in $Ga_{1-x}Mn_xAs$, while H acts as an amphoteric impurity [30], the decrease in carrier concentration can be a priori caused either by a compensation of the Mn acceptors by H donors or to a passivation of the Mn acceptors by the formation of electrically neutral Mn–H complexes. Even if the loss of conductivity and the accompanying loss of ferromagnetism upon hydrogenation in $Ga_{1-x}Mn_xAs$ were reported by several groups, the local microscopic structure of hydrogen in this material has remained an open question for a long time.

Different spectroscopic techniques have been used to investigate the possible formation of acceptor-hydrogen complexes in $Ga_{1-x}Mn_xAs$. Secondary ion mass spectrometry (SIMS) measurements of hydrogenated $Ga_{1-x}Mn_xAs$ samples showed that Mn and H concentrations are identical within experimental uncertainty, which suggested some association between the two atoms [28]. Infrared absorption (IR) spectra of hydrogenated and deuterated samples revealed absorption maxima at 2,143 and 1,546 cm^{-1}, respectively, which are compatible with the local vibrational As-H and As-D stretching modes. IR results were considered as an evidence for the formation of complexes involving manganese and hydrogen [27, 28], while the relative lattice location of the two atoms remained controversial; two main hypotheses were advanced (sketched in Fig. 5.4): in the first (Mn–H_{bc}), H is at bond-centered site in between the acceptor and a neighboring As atom, while in the second (Mn–H_{ab}) H is in antibonding configuration with respect to the same atoms.

DFT calculations privileged the first configuration, which minimizes the total energy and gives calculated local vibrational modes in agreement with the experi-

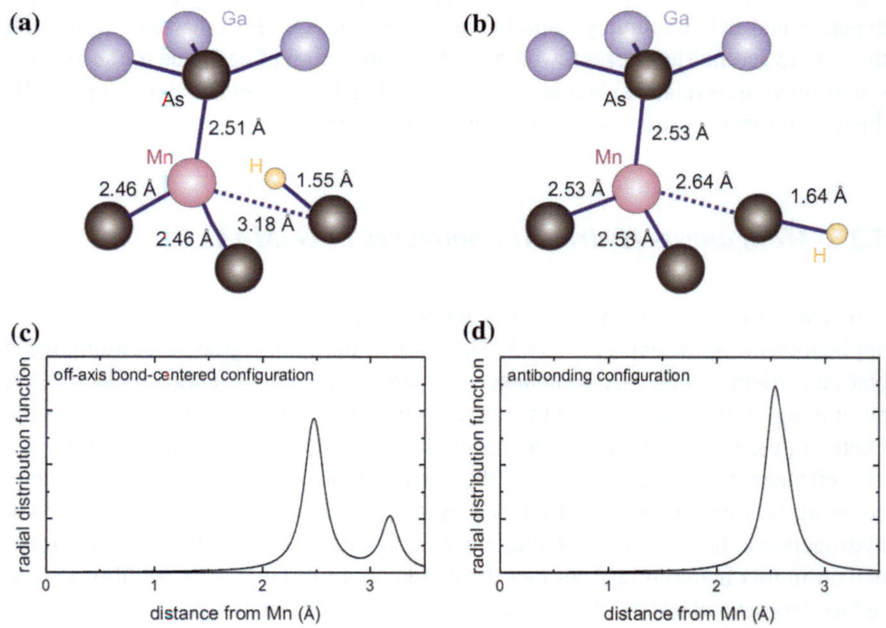

Fig. 5.4 **a** Off-axis bond-centered configuration (Mn–H_{bc}) and **b** antibonding configuration (Mn–H_{ab}) of the Mn–H complex with the bond lengths calculated by Goss and Briddon [31, 32]. **c** Expected radial distribution function for the Mn–H_{bc} configuration and **d** for the Mn–H_{ab} one. (Reprinted with permission from [33]. Copyright 2008 by the American Physical Society. http://prb.aps.org/abstract/PRB/v78/i23/e235208)

mental ones [31, 32]. Electron paramagnetic resonance (EPR), which is sensitive to the symmetry of the field of the Mn^{2+} atoms (from which information on the position of the H may be inferred), rather suggested the Mn–H_{ab} configuration [33]; however this technique only works at the dilute limit and cannot be used to study samples with Mn concentration equal to some percent since the spectra broaden due to the interaction of neighboring Mn^{2+} centers.[1]

Bihler et al. addressed the issue of Mn–H complexes in hydrogenated $Ga_{1-x}Mn_x$ As by Mn K-edge XAS [33]. Measurements were performed at the BM29 beamline of the European Synchrotron Radiation Facility (ESRF) in fluorescence mode on $0.03 < x < 0.05$ epilayers grown by low-temperature MBE [34].

Hydrogenation[2] was carried out via remote dc deuterium plasma operated at 0.9 mbar, with the sample heated to 170 °C. Annealed control samples were prepared in the same way, but with the hydrogen plasma switched off to study the influence of the temperature on the treatment. The authors measured two sample series with $x = 0.039$, for each series they investigated an as-grown piece, a hydrogenated piece, and a reference annealed piece obtained as detailed above.

EXAFS can give important information on the local structure of Mn–H complexes in $Ga_{1-x}Mn_xAs$ and help in discriminating between the two main models proposed. In fact, for the bond-centered complex configuration (Fig. 5.4a) theory predicts an extension of the accompanying Mn-As distance, resulting in a significant change in the radial distribution function of the Mn atom, which should show a double peak (Fig. 5.4c). In contrast, in the anti-bonding configuration (Fig. 5.4b), small changes of the distance distributions are predicted with respect to the pure substitutional Mn impurity in GaAs (Fig. 5.4d).

Figure 5.5a shows the Fourier transforms (FTs) of the Mn K-edge EXAFS spectra for an as grown, hydrogenated and annealed sample. Three clear peaks can be distinguished corresponding to the first three neighbors shells of the central Mn. It is evident that all the FTs are virtually identical except for a reduced magnitude of the nearest As-neighbor peak at (apparent) distance around 2.2 Å for the hydrogenated sample, which suggests a minor role of H in modifying the interatomic distances distribution. Based on that, a conventional data analysis (see Sect. 5.2.1) was employed, using a zincblende GaMnAs model cluster to generate theoretical backscattering amplitudes and phase shifts, and leaving the first three shells interatomic distances and DW factors free of varying in the fit. Figure 5.5b shows the values for structural parameters extracted from the best fits: the amplitude reduction in the FT after hydrogenation mentioned above is accounted for by the significant increase in the first shell DW factor of the hydrogenated sample. The second and third shell DW factors also increase, testifying to an increase in local structural disorder after hydrogenation. Besides, there is also a slight increase in the first shell distance

[1] Note that a Mn concentration exceeding 1 % is needed to achieve ferromagnetism in GaMnAs.

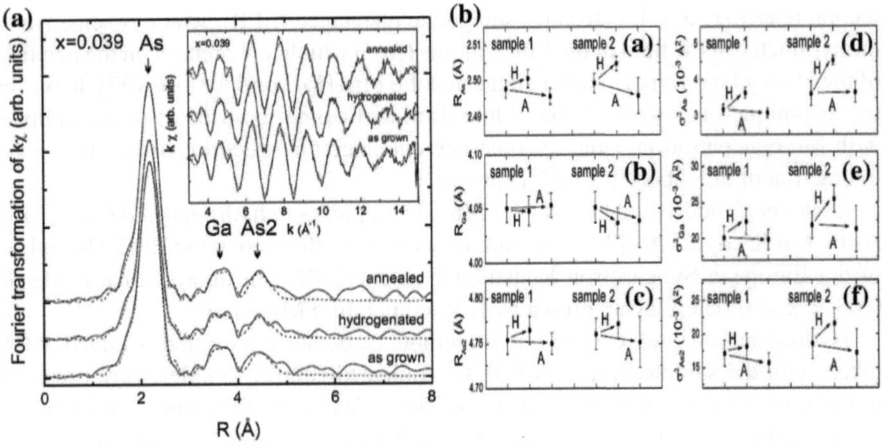

Fig. 5.5 **a** Comparison of the Fourier transforms of the Mn K-edge EXAFS of an as-grown, a hydrogenated, and an annealed reference sample. *Inset* shows the relative kχ(k) spectra. The *dashed curves* are fits to the measured data. **b** First, second and third shell (from *top* to *bottom*) Mn-centered interatomic distances and Debye-Waller factors for two $Ga_{1-x}Mn_xAs$ samples as determined from the fits to the spectra. (Reprinted with permission from [33]. Copyright 2008 by the American Physical Society. http://prb.aps.org/abstract/PRB/v78/i23/e235208)

upon hydrogenation, while all parameters of the annealed reference sample remain the same as those of the as-grown samples within the error bars.

What is very important to point out is that best fit for the hydrogenated sample is obtained by using the same model as for the H-free sample in which all nearest neighbors of Mn have the same distance: allowing one of the NNs to have a longer distance does not improve the fit quality. For the Mn–H_{bc} complex, one would expect an additional peak or shoulder on the higher distance side of the FT nearest neighbor peak, similar to the additional peak of the radial distribution shown in Fig. 5.4c. The fact that this was not observed ruled out the hypothesis of Mn–H_{bc} as the majority complex configuration in GaMnAs. In contrast, EXAFS results are in agreement with the one-peak structure of the radial distribution function predicted for the Mn–H_{ab} configuration (Fig. 5.4d). We refer to the same paper [33] for an explication of how Mn K-edge XANES, coupled to FMS simulations performed following the same strategy as in the previous section, confirms the EXAFS result. In summary, XAS supports the hypothesis of Mn–H_{ab} as the majority complex at relatively high Mn incorporation, despite the theoretical prediction of the Mn–H_{bc} one. The results reported in this section demonstrate the great complementarity between XAS and EPR, which, instead, better addresses the dilution limit.

5.3.3 Cobalt–Oxygen Vacancy Complexes in $Zn_{1-x}Co_xO$

$Zn_{1-x}Co_xO$, like $Ga_{1-x}Mn_xAs$ presented in the previous section, belongs to the family of dilute magnetic semiconductors (DMS). Among DMS, compounds based on ZnO (such as $Zn_{1-x}Co_xO$ and $Zn_{1-x}Mn_xO$, see also Chap. 16) are especially appealing since they exhibit ferromagnetism (FM) above room temperature [35] (i.e. in conditions favorable to the production of devices) and have low toxicity. Despite this great potential for spintronics applications, the origin of FM in dilute ferromagnetic oxides (DFO) has been mysterious for years.

In DFO the established theory of carrier-mediated magnetism [24] does not apply because the correlation between carriers and magnetism is not evident. Experimental and theoretical studies have attributed FM in these systems to a variety of origins including intrinsic defects, nonhomogeneity of the dopant spatial distribution, hydrogen contamination, or secondary phases. Recently, some theoretical works attributed an important role to oxygen vacancies (V_O) [36, 37]. In fact, while isolated V_O are deep donors, they can generate electronic states close to the conduction band when at complex with 3d impurities: FM is achieved when these states are occupied by electrons from a shallow donor. Despite a number of experiments attempted to reveal the presence of V_O in DFO, a convincing experimental demonstration of their existence was missing for a long time.

XAS, along with soft X-ray circular magnetic dichroism, was widely employed in the study of DFO, in particular to look for possible presence of small metallic Co clusters or nanoclusters, invisible to XRD, and other minority phases [38, 39] which may provide un unwanted spurious contribution to the magnetic properties of the samples.[3] A detailed discussion of the role of metallic impurities in semiconductors and the problematic of spurious phases will be given in Chaps. 13 and 16. In this section, instead, we focus on a more neglected aspect: the study of V_O-related defects in $Zn_{1-x}Co_xO$.

Some preliminary attempts to verify the existence of V_O in DFO via O, Co and Zn edge XAS at fixed incidence angle were not able to provide a clear answer to the question [40–42]. In a recent work, Ciatto et al. [43] exploited the polarization dependence of Co K-edge XANES to study the formation of V_O in wurzite single-phase $Zn_{1-x}Co_xO$ epilayers. Angle-dependent XANES measurements, performed at the SAMBA beam line of the SOLEIL synchrotron in fluorescence detection mode, were coupled with spin-polarized DFT calculations of V_O-containing structures. As described in Sect. 5.2.3, the relaxed interatomic distances calculated by DFT were used both as an input to simulate the Co K-edge XANES in FMS approach and to calculate the experimental amplitudes and phases for the analysis of the EXAFS region. Samples were measured both with the electric field vector (E) // c and with E $\perp$ c (c being the axis of symmetry of the hexagonal lattice). Figure 5.6 shows XANES

[3] Of course, the magnetism originating from spurious phases does not qualify the material as a magnetic semiconductor and should be eliminated or at least reduced in order to obtain good quality samples.

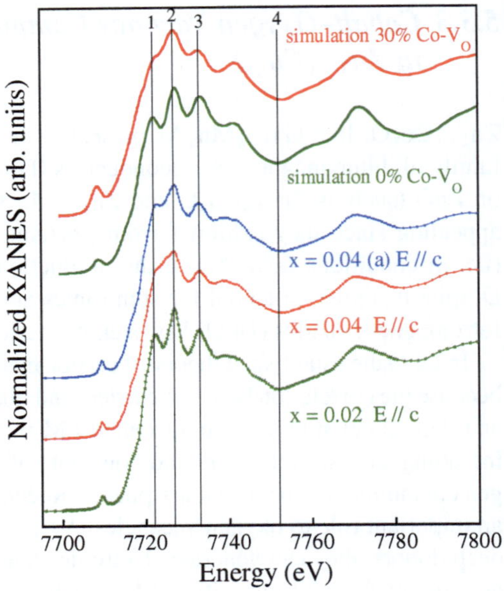

Fig. 5.6 Co K-edge XANES taken on $Zn_{1-x}Co_xO$ epilayers with E // c (*dotted line*) and FMS simulations (*continuous line*); *a* indicates "annealed in oxygen". Modified from [43]

taken on two selected samples with different Co concentration at E // c (green and red bottommost spectra): the valley between features 1 and 2, evident in the spectrum of the x = 0.02 sample, disappears in the spectrum of the x = 0.04 one.

Moreover, the amplitude of feature 3 is smaller for x = 0.04. Instead, for E ⊥ c, only minor changes are observed (not shown). The variations of the XANES for E // c go along with a change of magnetic properties: magnetization versus field (H) measurements (Fig. 5.7) show that saturation magnetization is remarkably higher for x = 0.04. For the wurtzite group and K-absorption edges, due to the directionality of the p-components of the final state, with E // c XANES is sensitive to variations of the electronic density along the c axis. Therefore, the experimental results can be explained by a defective local configuration which differs from a pure substitutional Co by a modification of the electronic charge distribution along c. One of the stable V_O configurations (according to DFT calculations) is the complex sketched in Fig. 5.8: in this case the O vacancy is located in a nearest-neighbor position with respect to a Co atom, and the Co and V_O positions define a direction parallel to the c axis. For the resulting complex (Co-V_O), a reduction in total energy with respect to the initial state (isolated Co plus simple V_O) is calculated. It turns out that this complex is also responsible for the appearance of empty levels close to the conduction band [43].

FMS XANES simulations are reported in Fig. 5.6 (continuous line): The spectrum of the x = 0.04 sample is reproduced well by replacing 30 % of substitutional Co atoms by Co-V_O complexes, while the spectrum of the x = 0.02 one is better reproduced without V_O. In order to confirm that FM is related to V_O, the x = 0.04 sample was annealed in oxygen to fill the V_O: the resulting XANES and magnetization curves are reported in Figs. 5.6 and 5.7, respectively (blue line). The valley

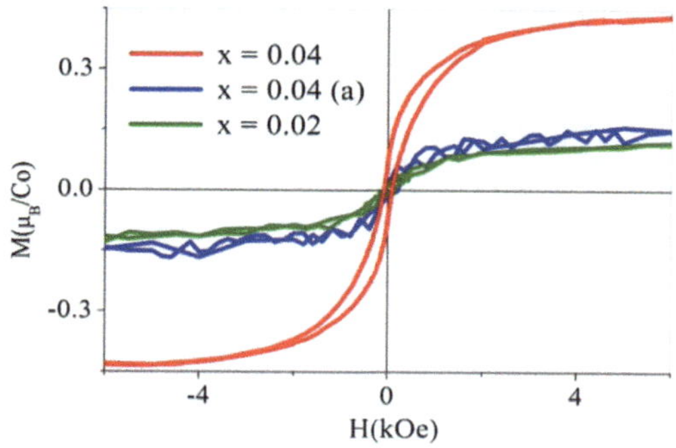

Fig. 5.7 Magnetic hysteresis cycle for the same epilayers as in the previous figure. Modified from [43]

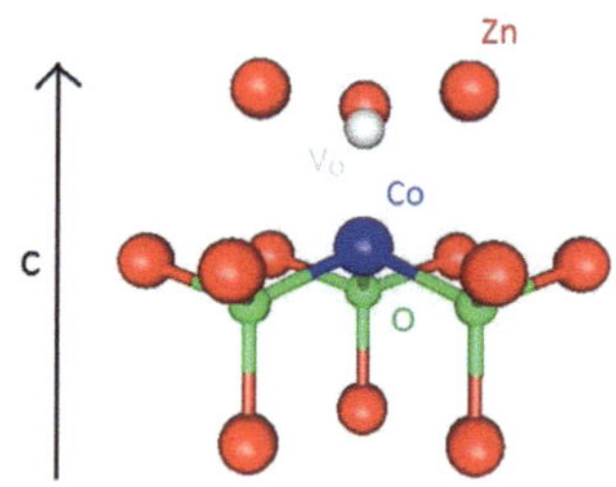

Fig. 5.8 Sketch of the Co-V_O complex used in the simulations

between features 1 and 2 reforms after annealing in oxygen; in parallel, saturation magnetization decreases to one-third of the original value. The observation of larger magnetic moments in the sample where a larger percentage of Co-V_O complexes are detected strongly suggested a V_O-related mechanism for FM. This could be provided by the formation of a partially filled donorlike band, realized by the concurrent presence of Co-V_O complexes and shallow donors.

The FTs of the Co K-edge EXAFS spectra taken on the same samples at the magic angle (~35° incidence angle) in the k-range 4.8–13 A^{-1} are shown in Fig. 5.9. The possibility of using the accurate values calculated by DFT for the atomic positions and distance distribution from the absorber, as anticipated in Sect. 5.2.3, allowed to minimize the number of fit variables and perform a fit weighting three different model structures: Co-V_O complexes (the relative paths of which are all multiplied by an amplitude reduction factor y), metallic Co (weighted by w), and substitutional Co (weighted by 1-y-w). In this case, the variables fitting the thermal part of the DW factors and the differences between actual and model distances were constrained to assume the same values for the Co-V_O complex and Co substitutional contributions, whereas the interatomic distances and DW factors of the metallic contribution, together with the S_0^2 many body factor, were fixed to the values extracted from the data analysis of an experimental Co foil. The fits performed following this procedure are superimposed to the data in Fig. 5.9 (circles).

The relative fraction of the different local environments extracted from the fit are shown in Table 5.1: while the error bars on y are relatively large due to the lower sensitivity of EXAFS to the presence of Co-V_O complexes with respect to XANES,

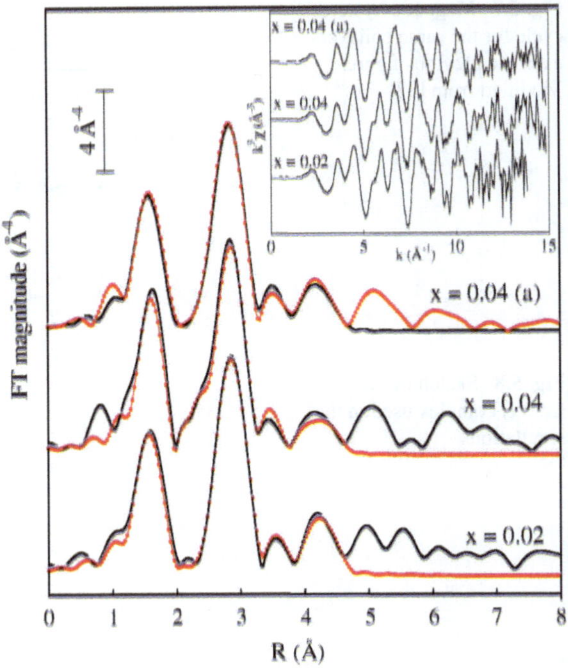

Fig. 5.9 Fourier transform of the Co K-edge EXAFS spectra (*continuous line*) and fits (*circle*) obtained weighting three Co local configurations, as explained in the text. *Inset* Background-subtracted EXAFS spectra. Modified from [43]

Table 5.1 Co-V_O complex fraction (y) and Co metal fraction (w) extracted from the fit of EXAFS data

Sample	Co-V_O (%)	Co metal (%)
X = 0.04 (a)	9 ± 14	0.3 ± 1.9
X = 0.04	18 ± 14	3.7 ± 2.1
X = 0.02	0 ± 14	0.1 ± 1.8

conversely, very small error bars are obtained when measuring the percentage of Co atoms involved in metallic parasitic phases (with respect to the total Co atoms).

The values extracted for *w* are rather small, testifying to the high quality of Co substitutional incorporation in the matrix, and enable quantitative considerations on the origin of RT FM: considering the Co atomic magnetic moment, Co metallic nanophases would still only be able to account for (at most) 13 % of the observed magnetism in the x = 0.04 sample. These results lead to discard the presence of Co nanoclusters, invisible to XRD, as the main source of FM in $Zn_{1-x}Co_xO$.

The results presented in this section provided direct evidence of the existence of cobalt-oxygen vacancy complexes in high-structural quality $Zn_{1-x}Co_xO$ epilayers having a specific spatial orientation, and that a larger percentage of these complexes lead to an enhancement of room temperature FM. This suggests an active role of these microstructures in the origin of the measured macroscopic magnetic moment, in agreement with recent theoretical models of DFO.

5.3.4 Erbium at Oxygen-Decorated Vacancies in (Er, O)-Doped Silicon

Doping silicon with erbium opens interesting perspectives for optoelectronic applications [44] since the $^4I_{13/2} - ^4I_{15/2}$ optical transition occurs at wavelength of 1.54 μm, close to one of the minima of absorption in silica optical fibers. This transition is normally forbidden by electric dipole selection rules, but when Er atoms are embedded in a matrix which renders their environment non centrosymmetric (as it may be the case of the inclusion in Si crystals). This possibility makes Er-doped Si interesting in the manufactory of devices which couple optical and electronic functionalities on the same platform. The main problem with this system from the technological applications standpoint is the low solubility of Er, which is around 2×10^{16} atoms/cm^2, concentration level which is insufficient for device fabrication. A viable route to increase the active ion concentration is provided by co-doping with O; in this way Er atoms bind to O ones and concentrations as high as 10^{19}Er atoms/cm^2 can be obtained without precipitation of Er silicates. A strong emission at 1.54 μm has been indeed observed in co-doped samples for Er/O ratio = 1:10 [45].

Different possible crystallographic sites have been theoretically proposed for Er in (Er, O) co-doped Si. According to the calculations of some groups the hexagonal interstitial site (H_{i}) would be more stable, while other studies suggested incorporation in tetrahedral interstitial sites (T_i) [46]. From the experimental standpoint both EPR [47] and Rutherford back-scattering spectroscopy (RBS)/channeling [48] indicated a coexistence of the two sites (H_i, T_i), in contrast they put a 5 % maximum limit to the incorporation in tetrahedral substitutional sites (T_s).

EXAFS is intuitively a very suitable technique to study the local environment of impurities in a matrix (see also Chap. 4), and the first applications of the technique to the present case revealed that Er is normally six fold coordinated to O atoms in doped Si, with Er–O distance of 2.24–2.26 Å [49, 50]. If these first attempts adopted the single scattering approximation in the data analysis and focused on the first atomic shells around Er, more recently D'Acapito et al. performed a multiple scattering analysis to extract more information on the higher coordination shells and bond angles distribution [51]. In this last work, the authors studied (Er, O)-codoped Si samples obtained both with MBE and ion implantation, by acquiring Er L3-edge spectra in fluorescence detection mode at the Gilda CRG beamline of the ESRF.

Figure 5.10 shows a typical absorption spectrum on a MBE-grown sample, while the inset shows the relative EXAFS Fourier Transform. The latter is characterized by the presence of a main peak just below 2 Å, related to the Er–O bond plus a second peak at apparent distance of 3.2 Å, originating from the second shell Er-Si coordination. The authors employed a conventional data analysis (Sect. 5.2.1), generating the theoretical signals from a crystallographic $Er_2Si_2O_7$ clusters, but then used the potentiality of the GNXAS code to better define the bond angles formed by the constituent atoms. Experimental data were reproduced using a first shell single scattering Er–O path with variable occupancy, distance and DW factor, and a Er–O–Si multiple scattering (MS) path described by the first shell variables plus the Si-O

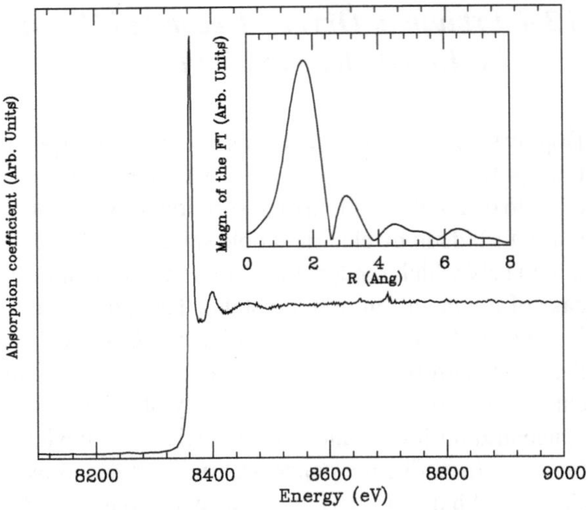

Fig. 5.10 X-ray absorption coefficient typically measured on (Er, O)-doped Si. *Inset* FT of the EXAFS data performed in the 2.4–8 Å range with a k^2 weight. (Reprinted with permission from [51]. Copyright 2004 by the American Physical Society. http://prb.aps.org/abstract/PRB/v69/i15/e153310)

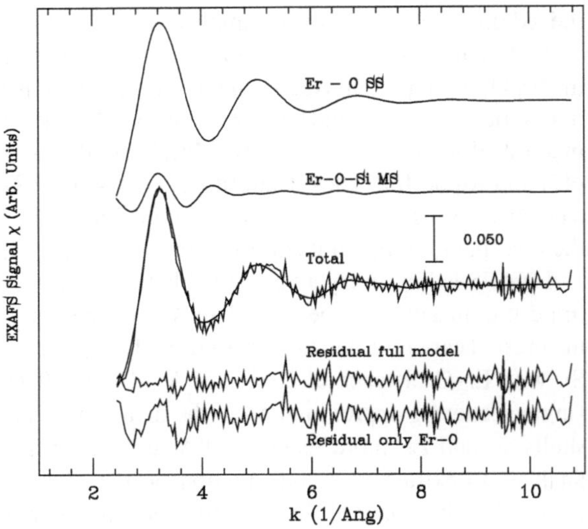

Fig. 5.11 Example of fit performed on one of the samples. The contributions from the Er–O and Er–O–Si atomic configurations are shown along with the experimental data and best fit (total). The two bottommost curves show the residual when using the full model or only the first Er–O shell in the fit. (Reprinted with permission from [51]. Copyright 2004 by the American Physical Society. http://prb.aps.org/abstract/PRB/v69/i15/e153310)

bond length and DW factor plus an angular $\theta_{Er-O-Si}$ parameter with associated DW factor. Second shell Er-Si bond length and DW factor, instead, were derived from the variables listed above. In order to limit the maximum number of variables some of the parameters (the angular DW factor and the O-Si distance) were defined for one of the samples and kept fixed in the analysis of the others.

Figure 5.11 shows an example of fit of the EXAFS spectrum for one of the samples. The fit is superposed to the experimental data, and the contributions of single and MS paths to the fit are illustrated in the uppermost part of the figure. The two bottommost curves represent the residual signals when the fits (obtained using the full model or the first shell only, respectively) are subtracted from the experimental

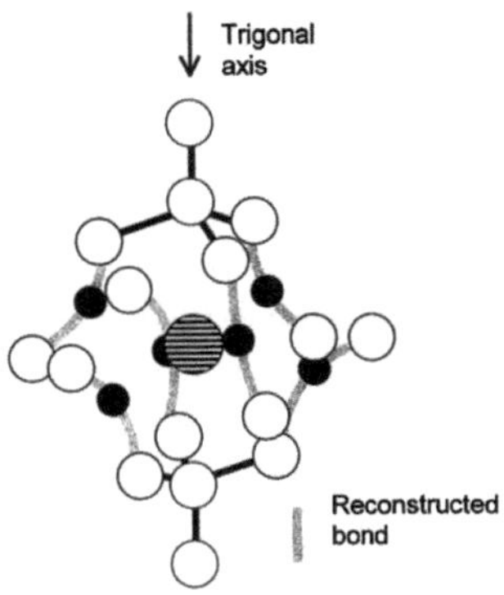

Fig. 5.12 Sketch of the local structure of Er located at the centre of an O-decorated ring hexavacancy. (Reprinted with permission from [52]. Copyright 2004 by the IOP. http://iopscience.iop.org/0953-8984/14/36/310)

data. The similarity of the first-shell residual and the MS signal is evident, as well as the improvement in residual minimization when the MS is employed. Hence, MS paths bring about a better agreement with data from which a refinement of the structural model can be obtained. If on the one hand the authors found expected values for the first shell structural parameters (coordination number around 5–6, and Er–O bond length around 2.24 Å), on the other hand they found appreciably longer Er-Si distances than the expected ones (3.6 Å vs. 2.5–3.0 Å), both for the H_i and T_i site hypothesis.

Actually, the extracted Er-Si distance is more similar to that expected for the substitutional site T_s; however, the fact that the T_s was ruled out by EPR and RBS data opened a controversial. Conciliation between the apparently different experimental results was provided by the hypothesis of formation of a specific defect complex in which Er atoms are located in a H_i site sitting in the middle of a O-decorated V_6 hexavacancy (Fig. 5.12) [52].

According to *ab initio* calculation [53], this vacancy has a center-to-silicon distance of 3.9 Å, sufficient to host a centered ErO_6 cluster in which the Er-Si distance is expected to be around 3.9 Å: this value is close to the estimation done by EXAFS. This result is yet another demonstration of the potential of this technique for the study of the three-dimensional structure of simple defect complexes in crystalline semiconductors; the application to the study of dopants in Si is discussed in Chap. 4.

5.4 Clustering and Anticlustering

In this section, we present examples of applications of XAS to the analysis of clustering/anticlustering phenomena in crystalline semiconductors. In case of GaAsBi alloys, Bi clustering can be detected and quantified by XAS. *Clustering* and *anticlustering* have here the meaning explained at Sect. 5.1 and a review of the vast literature about heterogeneous inclusions and cluster formation in semiconductors, even at the microscopic [54] and nanoscopic scale range [55] is out of the scope of this section.

5.4.1 Bismuth Clustering in GaAsBi Epilayers

In Sect. 5.3.1 we introduced the technological importance of dilute nitride semiconductors in the manufactory of long-wavelength infrared devices on GaAs substrates such as 1.3–1.55 μm emitters and 1 eV band-gap components for multijunction solar cells. Unluckily, there are several intrinsic drawbacks limiting their utilization in devices, such as the low electron mobility [56] and short carrier diffusion lengths, which led scientists to look for alternative semiconductor alloys. One of the novel routes proposed was the incorporation of bismuth into GaAs [57].

$GaAs_{1-x}Bi_x$ alloys grown by MBE present good structural quality and strong photoluminescence emission without need of annealing up to x = 0.1 [58]. Bi incorporation (at odds with N) perturbs the valence-band maximum instead of the conduction band minimum [59], which does not affect the electron mobility [56]. Moreover, Bi redshifts the band gap by more than four times as much as Sb or In enabling easier access to the useful wavelengths [60]. However, the origin of the properties of *dilute bismides* remains unclear: the giant reduction of the band gap is in disagreement with predictions [61] and the efficient PL emission is surprising, considering the low growth temperature which uses to produce defects [62].

It is well known that SRO affects the optical properties of dilute nitrides: for example, In-N preferential binding blue shifts the band gap of $In_xGa_{1-x}As_{1-y}N_y$ [63], and N (In) clustering induces a redshift in $GaAs_{1-x}N_x$ ($In_xGa_{1-x}N$) [2]. Similarly, in $GaAs_{1-x}Bi_x$, Bi impurities may correlate and affect the emission, and the presence of Bi clusters, similar to N, is likely to induce shallow bound states in the band gap.

The issue of clustering in $GaAs_{1-x}Bi_x$ epilayers grown by MBE on GaAs substrate has been addressed by Bi L2-edge EXAFS [64] in fluorescence mode, selecting the Bi $L\beta 1$ emission line. The experiment was performed at the BM29 beamline of the ESRF. Figure 5.13 shows the FTs of the EXAFS spectra (transformation range = 3.2–12.4 $Å^{-1}$) for three samples with different Bi concentration.

Comparing the FTs of samples A (x = 1.2 %) and B (x = 1.9 %), it is evident that the second shell peak is strongly damped in the latter, while the amplitude of the first and third shell peak is almost unchanged: since thermal disorder is supposed to be alike (the temperature was constant in the experiment), this suggests a different distribution of Bi and As atoms in the second shell of Bi. As for sample C (x = 2.4 %), the amplitude of the first shell Fourier peak is almost identical to that of sample A, while second and third shell peak amplitudes and shapes change, suggesting a different organization of the Bi neighbors.

In order to analyze these data, the VFF strategy described in Sect. 5.2.2 was used; this is particularly suitable in this case since the main models at test do not involve a deep modification in the structure symmetry but only a redistribution of the atoms in the V-group mixed atomic shell. In this approach, the spectrum of the lowest concentration sample (A), where Bi clustering is less probable, was used as a reference. Therefore, in the fit of sample A (Fig. 5.14, bottom), a random distribution of the anions was assumed and the best fit values of unique DW variables for each shell were determined from the minimization routine. In contrast, all interatomic distances were

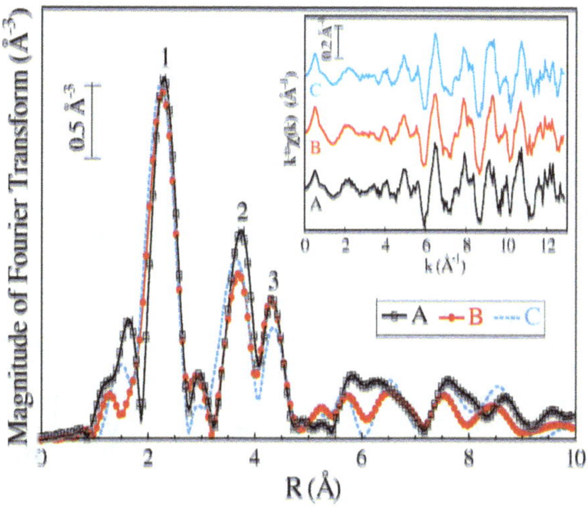

Fig. 5.13 Fourier transform of the Bi L2-edge EXAFS signal for three $GaAs_{1-x}Bi_x/GaAs$ epilayers with x = 0.012 (*A*), 0.019 (*B*) and 0.024 (*C*). The three main peaks are labelled. *Inset* Background subtracted $K^{*}\chi(K)$ spectra. (Reprinted with permission from [64]. Copyright 2008 by the American Physical Society. http://prb.aps.org/abstract/PRB/v78/i3/e035325)

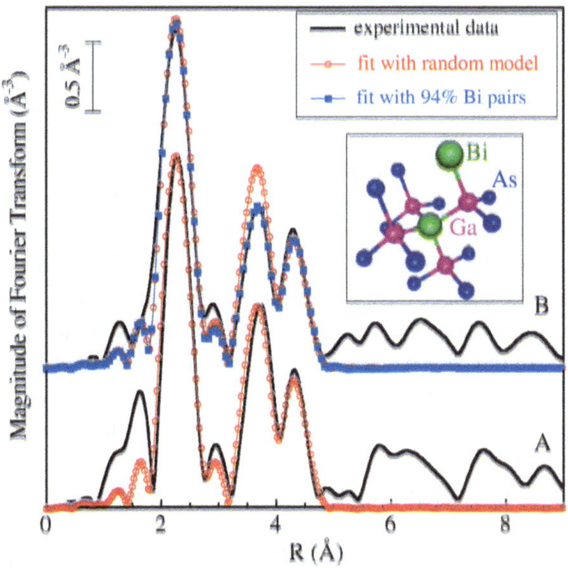

Fig. 5.14 Experimental and simulated Fourier transformed EXAFS spectra for samples *A* (*bottom*) and *B* (*top*). The fits are based on clusters with either a random distribution of Bi atoms or a predominance of Bi pairs. (Reprinted with permission from [64]. Copyright 2008 by the American Physical Society. http://prb.aps.org/abstract/PRB/v78/i3/e035325)

fixed to the VFF-calculated values: varying the distances in the fit to counterbalance possible systematic errors done by FEFF in the phase shifts calculations was not necessary. The DW factors determined from the analysis of sample A were fixed in the analysis of the other two samples.

As for sample B (Fig. 5.14, top), if the DW factors are fixed to the values determined for sample A, it is evident that the second shell peak cannot be reproduced well

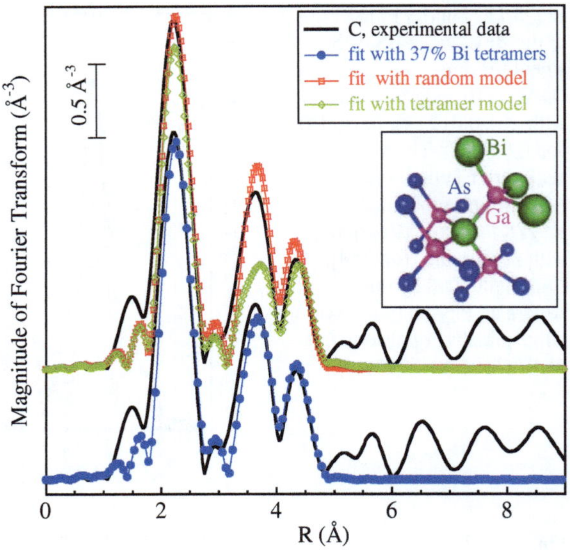

Fig. 5.15 Experimental Fourier-transformed EXAFS spectrum for sample *C* (*continuous black line*, *top* and *bottom*) and fits performed for clusters with a random anion distribution, a Bi tetramer (*see sketch in the inset*), and a combination of different local configurations as detailed in the text. (Reprinted with permission from [64]. Copyright 2008 by the American Physical Society. http://prb.aps.org/abstract/PRB/v78/i3/e035325)

by a random model (red open circles).[4] In contrast, experimental data for samples B and C can be fitted by considering a combination of three different local configurations: a next-nearest-neighbor Bi pair (inset of Fig. 5.14), a Bi tetramer (inset of Fig. 5.15), and a random Bi distribution. In the tetramer configuration, the central Bi atom is surrounded by three Bi next-nearest-neighbors, in a way that all Bi atoms are next nearest neighbors of each other. The procedure consists of simulating the EXAFS spectra by a weighted sum of simulated spectra for each configuration and using the relative weights as free variables in the fit.

In the case of sample B, an excellent best fit is obtained (Fig. 5.14, top, blue filled squares) with $94 \pm 17\,\%$ Bi pairs and $6 \pm 17\,\%$ Bi tetramers (0 % random). This result supports the formation of Bi pairs when the Bi concentration increases from 1.2 % to 1.9 %.

When a Bi pair forms, the damping of the second shell is caused by the broadening of the Bi-As distance distribution and not only by the presence of a Bi-Bi EXAFS path: according to VFF calculations, the Bi-As distances are spread over 0.086 for the pair versus 0.022 Å for the random model.

In the case of sample C, a fit with the random model (Fig. 5.15, top, red open squares) fails to reproduce the second and third shells of the experimental FT (continuous black line). In case of tetramer formation, the interatomic distances are strongly perturbed by the presence of the Bi next nearest neighbors in the second shell: according to the VFF calculation, the Bi-Ga first shell distances spread over 0.061 Å and the Bi-As second shell ones over 0.10 Å. The fit with the tetramer model (top,

---

[4] Increasing the second shell Bi-As DW factor by 25 % provides an apparently satisfactory fit; however, our aim is to go beyond this purely phenomenological procedure and to identify the actual configuration responsible for the signal damping.

olive open diamonds) is effective in reproducing the amplitude of the third shell and the region between the second and third shells where part of the Bi second shell signal is also found. In contrast, it fails in reproducing the amplitude of the whole second shell peak. By fitting with the same combination as for sample B, the best fit (bottom, blue filled circles) is obtained with $37 \pm 19\,\%$ Bi tetramers and $0 \pm 4\,\%$ Bi pairs (63 % random): such combination produces a good agreement with the experimental FT.[5] Other defects, such as interstitials or Bi heteroantisites, do not account for the modifications of the spectra [65].

It is important to note that in the present approach it is not necessary to make the DW factors vary in order to account for the different structural disorder in samples with different Bi concentration: once the correct SRO configuration is found, this accounts itself for the structural part of distances distribution broadening, while the thermal part is approximated by the DW of the reference sample A. A limitation of this approach is that it assumes that the damping of the signal in the mixed shell is entirely due to the realization of the specific SRO configurations: if other sources of distance broadening summed up, this would bring about an overestimation of the pair/tetramers concentrations extracted from the fits.

The results presented in this section gave evidence of an evolution from a random distribution of Bi anions to Bi pairing and clustering as the Bi concentration increases in $GaAs_{1-x}Bi_x$. As for N pairing in $GaAs_{1-x}N_x$ [2], it is probable that Bi clustering has an important role in the giant band gap bowing observed in $GaAs_{1-x}Bi_x$. Furthermore, Bi clusters could play the role of exciton trap in analogy to the mechanism proposed for $In_xGa_{1-x}N$ [66], accounting for the PL enhancement observed in dilute bismides. In fact, later studies have shown that the most efficient optical emission is observed in correspondence of the highest clustering [67].

5.4.2 Absence of Clustering in GaAsSbN and ZnSSe

In contrast with the case of Bi impurities in $GaAs_{1-x}Bi_x$, in several other ternary and quaternary alloys clustering phenomena do not occur or are very small. Even in this case, XAS can be very useful to verify the randomness of the material. One example is given by GaAsSbN alloys, which are also important semiconductors for applications in optoelectronics. When post-growth treated by thermal annealing, GaAsSbN alloys show a remarkable blueshift of the band gap as a function of the annealing time [68] which is not trivially due to N diffusion out of the active layer. Clustering/anticlustering of Sb or N atoms driven by thermal annealing could provide an explanation for the observed blue shift since their effect on the optical properties of III–V semiconductors has been predicted [2], as already anticipated.

The issue of Sb clustering in GaAsSbN and possible evolution as a function of growth conditions and annealing has been addressed by Sb K- and L3-edge XAS

[5] Note that agreement is better than that obtained using any other combination involving different SRO configurations, including other tetramers, triplets with Bi atoms set in a row or in a triangle.

performed in fluorescence mode at the BM29 and ID26 beamlines of the ESRF [69]. This enabled to evaluate the relative occupancy of As, N, and Sb in the second atomic coordination shell of Sb. Data analysis was performed using a VFF strategy similar to that adopted in the previous section, using GaAsSb reference samples to untangle the structural component of the DW factor related to the introduction of N atoms in the crystal from the thermal one. The results obtained [69] demonstrated that there is no Sb clustering in GaAsSbN and the anions are roughly randomly distributed in the mixed shell even if, in some case, a moderate trend over Sb-N preferential association is observed. They also showed that annealing does not affect appreciably the Sb local environment, therefore Sb clustering cannot be at the origin of the band gap blue shift. The PL blue shift upon annealing could be attributed to homogenization of initial nanometric spatial fluctuations of N concentration [70].

Similar results were found by Maruyama et al. [71] on a II–VI semiconductor alloy, ZnS_xSe_{1-x}, which is another interesting direct-wide gap system for optoelectronic applications. By using S K-edge EXAFS performed in fluorescence mode at the beamline (BL)-11B of the Photon Factory Synchrotron, the authors investigated the relative atom occupancy in the mixed second shell around S. By employing a conventional analysis strategy (Sect. 2.1) and Fourier filtering to isolate the second shell signal, they measured the ratio of the number of S atoms to that of Se ones and found the value of 0.057/0.942, in good agreement with the ratio between S and Se concentrations ($x = 0.06$, $1 - x = 0.94$). This result gave the indication that there is no S clustering in the alloy and that S and Se atoms are randomly distributed.

5.5 Summary

In this chapter we have provided several examples of application of XAS to the analysis of defect complexes and clustering in crystalline semiconductors of high technological interest. If EXAFS has been recognized, since its early development, to be a very important tool in the analysis of the local geometric arrangement of dopants and point defects in semiconductors, the extension to the study of composite defects is more recent. In fact, this required the enhanced sensitivity of XANES to the three-dimensional structure of the complexes and reliable simulations of the spectral region in which the multiple scattering expansion of the XAS cross section does not converge. The success of the method in the examples here reported depends on the high brilliance of third generation synchrotron sources which allowed the recording of high-quality spectra from dilute systems, the availability of reliable theory and computer codes able to accurately simulate the XAS spectra, and the trustworthiness of DFT and VFF methods in achieving accurate results on the microstructure of defect complexes and in mimicking SRO configurations.

Last but not least, the scientific cases presented in this chapter underline the strong complementarity of XAS with other spectroscopic techniques more traditionally employed in the study of defect complexes structure such as infrared absorption spectroscopy, Rutherford back-scattering spectroscopy/channeling, and electron paramagnetic resonance.

Acknowledgments I am grateful to all collaborators who took part in the different research projects presented in this chapter, in particular to A. Amore Bonapasta, F. Filippone, P. Alippi (ISM-CNR, Montelibretti, Italy) and F. Glas (LPN, Marcoussis, France) for fundamental theoretical support via calculations of the complex and cluster structures. I acknowledge the excellent support by the scientific and technical staff of the different beamlines at which the relevant experiments were carried out: BM29, BM08, and ID26 beamlines at the ESRF, the Samba beamline at SOLEIL, and the Aloisa beamline at ELETTRA. I thank F. D'Acapito for sending me Figs. 5.10 and 5.11, and J.D. Carey for agreeing in republishing Fig. 5.12. Measurements at the ESRF and SOLEIL were supported in part by the relative public user programs; experiments at ELETTRA were supported in part by the Research Infrastructure Action under the FP6 "Structuring the European Research Area" program.

References

1. J.M. Cowley, Phys. Rev. **77**, 669 (1950)
2. L. Bellaiche, A. Zunger, Phys. Rev. B **57**, 4425 (1998)
3. Y. Joly, Phys. Rev. B **63**, 125120 (2001)
4. F. Glas, Phys. Rev. B **51**, 825 (1995)
5. I. Anisimov, F. Aryasetiawan, A.I. Liechtenstein, J. Phys.: Condens. Matter **9**, 767 (1997)
6. X. Gonze, R. Stumpf, M. Scheffler, Phys. Rev. B **44**, 8503 (1991)
7. D. Vanderbilt. Phys. Rev. B **41**, 7892 (1990)
8. K.C. Pandey, A. Erbil, G.S. Cargill III, R.F. Boehme, D. Vanderbilt, Phys. Rev. Lett. **61**, 1282 (1988)
9. T. Maruyama, T. Ogawa, K. Akimoto, Y. Kitajima, Solid State Commun. **103**, 453 (1997)
10. M. Henini (ed.), *Dilute Nitrides* (Elsevier, Oxford, 2005)
11. F. Masia, A. Polimeni, G.B.H. von Högersthal, M. Bissiri, M. Capizzi, P.J. Klar, W. Stolz, Appl. Phys. Lett. **82**, 4474 (2003)
12. S. Sato, Y. Osawa, T. Saitoh, Jpn. J. Appl. Phys. Part 1, **36**, 2671 (1997)
13. J.F. Geisz, D.J. Friedman, Semicond. Sci. Technol. **17**, 769 (2002)
14. A. Polimeni, G.B.H.v. Högersthal, M. Bissiri, M. Capizzi, M. Fischer, M. Reinhardt, A. Forchel, Phys. Rev. B **63**, 201304(R) (2001)
15. A. Polimeni, G. Ciatto, L. Ortega, F. Jiang, F. Boscherini, F. Filippone, A. Amore Bonapasta, M. Stavola, M. Capizzi, Phys. Rev. B **68**, 085204 (2003)
16. G. Bisognin, D. De Salvador, A.V. Drigo, E. Napolitani, A. Sambo, M. Berti, A. Polimeni, M. Felici, M. Capizzi, M. Güngerich, P.J. Klar, G. Bais et al., Appl. Phys. Lett. **89**, 061904 (2006)
17. M. Felici, A. Polimeni, G. Salviati, L. Lazzarini, N. Armani, F. Masia, M. Capizzi, F. Martelli, M. Lazzarino, G. Bais, M. Piccin, S. Rubini et al., Adv. Mater. (Weinheim, Ger.) **18**, 1993 (2006)
18. A. Amore Bonapasta, F. Filippone, P. Giannozzi, Phys. Rev. B **68**, 115202 (2003) (and references therein)
19. F. Jiang, M. Stavola, M. Capizzi, A. Polimeni, A. Amore Bonapasta, F. Filippone, Phys. Rev. B **69**, 041309(R) (2004)
20. G. Ciatto, F. Boscherini, A. Amore Bonapasta, F. Filippone, A. Polimeni, M. Capizzi, Phys. Rev. B **71**, 201301(R) (2005)
21. G. Ciatto, F. Boscherini, A. Amore Bonapasta, F. Filippone, A. Polimeni, M. Capizzi, M. Berti, G. Bisognin, D. De Salvador, L. Floreano, F. Martelli et al., Phys. Rev. B **79**, 165205 (2009)
22. G. Ciatto, H. Renevier, M.G. Proietti, A. Polimeni, M. Capizzi, S. Mobilio, F. Boscherini, Phys. Rev. B **72**, 085322 (2005)

23. H. Ohno, Science **281**, 951 (1998)
24. T. Dietl, H. Ohno, F. Matsukura, J. Cibert, D. Ferrand, Science **287**, 1019 (2000)
25. Y. Ohno, D.K. Young, B. Beschoten, F. Matsukura, H. Ohno, D.D. Awschalom, Nature (London) **402**, 790 (1999)
26. S.T.B. Goennenwein, T.A. Wassner, H. Huebl, M.S. Brandt, J.B. Philipp, M. Opel, R. Gross, A. Koeder, W. Schoch, A. Waag, Phys. Rev. Lett. **92**, 227202 (2004)
27. R. Bouanani-Rahbi, B. Clerjaud, B. Theys, A. Lemaitre, F. Jomard, Physica B **340–342**, 284 (2003)
28. M.S. Brandt, S.T.B. Goennenwein, T.A. Wassner, F. Kohl, A. Lehner, H. Huebl, T. Graf, M. Stutzmann, A. Koeder, W. Schoch, A. Waag, Appl. Phys. Lett. **84**, 2277 (2004)
29. L. Thevenard, L. Largeau, O. Mauguin, A. Lemaitre, B. Theys, Appl. Phys. Lett. **87**, 182506 (2005)
30. C.G. Van de Walle, J. Neugebauer, Nature (London) **423**, 626 (2003)
31. A. Amore Bonapasta, F. Filippone, P. Giannozzi, Phys. Rev. B **72**, 121202R (2005)
32. J.P. Goss, P.R. Briddon, Phys. Rev. B **72**, 115211 (2005)
33. C. Bihler, G. Ciatto, H. Huebl, G. Martinez-Criado, P.J. Klar, K. Volz, W. Stolz, W. Schoch, W. Limmer, F. Filippone, A. Amore, Bonapasta, M.S. Brandt, Phys. Rev. B **78**, 235208 (2008)
34. W. Limmer, A. Koeder, S. Frank, V. Avrutin, W. Schoch, R. Sauer, K. Zuern, J. Eisenmenger, P. Ziemann, E. Peiner, A. Waag, Phys. Rev. B **71**, 205213 (2005)
35. P. Sharma, A. Gupta, K.V. Rao, F.J. Owens, R. Sharma, R. Ahuja, J.M. Osorio Guillen, B. Johansson, G.A. Gehring, Nat. Mater. **2**, 673 (2003)
36. J.M. Coey, M. Venkatesan, C.B. Fitzgerald, Nat. Mater. **4**, 173 (2005)
37. C.D. Pemmaraju, R. Hanafin, T. Archer, H.B. Braun, S. Sanvito, Phys. Rev. B **78**, 054428 (2008)
38. A. Ney, A. Kovács, V. Ney, S. Ye, K. Ollefs, T. Kammermeier, F. Wilhelm, A. Rogalev, R.E. Dunin-Borkowski, New J. Phys. **13**, 103001 (2011)
39. T.C. Kaspar, T. Droubay, S.M. Heald, M.H. Engelhard, P. Nachimuthu, S.A. Chambers, Phys. Rev. B **77**, 201303 (2008)
40. S. Krishnamurthy, C. McGuinness, L.S. Dorneles, M. Venkatesan, J.M.D. Coey, J.G. Lunney, C.H. Patterson, K.E. Smith, T. Learmonth, P.-A. Glans et al., J. Appl. Phys. **99**, 08M111 (2006)
41. S. Zhang, L. Zhang, H. Li, J. Li, Z. Jiang, W. Chu, Y. Huang, J. Wang, Z. Wu, J. Synchrotron Radiat. **17**, 600 (2010)
42. H.S. Hsu, J.C.A. Huang, Y.H. Huang, Y.F. Liao, M.Z. Lin, C.H. Lee, J.F. Lee, S.F. Chen, L.Y. Lai, C.P. Liu, Appl. Phys. Lett. **88**, 242507 (2006)
43. G. Ciatto, A.D. Trolio, E. Fonda, P. Alippi, A.M. Testa, A. Amore Bonapasta, Phys. Rev. Lett. **107**, 127206 (2011)
44. S. Coffa, A. Polman, R.N. Schwartz (ed.), *Rare Earth Doped Semiconductors II. Materials Research Society Symposium Proceedings*, vol. 402 (Materials Research Society, Pittsburgh, 1996)
45. S. Coffa, G. Franzo, F. Priolo, A. Polman, R. Serna, Phys. Rev. B **49**, 16313 (1994)
46. A.G. Raffa, P. Ballone, Phys. Rev. B **65**, 121309 (2002)
47. J.D. Carey, R.C. Barklie, J.F. Donegan, F. Priolo, G. Franzo, S. Coffa, Phys. Rev. B **59**, 2773 (1999)
48. M.B. Huang, X.T. Ren, Phys. Rev. B **68**, 033203 (2003)
49. D.L. Adler, D.C. Jacobson, D.J. Eaglesham, M.A. Marcus, J.L. Benton, J.M. Poate, P.H. Citrin, Appl. Phys. Lett. **61**, 2181 (1992)
50. M. Ishii, T. Ishikawa, T. Ueki, S. Komuro, T. Morikawa, Y. Aoyagi, H. Oyanagi, J. Appl. Phys. **85**, 4024 (1999)
51. F. D'Acapito, S. Mobilio, S. Scalese, A. Terrasi, G. Franzo, F. Priolo, Phys. Rev. B **69**, 153310 (2004)
52. J.D. Carey, J. Phys.: Condens. Matter **14**, 8537 (2002)
53. J.L. Hastings, S.K. Estreicher, P.A. Fedders, Phys. Rev. B **56**, 10215 (1997)
54. A. Wolska, M.T. Klepka, K. Lawniczak-Jablonska, J. Sadowski, A. Reszka, B.J. Kowalski, Radiat. Phys. Chem. **80**, 1026 (2011)

55. M. Opel, K.-W. Nielsen, S. Bauer, S.T.B. Goennenwein, J.C. Cezar, D. Schmeisser, J. Simon, W. Mader, R. Gross, Eur. Phys. J. B **63**, 437–444 (2008)
56. D.G. Cooke, E.C. Young, F.A. Hegmann, T. Tiedje, Appl. Phys. Lett. **89**, 122103 (2006)
57. S. Tixier, M. Adamcyk, T. Tiedje, S. Francoeur et al., Appl. Phys. Lett. **82**, 2245 (2003)
58. X. Lu, D.A. Beaton, R.B. Lewis, T. Tiedje, M.B. Whitwickal, Appl. Phys. Lett. **92**, 192110 (2008)
59. Y. Zhang, A. Mascarenhas, L.W. Wang, Phys. Rev. B **71**, 155201 (2005)
60. S. Francoeur, M.J. Seong, A. Mascarenhas, S. Tixier, M. Adamcyk, T. Tiedje, Appl. Phys. Lett. **82**, 3874 (2003)
61. A. Janotti, S.H. Wei, S.B. Zhang, Phys. Rev. B **65**, 115203 (2002)
62. K. Bertulis, A. Krotkus, G. Aleksejenko, V. Pacebutas, R. Andromavicius, G. Molis, Appl. Phys. Lett. **88**, 201112 (2006)
63. K. Kim, A. Zunger, Phys. Rev. Lett. **86**, 2609 (2001)
64. G. Ciatto, E.C. Young, F. Glas, J. Chen, R. Alonso, Mori, T. Tiedje, Phys. Rev. B **78**, 035325 (2008)
65. G. Ciatto, P. Alippi, A. Amore Bonapasta, T. Tiedje, Appl. Phys. Lett. **99**, 141912 (2011)
66. O. Brandt, K.H. Ploog, Nat. Mater. **5**, 769 (2006)
67. G. Ciatto, M. Thomasset, F. Glas, X. Lu, T. Tiedje, Phys. Rev. B **82**, 201304(R) (2010)
68. T. Kageyama, T. Miyamoto, S. Makino, F. Koyama, K. Iga, Jpn. J. Appl. Phys. Part 2 **38**, L298 (1999)
69. G. Ciatto, J.-C. Harmand, F. Glas, L. Largeau, M. Le Du, F. Boscherini, M. Malvestuto, L. Floreano, P. Glatzel, R. Alonso Mori, Phys. Rev. B **75**, 245212 (2007)
70. L. Grenouillet, C. Bru-Chevallier, G. Guillot, P. Gilet, P. Ballet, P. Duvaut, G. Rolland, A. Million, J. Appl. Phys. **91**, 5902 (2002)
71. T. Maruyama, T. Ogawa, K. Akimoto, Y. Kitajima, S. Ito, A. Ishibashi, Jpn. J. Appl. Phys. Part II **34**, L539–L542 (1995)

Chapter 6
Vibrational Anisotropy

Paolo Fornasini

Abstract The mean square relative displacement parallel to the bond direction (parallel MSRD) is directly measured by the EXAFS Debye-Waller exponent. For crystals, the comparison of the thermal expansions measured by EXAFS and by Bragg diffraction allows one to obtain also the perpendicular MSRD. The ellipsoid of relative thermal displacements is anisotropic with respect to the bond direction for all studied systems. The quantitative study of anisotropy gives original insights on the local lattice dynamics and can be exploited to investigate the vibrational mechanism of negative thermal expansion in crystals. The basic theory is here reviewed and the results of some recent experimental studies on crystals with the diamond-zincblende structure are critically compared.

6.1 Introduction

The comprehension of the effects of thermal disorder on extended X-ray absorption fine structure (EXAFS) has been progressively refined since the Seventies. In the first pioneering works [1, 2], the Debye-Waller exponent σ^2 was identified as the mean square relative displacement (MSRD) parallel to the bond direction and expressed in terms of eigenfrequencies and eigenvectors of the dynamical matrix of crystals. The difference between the Debye-Waller factors of EXAFS and of diffraction due to correlation effects was stressed and the correlated Debye and Einstein models were introduced. The relevance of asymmetry in the nearest neighbour distance distribution was early thereafter recognised [3]; a parametrised model based on the cumulant expansion was introduced [4, 5], which proved to be particularly effective when dealing with relatively weak thermal disorder. Various theoretical treatments were developed, based on different approches [6–15].

In the last years, the increase of experimental accuracy and the progress of analysis techniques led to the possibility of detecting the difference between the local thermal

P. Fornasini (✉)
Department of Physics, University of Trento, Povo (Trento), Italy
e-mail: paolo.fornasini@unitn.it

C.S. Schnohr and M.C. Ridgway (eds.), *X-Ray Absorption Spectroscopy of Semiconductors*, Springer Series in Optical Sciences 190,
DOI 10.1007/978-3-662-44362-0_6

expansions measured by EXAFS and by Bragg diffraction in crystals and to the consequent evaluation of the perpendicular MSRD [16, 17]. As a result, it is now possible to reconstruct the ellipsoid of relative thermal vibrations of the absorber and backscatterer atoms, which has been found to be disc-shaped with respect to the bond direction in all systems studied up to now. The degree of anisotropy of the relative ellipsoids, which depends on the difference of correlation parallel and perpendicular to the bond, gives original insights on the local lattice dynamics and has been exploited to study the local origin of negative thermal expansion in crystals [18, 19].

An updated introductory treatment of thermal effects in EXAFS spectra can be found in [20]. Here, after a synthetic theoretical introduction (Sect. 6.2), an account is given of recent experimental results concerning the effects of thermal vibrations in some selected semiconductors (Sect. 6.3). The difference between bond-lengths measured by EXAFS and Bragg diffraction is discussed in Sect. 6.4 and the applications to the study of negative thermal expansion in crystals is presented in Sect. 6.5.

6.2 Theory

The information available from a single-scattering analysis of EXAFS spectra is intrinsically one-dimensional. For a given coordination shell, EXAFS samples an effective distribution $P(r,\lambda) = \rho(r)\exp(-2r/\lambda)/r^2$, where $\rho(r)$ is the real distribution of inter-atomic distances and $\lambda(k)$ is the photo-electron mean free path (Chap. 1). For relatively weak thermal disorder, both distributions, real and effective, can be parametrised in terms of their cumulants [4, 5] C_i and C'_i, respectively. The relation between C_i and C'_i can be expressed by a recursion formula [21]; in practice, the difference is generally significant only for $i = 1$. The leading cumulants can be obtained by a careful analysis of EXAFS spectra, better if measured as a function of temperature [16, 17]. (Note that, due to a different convention, the cumulants of the real and effective distribution are elsewhere labelled as C^*_i and C_i, respectively [16, 17, 21]).

We review here the connection between the cumulants C_i of the one-dimensional real distribution of distances and the structural and dynamical properties of a three-dimensional crystalline system.

Let $\boldsymbol{R}_0$ be the distance between absorber and backscatterer atoms (a and b, respectively), ideally frozen at their rest positions, and let $\boldsymbol{u}_a$ and $\boldsymbol{u}_b$ be their instantaneous vibrational displacements with respect to the rest positions, due to zero-point plus thermal energy. The instantaneous inter-atomic vector distance $\boldsymbol{r}$ can be expressed as

$$\boldsymbol{r} = \boldsymbol{R}_0 + \Delta\boldsymbol{u}, \tag{6.1}$$

where $\Delta\boldsymbol{u} = \boldsymbol{u}_b - \boldsymbol{u}_a$ is the relative displacement. It is convenient [22] to consider the projections of $\Delta\boldsymbol{u}$ along the bond direction $\Delta u_\parallel$ and in the perpendicular plane

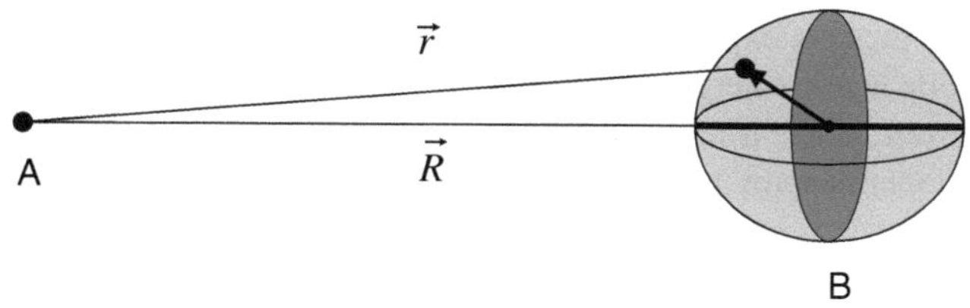

Fig. 6.1 Schematic representation of the relation between the equilibrium distance ***R*** and the instantaneous distance $\boldsymbol{r} = \boldsymbol{R} + \Delta\boldsymbol{u}$ between the absorber atom A and the backscatterer atom B. The ellipsoid of relative displacements is *light grey*. Equation (6.2) project the relative instantaneous displacement $\Delta\boldsymbol{u}$ (*arrow*) along the $\hat{R}$ direction (*thick line*) and in the perpendicular plane (*dark grey*)

$\Delta u_\perp$ (Fig. 6.1), defined by the following equations:

$$\Delta u_\parallel = \hat{R}_0 \cdot \Delta\boldsymbol{u}, \qquad \Delta u_\perp^2 = \Delta u^2 - \Delta u_\parallel^2 . \tag{6.2}$$

An EXAFS spectrum corresponds to sampling, by a large number of photoelectrons, a correspondingly large number of instantaneous configurations created by zero-point and thermal motions. The quantities measured by EXAFS are thus expressed as canonical averages.

6.2.1 Average Distance

The average inter-atomic distance, directly measured by the first EXAFS cumulant C_1, is [17]

$$C_1 = \langle r \rangle \simeq R_0 + \langle \Delta u_\parallel \rangle + \langle \Delta u_\perp^2 \rangle / 2R_0, \tag{6.3}$$

where $R_0 + \langle \Delta u_\parallel \rangle = R_c$ is the crystallographic distance, measured by Bragg diffraction experiments. In the harmonic approximation, $\langle \Delta u_\parallel \rangle = 0$ and there is no crystallographic thermal expansion. For an anharmonic crystal, the term $\langle \Delta u_\parallel \rangle$ increases with temperature, and accounts for the crystallographic expansion.

In terms of the average vector positions of the absorber and backscatterer atoms, $\langle \boldsymbol{r}_a \rangle$ and $\langle \boldsymbol{r}_b \rangle$, respectively, the average distance measured by EXAFS and the crystallographic distance have different expressions:

$$\langle r \rangle = \langle |\boldsymbol{r}_b - \boldsymbol{r}_a| \rangle \qquad R_c = |\langle \boldsymbol{r}_b \rangle - \langle \boldsymbol{r}_a \rangle| , \tag{6.4}$$

R_c being the distance between average atomic positions. Equation (6.3) shows that the average distance $\langle r \rangle$ is always larger than the crystallographic distance, owing to the presence of the perpendicular MSRD in the last term. Moreover, the difference increases with temperature, so that the thermal expansion measured by EXAFS is larger than the thermal expansion measured by Bragg diffraction or by dilatometric techniques. For nearest-neighbour atoms, the two distances $\langle r\rangle$ and R_c have been recently referred to as "true" and "apparent" bond-lengths, respectively, and their temperature dependencies as true and apparent bond expansions [23]. The possibility of directly measuring the true bond expansion by EXAFS represents the solution of an old crystallographic problem, which has been for a long time faced by approximate models, such as the riding model or the TLS model [24].

6.2.2 Parallel MSRD

The second cumulant C_2, say the variance σ^2 of the distribution of distances, corresponds to a good degree of accuracy to the mean value $\langle\Delta u_{\parallel}^2\rangle$, and is now generally referred to as parallel MSRD [25]:

$$C_2 = \sigma^2 = \langle \Delta u_{\parallel}^2 \rangle = \left\langle \left[\hat{R}\cdot(\boldsymbol{u}_b - \boldsymbol{u}_a)\right]^2 \right\rangle . \tag{6.5}$$

The parallel MSRD can be expanded into the sum of three terms [1]

$$\langle \Delta u_{\parallel}^2 \rangle \;=\; \langle (\hat{R}\cdot\boldsymbol{u}_b)^2\rangle \;+\; \langle (\hat{R}\cdot\boldsymbol{u}_a)^2\rangle \;-\; 2\,\langle (\hat{R}\cdot\boldsymbol{u}_b)(\hat{R}\cdot\boldsymbol{u}_a)\rangle . \tag{6.6}$$

The first two terms are the independent mean square displacements (MSD) of atoms a and b along the bond direction, which can be calculated from the atomic displacement parameters (ADP) obtained by the refinement of Bragg diffraction patterns [26]. The third term, the displacement correlation function (DCF), depends on the correlation of atomic motions. The correlation term DCF is relatively large for the first shell and significantly smaller for the second and outer shells, reducing to zero at large distances.

In the harmonic approximation, the parallel MSRD of a crystal is connected to the eigenvalues $\omega(\boldsymbol{q}, s)$ and eigenvectors $\boldsymbol{w}(\boldsymbol{q}, s)$ of the dynamical matrix by

$$\langle \Delta u_{\parallel}^2 \rangle \;=\; \frac{1}{\mathbf{N}} \sum_{\boldsymbol{q},s} \langle |Q(\boldsymbol{q}, s, t)|^2\rangle \left| \left(\frac{\boldsymbol{w}_b(\boldsymbol{q}, s) e^{i\boldsymbol{q}\cdot\boldsymbol{R}}}{\sqrt{m_b}} - \frac{\boldsymbol{w}_a(\boldsymbol{q}, s)}{\sqrt{m_a}} \right) \cdot \hat{R} \right|^2 \tag{6.7}$$

where $\boldsymbol{q}$ are the normal mode wavevectors, s are the branch indexes, $\mathbf{N}$ is the number of primitive cells and $Q(\boldsymbol{q}, s, t)$ is the normal coordinate of mode $(\boldsymbol{q}, s)$. The parallel MSRD peculiarly depends on the phase relations between eigenvectors, on their projections on the bond direction $\hat{R}$ and on the inter-cell phase relation $e^{i\boldsymbol{q}\cdot\boldsymbol{R}}$.

Equation (6.7) is of no practical use in EXAFS analyses. The temperature dependence of the parallel MSRD can be satisfactorily fitted to a correlated Debye Model [1, 2, 27]

$$\sigma_D^2 = \frac{3\hbar}{\omega_D^3 m} \int_0^{\omega_D} d\omega\, \omega \coth \frac{\hbar\omega}{2kT} \left[1 - \frac{\sin(q_D R)}{q_D R}\right], \tag{6.8}$$

where q_D is the radius of the Debye sphere and the Debye frequency ω_D is the only parameter of the model. The Debye frequency corresponds to a Debye temperature $\Theta_D = \hbar\omega_D/k_B$. For copper, as generally expected for monatomic Bravais crystals, the Debye temperatures of different coordination shells are reasonably similar [17] and in good agreement with the Debye temperatures of specific heat and of X-ray diffraction. For crystals with more than one atom per primitive cell, such as tetrahedral semiconductors, different Debye temperatures are expected for different coordination shells.

Alternatively, the temperature dependence of the parallel MSRD can be fitted to an Einstein model [2]:

$$\langle \Delta u_\parallel^2 \rangle = (\hbar/2\mu\omega_\parallel) \coth(\hbar\omega_\parallel/2kT) \tag{6.9}$$

where μ is the reduced mass of the absorber–backscatterer atomic pair and the Einstein frequency $\omega_\parallel$ is the only parameter of the model. The Einstein frequencies of different coordination shells are expected to be different also for monatomic Bravais crystals. The advantages of the Einstein model are its simplicity and the possibility of obtaining an effective force constant $k_\parallel = \mu(\omega_\parallel)^2$, which measures the strength of the bond between the absorber and backscatterer atoms embedded in the three-dimensional system. The force constant $k_\parallel$ refers to an effective pair potential that depends on the statistically averaged influence of all the other atoms and cannot be identified with the force constant of a single-bond potential [28].

The extent of correlation of the vibrations of absorber and back-scatterer atoms can be measured by a dimensionless function of temperature [29, 30]

$$\phi_\parallel(T) = \frac{\langle(\hat{R}\cdot \boldsymbol{u}_b)^2\rangle + \langle(\hat{R}\cdot \boldsymbol{u}_a)^2\rangle - \langle \Delta u_\parallel^2 \rangle}{2[\langle(\hat{R}\cdot \boldsymbol{u}_b)^2\rangle\langle(\hat{R}\cdot \boldsymbol{u}_a)^2\rangle]^{1/2}}. \tag{6.10}$$

A value $\phi_\parallel = 0$ corresponds to a completely uncorrelated motion of the two atoms. Values $\phi_\parallel = 1$ and $\phi_\parallel = -1$ correspond to atomic motions perfectly in phase and in opposition of phase, respectively.

6.2.3 Perpendicular MSRD

The perpendicular MSRD $\langle \Delta u_\perp^2 \rangle$ cannot be directly obtained from EXAFS spectra. It can however be calculated by inverting (6.3), provided $R_c = R_0 + \langle \Delta u_\parallel \rangle$ is known from Bragg diffraction measurements.

In the harmonic approximation, the perpendicular MSRD of a crystal can be connected to the eigenvalues $\omega(\boldsymbol{q}, s)$ and eigenvectors $\boldsymbol{w}(\boldsymbol{q}, s)$ of the dynamical matrix by an expression similar to (6.7) [31]. It is worth remembering that different dynamical matrices can exist, sharing the same eigenfrequencies but with different eigenvectors [32]. The reproduction of parallel and perpendicular MSRDs, experimentally obtained from EXAFS, represents a peculiar test for the phase relationships between eigenvectors obtained from model calculations or ab initio [33].

The temperature dependence of the perpendicular MSRD $\langle \Delta u_\perp^2 \rangle$ is reproduced by an Einstein model [31]

$$\langle \Delta u_\perp^2 \rangle = (\hbar/\mu\omega_\perp)\,\coth(\hbar\omega_\perp/2kT)\,, \tag{6.11}$$

the only difference with respect to (6.9) being that μ is substituted for 2μ to account for the different dimensionality (projection along a line and within a plane for parallel and perpendicular MSRD, respectively). Also the perpendicular Einstein frequency can be connected to an effective force constant $k_\perp = \mu(\omega_\perp)^2$. Finally, a perpendicular correlation function can be defined by substituting $\langle \Delta u_\perp^2 \rangle/2$ for $\langle \Delta u_\parallel^2 \rangle$ in the numerator of (6.10):

$$\phi_\perp(T) = \frac{\langle(\hat{R}\cdot \boldsymbol{u}_b)^2\rangle + \langle(\hat{R}\cdot \boldsymbol{u}_a)^2\rangle - \langle \Delta u_\perp^2 \rangle/2}{2[\langle(\hat{R}\cdot \boldsymbol{u}_b)^2\rangle\langle(\hat{R}\cdot \boldsymbol{u}_a)^2\rangle]^{1/2}}\,. \tag{6.12}$$

The division of $\langle \Delta u_\perp^2 \rangle$ by 2, here and in the following, projects the perpendicular MSRD along one direction.

6.2.4 Relative Vibrational Anisotropy

The knowledge of both parallel and perpendicular MSRDs allows one to reconstruct the ellipsoid of relative thermal vibrations of the pair of absorber and backscatterer atoms (assuming that the perpendicular vibrations are isotropic within the plane perpendicular to the bond). For all systems up to now investigated, the ellipsoid of relative vibrations has always been found to be anisotropic and disc-shaped, say $\langle \Delta u_\perp^2 \rangle/2 > \langle \Delta u_\parallel^2 \rangle$.

The degree of anisotropy of relative vibrations is measured by the ratio $\gamma = \langle \Delta u_\perp^2 \rangle/2\langle \Delta u_\parallel^2 \rangle$. Since parallel and perpendicular MSRDs are characterised by different Einstein frequencies, the ratio γ is generally temperature dependent (Sect. 6.3, Fig. 6.4). A temperature-independent measure of anisotropy, corresponding to the asymptotic behaviour of γ for $T \rightarrow \infty$, is the ratio of parallel to perpendicular effective force constants, $\xi = k_\parallel/k_\perp$ [34]. For perfectly isotropic relative vibrations, $\xi = 1$. The anisotropy of relative vibrations is a consequence of the different degree of correlation of atomic motion parallel and perpendicular to the bond.

6.3 Experimental Results on Vibrational Anisotropy

The comparison of the bond thermal expansions measured by EXAFS and by Bragg diffraction has been done for a number of different systems characterized by different structures: copper [17], germanium [16] and crystals with the structures of zincblende [19, 21, 35, 36], cuprite [26], delafossite [34]. Relative values of both parallel and perpendicular MSRDs with respect to a reference temperature have been obtained, according to (6.3) and (6.5), as:

$$\delta\langle \Delta u_{\parallel}^2 \rangle = \delta C_2, \quad \delta\langle \Delta u_{\perp}^2 \rangle = 2R_c(\delta C_1 - \delta R_c). \tag{6.13}$$

Absolute values of the MSRDs have been calculated by fitting Einstein correlated models (6.9) and (6.11) to the temperature dependence of the experimental data [31]. The MSRDs from EXAFS have been compared with the uncorrelated MSDs from diffraction, when available. Parallel MSRD are also shown as a function of temperature for ZnSe, GaN and InN in Chap. 3 and for Ge and InP in Chaps. 7 and 8, respectively.

6.3.1 The Case of CdTe

An example particularly suitable to illustrate the effects of thermal disorder on EXAFS of semiconductors is represented by CdTe, for which high quality EXAFS data were measured from liquid helium to room temperature at both the Cd and Te K edges [19]. The availability of data measured at two different edges, at different times and different synchrotron beamlines and the comparison of results obtained by different data analysis procedures allowed a self-consistent evaluation and discussion of the uncertainty of results [37].

The coefficient of apparent thermal expansion of the Cd–Te distance in CdTe, proportional to the lattice expansion, is negative below $T \simeq 65$ K, and above 150 K is nearly constant, $\alpha_{\text{app}} \simeq 4.3 \times 10^{-6}\,\text{K}^{-1}$ [38, 39]. The coefficient of true expansion, measured by EXAFS [19], is instead positive at all temperatures, $\alpha_{\text{tru}} \simeq 18 \times 10^{-6}\,\text{K}^{-1}$, much larger than α_{app} (Fig. 6.6, right).

The first-shell parallel MSRD $\langle \Delta u_{\parallel}^2 \rangle$ and the halved perpendicular MSRD $\langle \Delta u_{\perp}^2 \rangle/2$ of CdTe are shown in the left panel of Fig. 6.2, together with the sum of the MSDs of Cd and Te, $\langle (\hat{R} \cdot \boldsymbol{u}_{\text{Cd}})^2 \rangle + \langle (\hat{R} \cdot \boldsymbol{u}_{\text{Te}})^2 \rangle$, calculated from the diffraction data of [40]. The MSDs and MSRDs are well fitted by Einstein models (dashed lines in Fig. 6.2). The ratio of perpendicular to parallel MSRDs is a measure of anisotropy. By comparing the MSRDs with the sum of the MSDs, one can connect the anisotropy with the degree of correlation.

The MSRDs have the dimension of an area. They are suitable quantities for lattice dynamical interpretations, in view of their Einstein-like behaviour, which tends asymptotically to the classical linear behaviour at high temperatures. More

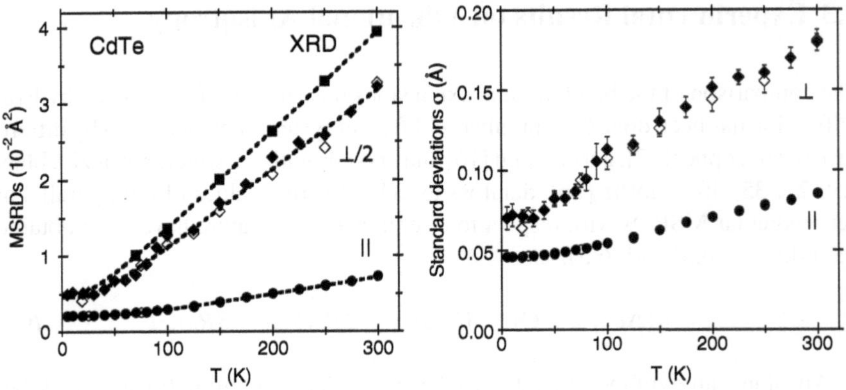

Fig. 6.2 *Left panel* Parallel MSRD (*circles*) and halved perpendicular MSRD (*diamonds*) for the first shell of CdTe; *solid* and *open* symbols refer to the Te and Cd edges, respectively; the *squares* are the sums of the Cd and Te MSDs from [40]. The *dashed lines* are the best-fitting Einstein models. *Right panel* Parallel and perpendicular standard deviations $\sigma_{\parallel} = [\langle \Delta u_{\parallel}^2 \rangle]^{1/2}$ and $\sigma_{\perp} = [\langle \Delta u_{\perp}^2 \rangle / 2]^{1/2}$

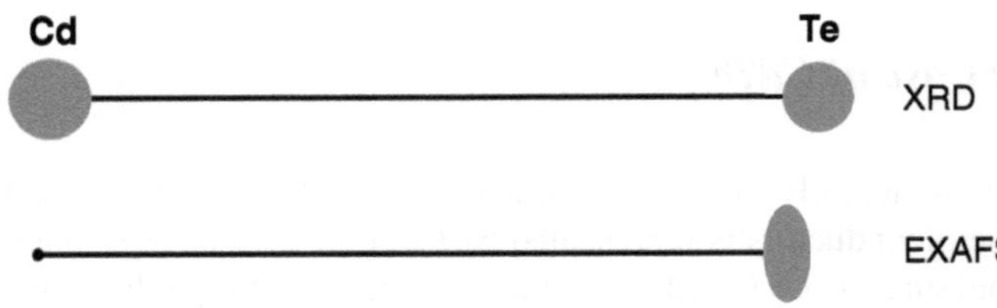

Fig. 6.3 Pictorial representation of the thermal ellipsoids for CdTe at 300 K. The inter-atomic distance is $R = 2.8$ Å. The standard deviations are $\sigma_{\mathrm{Cd}} = 0.15$ Å and $\sigma_{\mathrm{Te}} = 0.13$ Å for the MSDs and $\sigma_{\parallel} = 0.085$ Å and $\sigma_{\perp} = 0.18$ Å for the MSRDs

direct physical insight is given by the standard deviations $\sigma_{\parallel} = [\langle \Delta u_{\parallel}^2 \rangle]^{1/2}$ and $\sigma_{\perp} = [\langle \Delta u_{\perp}^2 \rangle / 2]^{1/2}$, which have the dimension of length and directly measure the size of the relative thermal ellipsoids. The standard deviations for the first shell of CdTe are shown in the right panel of Fig. 6.2: the non negligible extent of zero-point vibrations with respect to thermal effects is evident. Standard deviations can be visually represented in real space. A pictorial comparison of the information from diffraction and from EXAFS is given in Fig. 6.3, where the absolute and relative thermal ellipsoids for the temperature of 300 K are shown on the same scale of the inter-atomic distance.

The anisotropy, measured by the ratio $\gamma = \langle \Delta u_{\perp}^2 \rangle / 2 \langle \Delta u_{\parallel}^2 \rangle$, depends on temperature (Fig. 6.4, left panel). The high-T asymptotic value of γ is $\xi = k_{\parallel}/k_{\perp} = 4.6$ for CdTe. An alternative measure of anisotropy is the ratio of the standard deviations $\sigma_{\perp}/\sigma_{\parallel}$, which for CdTe tends asymtpotically to $\xi^{1/2} = 2.14$ (also shown in Fig. 6.4, left panel).

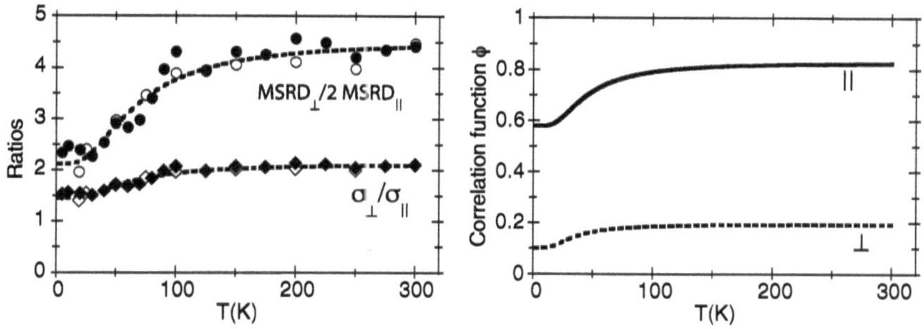

Fig. 6.4 *Left panel* anisotropy of the relative vibrations in CdTe measured by the ratios between the MSRDs (*circles*) and between the standard deviations (*diamonds*); *solid* and *open* symbols refer to Te and Cd K edges, respectively. *Right panel* correlation functions $\phi_{\parallel}(T)$ (*continuous line*) and $\phi_{\perp}(T)$ (*dashed line*) calculated from the Einstein models best fitting the MSDs and MSRDs for the first shell of CdTe

The anisotropy of relative vibrations depends on the different degree of correlation along the bond direction and within the perpendicular plane, measured by the $\phi(T)$ functions of (6.10) and (6.12), respectively. Since the experimental values of MSRD and MSD had been measured at different temperatures, the correlation functions $\phi(T)$ have been evaluated using the best-fitting Einsten models; the results are shown in Fig. 6.4 (right panel). The high-temperature asymptotic value for the parallel correlation, $\phi_{\parallel} = 0.82$, is significantly higher than the value for the perpendicular correlation, $\phi_{\perp} = 0.2$.

6.3.2 Comparison of Diamond and Zinblende Structures

The possibilities offered by the study of parallel and perpendicular MSRDs are highlighted by the comparison of different crystals with the diamond or zincblende structure.

6.3.2.1 First Coordination Shell

The nearest-neighbours effective force constants $k_{\parallel}$ and $k_{\perp}$ and the anisotropy parameter $\xi = k_{\parallel}/k_{\perp}$ are listed in Table 6.1 as a function of both the ionicity [41] and the ratio of the masses of the heavier to the lighter atom, M_h/M_l. Copper is added in the first column for comparison.

For the diamond-zincblende crystals, a clear correlation exists between ionicity and the effective force constants: both parallel and perpendicular force constants decrease when the ionicity increases (Fig. 6.5, top left). For the tetrahedral semiconductors, the parallel force constants (solid circles) are much larger than the perpendicular force constants (solid squares). For copper, the difference between parallel

Table 6.1 Parallel and perpendicular Einstein effective force constants k and anisotropy parameters $\xi = k_{\parallel}/k_{\perp}$ for the first shell of some selected systems, compared with ionicity and mass ratio

	Cu [17]	Ge [16]	GaAs [35]	InP [36]	CdTe [19]	CuCl [21]
Ionicity [41]	0	0	0.31	0.421	0.675	0.746
M_h/M_l	1	1	1.07	3.71	1.421	1.791
$k_{\parallel}$ (eV/Å^2)	3.2	8.5	7.02	6.65	3.71	1.4
$k_{\perp}$ (eV/Å^2)	2.6	2.9	1.85	1.11	0.81	0.26
$\xi = k_{\parallel}/k_{\perp}$	1.2	2.9	3.8	6.0	4.6	5.4

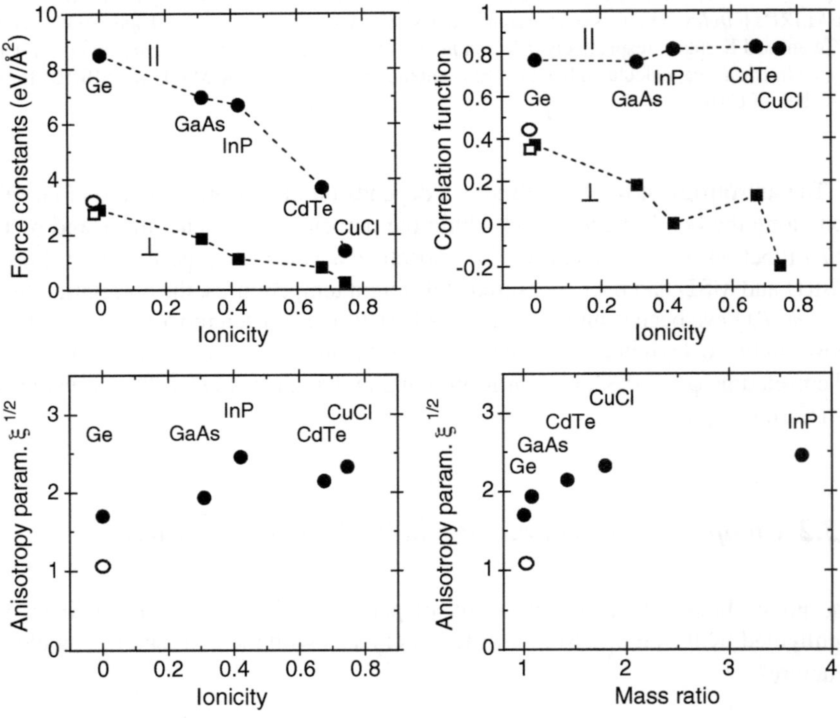

Fig. 6.5 Results for the first shell of crystals with the diamond or zincblende structure. *Top panels* parallel (*solid circles*) and perpendicular (*solid squares*) force constants (*left*) and correlation function at 300 K (*right*), plotted as a function of ionicity; the *dashed lines* are a guide to eye. *Bottom panels* anisotropy parameter $\xi^{1/2}$ as a function of both ionicity (*left*) and of mass ratio (*right*). In all panels, the open symbols are the values for Cu

and perpendicular force constant is instead quite small (open symbols). The first-shell parallel force constant is much larger in germanium than in copper. This difference, which decreases when the ionicity increases in the other tetrahedral semiconductors, can be attributed to the difference between the totally or partially covalent bonds of tetrahedral semiconductors and the metallic bond of copper. The first-shell perpendicular force constant is instead very similar in Ge and in Cu, and, for the other tetrahedral semiconductors, still decreases when ionicity increases. This behaviour can be qualitatively attributed to the open structure of semiconductors, which favours the perpendicular relative vibrations of neighbouring atoms in spite of the relative bond rigidity. The smaller lateral rigidity of the metallic bond with respect to the covalent bond is compensated by the closer packing of the fcc structure.

The anisotropy parameter $\xi^{1/2}$ exhibits an average positive trend as a function of ionicity (Fig. 6.5, bottom left). The scattering of data with respect to a smooth behaviour can be correlated to the different mass ratios M_h/M_l of the studied compounds: the larger the mass ratio, the larger the anisotropy (Fig. 6.5, bottom right). For copper, the relative thermal vibrations are only weakly anisotropic, $\xi^{1/2} \simeq 1.1$ (open circles).

The parallel and perpendicular correlation functions $\phi(T)$ evaluated for $T = 300$ K are shown in Fig. 6.5 (top right). The parallel correlation (solid circles) is stronger than the perpendicular correlation (solid squares), and quite independent of ionicity, while the perpendicular correlation exhibits an average decreasing trend. For copper, the parallel correlation (open circle) is only slightly larger than the perpendicular correlation (open square).

The anisotropic behaviour of relative atomic vibrations for some systems has been recently reproduced by the ab initio calculations performed by Vila et al. [33]. The calculated values of the asymptotic ratios

$$\lim_{T\to\infty} \langle \Delta u_\perp^2 \rangle / \langle \Delta u_\parallel^2 \rangle = 2\xi$$

are 2.36 for Cu at 500 K and 7.2 for diamond lattices at 600 K, in agreement with the experimental values: 2.2 for Cu, from 5.8 to 12 for tetrahedral semiconductors (Table 6.1).

6.3.2.2 Outer Coordination Shells

Further information can be obtained from the analysis of the parallel MSRD of the outer coordination shells.

The parallel effective force constants $k_\parallel$ and the Debye temperatures Θ_D for the different pairs of absorber-backscatterer atoms are listed in Tables 6.2 and 6.3, respectively. In the Debye model for the tetrahedral semiconductors, an extended first Brillouin zone has been considered.

For copper, the Debye temperatures of the different shells are very similar, while for the tetrahedral semiconductors the Debye temperatures are strongly dependent on the shell, the most significant difference being between the first and the second shell (Table 6.3). The differences of force constants between the different shells are still stronger for tetrahedral semiconductors, and not negligible even for copper (Table 6.2).

Table 6.2 Parallel effective force constants $k_{\parallel}$ (in eV/Å^2) for the first and outer shells of different crystals

Shell	Cu [17]	Ge [16]	GaAs [35]	InP [36]	CdTe [19]
1	3.20	8.15	7.02	6.65	3.71
2	1.97	2.48	2.30		0.91
			2.68	2.09	1.29
3	2.40	2.18	1.75	1.74	0.85
4	2.33				

For the 2nd shells of binary compounds, the first and second lines refer to the lightest pair (Ga–Ga, Cd–Cd) and to the heaviest pair (As–As, Te–Te, In–In), respectively

Table 6.3 EXAFS Debye temperatures (in K) for the first and outer shells of different crystals

Shell	Cu [17]	Ge [16]	GaAs [35]	InP [36]	CdTe [19]
1	328	460	402	416	228
2	283	299	230		114
			240	171	126
3	322	290	195	203	105
4	321				

For the 2nd shells of binary compounds, the first and second lines refer to the lightest pair (Ga–Ga, Cd–Cd) and to the heaviest pair (As–As, Te–Te, In–In), respectively

Table 6.4 Correlation functions $\phi_{\parallel}$ and $\phi_{\perp}$ at $T = 300$ K for different crystals

	Shell	Cu [17]	Ge [16]	GaAs [35]	InP [36]	CdTe [19]	CuCl [21]
$\phi_{\parallel}$	1	0.46	0.77	0.76	0.82	0.83	0.82
	2	0.14	0.18	0.32		0.33	
				0.44	0.42	0.41	
	3	0.24	0	0.14	0.37	0.27	
	4	0.23					
$\phi_{\perp}$	1	0.36	0.37	0.18	0	0.13	−0.2

For the 2nd shells of binary compounds, the first and second lines refer to the lightest pair (Ga–Ga, Cd–Cd) and to the heaviest pair (As–As, Te–Te, In–In), respectively

The values of the parallel and perpendicular correlation functions for the first shell and of the parallel correlation function for the outer shells, evaluated at $T = 300$ K, are listed in Table 6.4. For Cu and CdTe, good quality experimental MSD data are available in the literature [40, 42, 43]. For GaAs and Ge we refer to theoretical calculations [44, 45]. For InP, the perpendicular MSRD values are quite scattered, and the comparison is made with the experimental MSD data of [46]. For CuCl, the data are limited to the first shell, perpendicular MSRD values at 300 K are extrapolated from the values below 100 K and the comparison is made with an average of available MSD data [21]. In spite of the difficulty in assessing the accuracy of experimental and theoretical data, some qualitative properties are evident in Table 6.4. The first-shell parallel correlation is much stronger in tetrahedral semiconductors than in copper; the correlation is instead comparable for the outer shells.

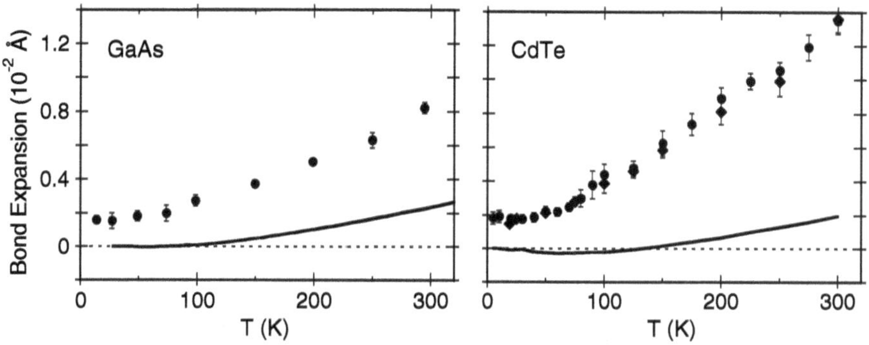

Fig. 6.6 First-shell true (*symbols*) and apparent (*continuous line*) bond expansions for GaAs (*left*) and CdTe (*right*)

Table 6.5 Difference $C_1 - R_c$ (10^{-3} Å) between the nearest-neighbours bond distance measured by EXAFS and by Bragg diffraction at low and room temperature

T(K)	Cu [17]	Ge [16]	GaAs [35]	InP [36]	CdTe [19]
10	1.3	1.5	1.6	1.2	1.8
300	3.8	3.9	5.8	9.5	11.5

The value for Ge at 300 K is erroneously quoted as 8×10^{-3} Å in Table 3 of [37]

6.4 True and Apparent Bond Expansion

Once the absolute values of the perpendicular MSRD of the first shell have been evaluated, one can attempt to gain a better quantitative evaluation of the difference between the true and apparent bond distances and thermal expansions, as per (6.3). In Fig. 6.6, the true and apparent bond expansions of GaAs and CdTe, evaluated with respect to the low-temperature crystallographic distance, are compared.

The values of the difference between true and apparent distances of some systems at low and room temperature are listed in Table 6.5. The differences depend on the extent of the perpendicular MSRD, and increase with the decreasing of the perpendicular force constant $k_\perp$ as well as with the increase of temperature. The discrepancy between true and apparent distances should be taken into account when calibrating EXAFS simulations against model systems of known crystallographic structure. When accuracies of the order of some 0.001 Å are sought, a calibration can be misleading if the effect of perpendicular vibrations is neglected.

6.5 Negative Thermal Expansion Crystals

The sensitivity of EXAFS to local dynamics, in particular to vibrational anisotropy, can be exploited to study the local mechanism at the origin of negative thermal expansion (NTE) in some crystals. A number of tetrahedral semiconductors (such as Si, Ge, GaAs, CdTe, InP, CuCl [47]) exhibit lattice NTE within a restricted low-temperature interval, and some more complex framework structures are characterised by lattice NTE within extended temperature intervals [48–50].

According to the phenomenological approach of Barron [23], the net lattice expansion is the result of a competition between two contributions: (a) a positive bond-stretching contribution due to the anharmonicity of the effective pair potential and (b) a negative contribution due to tension effects connected to atomic movements perpendicular to the bond. When the tension effects prevail over bond stretching, the crystal exhibits NTE.

Conventional techniques, such as dilatometry and Bragg diffraction, are sensitive only to the lattice thermal expansion, and cannot distinguish the bond stretching from the tension effects. These two effects can be distinguished by complementary EXAFS measurements: the temperature dependence of the average distance $\langle r \rangle$ directly measures the positive bond expansion due to the stretching effect, while the perpendicular MSRD is connected to the tension effect.

Several crystals with different structures and characterized by different NTE strengths have been recently investigated by EXAFS: diamond-zincblende structures [16, 19, 21, 35], cuprite structures [26], delafossite structures [34]. The lattice thermal expansion is isotropic with respect to the crystallographic axes in the cubic structures (diamond-zincblende and cuprite) and anisotropic in the delafossite structure, where it is negative along the c axis and positive in the perpendicular plane.

The EXAFS results confirm the possibility of distinguishing bond stretching from tension effects. For all considered systems, the bond thermal expansion, measured by the first EXAFS cumulant, is always positive; moreover, the stronger is the lattice NTE, the larger is the positive bond expansion. Within each family of isostructural compounds (diamond/zincblende, cuprite, delafossite) a stronger NTE corresponds to a smaller value of both parallel and perpendicular effective force constants as well as to a larger value of the relative vibrational anisotropy ξ.

According to EXAFS results, the bond stretching and tension effects can be separately quantified but are nevertheless in some way entangled. When the ionicity increases, the directional rigidity of the bond is reduced, the perpendicular effective force constant $k_{\perp}$ decreases and the perpendicular MSRD becomes larger. In a relatively open structure such as zincblende, larger vibrations perpendicular to the bond make possible also larger parallel vibrations (smaller parallel effective force constant $k_{\parallel}$) and a larger bond expansion.

It is of interest to compare the directional properties of the absolute atomic vibrations, measured by Bragg diffraction, with those of the relative vibrations, measured by EXAFS. The relative thermal ellipsoids are always anisotropic with respect to the bond direction, even for cubic systems where the atomic ellipsoids are spherical for symmetry reasons. In the layered delafossite structures, Cu atoms belonging to a layer are linearly coordinated to two O atoms along the c axis; the NTE along the c axis is connected to the anisotropy of the thermal ellipsoids of Cu atoms, whose vibrations are stronger within the plane normal to the c axis than along the c axis. Correspondingly, also the relative vibrations of the Cu–O atomic pair monitored by EXAFS are anisotropic, the ellipsoid being disc-shaped perpendicular to the

c axis [34]. The relative anisotropy measured by EXAFS for the Cu–O pair is however larger than the anisotropy of atomic Cu vibrations. It seems thus that the relative vibrational anisotropy is a more general property to be connected to NTE than the absolute atomic anisotropy.

References

1. G. Beni, P.M. Platzman, Phys. Rev. B **14**, 1514–1518 (1976)
2. E. Sevillano, H. Meuth, J.J. Rehr, Phys. Rev. B **20**, 4908–4911 (1979)
3. P. Eisenberger, G.S. Brown, Solid State Commun. **29**, 481–484 (1979)
4. G. Bunker, Nucl. Instr. Methods Phys. Res. **207**, 437–444 (1983)
5. E.D. Crozier, J.J. Rehr, R. Ingalls, in *Amorphous and Liquid Systems*, ed. by D.C. Koningsberger, R. Prins, X-ray Absorption, Chap. 9 (Wiley, New York, 1988), pp. 373–442
6. J.M. Tranquada, R. Ingalls, Phys. Rev. B **28**, 3520–3528 (1983)
7. M. Benfatto, C.R. Natoli, A. Filipponi, Phys. Rev. B **40**, 9626–9635 (1989)
8. A.I. Frenkel, J.J. Rehr, Phys. Rev. B **48**, 585–588 (1993)
9. T. Fujikawa, T. Miyanaga, J. Phys. Soc. Jpn. **62**, 4108–4122 (1993)
10. T. Miyanaga, T. Fujikawa, J. Phys. Soc. Jpn. **63**, 1036–1052 (1994)
11. N. Van Hung, J.J. Rehr, Phys. Rev. B **56**, 43–46 (1997)
12. T. Yokoyama, Phys. Rev. B **57**, 3423–3432 (1998)
13. A.V. Poiarkova, J.J. Rehr, Phys. Rev. B **59**, 948–957 (1999)
14. H. Katsumata, T. Miyanaga, T. Yokoyama, T. Fujikawa, T. Ohta, J. Synchrotron Radiat. **8**, 226–228 (2001)
15. S.A. Beccara, G. Dalba, P. Fornasini, R. Grisenti, F. Pederiva, A. Sanson, D. Diop, F. Rocca, Phys. Rev. B **68**, 140301 (2003)
16. G. Dalba, P. Fornasini, R. Grisenti, J. Purans, Phys. Rev. Lett. **82**, 4240–4243 (1999)
17. P. Fornasini, S. aBeccara, G. Dalba, R. Grisenti, A. Sanson, M. Vaccari, F. Rocca. Phys. Rev. B **70**, 174301 (2004)
18. S.A. Beccara, G. Dalba, P. Fornasini, R. Grisenti, F. Rocca, A. Sanson, Nucl. Instrum. Methods Phys. Res. B **200**, 237–241 (2003)
19. N. Abd el All, G. Dalba, D. Diop, P. Fornasini, R. Grisenti, O. Mathon, F. Rocca, B. Thiodjio Sendja, M. Vaccari. J. Phys.: Condens. Matter **24**, 115403 (2012)
20. P. Fornasini, e-J. Surf. Sci. Nanotech. **10**, 480–485 (2012).
21. M. Vaccari, R. Grisenti, P. Fornasini, F. Rocca, A. Sanson, Phys. Rev. B **75**, 184307 (2007)
22. P. Fornasini, J. Phys.: Condens. Matter **13**, 7859–7872 (2001)
23. G.D. Barrera, J.A. Bruno, T.H.K. Barron, N.L. Allan, J. Phys.: Condens. Matter **17**, R217–R252 (2005)
24. B.T.M. Willis, A.W. Pryor, *Thermal Vibrations in Crystallography* (Cambridge University Press, Cambridge, 1975)
25. P. Fornasini, F. Monti, A. Sanson, J. Synchrotron Radiat. **8**, 1214–1220 (2001)
26. A. Sanson, F. Rocca, G. Dalba, P. Fornasini, R. Grisenti, M. Dapiaggi, G. Artioli, Phys. Rev. B **73**, 214305 (2006)
27. W. Böhmer, P. Rabe, J. Phys. C Solid State Phys. **12**, 2465–2474 (1979)
28. N. Van Hung, P. Fornasini, J. Phys. Soc. Jpn. **76**, 084601 (2007)
29. C.H. Booth, F. Bridges, E.D. Bauer, G.G. Li, J.B. Boyce, T. Claeson, C.W. Chu, Q. Xiong, Phys. Rev. B **52**, R15745–R15748 (1995)
30. I.K. Jeong, R.H. Heffner, M.J. Graf, S.J.L. Billinge, Phys. Rev. B **67**, 104301 (2003)
31. M. Vaccari, P. Fornasini, J. Synchrotron Radiat. **13**, 321–325 (2006)
32. W. Cochran, Acta Cryst. A **27**, 556–559 (1971)
33. F.D. Vila, J.J. Rehr, H.H. Rossner, H.J. Krappe, Phys. Rev. B **76**, 014301 (2007)

34. S.I. Ahmed, G. Dalba, P. Fornasini, M. Vaccari, F. Rocca, A. Sanson, J. Li, A.W. Sleight, Phys. Rev. B **79**, 104302 (2009)
35. S.I. Ahmed, G. Aquilanti, N. Novello, L. Olivi, R. Grisenti, P. Fornasini, J. Chem. Phys. **139**, 164512 (2013)
36. C.S. Schnohr, P. Kluth, L.L. Araujo, D.J. Sprouster, A.P. Byrne, G.J. Foran, M.C. Ridgway, Phys. Rev. B **79**, 195203 (2009)
37. N. Abd el All, B. Thiodjio Sendja, R. Grisenti, F. Rocca, D. Diop, O. Mathon, S. Pascarelli, P. Fornasini. J. Synchrotron Radiat. **20**, 603–613 (2013)
38. T.F. Smith, G.K. White, J. Phys. C: Solid State Phys. **8**, 2031–2042 (1975)
39. S.I. Novikova, Soviet Phys. Solid State **2**, 2087–2089 (1961)
40. R.D. Horning, J.L. Staudenmann, Phys. Rev. B **34**, 3970–3979 (1986)
41. R.M. Martin, Phys. Rev. B **1**, 4005–4011 (1971)
42. C.J. Martin, D.A. O'Connor, J. Phys. C: Solid State Phys. **10**, 3521–3526 (1977)
43. J.T. Day, J.G. Mullen, R. Shukla, Phys. Rev. B **52**, 168–176 (1995)
44. J.S. Reid, Acta Cryst. A **39**, 1–13 (1983)
45. T. Soma, H. Matsuo, Phys. Status Solidi (b) **111**, K93–K97 (1982)
46. R. Saravanan, S.K. Mohanlal, K.S. Chandrasekaran, Z. Kristallogr. **200**, 7 (1992)
47. G.K. White, Contemp. Phys. **34**, 193–204 (1993)
48. T.A. Mary, J.S.O. Evans, T. Vogt, A.W. Sleight, Science **272**, 90–92 (1996)
49. W. Tiano, M. Dapiaggi, G. Artioli, J. Appl. Cryst. **36**, 1461–1463 (2003)
50. K.W. Chapman, P.J. Chupas, C.J. Kepert, J. Am. Chem. Soc. **127**, 15630–15636 (2005)

Part II
Disordered Semiconductors

Chapter 7
Amorphous Group IV Semiconductors

Mark C. Ridgway

Abstract The amorphous Group IV semiconductors are of technological significance and scientific importance. As an example of the former, amorphous Si has widespread use in thin-film transistors while, as an example of the latter, structural determinations of amorphous Ge are invaluable in assessing the validity of Continuous Random Network models. The application of XAS to the study of the elemental and binary amorphous Group IV semiconductors is now described and we demonstrate XAS is an ideal technique to study the structural and vibrational properties of these materials. A commonality in amorphous phase structure is apparent including an increase in disorder and bondlength and a decrease in coordination number relative to the crystalline phase.

7.1 Introduction

Crystalline semiconductors are used extensively in the fabrication of electronic and photonic devices. As Chaps. 2–6 readily demonstrated, the application of XAS to these materials has been crucial for the understanding of their structural and vibrational properties. While the use of amorphous semiconductors is not as widespread as that of their crystalline counterparts, important applications certainly exist such as thin-film transistors (TFTs) formed from amorphous Si for use in display screens. Short-range order (SRO) is intrinsic to both crystalline and amorphous materials and, as a probe of SRO, XAS is thus equally applicable to both phases. In this chapter, we now review the application of XAS to the amorphous Group IV semiconductors including Si, Ge, SiC and $Si_{1-x}Ge_x$. For brevity, the hydrogenated Group IVs are not covered.

M.C. Ridgway (✉)
Department of Electronic Materials Engineering, Australian National University, Canberra, ACT 0200, Australia
e-mail: mark.ridgway@anu.edu.au

C.S. Schnohr and M.C. Ridgway (eds.), *X-Ray Absorption Spectroscopy of Semiconductors*, Springer Series in Optical Sciences 190,
DOI 10.1007/978-3-662-44362-0_7

7.2 Structure of Amorphous Semiconductors

The structure of a covalently-bonded amorphous solid was first considered in detail by Zachariasen [1] with the Continuous Random Network (CRN) model. Comparing crystalline and amorphous phases, the CRN model necessitates that the amorphous phase retain the same coordination number (CN) and bondlength without the introduction of dangling bonds. Following both Zachariasen [1] and Zallen [2], Fig. 7.1 shows photographs of two-dimensional ball-and-stick models of the crystalline and amorphous phase. The periodicity leading to long-range order (LRO) is readily apparent on the left yet by definition necessarily absent on the right. The CN (in this case 3 atoms) and bondlength are the same for both phases. The bondangle is 120° for the crystalline phase and distributed about 120° for the amorphous phase. The ring size is six for the crystalline phase and distributed about six for the amorphous phase. Clearly the crystalline and amorphous phases share common SRO.

For an amorphous semiconductor, Polk [3] was an early pioneer in constructing a three-dimensional CRN with four-fold coordinated atoms that satisfied the Zachariasen criteria. Significantly, the bondstrain did not increase as a function of CRN size demonstrating such an approach was indeed appropriate. By enabling a small variation in the bondlength, the bondstrain was reduced yielding a decrease in the width of the bondangle distribution. The ring sizes were distributed over five-, six- and seven-fold. The structural disorder in such a model was largely contained in the bondangle distribution as consistent with the energetically more favourable bond bending over bond stretching in covalently-bonded semiconductors (see Chap. 2).

Following on from the early work of Polk, more recent, computationally-intensive examples of the construction of CRNs appropriate to amorphous semiconductors are those of Djordjevic et al. [4] and Barkema and Mousseau [5]. Figure 7.2 shows a ball-and-stick model of tetrahedral amorphous diamond from Djordjevic et al. [4]. Surprisingly, the authors reported "there are no major differences between [the structures of] amorphous C and amorphous Si". Barkema and Mousseau [5] utilized a modified version of the Wooten-Winer-Weaire algorithim to create 1,000-atom

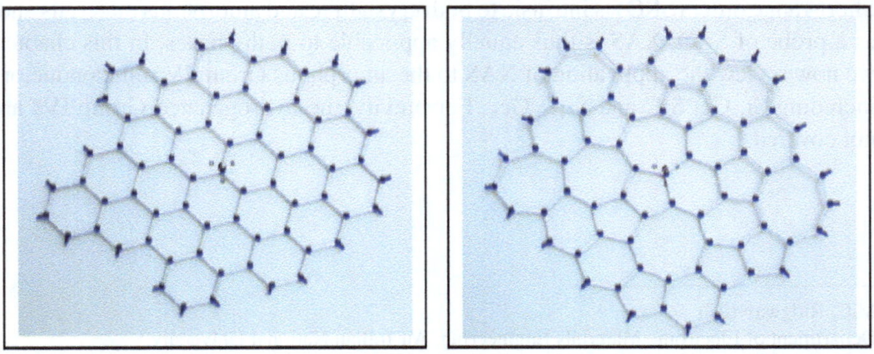

Fig. 7.1 Illustrative ball-and-stick models of (*left*) crystalline and (*right*) amorphous structures

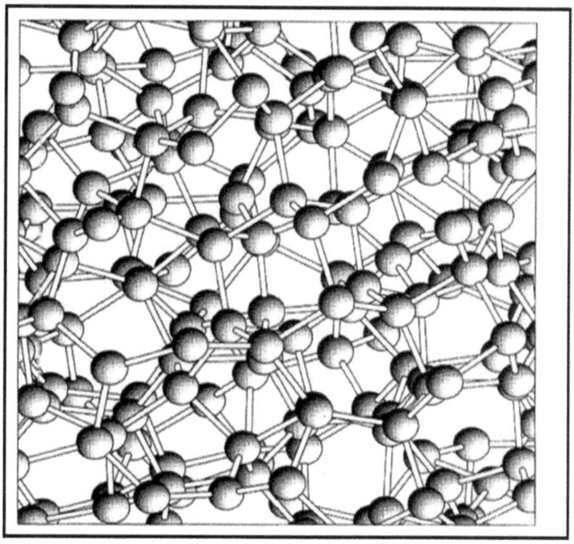

Fig. 7.2 Computer-generated model of a tetrahedral amorphous diamond structure. (Reprinted figure with permission from Djordjevic et al. [4]. Copyright (1995) by the American Physical Society. http://link.aps.org/abstract/PRB/v52/p5685)

configurations with bondangle distributions consistent with experimental data and an average coordination number of four atoms. As described below, XAS has been used extensively to assess the validity of many structural models. The current section has been purposefully general and historical. The subsequent presentation for the amorphous Group IV semiconductors will include brief descriptions of current, materials-specific structural models for comparison with experimental results, including both XAS and complementary non-XAS derived data.

7.3 XAS of Amorphous Semiconductors

A commonality will be readily apparent in the XAS spectra of the amorphous semiconductors that follow, the origin of which is now illustrated. Figure 7.3 shows time-lapsed photographs of the two-dimensional crystalline and amorphous models depicted in Fig. 7.1 as they are rotated about a central atom which we nominate as the absorber. As above, SRO is intrinsic to the two structures, yielding the well-defined, inner-most ring that appears when either model is rotated. In contrast, LRO is present only in the crystalline model. Therein, the atoms that comprise a given ring have the same radial distance to the absorber, yielding the additional well-defined rings of greater radii upon rotation. LRO is absent in the amorphous model and upon rotation the rings beyond the first now appear progressively blurred. Such smearing results from more broadly distributed or, equivalently, non-unique radial distances to the absorber.

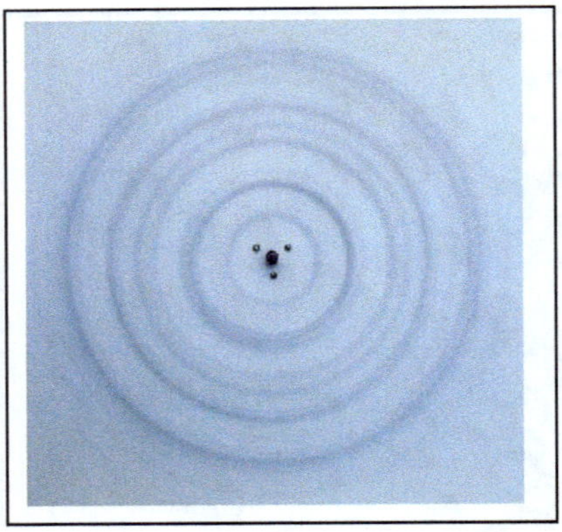

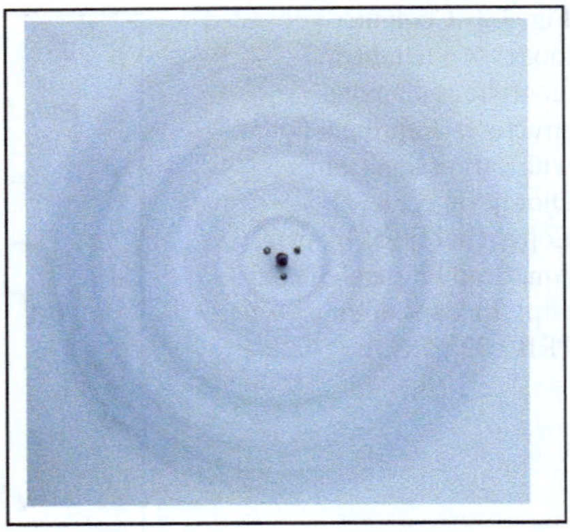

Fig. 7.3 Rotating ball-and-stick models of the (*left*) crystalline and (*right*) amorphous structures shown in Fig. 7.1

In an XAS experiment of an amorphous semiconductor, the smearing illustrated above typically manifests as a dramatic reduction in the scattering contribution from atoms beyond the nearest neighbour (NN) due to the presence of structural and/or thermal disorder. Both forms of disorder lead to non-unique radial distances to the absorber and, as a consequence, coherency of the scattered wave is lost and the phase shift between outgoing and scattered spherical waves at the absorber is non-unique. This is well illustrated in Fig. 7.4 which compares XAS spectra of crystalline and amorphous Ge [6]. In Fig. 7.4a, the EXAFS spectrum is single frequency for the amorphous phase due to scattering from only the NN shell and multi-frequency for the crystalline phase due to the superposition of scattering contributions from the NN shell and beyond, including second and third nearest neighbour (2NN and 3NN, respectively) shells. Consequently, the Fourier-transformed (FT) spectra of Fig. 7.4b exhibit a single peak for the amorphous phase and multiple peaks for the crystalline phase. The presence of increased disorder in the amorphous sample results

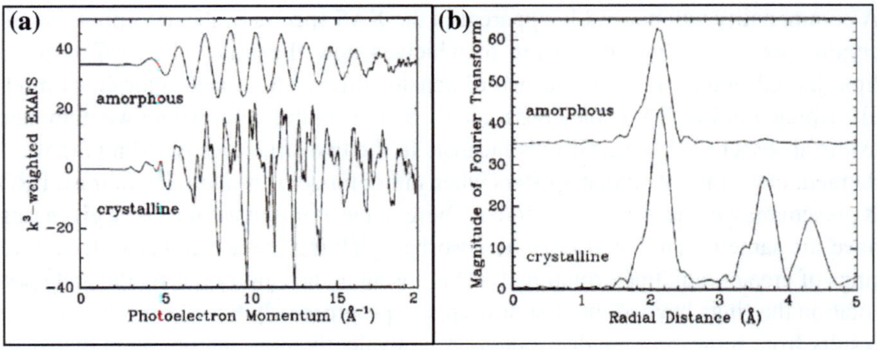

Fig. 7.4 **a** EXAFS and **b** FT EXAFS spectra comparing crystalline and amorphous Ge. (Reprinted figure with permission from [6]. Copyright (2000) by the American Physical Society. http://link.aps.org/abstract/PRB/v61/p12586)

in a reduction in amplitude of the NN peak and the entire loss of the 2NN and 3NN peaks apparent in the crystalline sample. Clearly, there is sufficient disorder to completely damp out scattering contributions in the amorphous material from beyond the NN. This feature typifies XAS spectra of amorphous semiconductors. None-the-less, confirmation with a complementary technique such as transmission electron microscopy (TEM) coupled with electron diffraction (ED) is always recommended to rule out the presence of a small crystalline fraction at a concentration below the sensitivity of XAS.

7.4 Preparation of Amorphous Group IV Semiconductor Samples for XAS

Careful preparation of samples, whether crystalline or amorphous, for an XAS experiment is essential to extract data of the highest quality. The influences of total thickness, thickness uniformity, homogeneity and continuity are well known. Common techniques for the preparation of amorphous semiconductor XAS samples include evaporation, sputtering and ion implantation all of which have characteristic advantages/disadvantages. For example, evaporation and sputtering are efficient means of depositing thin, amorphous layers onto X-ray transparent material such as Kapton. Thereafter samples can be cut and stacked to achieve the optimum thickness. However, pin-holes (evaporation) and voids (sputtering) are common artefacts. While ion implantation can circumvent these problems, this process and the subsequent separation of an amorphised layer from an underlying bulk crystalline substrate is of greater complexity. None-the-less, many material-specific ion implantation protocols have been developed [7] with the goal of producing the ideal XAS sample.

Subsequent to either deposition or amorphisation, the amorphous-phase structure is likely to be in an energetic state (herein termed "unrelaxed") above that of the minimum-energy configuration. However, the intrinsic, minimum-energy configuration (herein termed "relaxed") can be approached and ideally attained via thermally-induced relaxation which reduces point-defect and dangling-bond concentrations. Annealing temperatures and times sufficient for relaxation, yet insufficient for recrystallization, require experimental determination for a given material. As illustrated below, this process has been investigated and successfully applied to a variety of amorphous semiconductors and XAS has proven invaluable as a tool in identifying the atomic-scale processes operative during structural relaxation.

7.5 Amorphous Group IV Semiconductors

7.5.1 Amorphous Si (a-Si)

Crystalline Si forms the basis of the vast majority of electronic devices in use today. During the fabrication process, a-Si is commonly formed or deposited via ion

Table 7.1 Structural parameters of c- and a-Si showing bondangle distribution mean ϑ, RMS bondangle distribution mean $\Delta\vartheta$, fraction of atoms CN_m with $CN\, m$ and average coordination number CN_{ave}

	Crystalline	Stich et al. [8]	Kim and Lee [9]	Cooper et al. [10]	Mousseau and Barkema [11]	Bernstein et al. [12]
ϑ (deg)	109.5	108.32	106.7	107.38	109.32	109.1
$\Delta\vartheta$ (deg)	0	15.5	16.3	15.2	9.74	11
CN_3 (%)	0	0.2	3.2	4.7	3.5	0.93
CN_4 (%)	100	96.6	82.8	92.6	96.0	97
CN_5 (%)	0	3.2	12.5	1.6	0.5	1.9
CN_6 (%)	0	0	1.68		0	0.46
$CN_{ave}(atoms)$	4	4.03	4.28	3.96		

implantation or sputtering, respectively. Given the physical properties of the amorphous phase can influence device performance, the structure of a-Si has been examined in detail both theoretically and experimentally with a multitude of calculation and characterization methods brought to bear on this material of such technological and scientific significance.

Theoretical determinations of the atomic-scale structure of a-Si are numerous with molecular dynamics (MD), density functional theory (DFT) and reverse Monte Carlo (RMC) the most frequently utilised calculation methodologies. In general, the majority of studies report a-Si exhibits near four-fold coordination with increases in the width of the bondlength, bondangle and ring-size distributions. Table 7.1 compares structural parameters of a relaxed CRN models of a-Si determined by a variety of authors [8–12].

Table 7.1 demonstrates SRO is indeed maintained in a-Si with the majority of atoms retaining four-fold coordination and an average CN of very close to four atoms. As above, the broadening of the bondangle distributions is readily apparent.

A notable non-XAS determination of the structure of a-Si is that of Laaziri et al. [13] who reported a high-energy X-ray diffraction (XRD) study of the radial distribution function (RDF) of a-Si formed by Si ion implantation of a crystalline Si substrate. Figure 7.5 shows a high-resolution RDF of a-Si before and after relaxation. The latter introduces a small increase (non-perceptible in the figure) in the CN from 3.79 ± 0.01 to 3.88 ± 0.01 atoms while the bondlength remains unaltered at ~2.35 Å. A greater change upon relaxation is apparent in the 2NN peak at ~3.8 Å plus that at ~4.7 Å, the latter attributed to dihedral bond ordering. Figure 7.5 clearly demonstrates two advantages of XRD relative to XAS when performed over a sufficiently large range of scattering vector (in this case to 60 $Å^{-1}$): the CN can be determined with great accuracy and structure in the amorphous phase beyond the NN can be characterized.

While the performance of XRD experiments such as those of reference 13 is complex in terms of sample preparation, measurement and analysis, the execution

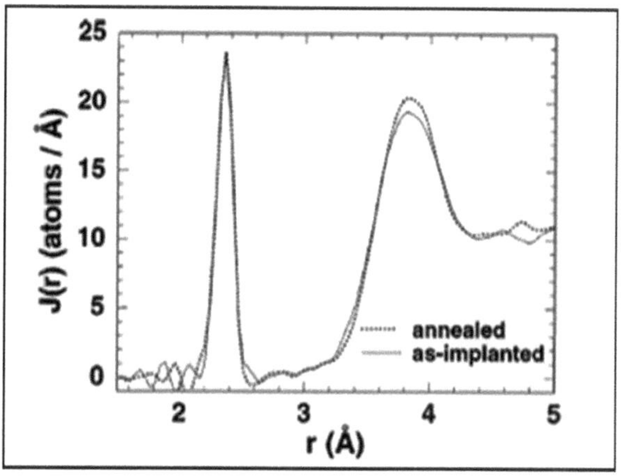

Fig. 7.5 High-resolution RDF of a-Si determined by XRD before ("as-implanted") and after ("annealed") relaxation. (Reprinted figure with permission from [13]. Copyright (1999) by the American Physical Society. http://link.aps.org/abstract/PRL/v82/p3460)

of an XAS experiment at the Si K-edge is also complicated by several factors. The edge energy of 1.839 keV necessitates a sample end-station with a vacuum or He environment. Unfortunately, the Si K-edge is intermediate between the energy ranges typically covered by soft and hard X-ray beamlines and as a consequence appropriate "tender" beamlines are scarce relative to their common hard X-ray counterparts. Also, the reduced X-ray penetration depth at the Si K-edge effectively precludes transmission measurements and requires removal of the native SiO_2 surface layer immediately prior to the experiment. Finally, data analysis is complicated by a KL multi-electron excitation 0.124 keV beyond the K-edge. A protocol to account for the multi-electron excitation has been described by Wakagi et al. [14].

Examples of experimental XAS studies of a-Si include Filipponi et al. [15], Wakagi et al. [14], Glover et al. [16] and Glover [17] with samples prepared by either sputtering [14, 15] or Si ion implantation [16, 17]. Figure 7.6 shows spectra from Glover et al. [16] including XANES, EXAFS and Fourier-transformed data, all as a function of ion fluence. The crystalline-to-amorphous phase transformation is readily apparent with the loss of multiple-scattering contributions in the XANES spectra, the dominant single-frequency oscillation in the EXAFS spectra and the loss of scattering contributions from beyond the NN in the FT spectra. In the latter, scattering from next NNs is absent for ion fluences of $1 \times 10^{15}/cm^2$ and beyond, consistent with amorphous-phase formation.

The data extracted from Fig. 7.6 was analysed with the Cumulant Method as described in Chap. 1 and the resulting structural parameters are shown in Fig. 7.7. Within experimental error, no ion fluence dependence is apparent beyond the amorphisation threshold of $1 \times 10^{15}/cm^2$ with the CN less than four (3.78 ± 0.16 atoms), the bondlength (2.351 ± 0.005 Å) unchanged relative to crystalline material and the Debye-Waller Factor (DWF) (0.0048 ± 0.0005 Å^2) increased significantly as consistent with the disorder inherent in the amorphous phase. Note that the third cumulant (C3), the third moment of the inter-atomic distance distribution, is non-zero, indicative of an asymmetric deviation from a Gaussian inter-atomic distance

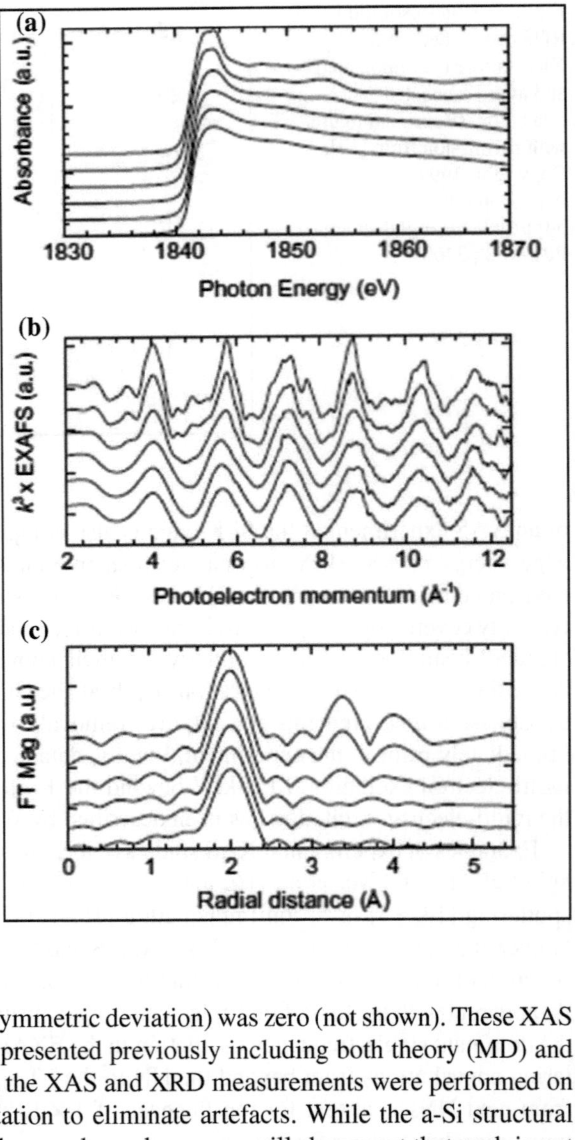

Fig. 7.6 **a** XANES, **b** EXAFS and **c** FT EXAFS spectra **c** for Si as a function of ion fluence. (Reprinted from [16], Copyright (2003), with permission from Elsevier)

distribution. In contrast, C4 (symmetric deviation) was zero (not shown). These XAS results agree well with those presented previously including both theory (MD) and experiment (XRD). Note that the XAS and XRD measurements were performed on a-Si prepared by ion implantation to eliminate artefacts. While the a-Si structural parameters exhibited no ion fluence dependence, we will show next that such is not the case for a-Ge due to an ion-induced amorphous-to-porous transformation.

7.5.2 Amorphous Ge (a-Ge)

The first transistor was fabricated from Ge and this material retains intrinsic advantages over Si in terms of hole mobility. Historically, the integration of Ge into electronic devices was impeded by an electrically-poor, water-soluble oxide but with

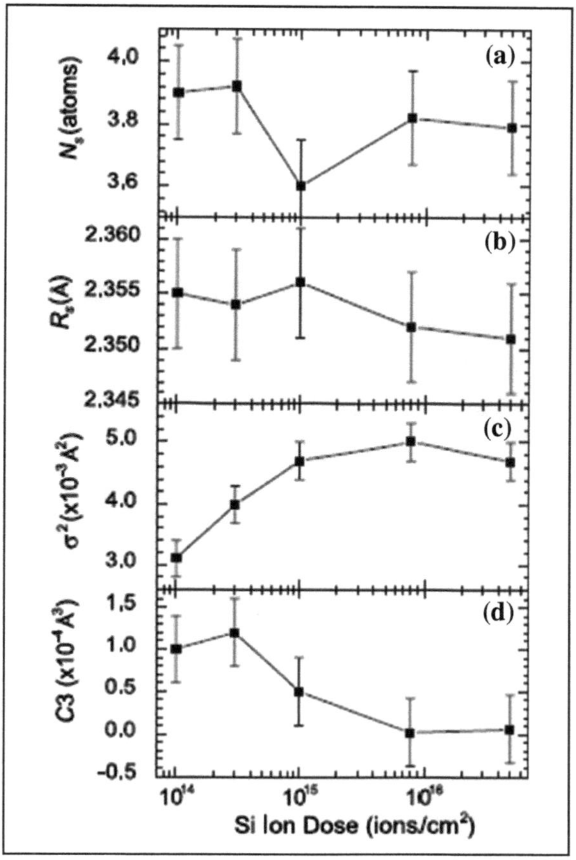

Fig. 7.7 Structural parameters for a-Si including **a** CN, **b** bondlength, **c** DWF and **d** C3 as a function of ion fluence. (Reprinted from [16], Copyright (2003), with permission from Elsevier)

the recent advent of alternative dielectrics the use of Ge is now increasing rapidly. For example, the channel region for the p-FETs in future CMOS devices is likely to be formed from elemental Ge (with that for n-FETs formed from $In_xGa_{1-x}As$). Knowledge of the atomic-scale structure of a-Ge is clearly of absolute necessity.

Like a-Si, disorder in a-Ge is entirely structural. Theoretical predictions of the structure of a-Ge have progressed significantly since the ball-and-stick model of Polk and MD is now the most common calculation methodology. As an example, we show the ab initio MD simulations of Kresse and Hafner [18] where a-Ge was produced by a rapid quench from the melt. Structural parameters are listed in Table 7.2.

Clearly the bondangle distribution is broad and the CN is less than four. The nature of the bonding defects is apparent in Fig. 7.8 which shows the electron density for three-, four- and five-fold coordinated atoms in a-Ge. The bondangle distribution means for the three- and five-fold coordinated atoms (100.3° and 103.7°, respectively) are significantly lower than the tetrahedral value while the RMS widths (16.6° and 25.7°, respectively) are broader. Furthermore, the bondlength (2.58 Å) associated with the five-fold coordinated atom is considerably higher than the tetrahedral

Table 7.2 Structural parameters of c- and a- Ge showing bondangle distribution mean ϑ, RMS bondangle distribution mean $\Delta\vartheta$ and fraction of atoms CN_m with CN m [18]

	Crystalline	Relaxed amorphous
ϑ (deg)	109.5	108.5
$\Delta\vartheta$ (deg)		14.9
CN_3 (%)	0	6.3
CN_4 (%)	100	90.6
CN_5 (%)	0	3.1

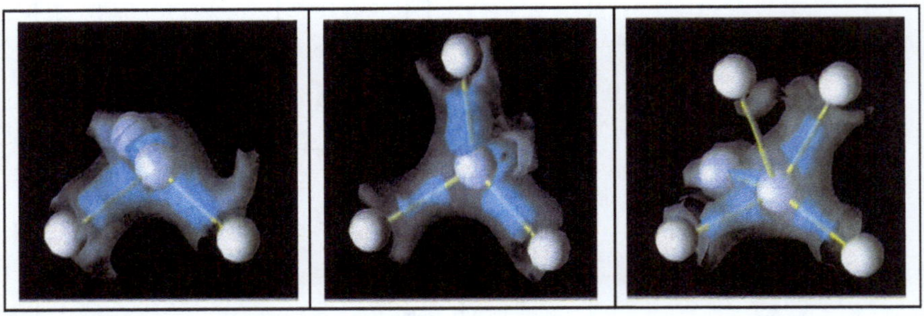

Fig. 7.8 Electron density plots for three-, four- and five-fold coordinated atoms in a-Ge. (Reprinted figure with permission from [18]. Copyright (1994) by the American Physical Society. http://link.aps.org/abstract/PRB/v49/p14251)

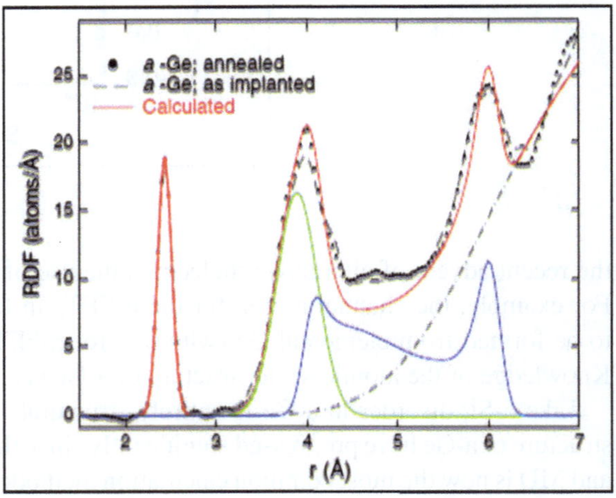

Fig. 7.9 High-resolution RDF of a-Ge before (" as-implanted") and after ("annealed") relaxation. (Reprinted figure with permission from [19]. Copyright (2012) by the American Physical Society. http://link.aps.org/abstract/PRL/v108/p255501). The calculated curve (*red*) is the sum of contributions from NN, 2NN (*green*) and 3NN (*blue*) plus the average density convoluted with an error function (*dashed*)

value (2.46 Å). As demonstrated below, these theoretical predictions will aid in the interpretation of XAS data.

Prior to describing an XAS example, Fig. 7.9 shows a RDF determined from XRD measurements [19]. The CNs associated with the NN and 2NN are 3.68 ± 0.02 and 10.2 ± 0.3 atoms, respectively, while their bondlengths are 2.429 ± 0.001 and 3.890 ± 0.006 Å, respectively. The magnitude of the 2NN CN is consistent with a 3NN contribution. The authors concluded that only first-principles-based models that calculate the electron density, such as those shown above, are appropriate for the identification of 2NN and 3NNs. The significant difference between spectra in the range 4.5–5.0 Å before and after relaxation was attributed to dihedral ordering. The relaxation process was considered a combination of defect annealing enabling the expansion of once-compressed tetrahedrons and a reduction in the width of the bondangle distribution.

Ge was not only the basis of the first transistor but also of some of the first quantitative XAS measurements [20] and since that time a-Ge has been studied extensively with XAS. Recent examples include Wakagi et al. [21], Dalba et al. [22, 23], Filipponi et al. [24], Ridgway et al. [6, 25], Glover et al. [26] and Glover [17]. We now focus on the latter four reports where samples were once again fabricated by ion implantation to inhibit artefacts. Figure 7.10 shows the NN bondlength and C3 for ion implanted a-Ge as a function of ion fluence and compares structural parameters before and after relaxation. Data analysis was performed with the Cumulant Method. Note that all ion fluences were in excess of the amorphisation threshold and thus Fig. 7.10 illustrates

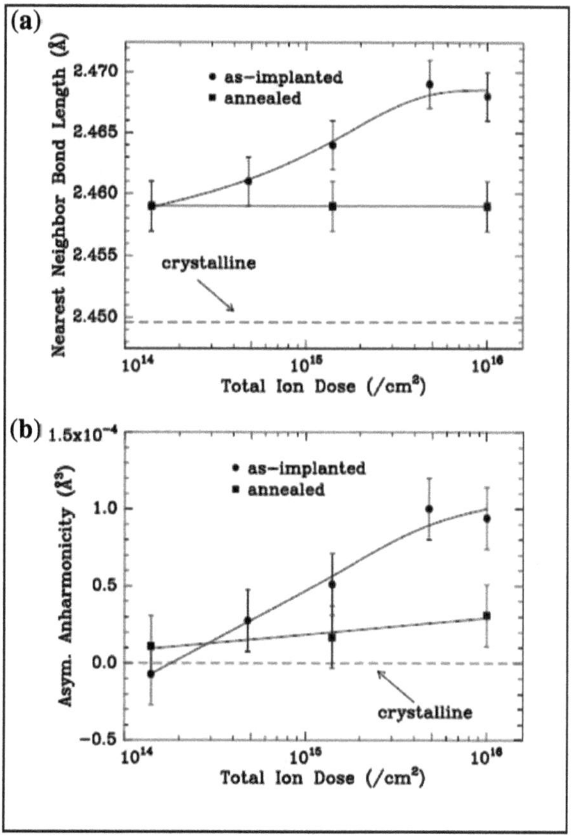

Fig. 7.10 **a** NN bondlength and **b** C3 for a-Ge as a function of ion fluence comparing unrelaxed ("as-implanted") and relaxed ("annealed") samples. (Reprinted from [25], Copyright (2001), with permission from Elsevier)

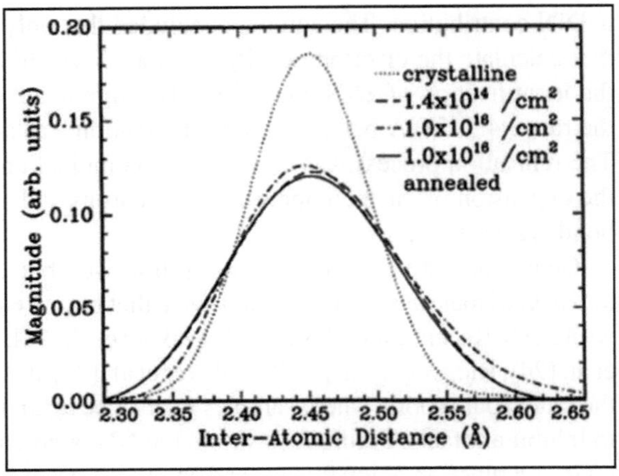

Fig. 7.11 Reconstructed inter-atomic depth distribution for a-Ge as a function of ion fluence before and after ("annealed") relaxation. (Reprinted from [25], Copyright (2001), with permission from Elsevier)

the evolution of the amorphous phase structure (and not the crystalline-to-amorphous phase transformation). The subtle yet discernable changes in bondlength and C3 readily demonstrate not only the power of this form of analysis but also the technique in general. The ion-fluence-dependent trends apparent in Fig. 7.10 were attributed to the implantation-induced increase in the fraction of defective atomic configurations in the amorphous phase. Most importantly, Fig. 7.10 demonstrates that upon thermal annealing, both the bondlength and C3 relax to ion-fluence-independent values. Equivalently, relaxation induces a common atomic-scale structure in a-Ge formed by ion implantation independent of the implantation parameters. This common structure is best illustrated with the reconstructed inter-atomic depth distribution shown in Fig. 7.11. Clearly, in an unrelaxed state, the distributions for the two ion fluences differ significantly, most notably in asymmetry. Upon relaxation, differences are effectively indiscernible. (Note that the distributions for the low ion fluence sample before and after relaxation were near identical.)

As noted in Chap. 1, XAS is an effective means of probing both the structural and vibrational properties of material, the latter achieved via measurement-temperature-dependent experiments performed over a range of ~300 °C or more. Typically, the first four cumulants are determined then fit with a correlated Einstein Model. The separate contributions of structural (static) and thermal disorder can then be identified. This form of analysis has been applied to a-Ge with examples that include Crozier and Seary [27], Dalba et al. [22, 23] and Filipponi et al. [24]. Figure 7.12 from Dalba et al. [23] compares the temperature dependence of C2, the second moment of the inter-atomic depth distribution, for c- and a-Ge while Table 7.3 compares Einstein frequencies from several studies. The slightly higher Einstein frequency measureable in c-Ge is indicative of stiffer bonding in the crystalline phase. Measurement-temperature-dependent experiments of c-Ge are also presented in Chap. 6.

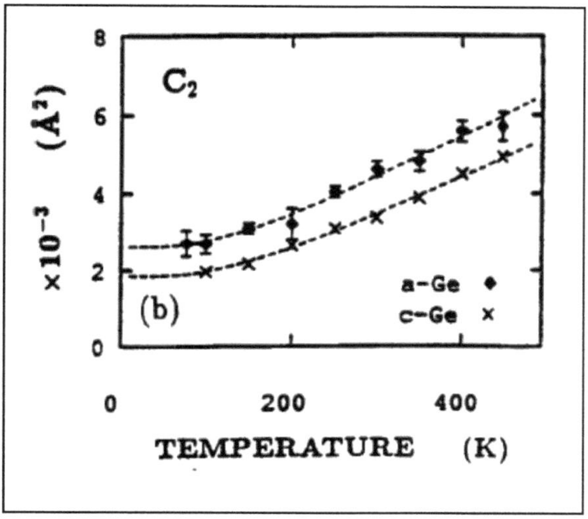

Fig. 7.12 Measurement-temperature-dependence of C2 for c- and a-Ge. (Reprinted figure with permission from [22]. Copyright (1995) by the American Physical Society. http://link.aps.org/abstract/PRB/v52/p11034)

Table 7.3 Einstein frequency comparison of c- and a-Ge where * indicates an average of two values

	Einstein frequency (THz)
Crystalline Ge [22]	7.50 ± 0.2
Amorphous Ge/Evaporated [22]	6.90 ± 0.2
Amorphous Ge/Sputtered [23]	7.36 ± 0.2*

7.5.3 Amorphous SiC (a-SiC)

We now move from mono- to multi-elemental Group IV semiconductors and consider the two examples of SiC and $Si_{1-x}Ge_x$. The former has excellent thermal conductivity which enables applications in high-power, high-voltage electronic devices. Crystalline SiC is also a common substrate for the growth of GaN for use in photonic devices such as LEDs and is found in the cubic zincblende structure (3C/β-SiC) and polytypes of the hexagonal wurtzite structure (6H/α-SiC). Bonding is purely heteropolar with separate Si and C sub-lattices. For the amorphous phase, we will show that XAS is ideally suited to identify both heteropolar (Si–C) and homopolar (Si–Si) forms of bonding, the latter commonly referred to as chemical disorder. For the crystalline phase, XAS measurements are presented in Chap. 3.

Yuan and Hobbs [28] used MD simulations to investigate the influence of chemical disorder on the stability of structural disorder during the amorphisation of SiC by ion implantation. Their variable parameter was chemical disorder χ where χ is the ratio of C–C to C–Si bonds. For crystalline SiC, $\chi = 0$ given the absence of homopolar bonds. Yuan and Hobbs identified a chemical disorder threshold of 0.3-0.4 for the amorphisation of SiC, a result consistent with earlier modelling [29] that demonstrated structural disordering in the absence of chemical disordering could

Fig. 7.13 RDF of a-SiC comparing unrelaxed ("as-irradiated") and relaxed (800 °C-annealed) samples. (Reprinted figure with permission from [30]. Copyright (2002) by the American Physical Society. http://link.aps.org/abstract/PRL/v89/p55502)

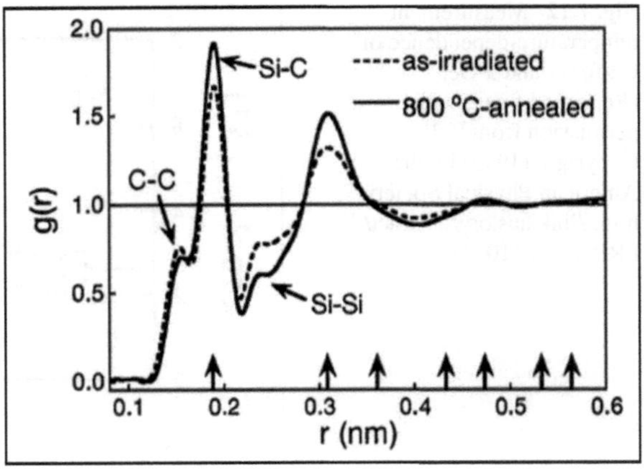

not be achieved. These authors [28] concluded that amorphisation of SiC by ion implantation was enabled by the presence of homopolar bonding created by anti-site replacement events and Frenkel-type defects.

Ishimaru et al. [30] reported a structural determination of a-SiC using ED. Figure 7.13 shows the experimentally-determined RDF of a-SiC formed by ion implantation. Both heteropolar and homopolar bonding in the NN shell are readily apparent with bondlengths of 1.51 Å (C–C), 1.88 Å (Si–C) and 2.38 Å (Si–Si). The C–C bondlength is consistent with that of diamond and graphene while the Si–C bondlength is equal to that in the crystalline phase. Structural relaxation induces a decrease in homopolar bonding with a concomitant increase in heteropolar bonding. The greater reduction in Si–Si bonding relative to C–C bonding is consistent with the greater bond energy of the latter. Figure 7.13 readily demonstrates that one process operative during structural relaxation of a-SiC is a reduction in homopolar bonding while a second is potentially point defect annihilation [30].

The complexities of XAS measurements in the soft X-ray energy range may be a contributing factor to a reduced number of experimental reports for a-SiC. None-the-less, examples include Kaloyeros et al. [31] and Bolse et al. [32] and we now use an example from the latter. Figure 7.14 shows FT EXAFS spectra measured at the Si K-edge illustrating the crystalline-to-amorphous transformation induced in α-SiC by Na ion implantation. As the ion fluence increases, note the progressive decrease in NN (C) and 2NN (Si) amplitudes due to enhanced structural disorder. Upon comparison with the a-Si spectrum (bottom), a Si–Si contribution at a radial distance intermediate between the NN and 2NN peaks in the c-SiC spectrum is apparent and representative of homopolar bonding within the NN shell of a-SiC. At the highest ion fluence, scattering contributions from beyond the NN shell are not observed as consistent with amorphous phase formation. Supported by Complementary Raman measurements, the authors concluded that the NN peak is the superposition of heteropolar (Si–C) and homopolar (Si–Si) bonding contributions. Once again, theory (MD) and experiment (Ed and XAS) agree well and verify the

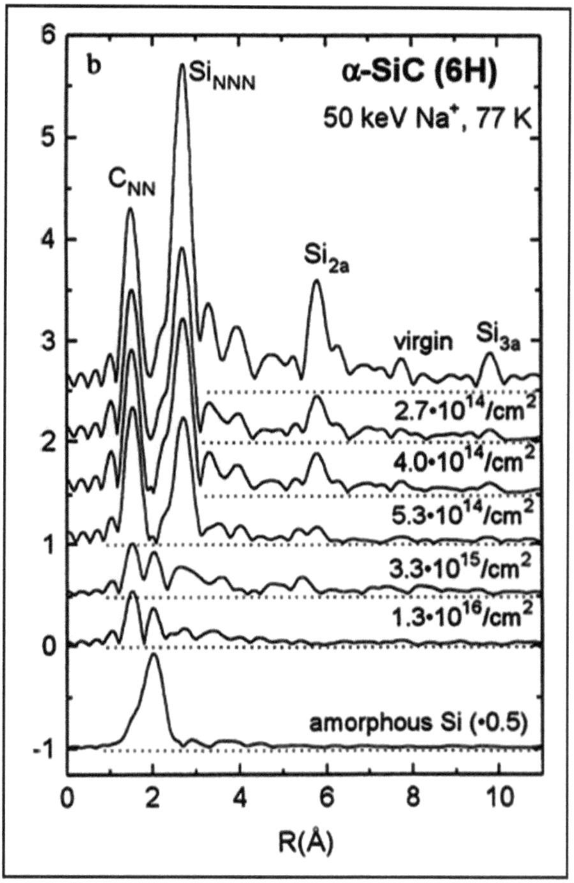

Fig. 7.14 FT EXAFS spectra of the crystalline-to-amorphous phase transformation in α-SiC as a function of ion fluence. (Reprinted from [32], Copyright (1999), with permission from Elsevier)

presence of homopolar bonding in a-SiC. Note that the amorphous phase was formed by ion implantation for both the theoretical and experimental components.

7.5.4 Amorphous $Si_{1-x}Ge_x$ (a-$Si_{1-x}Ge_x$)

Crystalline $Si_{1-x}Ge_x$ alloys have widespread use in electronic devices and often comprise the source/drain regions in FETs. The material is deposited in either a strained or relaxed state with the latter then serving as a template upon which strained Si is deposited. Si and Ge are completely miscible across the entire $Si_{1-x}Ge_x$ composition range and thus c-$Si_{1-x}Ge_x$ alloys, including the c-$Si_{0.5}Ge_{0.5}$ stoichiometry, lack separate Si and Ge sub-lattices in contrast to c-SiC as discussed above. As a consequence, all NN and next NN shells surrounding the absorbing atom in an

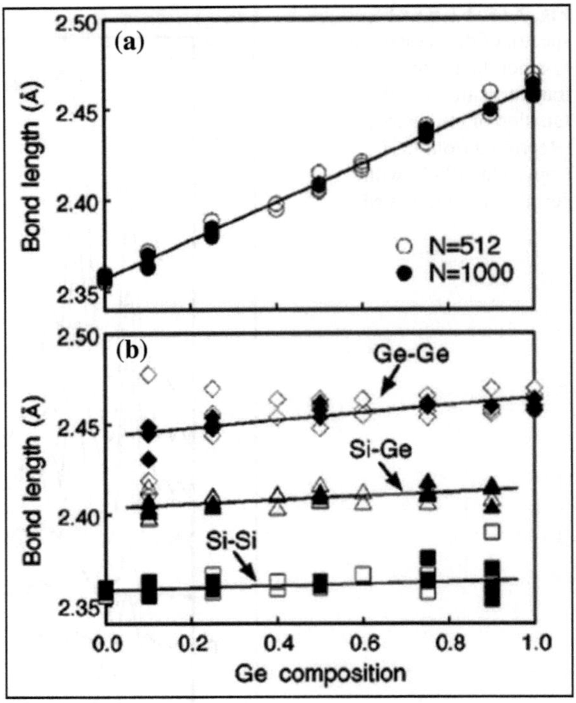

Fig. 7.15 **a** Total and **b** partial (Ge–Ge, Si–Ge and Si–Si) bondlengths in a-$Si_{1-x}Ge_x$ as a function of Ge composition derived from MD simulations. (Reprinted figure with permission from [33]. Copyright (2003) by the American Physical Society. http://link.aps.org/abstract/PRB/v68/p235207)

XAS experiment of c- and a-$Si_{1-x}Ge_x$ are multi-elemental. For c-$Si_{1-x}Ge_x$, XAS measurements and theoretical calculations are presented in Chap. 2.

Ishimaru et al. [33] performed MD simulations designed in part to assess whether the a-$Si_{1-x}Ge_x$ alloys exhibited phase separation, chemical ordering or random mixing within the NN shell. Indeed, the latter was confirmed with Si–Si, Si–Ge and Ge–Ge configurations apparent. As the Ge fraction increased, the mean and width of the bondangle distribution decreased and increased, respectively, suggesting higher Ge fractions yielded higher network distortion. Figure 7.15 shows total and partial bondlengths as a function of Ge fraction [33]. The total bondlength increases smoothly as a function of Ge fraction, from elemental Si to elemental Ge, in a Vegard-like manner while the partial bondlengths (Si–Si, Si–Ge and Ge–Ge) differ in magnitude in a Bragg-Pauling-like manner and exhibit only weak composition dependence (see also Chap. 2).

Non-XAS measurements of the structure of a-$Si_{1-x}Ge_x$ include XRD determinations of the RDF [33] (for $Si_{0.5}Ge_{0.5}$ only) and Raman measurements of the Si–Si, Si–Ge and Ge–Ge vibrational modes [34, 35]. A linear dependence on Ge fraction was reported for the Si–Si and Si–Ge frequencies [35]. Both the XRD and Raman results were consistent with a random mixing of the two elements within the NN shell.

Examples of XAS measurements for the a-$Si_{1-x}Ge_x$ alloys include Minomura et al. [36] and Ridgway et al. [37] and results from the latter are now presented.

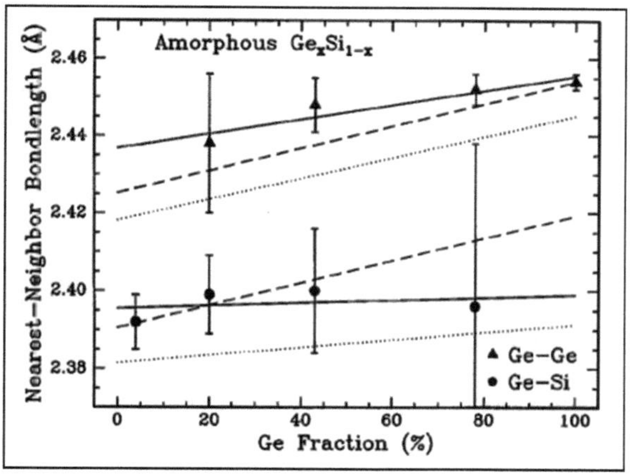

Fig. 7.16 Ge–Ge and Ge–Si bondlengths as a function of Ge composition for $Si_{1-x}Ge_x$ alloys. Symbols are data for the amorphous sample and *solid* and *dotted lines* represent linear fits for amorphous and crystalline samples, respectively. *Dashed line* represents a theoretical prediction using a** = 0.707 [38]. (Reprinted figure with permission from [37]. Copyright (1999) by the American Physical Society. http://link.aps.org/abstract/PRB/v60/p10831)

To enable transmission measurements, a sophisticated sample preparation protocol was developed involving deposition of c-$Si_{1-x}Ge_x$ on Si-on-Insulator substrates, amorphisation by Si ion implantation followed by separation of the a-$Si_{1-x}Ge_x$ layer from the c-Si substrate by selective chemical etching. Figure 7.16 shows Ge K-edge measurements of the Ge–Si and Ge–Ge NN bondlengths as a function of Ge fraction, comparing amorphous and crystalline alloys. In general, the bondlengths in the amorphous phase exceed those in the crystalline phase, consistent with previous results for a-Si and a-Ge. Despite the error bars, the composition dependence of the Si–Ge bondlength appears weaker than that of Ge–Ge, entirely consistent with the MD simulation of Ishimaru et al. [33]. Also included in Fig. 7.16 is a theoretical prediction from Mousseau and Thorpe [38] based on a topological rigidity parameter a** calculation of 0.707. This parameter has values of 0 and 1 at the Vegard and Bragg-Pauling limits, respectively. In the Vegard limit, $Si_{1-x}Ge_x$ alloys exhibit one, composition-*dependent* bondlength while, in contrast, three, composition-*independent* bondlengths result in the Bragg-Pauling limit (see also Chap. 2). Figure 7.16 demonstrates a mixed character is intrinsic to a-$Si_{1-x}Ge_x$ alloys. The magnitude of the topological rigidity parameter however indicates Bragg-Pauling character dominates as consistent with the energetically more favourable bond bending over bond stretching in covalently-bonded semiconductors. Note also the similarity in slope for a given bond when comparing amorphous and crystalline material, consistent with a phase-independent prediction for the topological rigidity parameter [38]. This XAS determination of the structure of a-$Si_{1-x}Ge_x$ yields excellent agreement with MD simulations [33] and first-principles calculations [38] and furthermore readily demonstrates the capabilities of XAS as an analytical technique.

7.6 Summary

This chapter has sought to review XAS measurements of the amorphous Group IV semiconductors with the inclusion of complementary theoretical and non-XAS measurements for comparison. The applicability of XAS to both crystalline and amorphous materials is clearly advantageous and enables detailed comparison of the structure of the two phases. In general, there is a commonality in the structure of the amorphous Group IV semiconductors with an increase in disorder and bondlength and a decrease in coordination number relative to the crystalline phase. Within the amorphous phase, structural, thermal and, with the appropriate materials system (SiC), chemical forms of disorder are readily discernable and quantifiable. This ability to probe both structural and vibrational properties makes XAS an invaluable tool in the study of the amorphous Group IV semiconductors.

Acknowledgments The author gratefully acknowledges C.J. Glover, G.J. Foran and K.M. Yu for significant contributions, fruitful collaborations and memorable times at XAS beamlines around the world in our pursuit of the atomic-scale structure of the amorphous Group IV semiconductors. The author thanks the Australian Research Council, Australian Synchrotron Research Program and Australian Synchrotron for financial support. Sample fabrication was enabled through access to the ACT Node of the Australian National Fabrication Facility.

References

1. W.H. Zachariasen, J. Am. Chem. Soc. **54**, 3841 (1932)
2. R. Zallen, *The Physics of Amorphous Solids* (Wiley, Brisbane, 1983)
3. D.E. Polk, J. Non-Cryst. Solids **5**, 365 (1971)
4. B.R. Djordjevic, M.F. Thorpe, F. Wooten, Phys. Rev. B **52**, 5685 (1995)
5. G.T. Barkema, N. Mousseau, Phys. Rev. B **62**, 4985 (2000)
6. M.C. Ridgway, C.J. Glover, K.M. Yu, G.J. Foran, C. Clerc, J.L. Hansen, A. Nylandsted Larsen, Phys. Rev. B **61**, 12586 (2000)
7. S. Decoster, C.J. Glover, B. Johannessen, R. Giulian, D.J. Sprouster, P. Kluth, L.L. Araujo, Z. Hussain, C. Schnohr, H. Salama, F. Kremer, K. Temst, A. Vantomme, M.C. Ridgway, J. Synchrotron Radiat. **20**, 426 (2013)
8. I. Stich, R. Car, M. Parrinello, Phys. Rev. B **44**, 11092 (1991)
9. E. Kim, Y.H. Lee, Phys. Rev. B **49**, 1743 (1994)
10. N.C. Cooper, C.M. Goringe, D.R. McKenzie, Comput. Mater. Sci. **17**, 1 (2000)
11. N. Mousseau, G.T. Barkema, Phys. Rev. B **61**, 1898 (2000)
12. N. Bernstein, J.L. Feldman, M. Fornari, Phys. Rev. B **74**, 205202 (2006)
13. K. Laaziri, S. Kycia, S. Roorda, M. Chicoine, J.L. Robertson, J. Wang, S.C. Moss, Phys. Rev. Lett. **82**, 3460 (1999)
14. M. Wakagi, K. Ogata, A. Nakano, Phys. Rev. B **50**, 10666 (1994)
15. A. Filipponi, F. Evangelisti, M. Benfatto, S. Mobilio, C.R. Natoli, Phys. Rev. B **40**, 9636 (1989)
16. C.J. Glover, G.J. Foran, M.C. Ridgway, Nucl. Instrum. Methods B **199**, 195 (2003)
17. C.J. Glover, Ph.D. thesis, Australian National University, 2002
18. G. Kresse, J. Hafner, Phys. Rev. B **49**, 14251 (1994)
19. S. Roorda, C. Martin, M. Droui, M. Chicoine, A. Kazimirov, S. Kycia, Phys. Rev. Lett. **108**, 255501 (2012)
20. D.E. Sayers, E.A. Stern, F.W. Lytle, Phys. Rev. Lett. **27**, 1204 (1971)

21. M. Wakagi, A. Maeda, Phys. Rev. B **50**, 14090 (1994)
22. G. Dalba, P. Fornasini, M. Grazioli, F. Rocca, Phys. Rev. B **52**, 11034 (1995)
23. G. Dalba, P. Fornasini, R. Grisenti, F. Rocca, I. Chambouleyron, C.F.O. Graeff, J. Phys.: Condens. Matter **9**, 5875 (1997)
24. A. Filipponi, A. Di Cicco, Phys. Rev. B **51**, 12322 (1995)
25. M.C. Ridgway, C.J. Glover, I.D. Desnica-Frankovic, K. Furic, K.M. Yu, G.J. Foran, C. Clerc, J.L. Hansen, A. Nylandsted Larsen, Nucl. Instrum. Methods B **21**, 175–177 (2001)
26. C.J. Glover, M.C. Ridgway, I.D. Desnica-Frankovic, K.M. Yu, G.J. Foran, C. Clerc, J.L. Hansen, A.N. Larsen, Phys. Rev. B **61**, 073204 (2001)
27. E.D. Crozier, A.J. Seary, Can. J. Phys. **59**, 877 (1981)
28. X. Yuan, L.W. Hobbs, Nucl. Instrum. Methods B **191**, 74 (2002)
29. C.E. Jesurum, V. Pulim, L.W. Hobbs, J. Nucl. Mater. **253**, 87 (1998)
30. M. Ishimaru, I-T Bae, Y. Hirotsu, S. Matsumura, K.E. Sickafus. Phys. Rev. Lett. **89**, 055502 (2002)
31. A.E. Kaloyeros, R.B. Rizk, J.B. Woodhouse, Phys. Rev. B **38**, 13099 (1988)
32. W. Bolse, Nucl. Instrum. Methods B **148**, 83 (1999)
33. M. Ishimaru, M. Yamaguchi, Y. Hirotsu, Phys. Rev. B **68**, 235207 (2003)
34. N.J. Shevchik, J.S. Lannin, J. Tajeda, Phys. Rev. B **7**, 3987 (1973)
35. J. Olivares, P. Martin, A. Rodriguez, J. Sangrador, J. Jimenez, T. Rodriguez, Thin. Sol. Films **358**, 56 (2000)
36. S. Minomura, K. Tsuji, M. Wakagi, T. Ishidate, K. Inoue, M. Shibuya, J. Non-Cryst. Solids **541**, 59–60 (1983)
37. M.C. Ridgway, C.J. Glover, K.M. Yu, G.J. Foran, C. Clerc, J.L. Hansen, A.N. Larsen, Phys. Rev. B **60**, 10831 (1999)
38. N. Mousseau, M.F. Thorpe, Phys. Rev. B **46**, 15887 (1992)

Chapter 8
Amorphous Group III–V Semiconductors

Mark C. Ridgway

Abstract Structural disorder is common to the amorphous Group IV and III–V semiconductors as manifested by an increase in bondlength and Debye-Waller factor and a decrease in coordination number relative to the crystalline phase. A second component of disorder, unique to compound semiconductors, is chemical disorder in the form of homopolar bonding. In this chapter, the application of XAS to the characterisation of both structural and chemical disorder in the amorphous Group III–V semiconductors is described with an emphasis on the identification and quantification of homopolar bonding. We show chemical disorder is characteristic of these materials, comprising ∼5–15 % of all bonds, and also demonstrate that XAS is an ideal technique for such studies.

8.1 Introduction

While in a crystalline form the Group III–V semiconductors have multiple applications in electronic and photonic device fabrication, in an amorphous form their technological usage is much reduced. None-the-less, the amorphous Group III–V semiconductors have been studied extensively with XAS. The driving force behind such investigations has largely been scientific as opposed to technological with assessments of the validity of structural models as a significant focus. Compared to their elemental amorphous Group IV semiconductor counterparts, disorder in the binary amorphous Group III–V semiconductors is more complex. Specifically, disorder may now have both structural and chemical components, the latter in the form of homopolar bonding (which is forbidden in the crystalline phase). As we discuss below, the application of XAS has been crucial for the identification and quantification of chemical disorder. In this chapter, we now review XAS of the amorphous Group III–V semiconductors including both Ga- and In-based materials. Such a wide scope of materials can never be all inclusive and thus the amorphous ternary and quaternary Group III–Vs are not covered.

M.C. Ridgway (✉)
Department of Electronic Materials Engineering, Australian National University, Canberra, ACT 0200, Australia
e-mail: mark.ridgway@anu.edu.au

C.S. Schnohr and M.C. Ridgway (eds.), *X-Ray Absorption Spectroscopy of Semiconductors*, Springer Series in Optical Sciences 190,
DOI 10.1007/978-3-662-44362-0_8

8.2 Structure of Amorphous Group III–V Semiconductors

In Chap. 7, the Continuous Random Network (CRN) model of Zachariasen [1] was introduced followed by the three-dimensional CRN of four-fold coordinated atoms constructed by Polk [2]. The Polk model, with a ring-size distribution spanning five-, six- and seven-fold, is clearly applicable to the elemental amorphous Group IV semiconductors. Disorder is purely structural and largely contained in the bondangle distribution as consistent with the energetically more favourable bond bending over bond stretching in covalently-bonded semiconductors (see Chap. 2). However, the presence of odd-membered rings inhibits the applicability of the Polk model to the binary amorphous Group III–V semiconductors given homopolar bonding is then mandated. The non-zero ionicity of the Group III–V semiconductors is such that homopolar bonding comes at the cost of Coulomb energy while restricting the ring-size distribution to only even-membered rings comes at the cost of elastic energy (as the potential configurations to minimize distortion in the bondangle and bondlength distributions are then limited). While a CRN similar to that of Polk but without odd-membered rings was successfully constructed by Connell and Temkin [3], it is the balance between the Coulomb and elastic energies that will ultimately determine the structure of the amorphous Group III–V semiconductors [4].

A more recent example of the now computer-generated construction of an amorphous Group III–V semiconductor model is that of Lewis et al. [5] who utilised molecular dynamics (MD) coupled with a rapid quench from the melt. Figure 8.1 shows the resulting structures for crystalline and amorphous InP (c-InP and a-InP, respectively). Upon close inspection, homopolar bonding is visible in the latter. As described below, XAS has been used extensively to not only identify but quantify the extent of homopolar bonding in amorphous Group III–V semiconductors and as such has been instrumental in assessing the validity of structural models including the pioneering works of both Polk and Connell-Temkin. While the current section

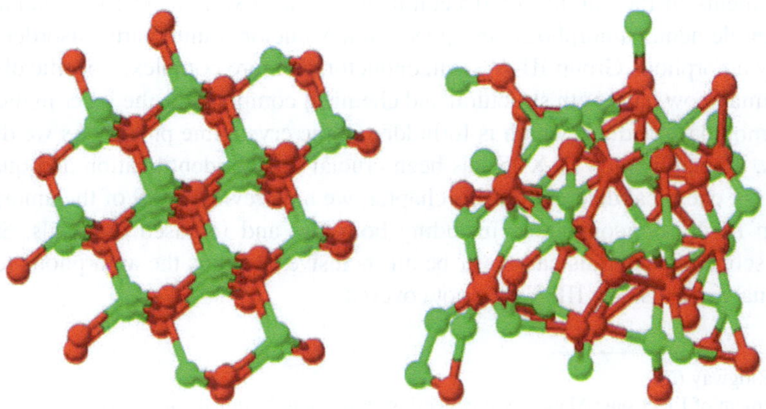

Fig. 8.1 The atomic-scale structure of **a** c-InP and **b** a-InP. Figure drawn with data from [5]

has been intentionally general, the subsequent presentation for the amorphous Group III–V semiconductors will include brief descriptions of current, materials-specific structural models for comparison with experimental results, including both XAS and complementary non-XAS derived data.

8.3 Preparation of Amorphous Group III–V Semiconductor Samples for XAS

As noted in Chap. 7, careful preparation of samples for an XAS experiment is essential to extract the best data possible. Relative to their Group IV counterparts, the preparation of amorphous Group III–V semiconductor samples is more fraught with difficulty. Historically, samples were prepared by depositing amorphous layers onto X-ray transparent material using flash evaporation. Unfortunately, the volatility of the Group V component often led to significant deviations from the desired stoichiometry and differences in experimental results from group to group. More recently, ion implantation has been used to negate such artefacts. Typically, a surface layer of several microns in thickness is amorphized with a multi-energy, multi-fluence ion implantation protocol and then separated from an underlying bulk crystalline substrate via selective chemical etching. A variety of material-specific ion implantation protocols are described in [6].

The non-equilibrium nature of the deposition and amorphization processes is such that the amorphous-phase structure will initially form in an energy state higher than that of the minimum-energy configuration. (We call the high-energy state "unrelaxed"). However, thermal annealing can effectively reduce point-defect, dangling-bond and homopolar bonding fractions within the amorphous phase such that the intrinsic, minimum-energy configuration can be approached. (We call the minimum-energy state "relaxed"). The annealing conditions required to attain the relaxed amorphous-phase structure are material-specific and great care must be exercized so as to induce relaxation but not recrystallization.

8.4 Amorphous Ga-Based Group III–V Semiconductors

8.4.1 Amorphous GaN (a-GaN)

Crystalline GaN has both electronic and photonic applications where the former includes high-power, high-frequency and high-temperature devices. The direct bandgap of GaN yields far brighter emission than that of the indirect bandgap SiC and the development of blue GaN LEDs has enabled the realisation of full-colour displays and white LEDs. MD simulations of a-GaN identified a large, state-free optical gap and weakly-delocalised band tails prompting suggestions of potential applications

Table 8.1 Structural parameters for a-GaN from two different models of Cai and Drabold comparing the fraction of homopolar bonding (N–N and Ga–Ga), the fraction of N_m and Ga_m atoms with CN m and the average CN for N and Ga [8]

Model	N–N (%)	Ga–Ga (%)	N_3 (%)	N_4 (%)	Ga_3 (%)	Ga_4 (%)	Ga_5 (%)	CN_N (atoms)	CN_{Ga} (atoms)
c-GaN	0	0	0	100	0	100	0	4	4
a-GaN LDA I	0	0	3	97	6	91	3	3.97	3.97
a-GaN LDA II	0	0	9	91	9	91	0	3.91	3.91

for the amorphous phase [7]. Clearly, accurate atomic-scale determinations of the structure of a-GaN are thus of critical importance and both theoretical and experimental examples are now presented.

Cai and Drabold examined the properties of a-GaN with first-principles calculations [8]. Given the high ionicity of GaN, homopolar bonding is expected to be minimal in this materials system and indeed this is borne out theoretically. Table 8.1 shows structural components for several different means of modelling. Short-range order is clearly retained in the amorphous phase with both elemental components having coordination numbers (CNs) near four-fold. The most abundant configurational defect for both Ga and N is a three-fold coordinated atom. The bondangle distributions for Ga–N–Ga and N–Ga–N (not shown) are centred in close proximity to the tetrahedral angle. The authors of [8] note that their "work focuses primarily on ideal a-GaN … [and] ion-bombarded samples are … likely to exhibit far more disorder." Indeed the MD simulations of Nord et al. [9] for the structure of a-GaN produced by ion implantation identified the formation of both long, weak Ga–Ga bonds (though not solid Ga metal) as the precursor to amorphization plus N_2 dimers in segregated gaseous N. Odd-membered rings (three- and five-fold) were dominant in the ring-size distribution as consistent with a significant homopolar bonding fraction.

Ishimaru et al. [10] reported electron diffraction (ED) results for ion-implanted GaN samples. Given c-GaN is highly radiation-resistant, ion implantation was performed at low temperature (60 K) to successfully stabilize the amorphous phase. Nitrogen-filled bubbles were readily apparent with TEM though metallic Ga precipitates were not observed. Figure 8.2 compares the pair distribution function (PDF) of a- and nanocrystalline (nc-) GaN. The dominant peaks at 1.9 and 3.2 Å are consistent with the nearest neighbour (NN) and second nearest neighbour (2NN) bondlengths in the heteropolar-only crystalline phase while in the a-GaN spectrum the peaks of lesser magnitude at 1.5 and 2.6 Å are, respectively, homopolar N–N and Ga–Ga bonding. Samples in this study were not subjected to any form of thermally-induced structural relaxation which, as the authors noted, could potentially have led to a reduction in homopolar bonding fraction.

To our knowledge, XAS measurements for stoichiometric a-GaN in the absence of a large fraction of stabilising impurities such as O have not been reported. Experiments have been performed for ion-implanted GaN with large amorphous and small

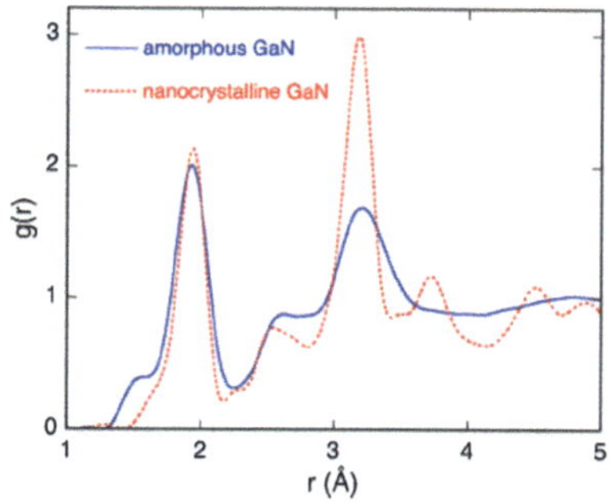

Fig. 8.2 PDF of a- and nc-GaN. (Reproduced with permission from [10], Copyright 2011, AIP Publishing LLC)

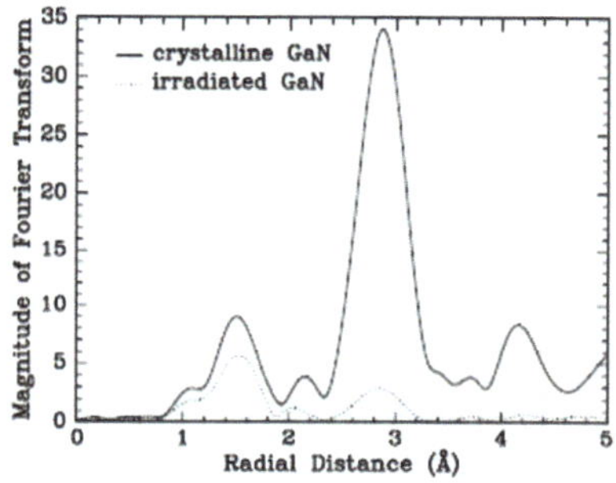

Fig. 8.3 FT EXAFS spectra for c- and implanted ("irradiated") GaN. (Reprinted from [11], Copyright (2006), with permission from Elsevier)

nanocrystalline components at both the Ga (Ridgway et al. [11]) and N (Katsikini et al. [12] and Deenapanray et al. [13]) K-edges. Figure 8.3 shows Fourier-transformed (FT) EXAFS spectra recorded at the Ga K-edge comparing implanted and c-GaN [11]. TEM revealed the presence of both N_2-filled bubbles and GaN nanocrystallites interspersed within an amorphous matrix with the nanocrystallite fraction estimated at $\sim$10 %. Comparing the FT EXAFS spectra, N nearest neighbour (NN) and Ga second nearest neighbour (2NN) peaks are apparent for both implanted and c-GaN. The Ga 2NN peak for the implanted sample is due in part or in whole to the presence of nanocrystallites. Unambiguous evidence for a Ga component in the NN shell was lacking suggesting the stoichiometric imbalance associated with N_2-filled bubble formation was not accommodated by Ga–Ga homopolar bonding in the amorphous phase or metallic Ga precipitation. As Table 8.2 demonstrates, it is a significant reduction in CN_{Ga} that accommodates the stoichiometric imbalance. Measurements at the N K-edge complement those performed at the Ga K-edge. For implanted GaN, Katsikini et al. [12] identified the presence of the N split-interstitial and, at very high Si ion fluences sufficient to induce significant non-stoichiometry, N dangling bonds.

While the two theoretical studies presented above differ considerably, such differences can be understood by the extent of (computer-generated) structural relaxation. Comparing the experimental ED and XAS studies, differences are also apparent and again are potentially the result of the extent of (thermally-induced) structural relaxation. As the majority of authors comment, sufficiently relaxed a-GaN samples should ideally have a common atomic-scale structure.

Table 8.2 Structural parameters determined at the Ga K-edge for c- and implanted GaN including bondlength, Debye-Waller Factor (DWF), CN and Ga-Ga bonding fraction [11]

	NN bondlength (Å)	NN DWF ($Å^2$)	CN_{Ga} (atoms)	Ga–Ga fraction (%)
Crystalline GaN	1.955 ± 0.018	0.0051 ± 0.0024	4 (fixed)	0
Implanted GaN	1.952 ± 0.004	0.0055 ± 0.0009	2.61 ± 0.23	0

8.4.2 Amorphous GaP (a-GaP)

Crystalline GaP finds applications in red, orange and green LEDs and as a component in multi-junction solar cells. Applications of a-GaP are limited but include diodes [14]. None-the-less, the atomic-scale structure of a-GaP has been studied in detail experimentally for comparison with other amorphous III–V materials.

To our knowledge, theoretical calculations of the structure of a-GaP have not been reported. Such calculations would clearly be of value given GaP has an ionicity (0.37), and thus potentially chemical disorder, intermediate between that of the well-studied GaN (0.50) and GaAs (0.31).

A recent determination of the structure of a-GaP using a technique with sufficient sensitivity to provide a meaningful comparison with XAS is unfortunately lacking and thus we proceed directly to XAS measurements. As above, a-GaP has been well studied with examples that include Theye et al. [15], Udron et al. [16], Elgun et al. [17, 18] and Ridgway et al. [11, 19]. Note these three groups utilised flash evaporation, sputtering and ion implantation, respectively, for the preparation of their amorphous phase samples. In [15, 18] measurements were performed at both the Ga and P K-edges. However, their conclusions differed dramatically: chemical disorder was reported by Theye et al. and Udron et al. yet not by Elgun et al. The Ga K-edge measurements of Ridgway et al. [11], as shown in Fig. 8.4, were consistent with chemical disorder. Scattering beyond the NN is not visible, typical of amorphous semiconductors. Table 8.3 compares the structural parameters of a-GaP determined by the three groups. The differences may well reflect the different sample preparation protocols.

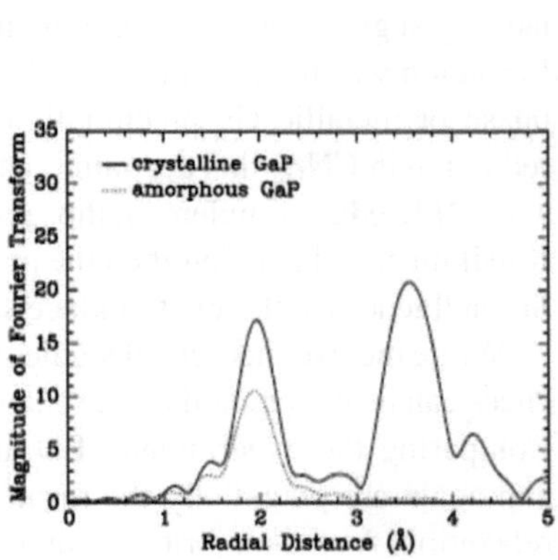

Fig. 8.4 FT EXAFS spectra for c- and a-GaP. (Reprinted from [11], Copyright (2006), with permission from Elsevier)

Table 8.3 Structural parameters determined at the Ga and P K-edges for a-GaP [15–19]

Ga K-edge	Ga-P bondlength (Å)	Ga–Ga bondlength (Å)	CN_{Ga} (P atoms)	CN_{Ga} (Ga atoms)
Theye [15], Udron [16]	2.37 ± 0.01	2.42 ± 0.03	3.2 ± 0.3	0.8 ± 0.3
Elgun [17, 18]	2.33 ± 0.02	Not reported	3.9 ± 1.4	Not reported
Ridgway [11, 19]	2.378 ± 0.025	Not reported	3.43 + 0.75	0.43 ± 0.4
P K-edge	P–Ga bondlength (Å)	P–P bondlength (Å)	CN_P (Ga atoms)	CN_P (P atoms)
Theye [15], Udron [16]	2.37 ± 0.01	2.15 ± 0.01	2.9 ± 0.3	0.8 ± 0.3
Elgun [17, 18]	2.31 ± 0.01	Not reported	4.0 ± 0.8	Not reported

8.4.3 Amorphous GaAs (a-GaAs)

Crystalline GaAs is the most widely used compound semiconductor in photonic devices (laser diodes, for example) but also has common usage in electronic devices (monolithic microwave integrated circuits, for example). GaAs is often alloyed with Al and In as the ternary III–V alloys $Al_xGa_{1-x}As$ and $In_xGa_{1-x}As$ to tune the alloy bandgap to application-specific requirements. Amorphous GaAs has limited applications though solar cells have been suggested as one potential use [20].

Theoretical predictions of the atomic-scale structure of a-GaAs are numerous with examples that include Fois et al. [21], Molteni et al. [22, 23], Seong and Lewis [24], Mousseau and Lewis [4, 25] and Ebbsjo et al. [26]. With one exception, there is general agreement that homopolar bonding is present at a level of <14 % of all bonds. Table 8.4 lists structural parameters for a-GaAs from selected models wherein a broadening of the CN, bondlength and bondangle distributions is apparent. Using an alternative inter-atomic potential, Ebbsjo et al. [26] created a CRN comprised of only even-membered rings (four-, six- and eight-fold) and as a consequence chemical disorder was zero. Their calculated static structure factor was in good agreement with the XRD measurements of Udron et al. [27] though there is an appreciable spread in the bondangle distribution.

The atomic-scale structure of a-GaAs has been studied by XRD with examples that include Shevchik and Paul [28], Temkin [29] and most recently Udron et al. [27]. Using unrelaxed samples of $Ga_{0.48}As_{0.52}$ formed by flash evaporation, Udron et al. [27] concluded a-GaAs was chemically ordered with ~10 % of As atoms having three-fold coordination, the natural coordination state of elemental As.

A XAS measurement of a-GaAs is intrinsically complicated by the similarity in scattering strengths of the two constituents which can inhibit an unambiguous determination of both the presence and fraction of homopolar bonding. None-the-less, a-GaAs has been well studied with XAS, examples include Del Cueto and Shevchik [30], Theye et al. [15], Udron et al. [27], Baker et al. [31], Ridgway et al. [11, 19, 32] and Glover [33]. We now focus on the results of Glover, the most sophisticated of the analyses reported thus far.

Table 8.4 Structural parameter predictions for a-GaAs showing the fraction of atoms CN_m with CN of m, the average CN, the fraction of wrong bonds and the RMS bondangle deviation

	Fois et al. [21]	Molteni et al. [22, 23]	Seong and Lewis [24]	Mousseau and Lewis [4, 25]	Mousseau and Lewis [4, 25]	Ebbsjo et al. [26]
CN_3 (%)	21.9	14	24.2	4.6	5.1	8.9
CN_4 (%)	78.1	66	59.8	95.4	94.4	90.6
CN_5 (%)	0	18	12.9	0	0.5	0.5
CN_6 (%)	0		2.4	0	0	0
CN_7 (%)	0		0.7	0	0	0
$CN_{average}$ (atoms)	3.83	4.09	3.94	3.95	3.95	3.92
Wrong bonds (%)	10.0	12.9	12.2	14.1	3.9	0
$\Delta\theta$ (°)		17.0	17.0	11.0	10.8	21–22

The assumption of a Gaussian inter-atomic distance distribution in the presence of significant structural disorder can yield large errors in structural parameter determinations. However, given the aforementioned similarity in scattering strengths for Ga and As, the Cumulant Method has been applied for the analysis of a-GaAs under the reasonable assumption that the first shell surrounding either a Ga or As atom can be treated as monoelemental even in the presence of homopolar bonding. Using this approach, Glover [33] determined the first four cumulants (C1, C2, C3 and C4, respectively) of the inter-atomic distance distributions for both Ga and As absorbers in a-GaAs, thus accounting for both asymmetric (C3) and symmetric (C4) deviations from a Gaussian distribution. Samples were prepared by ion implantation with equal ion fluences of Ga and As ions to maintain stoichiometry. Figure 8.5 shows reconstructed inter-atomic distance distributions at both edges from which it is apparent that the spectra are asymmetric yet differ, primarily as a consequence of positive and negative values of C4 for Ga and As absorbers, respectively. Whilst some ion-fluence-dependent evolution was apparent for an As absorber, such was not the case for a Ga absorber. Supported by simulations, Glover postulated that this evolution was potentially the result of increasing chemical disorder.

Glover also examined the influence of thermal annealing and, typical of the amorphous semiconductors, structural relaxation was observable in a-GaAs. While the amplitude-dependent parameters (CN, DWF and C4) did not change upon annealing to 200 °C, the phase-dependent parameters (bondlength and C3) did. Table 8.5 compares the structural parameters of c-GaAs and relaxed a-GaAs. As expected, the bondlengths and DWFs for a-GaAs exceeded those for c-GaAs with the CNs decreasing below four atoms. Differences between the parameters measured for Ga and As absorbers are apparent. The lesser value of CN_{As} relative to CN_{Ga} was attributed to a greater fraction of three-fold coordinated As atoms and was consistent with theo-

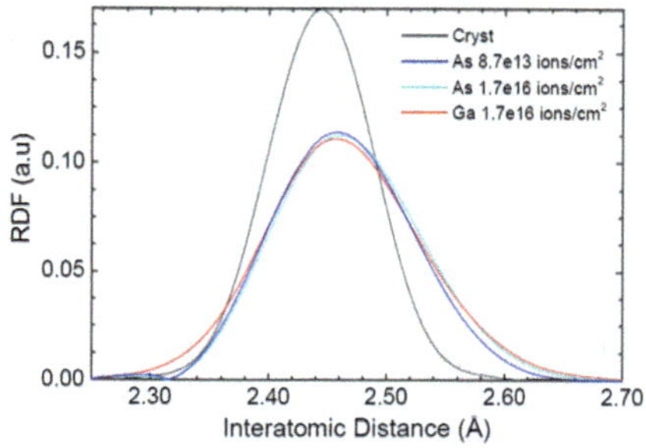

Fig. 8.5 Reconstructed inter-atomic distance distributions for a-GaAs measured at the Ga (*red*) and As (*blue* and *turquoise*) edges as a function of total ion fluence [33]. No ion fluence dependence at the Ga edge was measureable hence only one spectrum is shown. A c-GaAs spectrum (*black*) is included for comparison

Table 8.5 Structural parameters for c- and relaxed a-GaAs [33]

	c-GaAs/Ga edge	c-GaAs/As edge	a-GaAs/Ga edge	a-GaAs/As edge
CN (atoms)	4 (fixed)	4 (fixed)	3.79 ± 0.1	3.60 ± 0.1
Bondlength (Å)	2.448 (fixed)	2.448 (fixed)	2.467 ± 0.002	2.470 ± 0.002
DWF ($x10^{-3}$ Å^2)	1.8 (fixed)	1.8 (fixed)	4.4 ± 0.6	3.2 ± 0.6
C3 ($x10^{-4}$ Å^3)	0	0	0.6 ± 0.3	0.7 ± 0.3
C4 ($x10^{-5}$ Å^4)	0	0	0.6 ± 0.2	-0.8 ± 0.2

rectical predictions [4, 25]. Clearly greater disorder is associated with a Ga absorber than an As absorber. While non-quantified, Glover also concluded that chemical disorder *must* be present in a-GaAs.

8.4.4 Amorphous GaSb (a-GaSb)

GaSb in a crystalline form has a broad range of applications across the electronics (high-speed transistors and microwave devices), photonics (IR detectors, LEDs and lasers) and photovoltaic (thermophotovoltaic generators) sectors. At present, we are unaware of applications of GaSb in an amorphous form. None-the-less, fundamental studies of the atomic-scale structure of a-GaSb are of value to validate future theoretical models and for comparison with other amorphous Group III–V semiconductors.

While theoretical predictions of the structure of a-GaSb are lacking (to the best of our knowledge), the low ionicity of this material leads one to intuitively anticipate an above average wrong bonding fraction. A variety of characterisation techniques have been applied for experimental determinations of the structure of a-GaSb. An early report on a-GaSb (and other amorphous semiconductors) by Shevchik and Paul [28] using samples produced by sputtering and characterized with XRD concluded a CRN model was applicable to this material and the presence of five-fold rings implied the presence of wrong bonding. More recently, a-GaSb was prepared by Barkalov et al. [34] using solid-state amorphization of metastable high-pressure phases heated at atmospheric pressure. Their neutron diffraction (ND) measurements at atmospheric pressure yielded mean NN bondlength and CN values of 2.66 Å and 3.88 atoms, respectively, and a bondangle of 108.6°. The RDF of a-GaSb derived by Barkalov et al. from ND exhibited significantly less spread in the NN bondlength distribution than that extracted from XRD by Shevchik and Paul leading the former authors to conclude that samples prepared by solid-state amorphization were more chemically ordered than those prepared by sputtering. Samples of a-GaSb have also been produced in-situ in an electron microscope by Yasuda and Furuya [35] using high-energy (2 MeV) electron irradiation of c-GaSb. These authors reported chemical and structural disordering became more difficult to achieve as the irradiation temperature increased. Like SiC (see Chap. 7), amorphous phase formation in a-GaSb appears to proceed via the production of chemical disorder that subsequently stabilizes structural disorder. Dias da Silva et al. [36] brought multiple analytical techniques to bear on a-GaSb and from IR and Raman measurements concluded that the thermally-induced relaxation of a-GaSb proceeded via the reduction in both structural and chemical disorder.

While the advantages of ion implantation for the preparation of amorphous XAS samples have been much extolled in this chapter, this technique is not the most appropriate for a-GaSb (and a-InSb). Ion implantation of the antimonides induces not only a crystalline-to-amorphous transformation but also a concomitant continuous-to-porous transformation [37, 38], the latter resulting in a dramatic increase in surface area. In a-GaSb samples, oxidation of the surface/near-surface Ga component yields not only a significant oxide fraction but Sb–Sb homopolar bonding within the oxide layer [37]. This Sb–Sb homopolar bonding cannot be differentiated from that in the sub-surface a-GaSb and thus inhibits quantification of chemical disorder within the amorphous phase [37, 39].

XAS measurements of stoichiometric a-GaSb are scarce, potentially the result of non-stoichiometric samples produced by flash evaporation and/or the porosity produced by ion implantation. We thus focus on the report of Sapelkin et al. [40] using samples produced by solid-state amorphization. Figure 8.6 shows isolated EXAFS and FT EXAFS spectra of a-GaSb at both the Ga and Sb edges. Beating is readily apparent in the Ga edge EXAFS spectrum consistent with a mixed first shell which is indeed borne out with CN_{Ga} comprised of 3.4 ± 0.3 Sb atoms and 0.7 ± 0.5 Ga atoms. At the Sb edge, wrong bonding was "close to zero" but as the authors note the similarity in Ga-Sb and Sb–Sb bondlengths impedes quantification. Relative to c-GaSb, an increase in bondlength was measureable at both edges for a-GaSb. Examination of the FT EXAFS spectrum recorded at the Sb edge however reveals the presence of a 2NN scattering contribution which is atypical of the amorphous semiconductors and prompts the question as to the presence of a small but non-negligible crystalline component within the sample. These suspicions are further bolstered by a Cumulant Method analysis of this spectrum wherein the authors report C3 and C4 values "of amorphous (and crystalline) GaSb are close to zero".

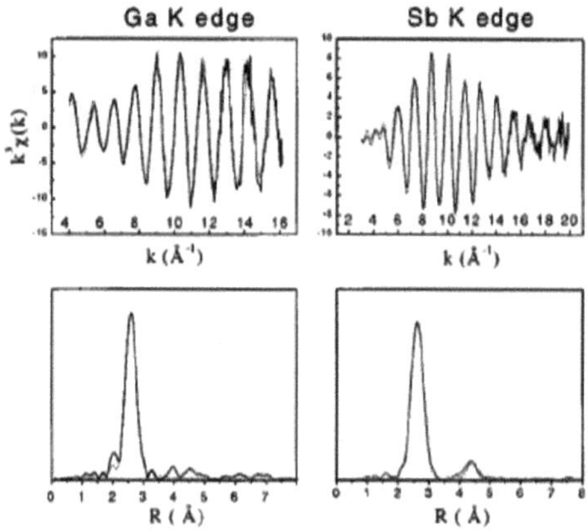

Fig. 8.6 (*Top*) isolated EXAFS and (*bottom*) FT EXAFS spectra for a-GaSb measured at 78 K (Ga) and 80K (Sb). (Reprinted figure with permission from [40]. Copyright (1997) by the American Physical Society. http://link.aps.org/abstract/PRB/v56/p11531). *Solid* and *dotted lines* are data and fits, respectively.

8.5 Amorphous In-Based Group III–V Semiconductors

8.5.1 Amorphous InN (a-InN)

Crystalline and poly-crystalline InN have a broad range of applications in electronics (high-power, high-frequency transistors), photonics (LEDs and laser diodes) and photovoltaics (solar cells). For the latter, InN is typically alloyed with GaN to yield a band-gap best matching the solar spectrum. XAS studies of crystalline $In_xGa_{1-x}N$ alloys are presented in Chap. 3. Reports on the atomic-scale structure of a-InN are scant but do include several theoretical and experimental contributions as discussed below.

Cai and Drabold [41] performed Density Functional Theory (DFT) calculations to generate structural models of a-InN. Table 8.6 lists the structural components garnered from a 250-atom model. Overall, short-range order is clearly maintained in the amorphous phase with both In and N atoms largely retaining four-fold coordination though a non-negligible fraction are either under- or over-coordinated. The bondangle distributions are centred close to the tetrahedral angle though are broad with FWHM of 28 and 34° for In-N-In and N-In-N, respectively. Partial pair correlation functions are shown in Fig. 8.7 [41]. Note the small N–N homopolar bonding contribution at a radial distance of ~1.5 Å. The authors concluded that a-InN was largely chemically ordered as supported by the very small fraction of odd-membered rings.

Amorphous InN has been produced experimentally by high-pressure cycling [42]. Raman Spectroscopy measurements showed a similarity in peak positions for samples of crystalline wurtzite and "recovered material" (considered amorphous based on XRD measurements). This led the authors [42] to suggest that short-range order was indeed maintained in amorphous InN.

To our knowledge, an XAS measurement of a-InN has surprisingly not been reported. If sufficiently present, chemical disorder should be readily identifiable given

Table 8.6 Structural parameters for a 250-atom model of a-InN [41]

	N_3 (%)	N_4 (%)	N_5 (%)	In_2 (%)	In_3 (%)	In_4 (%)	In_5 (%)	CN_N (atoms)	CN_{In} (atoms)
c-InN	0	100	0	0	0	100	0	4	4
a-InN	8.8	87.2	4	0.8	8	88	3.2	3.95	3.94

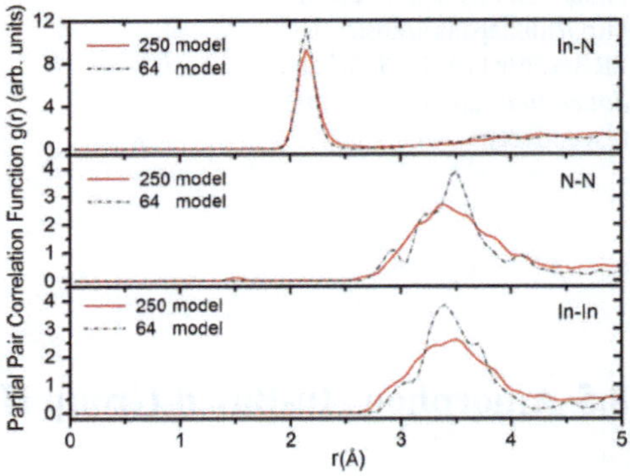

Fig. 8.7 Partial pair correlation functions for 250- and 64-atom models of a-InN. (Reprinted figure with permission from [41]. Copyright (2009) by the American Physical Society. http://link.aps.org/abstract/PRB/v79/p195204)

the anticipated differences in heteropolar and homopolar bondlengths and scattering strengths of In and N. Such a measurement would be of much value in assessing the validity of the theoretical model described above.

8.5.2 Amorphous InP (a-InP)

Crystalline InP has numerous applications in the electronics (high-frequency devices), photonics (laser diodes) and photovoltaics (solar cells) industries. Regarding photonics, READE [43] states "InP can provide all-in-one integrated functionality that includes light generation, detection, amplification, high-speed modulation and switching, as well as passive splitting, combining and routing. The same material can be used to make high-speed modulators, switches, amplifiers and detectors, or just passive waveguides for interconnecting these diverse devices." Given such widespread usage, investigations of the structure and properties of both c- and a-InP are clearly warranted.

Lewis et al. [5] used MD to examine the structure of a-InP with the amorphous phase attained via a quench from the melt. Partial and total radial distribution functions (RDFs) are shown in Fig. 8.8 and readily demonstrate the presence of homopolar bonding with P–P and In–In correlations apparent at 2.19 and 2.81 Å. The difference in homopolar bondlengths is consistent with the difference in size of the two elemental components and the proportion of homopolar bonding is 8.4 %. The heteropolar bondlength in the amorphous phase (2.51 Å) exceeds that in the crystalline phase (2.45 Å), a common trend for Group III–V semiconductors as we have seen

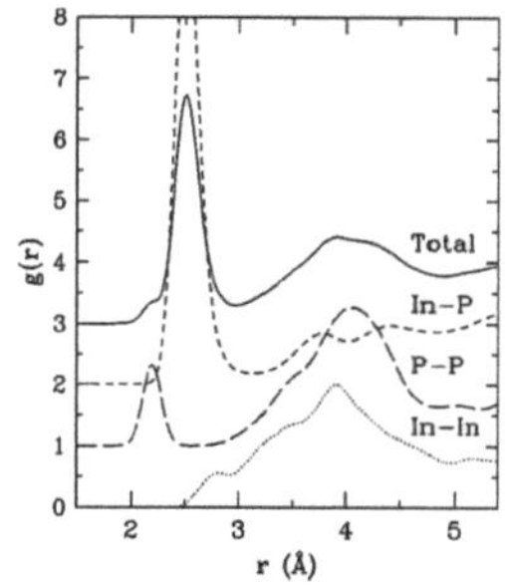

Fig. 8.8 Partial and total RDFs for a-InP. (Reprinted figure with permission from [5]. Copyright (1998) by the American Physical Society. http://link.aps.org/abstract/PRB/v57/p1594)

Table 8.7 Comparison of partial and total CNs for c- and a-InP [4]

	CN_{InIn} (atoms)	CN_{InP} (atoms)	CN_{PP} (atoms)	CN_{In} (atoms)	CN_{P} (atoms)	$CN_{average}$ (atoms)
c-InP	0	4	0	4	4	4
a-InP	0.34	3.91	0.38	4.25	4.29	4.27

previously. Table 8.7 compares NN CNs for c- and a-InP. Surprisingly, both elemental components in a-InP have CNs $>$ 4 atoms which thus explains the abnormal increase in density upon amorphization (despite the aforementioned increase in bondlength). Lewis et al. concluded that while heteropolar bonding is clearly dominant in a-InP, a significant fraction of homopolar bonding (and odd-membered rings) is indeed present.

Early measurements of the structure of a-InP using XAS (for example, Flank et al. [44], Udron et al. [45], Bayliss et al. [46] and Baker et al. [47–49]) and alternative techniques such as ED (Gheorghiu et al. [50]) were complicated by non-stoichiometric samples. None-the-less, the presence of homopolar bonding was typically reported. More recently, the application of ion implantation for XAS sample preparation eliminated non-stoichiometry-related artefacts and, as we demonstrate below, enabled detailed and exacting studies of the structural and vibrational properties of a-InP by Glover et al. [51], Glover [33], Azevedo et al. [52, 53], Ridgway et al. [19], Schnohr et al. [54, 55] and Schnohr [56]. Given the difference in scattering strengths of the two components, this material is ideal for the study of homopolar bonding and hence an assessment of the validity of various theoretical models. As a consequence, a-InP is now arguably the most thoroughly studied amorphous Group III–V semiconductor using XAS.

Glover et al. [33, 51] combined heterostructure growth, selective chemical etching and ion implantation to produce stoichiometric a-InP for XAS analysis. Figure 8.9 shows a FT and back FT EXAFS spectra recorded at the In K-edge with the fitted In-P and In–In scattering contributions shown. CN_{InP}, CN_{InIn} and CN_{In} values of 3.56, 0.60 and 4.16 atoms, respectively, were determined in addition to an increase in heteropolar bondlength upon amorphization. The homopolar fraction was $14 \pm 4\,\%$. Such results agree well with the theoretical predictions of Lewis et al.

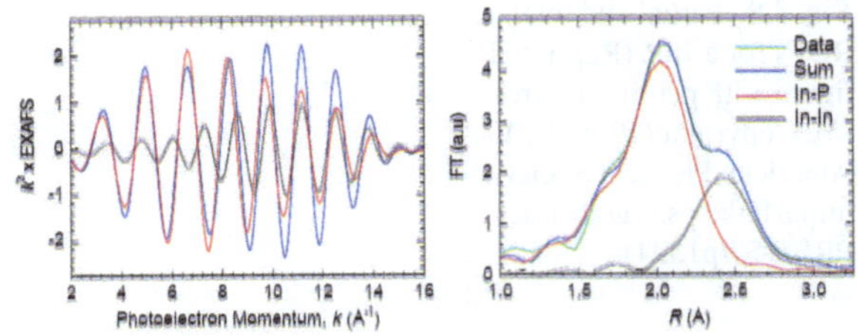

Fig. 8.9 FT EXAFS (*right*) and back FT EXAFS (*left*) spectra for a-InP recorded at the In K-edge, showing the separate In-P (*red*) and In–In (*black*) scattering contributions [33]

[5]. For ion fluences *above* that required for amorphization, an ion-fluence *dependence* was apparent in selected structural parameters, most notably the heteropolar bondlength that increased with an increase in ion fluence (not shown) [33]. This demonstrates the need for thermally-induced structural relaxation to attain the minimum-energy, ion-fluence-*independent* configuration of the amorphous phase. Note that Glover [33] also performed complementary measurements at the P K-edge where P–P homopolar bonding was readily apparent. Surprisingly, the P–P homopolar bonding fraction was approximately twice that measured for In–In at the In–K edge.

Azevedo et al. [52, 53] studied the subtle, temperature-dependent, structural-parameter variations in a-InP associated with the thermally-induced relaxation process. Figure 8.10 shows the partial CN, bondlength and DWF as a function of annealing temperature. Recrystallization of the amorphous phase, as inferred from the appearance of a 2NN scattering contribution in the FT EXAFS spectrum (not shown), began at a temperature between 200 and 237 °C. For temperatures $\leq$200 °C, Fig. 8.10 demonstrates the amorphous phase becomes more "crystalline-like" as the annealing temperature increases. Equivalently, CN_{InP} increases (approaches 4 atoms) at the expense of CN_{InIn} (approaches 0 atoms) while both the bondlength and DWF decrease. Structural relaxation in a-InP thus involves the reduction of chemical *and* structural disorder. Most importantly, Azevedo et al. clearly demonstrated that chemical disorder in the form of homopolar bonding is characteristic of the minimum-energy, structurally-relaxed configuration of a-InP and is thus not a preparation-specific artefact.

Unlike most semiconductors, crystalline InP can be rendered amorphous by either elastic or inelastic ion-solid interactions, the former characterized by ballistic atomic displacements while the latter consists of atomic excitation and ionization. Schnohr et al. [54, 56] compared the atomic-scale structure of a-InP amorphized by these two processes, examining both unrelaxed and relaxed states. Despite the vastly different means of forming the amorphous phase, no significant structural differences were measureable as attributed to a common melt and quench process governing amorphization.

Schnohr et al. [55, 56] subsequently extended the application of XAS to amorphous semiconductors well beyond the norm, performing a detailed analysis of the vibrational properties of a-InP using a correlated Einstein model and measurement-temperature-dependent experiments. Figure 8.11 shows the NN DWF for the heteropolar In-P bond in a-InP with results for c-InP included for comparison. The

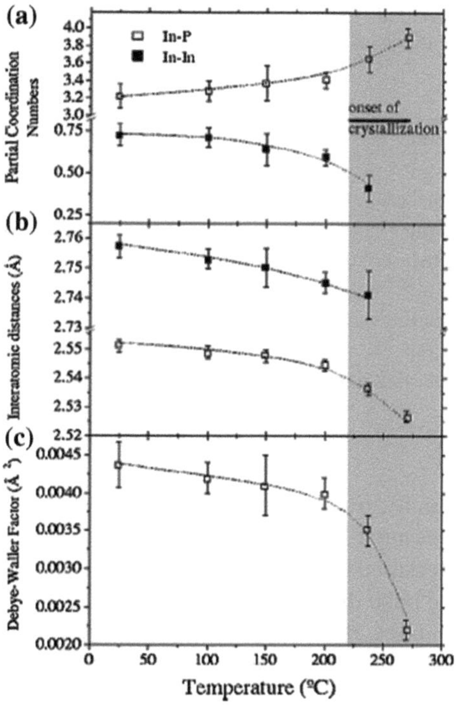

Fig. 8.10 Partial CNs, bondlength and DWF for a-InP as a function of thermal annealing temperature. (Reprinted figure with permission from [52]. Copyright (2003) by the American Physical Society. http://link.aps.org/abstract/PRB/v68/p115204)

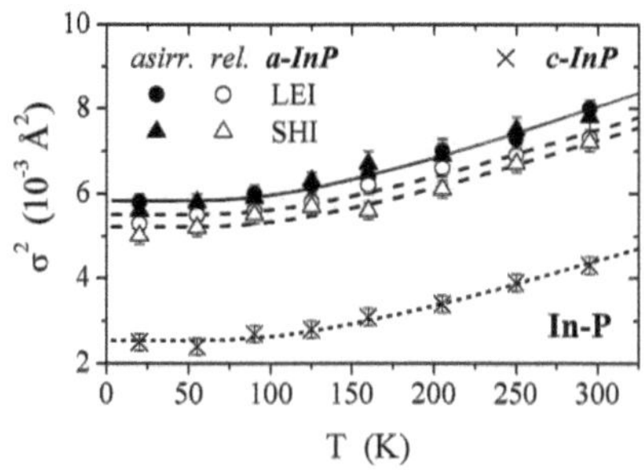

Fig. 8.11 DWF as a function of measurement temperature for a-InP formed by elastic (LEI) and inelastic (SHI) processes in unrelaxed (asirr) and relaxed (rel) states. For comparison, c-InP is included (*bottom*). (Reprinted figure with permission from [55]. Copyright (2009) by the American Physical Society. http://link.aps.org/abstract/PRB/v79/p195203)

temperature dependence for the two phases is clearly similar. As above, the relaxed samples exhibit a slightly lower, more crystalline-like DWF relative to unrelaxed samples. The offset in the c-InP and a-InP data is the result of differences in structural disorder which is greater in the amorphous material. Table 8.8 lists a selection of parameters. The Einstein temperatures of the crystalline and amorphous phases are similar, as above, though slightly higher in the former as consistent with stiffer bonding. Table 8.8 thus readily demonstrates that while thermal disorder in the two phases is comparable, structural disorder is much greater in the amorphous phase.

Table 8.8 Einstein temperatures (θ_E) and DWFs for a-InP formed by elastic and inelastic processes, in unrelaxed and relaxed states [55]

	$\theta_{E\ In-P}$ (K)	DWF_{In-P} (static) (10^{-3} Å^2)	$\theta_{E\ In-In}$ (K)	DWF_{In-In} (static) (10^{-3} Å^2)
c-InP	392 ± 8	0.0 ± 0.2		
a-InP (inelastic/unrelaxed)	370 ± 10	3.1 ± 0.2	260 ± 20	3.9 ± 0.3
a-InP (inelastic/relaxed)	382 ± 10	2.6 ± 0.2	240 ± 20	3.9 ± 0.3
a-InP (elastic/unrelaxed)	369 ± 10	3.1 ± 0.2	260 ± 20	3.0 ± 0.3
a-InP (elastic/relaxed)	384 ± 10	2.9 ± 0.2	270 ± 20	3.1 ± 0.3

Finally, from a consideration of the *uncorrelated* mean square displacement (MSD) measured by XRD and the *correlated* mean square relative displacement (MSRD) measured by EXAFS, Schnohr et al. [55] suggested that atomic vibrations parallel to the bond direction in a-InP were, like c-InP, correlated. Such behaviour is consistent with the approximately six-to-seven fold difference in the force constants for bond stretching and bond bending [57, 58] as discussed in detail in Chap. 6.

8.5.3 Amorphous InAs (a-InAs)

The small bandgap of c-InAs is such that applications of this material include IR detectors and terahertz radiation sources. In the former, advantages are achieved by operation in the photovoltaic mode where the low-frequency 1/f noise characteristic of materials such as HgCdTe is inhibited. We are unaware of specific applications of a-InAs.

Theoretical considerations of the structure of a-InAs include those of O'Reilly and Robertson [59] and more recently Wang et al. [60]. The DFT calculations performed by the latter authors yielded the structural parameters listed in Table 8.9. Again, four-fold coordination is retained in the amorphous phase with a small fraction of atoms either under- or over-coordinated.

Early non-XAS experimental studies of a-InAs focussed on the applicability of the CRN model and the identification of homopolar bonding. These included the

Table 8.9 CN and the fractions of three-, four- and five-fold coordinated atoms in a-InAs [60]

	$CN_{average}$ (atoms)	CN_3 (%)	CN_4 (%)	CN_5 (%)
c-InAs	4	0	100	0
a-InAs	3.99	1.56	97.66	0.78

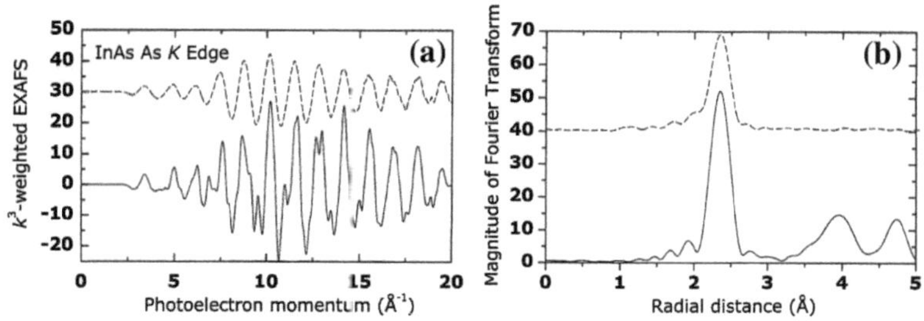

Fig. 8.12 **a** EXAFS and **b** FT EXAFS spectra comparing c- and a-InAs (*solid* and *dashed lines*, respectively). (Reprinted from [63]. Copyright (2002), with permission from Elsevier)

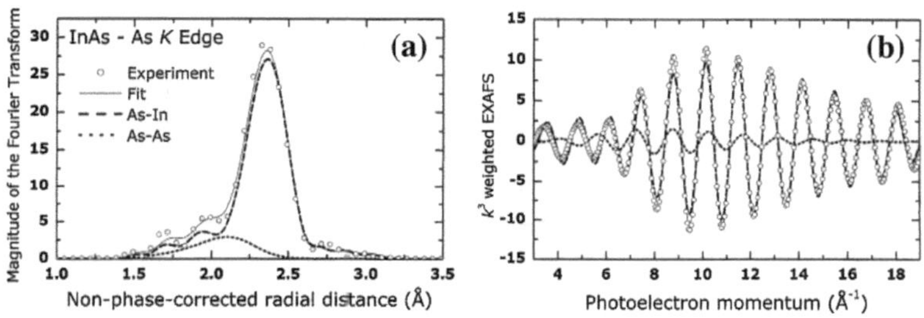

Fig. 8.13 **a** FT EXAFS and **b** back FT EXAFS spectra for a-InAs showing the separate As–In (*dashed line*) and As–As (*dotted line*) scattering contributions. (Reprinted from [63], Copyright (2002), with permission from Elsevier)

XRD and Raman measurements of Shevchik [61] and Wihl et al. [62], respectively. The latter authors reported the presence of As–As bonding.

More recently, Azevedo et al. [53, 63] have performed XAS measurements of a-InAs formed by ion implantation. This material is well suited to such analysis given both the In and As K-edges are generally readily accessible. Figure 8.12a [63] shows EXAFS spectra measured at the As K-edge comparing c- and a-InAs. For the latter, the single frequency oscillation is consistent with scattering from a single shell only and this is borne out in the FT EXAFS spectrum of Fig. 8.12b. As seen throughout this chapter, this trait is typical of amorphous semiconductors. In the FT and back FT spectra of Fig. 8.13a and b, respectively, the separate As–In and As–As contributions and the resulting fits are indicated.

Structural parameters extracted from analysis at both the As and In K-edges are listed in Table 8.10 [63]. The heteropolar bondlength increases upon amorphization. As anticipated from the two different frequencies visible in Fig. 8.13b, different homopolar bondlengths are measureable. Again, short-range order is retained in the amorphous phase with near four-fold coordination for both In and As atoms. Comparing the DWFs of c- and a-InAs, that of a-InAs increases by a factor of two upon amorphization. The disorder associated with homopolar bonds clearly exceeds that of heteropolar bonds by a similar factor. This combination of structural and chemical

Table 8.10 Structural parameters for c- and a-InAs [63]

	Crystalline InAs	Amorphous InAs
Bondlength In–As (Å)	2.617 ± 0.006	2.626 ± 0.006
Bondlength In–In (Å)		2.86 ± 0.04
Bondlength As–As (Å)		2.45 ± 0.02
CN_{InAs} (atoms)	4	3.3 ± 0.2
CN_{AsAs} (atoms)	0	0.5 ± 0.1
CN_{InIn} (atoms)	0	0.7 ± 0.2
DWF In–As (Å^2)	0.0019 ± 0.0002	0.0033 ± 0.0003
DWF In–As (Å^2)		0.007 ± 0.001
DWF In–In (Å^2)		0.008 ± 0.001

disorder makes a-InAs a proto-typical example of an amorphous III–V semiconductor while the ready identification of both forms of disorder again demonstrates the applicability and value of XAS to these materials systems.

8.5.4 Amorphous InSb (a-InSb)

Like c-GaSb, the small bandgap of c-InSb yields applications as IR detectors for use in thermal-imaging cameras and missile guidance systems while the high electron mobility enables applications that include transistors with high switching speed and low power consumption. A potential use of a-InSb may be in optical discs to achieve resolution beyond the diffraction limit [64].

Theoretical predictions of the atomic-scale structure of a-InSb include DFT calculations by Wang et al. [60, 65] and MD simulations by Rino et al. [66]. Figure 8.14 shows the total and partial pair correlation functions for a-InSb reported by Wang et al. [65]. In–In and Sb–Sb homopolar bonding is readily apparent at radial distances comparable to those of the pure metals. The fraction of homopolar bonds was estimated at ∼13 % with bondlengths $R_{Sb-Sb} > R_{In-Sb} > R_{In-In}$. Table 8.11 compares structural parameters derived from DFT [60, 65] and MD [66] plus an early

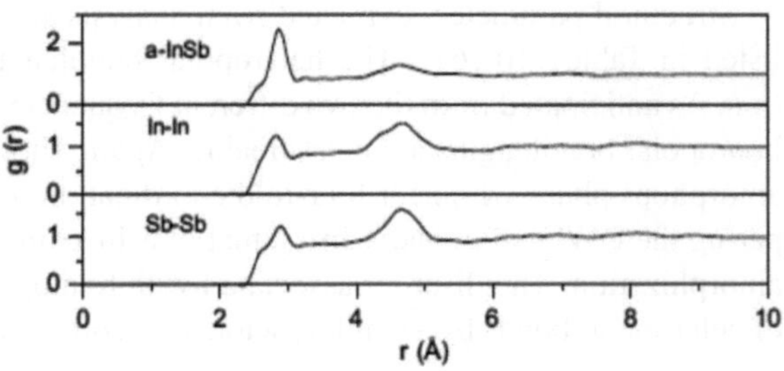

Fig. 8.14 Total and partial pair correlation functions for a-InSb. (Reprinted from [65], Copyright (2010), with permission from Elsevier)

Table 8.11 Structural parameters for ε-InSb

	$CN_{average}$ (atoms)	CN_3 (%)	CN_4 (%)	CN_5 (%)
c-InSb	4	0	100	0
a-InSb, Wang [60, 65]	3.97	2.73	97.27	0
a-InSb, Rino [66]	4.23			
a-InSb, Shevchik [28]	3.82			

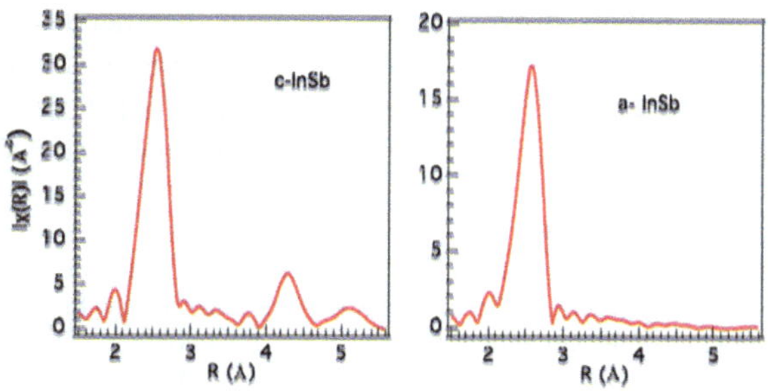

Fig. 8.15 FT EXAFS spectra measured at the In K-edge comparing (*left*) c- and (*right*) a-InSb. (Reproduced with permission from [67], Copyright 2010, AIP Publishing LLC)

experimental XRD [28] result. Again, near four-fold coordination is retained in the amorphous phase of a-InSb.

In addition to the XRD measurement referred to above, other non-XAS determinations of the structure of a-InSb include Raman measurements by Wihl et al. [62]. Therein, homopolar bonding (Sb–Sb) in a-InSb was identified (though not quantified).

To our knowledge, the only XAS measurements of the structure of a-InSb are those of Krbal et al. [67]. Such an experiment is hampered by the same difficulties experienced for a-GaAs. Though the K-edges of both absorbers are readily accessible, their similarity in scattering strength impedes determinations of the homopolar bonding fraction. The FT EXAFS spectra measured at the In K-edge and shown in Fig. 8.15 are once again characteristic of a crystalline and amorphous semiconductor comparison. While scattering from multiple shells is apparent for the crystalline phase, scattering is limited to a single shell for the amorphous phase. Near four-fold coordination was retained in the amorphous phase and the bondlength was greater than that of the crystalline phase. Superior fits were achieved with the addition of a 5–10 % homopolar bonding component, indeed suggesting the presence of In–In and Sb–Sb bonding. Comparing crystalline and amorphous material, the authors reported a higher density for the latter, an atypical result for semiconductors (excluding a-InP) and an unanticipated result given the CN and bondlength data. The higher density for the amorphous phase was attributed to the presence of homopolar bonding. We note that Krbal et al. also reported Einstein temperature determinations ($\theta_E^{crystalline} = 236 \pm 3$ K and $\theta_E^{amorphous} = 221 \pm 2$ K) that were consistent with those of other Group III–V semiconductors presented in this chapter where $\theta_E^{crystalline} > \theta_E^{amorphous}$ as attributed to weaker bonding in the amorphous phase.

8.6 Summary

XAS measurements of the amorphous Group III–V semiconductors have been presented with the inclusion of additional theoretical and experimental results for comparison. Disorder in these materials is both structural and chemical. The former yields an increase in bondlength and Debye-Waller factor and a decrease in coordination number relative to the crystalline phase while the latter, in the form of homopolar bonding, comprises ~5–15 % of all bonds. While thermally-induced relaxation reduces both structural and chemical disorder as the amorphous-phase structure approaches the minimum-energy configuration, chemical disorder appears intrinsic to these materials. With the capability of characterizing both forms of disorder, XAS is clearly well suited for the study of the amorphous Group III–V semiconductors.

Acknowledgments The author gratefully acknowledges C.J. Glover, C.S. Schnohr, G. de M. Azevedo, G.J. Foran, K.M. Yu, S.M. Kluth and P. Kluth for significant contributions, fruitful collaborations and memorable times at XAS beamlines around the world in our pursuit of the atomic-scale structure of the amorphous Group III–V semiconductors. The author thanks the Australian Research Council, Australian Synchrotron Research Program and Australian Synchrotron for financial support. Sample fabrication was enabled through access to the ACT Node of the Australian National Fabrication Facility.

References

1. W.H. Zachariasen, J. Am. Chem. Soc. **54**, 3841 (1932)
2. D.E. Polk, J. Non-Cryst. Solids **5**, 365 (1971)
3. G.A.N. Connell, R.J. Temkin, Phys. Rev. B **9**, 5323 (1974)
4. N. Mousseau, L.J. Lewis, Phys. Rev. B **56**, 9461 (1997)
5. L.J. Lewis, A. De Vita, R. Carr, Phys. Rev. B **57**, 1594 (1998)
6. S. Decoster, C.J. Glover, B. Johannessen, R. Giulian, D.J. Sprouster, P. Kluth, L.L. Araujo, Z. Hussain, C. Schnohr, H. Salama, F. Kremer, K. Temst, A. Vantomme, M.C. Ridgway, J. Synchrotron Radiat. **20**, 426 (2013)
7. P. Stumm, D.A. Drabold, Phys. Rev. Lett. **79**, 677 (1997)
8. B. Cai, D.A. Drabold, Phys. Rev. B **84**, 075216 (2011)
9. J. Nord, K. Nordlund, J. Keinonen, Phys. Rev. B **68**, 184104 (2003)
10. M. Ishimaru, Y. Zhang, X. Wang, W.-K. Chu, W.J. Weber, J. Appl. Phys. **109**, 043512 (2011)
11. M.C. Ridgway, S.E. Everett, C.J. Glover, S.M. Kluth, P. Kluth, B. Johannessen, Z.S. Hussain, D.J. Llewellyn, G.J. Foran, G.M. de Azevedo, Nucl. Instrum. Methods B **250**, 287 (2006)
12. M. Katsikini, F. Pinakidou, E.C. Paloura, W. Wesch, Appl. Phys. Lett. **82**, 1556 (2003)
13. P. Deenapanray, M. Petravic, K.-J. Kim, B. Kim, G. Li, Appl. Phys. Lett. **83**, 4948 (2003)
14. H. Kubota, T. Matsumoto, T. Hirayu, T. Fujiyoshi, R. Miyagawa, K. Miyahara, M. Onuki, Sol. Energy Mater. Sol. Cells **35**, 353 (1994)
15. M.-L. Theye, A. Gheorghiu, H. Launois, J. Phys. C: Solid State Phys. **13**, 6569 (1980)
16. D. Udron, A.-M. Flank, A. Gheorghiu, P. Lagarde, M.-L. Theye, Philos. Mag. Lett. **59**, 9 (1989)
17. N. Elgun, S.J. Gurman and E.A. Davis, J. Phys.: Cond. Matter **4**, 7759 (1992)
18. N. Elgun, S.J. Gurman and E.A. Davis, J. Phys.: Cond. Matter **12**, 4723 (2000)
19. M.C. Ridgway, G. de M. Azevedo, C.J. Glover, K.M. Yu, G.J. Foran, Nucl. Instrum. Methods B **199**, 235 (2003)
20. A. Madan, M.P. Shaw, *The Physics and Applications of Amorphous Semiconductors* (Elsevier, 1988), p. 166

21. E. Fois, A. Selloni, G. Pastore, Q.-M. Zhang, R. Car, Phys. Rev. B **45**, 13378 (1992)
22. C. Molteni, L. Colombo, L. Miglio, Europhys. Lett. **24**, 659 (1993)
23. C. Molteni, L. Colombo, L. Miglio, Phys. Rev. B **50**, 4371 (1994)
24. H. Seong, L.J. Lewis, Phys. Rev. B **53**, 4408 (1996)
25. N. Mousseau, L.J. Lewis, Phys. Rev. Lett. **78**, 1484 (1997)
26. I. Ebbsjo, R.K. Kalia, a. Nakano, J.P. Rino, P. Vashishta, J. Appl. Phys. **87**, 7708 (2000)
27. D. Udron, M.-L. Theye, D. Raoux, A.-M. Flank, P. Lagarde, J.-P. Gaspard, J. Non-Cryst. Solids **131**, 137–138 (1991)
28. N.J. Shevchik, W. Paul, J. Non-Cryst. Solids **13**, 1 (1973)
29. R.J. Temkin, Solid State Commun. **15**, 1325 (1974)
30. J.A. Del Cueto, N.J. Shevchik, J. Phys. C **11**, L829 (1978)
31. S.H. Baker, M.I. Manssor, S.J. Gurman, S.C. Bayliss, E.A. Davis, J. Non-Cryst. Solids **144**, 63 (1992)
32. M.C. Ridgway, C.J. Glover, G. Foran, K.M. Yu, J. Appl. Phys. **83**, 4610 (1998)
33. C.J. Glover, Ph.D. thesis, Australian National University, 2002
34. O.I. Barkalov, A.I. Kolesnikov, V.E. Antonov, E.G. Ponyatovsky, U. Dahlborg, M. Dahlborg, A. Hannon, Phys. Status Solidi B **198**, 491 (1996)
35. H. Yasuda, K. Furuya, Philos. Mag. A **80**, 2355 (2000)
36. J.H. Dias da Silva, J.I. Cisneros, M.M. Guraya, G. Zampieri, Phys. Rev. B **51**, 6272 (1995)
37. S.M. Kluth, B. Johannessen, P. Kluth, C.J. Glover, G.J. Foran, M.C. Ridgway, Nucl. Instrum. Methods B **238**, 264 (2005)
38. S.M. Kluth, B. Johannessen, P. Kluth, C.J. Glover, G.J. Foran, M.C. Ridgway, Nucl. Instrum. Methods B **242**, 640 (2006)
39. P. Kluth, S.M. Kluth, B. Johannessen, C.J. Glover, G.J. Foran, M.C. Ridgway, J. Appl. Phys. **110**, 113528 (2011)
40. A.V. Sapelkin, S.C. Bayliss, A.G. Lyapin, V.V. Brazhkin, A.J. Dent, Phys. Rev. B **56**, 11531 (1997)
41. B. Cai, D.A. Drabold, Phys. Rev. B **79**, 195204 (2009)
42. S.V. Ovsyannikov, V.V. Shchennikov, A.E. Karkin, A. Polian, O. Briot, S. Ruffenach, B. Gill, M. Moret, Appl. Phys. Lett. **97**, 032105 (2010)
43. http://www.reade.com/products/36-phosphide-compounds-powder/261-indium-phosphide-indiumiii-phosphide-inp-indium-phosphide-crystalline-indium-phosphide-lump-indium-phosphide-pieces-cas22398-80-7
44. A.-M. Flank, P. Lagarde, D. Udron, S. Fisson, A. Gheorghiu, M.-L. Theye, J. Non-Cryst. Solids **435**, 97 (1987)
45. D. Udron, A.-M. Flank, P. Lagarde, D. Raoux, M.-L. Theye, J. Non-Cryst. Solids **150**, 361 (1992)
46. S.C. Bayliss, S.H. Baker, J.S. Bates, S.J. Gurman, R. Bradley, E.A. Davis, J. Non-Cryst. Solids **164**, 143 (1993)
47. S.H. Baker, S.C. Bayliss, S.J. Gurman, N. Elgun, J.S. Bates, E.A. Davis, J. Phys.: Condens. Matter **5**, 519 (1993)
48. S.H. Baker, S.C. Bayliss, S.J. Gurman, N. Elgun, B.T. Williams, E.A. Davis, J. Non-Cryst. Solids **169**, 111 (1994)
49. S.H. Baker, S.C. Bayliss, S.J. Gurman, N. Elgun, E.A. Davis, J. Phys.: Condens. Matter **8**, 1591 (1996)
50. A. Gheorghiu, M. Ouchene, T. Rappeneau, M.-L. Theye, J. Non-Cryst. Solids **59–60**, 621 (1983)
51. C.J. Glover, M.C. Ridgway, K.M. Yu, G.J. Foran, T.W. Lee, Y. Moon, E. Yoon, Appl. Phys. Lett. **74**, 1713 (1999)
52. G.M. de Azevedo, C.J. Glover, M.C. Ridgway, K.M. Yu, G.J. Foran, Phys. Rev. B **68**, 115204 (2003)
53. G.M. de Azevedo, C.J. Glover, K.M. Yu, G.J. Foran, M.C. Ridgway, Nucl. Instrum. Methods B **206**, 1024 (2003)
54. C.S. Schnohr, P. Kluth, A.P. Byrne, G.J. Foran, M.C. Ridgway, Phys. Rev. B **77**, 073204 (2008)

55. C.S. Schnohr, P. Kluth, L.L. Araujo, D.J. Sprouster, A.P. Byrne, G.J. Foran, M.C. Ridgway, Phys. Rev. B **79**, 195203 (2009)
56. C.S. Schnohr, Ph.D. thesis, Australian National University, 2009
57. R.M. Martin, Phys. Rev. B **1**, 4005 (1970)
58. A.-B. Chen, A. Sher, Phys. Rev. B **32**, 3695 (1985)
59. E.P. O'Reilly, J. Robertson, Phys. Rev. B **34**, 8684 (1986)
60. L. Wang, X. Chen, W. Lu, Y. Huang, J. Zhao, Solid State Commun. **149**, 638 (2009)
61. N.J. Shevchik, Phys. Rev. Lett. **31**, 1245 (1973)
62. M. Wihl, M. Cardona, J. Tauc, J. Non-Cryst. Solids **8–10**, 172 (1972)
63. G.M. de Azevedo, M.C. Ridgway, K.M. Yu, C.J. Glover, G.J. Foran, Nucl. Instrum. Methods B **190**, 851 (2002)
64. G. Pilard, C. Fery, L. Pacearescu, H. Hoelzemann, S. Knappmann, Jpn. J. Appl. Phys. **48**, 03064 (2009)
65. L. Wang, X.S. Chen, Y. Huang, W. Lu, J.J. Zhao, Phys. B **405**, 2481 (2010)
66. J.P. Rino, D.S. Borges, S.C. Costa, J. Non-Cryst. Solids **348**, 17 (2004)
67. M. Krbal, A.V. Kolobov, B. Hyot, B. Andre, P. Fons, R.E. Simpson, T. Uruga, H. Tanida, J. Tominaga, J. Appl. Phys. **108**, 023506 (2010)

Chapter 9
Semiconductors Under Extreme Conditions

Andrea Di Cicco and Adriano Filipponi

Abstract The importance of exploring physical properties and phase transitions of semiconductors under extreme conditions of high pressure P and/or high temperature T are illustrated using basic thermodynamic concepts. The specific role of XAS in the investigation of amorphous or liquid systems is addressed with emphasis on its unique sensitivity to the short range order. Current techniques for the performance of high-pressure and/or high temperature XAS measurements are briefly reviewed. Examples of XAS results on simple systems such as liquid and amorphous selenium and germanium are discussed.

9.1 Introduction

The richness and variety of polymorphic phase diagrams displayed by several elemental and compound semiconductors is a manifestation of the complexity of the interatomic interactions. Understanding the properties and the nature of the various existing stable or metastable condensed matter states deserves therefore a particular interest and importance.

The relevant parameters that are more easily controllable from the exterior to define the sample thermodynamic state are the temperature T and hydrostatic pressure P. Here we assume that the sample is in thermodynamic equilibrium which implies that all thermodynamic parameters are well defined and do not vary in time

A. Di Cicco (✉)
Physics Division, School of Science and Technology, Università di Camerino,
Via Madonna delle Carceri 9, 62032 Camerino, Italy
e-mail: andrea.dicicco.@unicam.it

A. Filipponi
Dipartimento di Scienze Fisiche e Chimiche, Università degli Studi dell'Aquila,
67100 Coppito, AQ, Italy
e-mail: adriano.filipponi@aquila.infn.it

C.S. Schnohr and M.C. Ridgway (eds.), *X-Ray Absorption Spectroscopy of Semiconductors*, Springer Series in Optical Sciences 190,
DOI 10.1007/978-3-662-44362-0_9

since the statistical distributions for the microscopic variables have reached a stationary equilibrium profile. In this conceptual framework stability issues are conveniently discussed in terms of the Gibbs free energy that for (P, V, T) thermodynamic systems is defined as:

$$G(T, P) = U(T, V) - TS(T, V) + PV(T, P) \tag{9.1}$$

where $U(T, V)$ is the internal energy, $S(T, V)$ is the entropy of the thermodynamic system and $V(T, P)$ is the volume expressed in terms of the independent equilibrium state variables through the equation of state. The total differential of G is $dG = -SdT + VdP$. The Gibbs free energy has the property that $\Delta G \leq 0$ in spontaneous processes occurring in contact with an environment at constant (T, P). If two or more competing phases exist for a given system the corresponding Gibbs free energies $G_i(T, P)$ should be compared to establish which phase is energetically favored and stable. For the stable phase i: $G_i(T, P) < G_j(T, P), \forall j \neq i$ and all other phases are metastable. By performing the above comparison as a function of T and P one obtains information on the phase diagram of the substance and in particular important indications on the boundaries of the stability regions for each phase. The variation of one or both thermodynamic state parameters through one of these boundaries induces a phase transition in the sample.

A qualitative understanding of the role of the thermodynamic state parameters T and P for determining phase transition boundaries can be appreciated looking at (9.1). By increasing the temperature parameter T the term $-TS$ contributes to lowering G in a way that is more effective for the states with a greater entropy S. Similarly the increase of pressure P favors the phases with a smaller specific volume through the term $+PV$. Thus the change of the state parameters induces phase transitions which allow one to access states of greater entropy, when T is increased, or greater atomic packing when P is increased.

The above considerations explain the importance of the experimental capability to vary the environmental parameters T and P in a sufficiently wide range. This opens up a two dimensional world with respect to the standard experiments at ambient conditions ($T \simeq 300$ K, $P \simeq 0.1$ MPa). The existence, nature, characteristics and stability (or metastability) boundaries of the different existing phases provide important information on the nature of the interaction and improves our understanding of the system and prediction capabilities. This (T, P) range usually involves the so called extreme thermodynamic conditions which are the subject of the present chapter.

For condensed phases where the volume is mostly determined by the atomic sizes, a useful P range to induce phase transitions extends to at least several tens (if not hundreds) of GPa, which is a range reachable with current high-pressure experimental techniques. Semiconducting phases are often associated with relatively open crystal structures in which the atoms are linked by strongly directional covalent bonds. In these cases a relatively moderate pressure in the few GPa range may induce transitions towards slightly more compact structures possibly with a lower gap or metallic nature. The entropic difference between solid and liquid phases establishes the melting line of the solid phases of the substance which can be reached by suitably tuning the temperature parameter.

In general higher temperatures favor higher entropy states associated with a larger degree of disorder. These include also higher disordered crystalline or amorphous covalent structures. Disordered structures are often metastable states generated by vapor deposition or high pressure amorphization processes where the limited diffusion prevents the system to reach the stable crystalline state and a quenched disorder is obtained. As a result elemental or compound semiconductor often display highly polymorphic phase diagrams with a variety of stable and metastable phases occurring in the experimentally accessible thermodynamic state range.

Investigating the physical properties of the observed phases provides basic fundamental information. This knowledge usually includes structural, dynamical and electronic properties. X-ray diffraction clearly represents a fundamental technique to probe structural properties and is particularly suited for single crystal samples or crystalline powder where the long range crystalline order produces Bragg reflections, but also for disordered phases where the diffraction pattern associated with the structure factor provides information on the radial distribution function. In this context a major role is also played by the X-ray absorption spectroscopy (XAS) [1]. The X-ray absorption cross-section at and above the main core hole absorption edges contains important information on: (1) the local structural order around the photoabsorber atomic species in terms of short range properties of the radial distribution function and possibly from higher order distribution functions through the multiple scattering effects; (2) the transition cross-section to low-lying unoccupied electronic states revealing changes in their density and matrix elements.

The main relevant characteristics of this spectroscopy in this context are: (a) conceptually similar experiments and data types are collected independently of the sample state: (the substantial blindness to long range crystalline order makes crystalline and disordered phases data relatively similar and directly comparable); (b) The atomic selectivity allows one to probe the sample properties even if it is confined in a cell and the cell just contributes with a smooth absorption background. In the case of compounds/alloys, also in the diluted limit, the element selectivity allows one to measure structural properties around minority atomic species (also in the 1 % or lower concentration range) where all other structural techniques will fail.

For all the above reasons XAS has played and will play an important specific role in the investigation of amorphous or liquid systems [2] for its sensitivity to the short range structural order including potential information on triplet correlations and sensitivity to important electronic properties. In the following paragraphs we shall briefly review some of the techniques used to obtain XAS spectra under extreme conditions and a few selected examples of applications to elemental disordered semiconductors. The necessary short account here contained is mainly related to the direct experience of the authors and can not be exhaustive. Several other important XAS applications to semiconducting materials under pressure can be found in the cited literature and references there contained.

9.2 Experimental Set-Ups at Scanning Energy Beamlines

Energy scanning X-ray Absorption Spectroscopy beamlines still represent the option which guarantees the collection of spectra with the best signal to noise ratio and energy resolution performances. The best and most stable instruments are those which exploit a continuous spectrum synchrotron radiation source such as the one generated by a bending magnet or a wiggler. The energy scan is performed by rotating a double crystal monochromator and the choice of the Bragg reflection determines the flux-resolution trade-off. X-ray mirrors are often used to improve focalization conditions and for harmonic rejection purposes together with possible crystals detuning. Two fundamental aspects involve the set-ups usable for extreme conditions: (1) the beamline has to be compatible with sample environment (cell/devices) able to generate the required extreme conditions; (2) the beamline has to be equipped with suitable ancillary techniques for the required sample characterization.

High-temperature high-vacuum experiments are relatively easy to perform with a variety of resistive heating furnaces [2, 3] and the only concerns regard T homogeneity and possible sample reaction/contamination. High pressure conditions on the other hand present considerable experimental challenges. In order to reach ideal hydrostatic conditions the sample has to be confined in an environment made of a pressure transmitting medium upon which external forces are acted in the presence of suitable X-ray windows. Both windows and pressure transmitting medium materials (in addition to the sample) contribute to the absorption level and may reduce considerably the useful flux increasing the relative harmonic contamination.

However, the most important limitation regards the usage of X-ray windows. When a single crystal window, such as diamond, is inserted in the beam at a fixed orientation a sequence of Bragg reflections will occur at discrete energy values which satisfy the Ewald sphere construction. As result a considerable part of the beam intensity will be deviated from the transmitted beam direction for a set of discrete energy values through the scan, appearing as additional narrow absorption peaks in the spectrum. Apart from relatively narrow low energy regions of usability, absorption spectra in the presence of single crystal windows are spoiled by the presence of these artifacts. As a result diamond anvils cells are hardly usable on these beamlines and suitable experimental techniques based on polycrystalline confinement gasket materials are recommended. Recent advances in the usage of nanocrystalline diamonds and the possibility of extreme focusing in XAS beamlines could soon overcome some of those drawbacks (see for example [4–6]).

The most suitable technique for moderate pressure large volume investigations was developed as a modification of the original toroidal gasket technique [7] and exploits the Paris-Edinburgh press [8] with the usage of doubly conical gaskets as first employed at LURE [9] and successively further developed at ESRF-BM29 [10, 11]. A typical sample assembly for high-pressure XAS measurements using those devices is shown in Fig. 9.1. The reduced gasket lateral size is required to increase X-ray transmission and limits the maximum pressure to about 6–8 GPa. Gaskets are manufactured from sintered rods of a Boron-epoxy mixture which guarantees

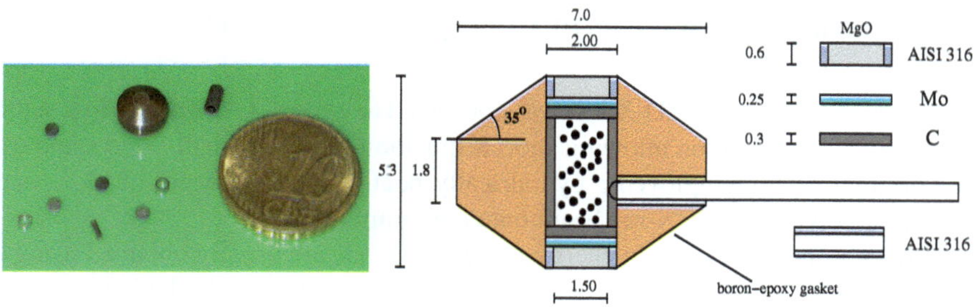

Fig. 9.1 Schematic drawing of the (cylindrically symmetric) sample section showing a 7 mm diameter boron-epoxy gasket, graphite heater (C), and its internal cavity filled with a sample. The dimensions in mm refer to the parts as assembled prior to pressurization. The real parts are shown in the picture

a sufficient strength and low X-ray absorption. The sample is placed in a mm-size internal cylindrical cavity surrounded by a thin graphite cylinder used for resistive heating. The available volume exceeds the one normally required for X-ray absorption of bulk specimens which therefore have to be diluted in powder form with a soft low-absorbing inert material such as for instance boron-nitride powder. Pressure is applied along the vertical axis by two anvils with WC (tungsten carbide) insets. The gasket progressively flows and widens with pressure up to about twice its original size still maintaining a gap of a few tenths of mm between the anvils for the X-ray probe. The low energy limit of this technique is typically the Cu K-edge, achievable with small gaskets and focalized beam. Several fourth period elements, semiconducting or involved in compound semiconductors, can be conveniently investigated.

Experiments under extreme conditions also require additional surrounding information and in particular independent probes of the thermodynamic sample state (P, T). For the high pressure range this is a known issue related with the definition of the pressure scale and development of suitable calibration techniques. The coupling of P and T and the difficulty of an in-situ implementation of the ruby fluorescence method make this a relatively complicated issue. Thermocouples can often be used for the temperature measurement. Knowledge of the thermodynamic parameters is moreover not always sufficient to establish unambiguously the sample state in particular when metastable states occur. Crossing a phase transition boundary by increasing or decreasing pressure often results in an hysteresis loop owing to the slow crystallization kinetic of the new phase especially at low temperature. An undercooled liquid state may persist well below the equilibrium melting temperature if crystalline nucleation is hampered by a large liquid/crystal interface energy. In all these cases a simultaneous diffraction check of the nature of the crystalline sample components is recommended [10, 11]. While taking XAS spectra at selected temperatures does not always allow a clear identification of the sample phase, on the other hand the phase sensitivity of XAS can be exploited by monitoring the sample absorption at fixed energy on a significant spectral feature during continuous temperature scans [12] where discrete jumps at first order phase transitions allow one to pin the transition temperatures very effectively.

9.3 Experimental Set-Ups at Energy-Dispersive Beamlines

Energy-dispersive XAS (ED-XAS) was conceived to collect XAS spectra in a static geometrical configuration allowing short acquisition times on small-sized samples. The original design employed a cylindrically bent triangular crystal to focus and disperse a polychromatic X-ray beam onto the sample [13].

X-ray absorption spectra in energy-dispersive mode are usually obtained using a linear photodiode array detector or a CCD measuring successively the incident and transmitted flux of photons. The beam position (pixels) is successively correlated to energy by means of standards.

Several synchrotron radiation facilities have developed ED-XAS dedicated beamlines (see [14] and references therein) and the method was applied quite successfully in various fields including studies under high-pressure conditions. In fact, ED-XAS was soon found to be well suited for studies at extreme conditions of pressure using a diamond-anvil cell (DAC) [15, 16], where the strongly focusing crystal and the absence of movements provide the necessary small and stable focal spot (see Fig. 9.2). Bragg reflections from the diamond anvils, always affecting the quality of the XAS spectra, can be usually removed from the energy range of interest by suitable positioning of the DAC. The pressure range that can be covered has a natural upper limit defined by the size of the diamond culet, which is in turn limited by the size of the focal spot.

Several improvements in the optical scheme (see for example [14] and references therein) and in the DAC technology (diamonds of improved shape, drilled culets,

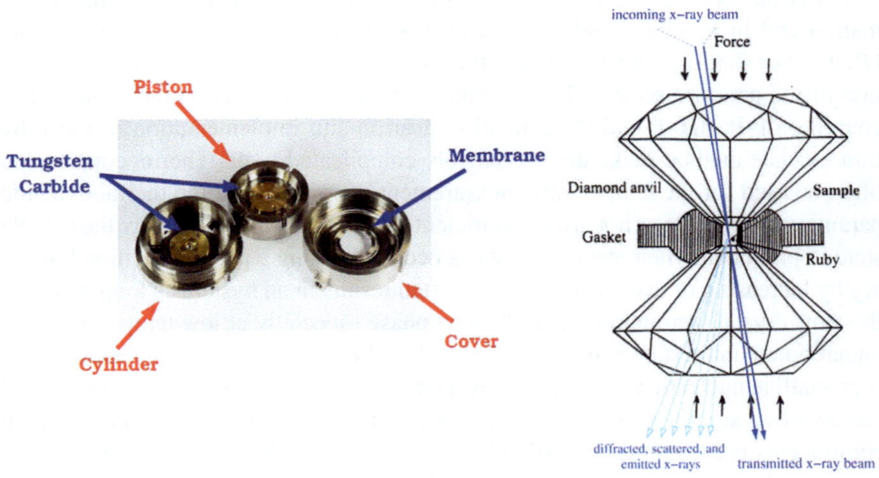

Fig. 9.2 *Left* photo of a membrane-DAC. Pressure is generated inflating a gas (helium) into the membrane through a capillary. *Right* typical set-up for X-ray measurements under pressure: an incoming X-ray beam is transmitted through the diamonds (anvils) and sample region so that transmitted and scattered X-rays (shown only for selected angles for clarity) can be measured. Depending on the maximum pressure achievable, sample size is in the 10–150 μm range

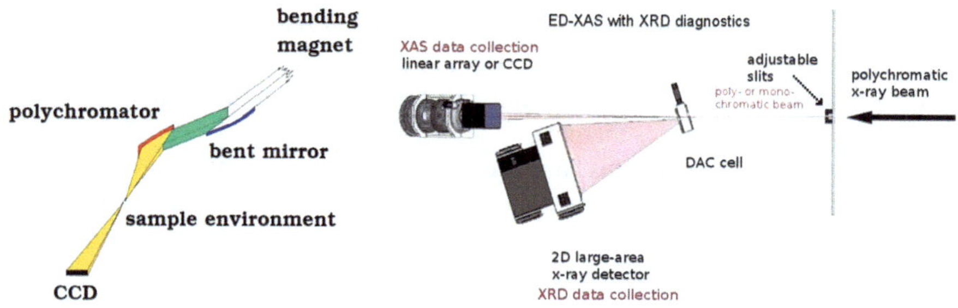

Fig. 9.3 *Left* sketch of a typical ED-XAS beamline. The *white* beam coming from a storage ring bending magnet is vertically focused by a bent mirror and dispersed bending a crystal. The sample is placed at the focal point and the transmitted beam is collected by a position-sensitive detector. *Right* sketch of a ED-XAS set-up including in-situ XRD diagnostics. The slit can be adjusted to provide a quasi-monochromatic beam used to collect X-ray diffraction at the selected energy

nanocrystalline diamonds) [4, 17] allowed to reach pressures up to the Mbar regime and extend XAS measurements even at relatively low photon energies (<7 keV).

The conventional optical scheme for ED-XAS is shown in Fig. 9.3. The spectrometer is installed on a divergent synchrotron source. A bending magnet, for example, naturally yields the necessary horizontal divergence (typically a few mrads) for obtaining a wide energy dispersion. In this way, using reasonable radii of curvature of the polychromator crystal, it is possible to collect XAS spectra extended hundreds of eV above a selected core edge. XRD patterns collected in the same conditions, without removing the DAC cell, are a useful, often mandatory, diagnostics for high-pressure XAS measurements. The set-up shown in Fig. 9.3, using a large-area X-ray detector (typically an image-plate) and an adjustable slit selecting poly- or quasi mono-chromatic incident beams, is implemented in several beamlines and can be easily used for this purpose (see for example [18] and references therein).

9.4 XAS of Amorphous and Liquid Se at High Pressures

Pure selenium is a semiconductor both in its stable solid and liquid phases at ambient pressure, a property shared with only few other materials. It has peculiar structural properties as it belongs to a class of polymorphic systems (S, Se, Te) where the interplay between strong covalent and metallic bonding for varying external conditions is particularly important. Polymer-like chains can be found in different polymorphs of these elemental substances in their various aggregation states (crystalline, liquid, amorphous).

The application of external pressure in a class of substances including elemental semiconductors (like Se, Si, Ge) results in unusual and interesting phenomena, such as pressure-induced polyamorphism, pressure-induced amorphization or nucleation of crystalline seeds. The nature of the behavior of amorphous solids under pressure is complicated by their own metastability and presence of multiple disordered structures, by the appearance of stable and metastable crystalline phases, and by the possibility of irreversible relaxation of their properties and structure (see for example [19] and references therein).

Selenium and related systems have been subject of many scientific investigations, including XAS, aimed to elucidate the structural properties of the different phases obtained at high pressures and temperatures. Several experiments have been recently performed on amorphous Se (a-Se) at high pressure [20–22] and on liquid Se (l-Se) at high temperature [23–25], with the aim of understanding their peculiar behaviour.

Amorphous Se (a-Se) has been studied by XAS at room pressure by several groups, discussing local structure and photo-crystallization properties (see [26] and references therein). XAS data collected up to about 10 GPa [20] have shown that the Se–Se covalent bond length in a-Se slightly increases with pressure, changing dramatically from about 2.36–2.40 Å just near the crystallization threshold (around 10 GPa). Further high-pressure Raman and X-ray diffraction experiments [21] on a-Se are compatible with those observations, reporting also evidence of the presence of various crystalline phases in the range of pressures 10.3–24.4 GPa. The nucleation of competing crystalline phases at 10.4 GPa was also studied in details using time-resolved X-ray diffraction (XRD) and tomography [22].

Several XAS experiments have been performed on liquid Se (l-Se) using different techniques covering a wide range of pressures and temperatures. The first experiments were carried out using polycrystalline sapphire cells allowing variable sample thickness (from ~30 μm or less for l-Se up to 20 mm for Se vapours) [23, 24, 27]. Pressures and temperatures up to about 2 kbar and 2,000 K can be reached using these devices. XAS structural results indicated an apparent shortening of the bondlength distance for increasing temperature, especially at the transition to the liquid phase (from ~2.36 to ~2.33 Å upon melting). A specific XAS study of the semiconductor-to-metal transition near the critical point of fluid Se (1,893 K, 385 bar) was also carried out using this cell. Structural results using the cumulant expansion method indicated a remarkable reduction of the interatomic distance along the chains in the metallic disordered phase (down to 2.24 Å), and a reduction of the coordination number at high temperatures. More recently, XAS experiments aimed to measure l-Se at high pressure [25, 28, 29] were performed using large-volume cells (MAX90 and Paris-Edinburgh) in an extended range of pressures (up to ~8 GPa) and temperatures (up to ~1,200 K). The authors reported a modification of the two-fold chain structure associated with the covalent bonds at high pressures, possibly related to a metalization process. As correctly stated in the original papers, deduction of the structural parameters of these disordered phase is quite difficult using standard EXAFS data-analysis methods.

Very recently, Se K-edge XAS experiments under pressure on l-Se and a-Se were performed at the XAFS beamline at ELETTRA (Trieste, Italy) and at the ODE beamline at Soleil (Saclay, France), with the aim of investigating the structure of stable and metastable disordered states by accurate data-analysis of high-quality data. These XAS data, complemented by Raman scattering experiments under pressure

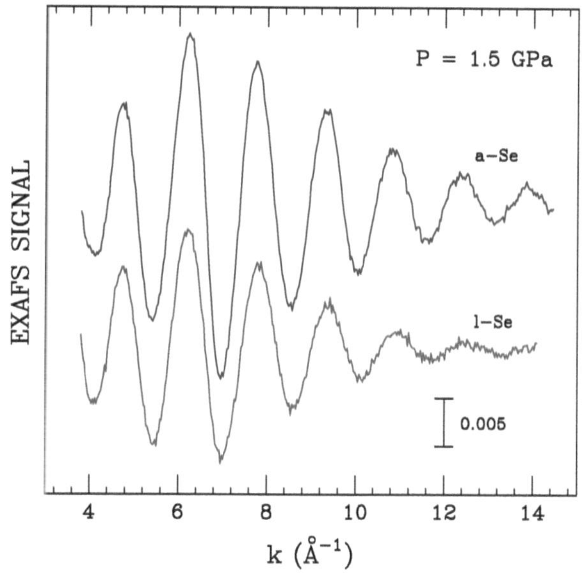

Fig. 9.4 Experimental K-edge XAS spectra of liquid (l-Se, T = 720 K) and amorphous (a-Se, T = 300 K) selenium, collected for an external pressure of 1.5 GPa with a Paris-Edinburgh press (see text)

[30], have shown significant effects of metastability of various phases, including nucleation of crystalline structures from a-Se and undercooled l-Se [31].

The a-Se and l-Se Se K-edge XAS structural signals reported in Fig. 9.4, obtained at the XAFS beamline at ELETTRA [32] with a Paris-Edinburgh high-pressure cell at 1.5 GPa, show clearly a single leading oscillation associated with the first-neighbour distribution. Those experiments included XRD diagnostics and the execution of single-energy scans [12], confirming the Se melting curve and showing the possibility of measuring deeply undercooled Se. Looking at Fig. 9.4 we notice that the main modification of the XAS signal is an amplitude reduction and a slight phase difference for l-Se. The inter-chain signal is certainly very weak both in the solid and in the liquid disordered phase, but modifications of the first-neighbour distribution can be measured quite accurately by XAS.

Useful short-range structural information can be obtained by application of the GNXAS data-analysis method [33–35]. The comparison of experimental and calculated best-fit XAS signals of l-Se at 1.5 GPa is shown in Fig. 9.5. The agreement with the best-fit signal obtained by using an asymmetric Γ-like first-neighbor distribution is excellent, as shown by both the raw data (left) and Fourier transform (right). The use of this distribution containing only an additional skewness parameter (as compared to the Gaussian distribution) was shown to be very effective for solids and liquids even at moderate temperatures [36–38]. In this particular case, XAS structural refinement can be safely limited to the first-neighbors as shown by the single peak observed in the Fourier transform of Fig. 9.5. The best-fit structural parameters of the Γ-like first-neighbour distribution resulted to be $R \sim 2.396$ Å (average distance), $\sigma^2 \sim 0.0105$ Å^2 (distance variance), and $\beta \sim 0.7$ (skewness), assuming a fixed coordination number $N = 2$. However, individual parameters are useful only

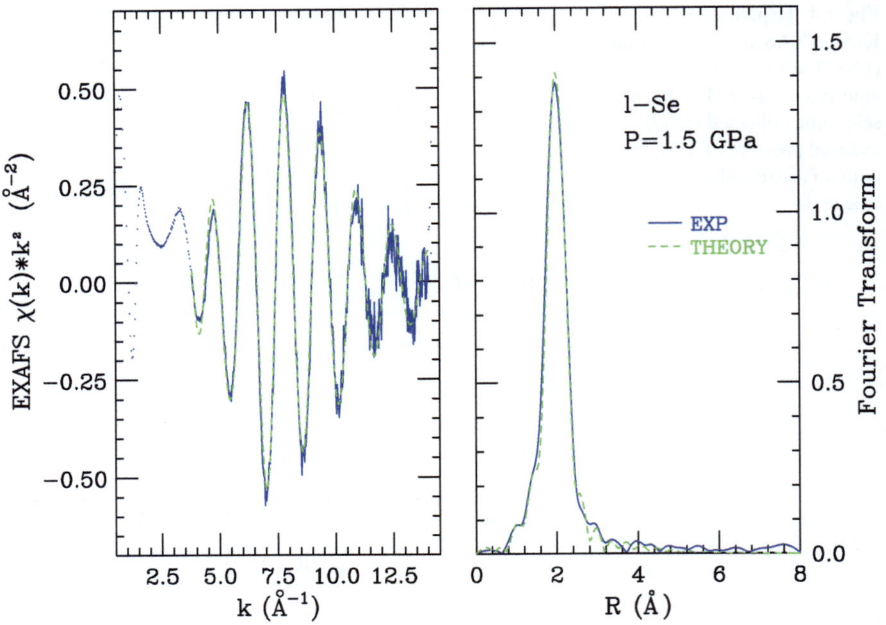

Fig. 9.5 Comparison between Se K-edge experimental (*blue*) and best-fit calculated (*dashed, green*) XAS signals of l-Se. The *left* hand side panel reports the raw k^2 EXAFS data. The *right* hand side panel shows the Fourier Transform in the entire fitting range

to define the shape of the first-neighbor distribution that is measured very accurately by XAS at short distances. As a matter of fact, the pair (or higher order) distribution function is a meaningful physical quantity that can be reconstructed on the basis of the XAS structural refinement, and easily compared with other experimental or theoretical results.

The pair distribution function $g(R)$ resulting from the XAS refinement is compared in Fig. 9.6 (right hand side panel) with the results of time-of-flight neutron diffraction (ND) measurements (measured at lower temperatures T = 538 K and room pressure) [39]. The first peak of the $g(R)$ shows remarkable differences with the same curve obtained by neutron diffraction: it is narrower and more skewed, reaching a maximum of about 3.8 instead of 2.6, although the average peak position is roughly the same. However, as shown in Fig. 9.6 (left hand side panel) the structure factor $S(k)$ reconstructed by XAS results is fully compatible with the original $S(k)$ neutron diffraction data [39]. This result shows that present XAS structural results are reliable and match previous ND determinations. We have also verified that the first peak of the $g(R)$ obtained by ND can not reproduce our XAS results, while the latter are obviously blind to the medium-range order information contained in the original $g(R)$. This confirms previous results (see for example [40]) showing both the agreement with original $S(k)$ data and the higher sensitivity of XAS to the very short-range structure. The unique XAS sensitivity to the local structure and bonding properties is then shown to be a very useful tool to investigate the transformation occurring in disordered semiconductors at high pressure.

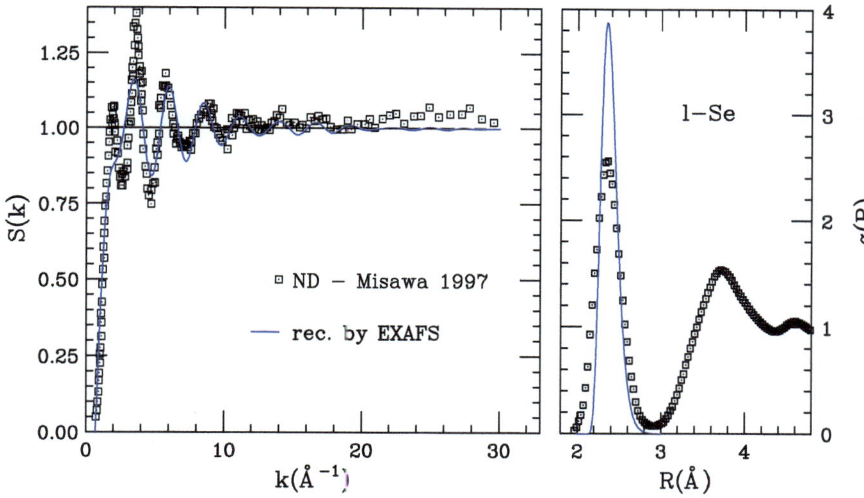

Fig. 9.6 *Left* hand side panel structure factor obtained by neutron diffraction measurements [39] (*squares*) compared with the S(k) reconstructed by XAS. *Right* hand side panel pair distribution function $g(R)$ obtained by neutron diffraction [39] (*squares*) compared with the first-neighbour distribution reconstructed by XAS (*blue*)

9.5 The Physics of Ge and Related Systems at High *P* and High *T*

Elemental Germanium is a fourth group semiconducting element characterized by a rich polymorphic phase diagram [41]. The semiconducting phases include the diamond structure stable at ambient conditions and up to about 8 GPa and to the melting line with an anomalous $dT_m(P)/dP < 0$ behavior owing to the lower liquid density and a number of crystalline metastable phases found upon pressure decompression from the β-Sn metallic phase. Metastable amorphous Ge (a-Ge) can be prepared by a variety of vapor deposition techniques. It is stable up to a temperature of about 400–500 K where nucleation to crystalline Ge (c-Ge) takes place. Retaining a continuous random network structure with tetrahedral bonding a-Ge is a prototype for a wide class of amorphous semiconductors. The Ge phase diagram with the stability regions of the various phases up to moderate pressures is reproduced in Fig. 9.7. The Ge liquid phase retains a partial ability to form covalent bonds in the melt resulting in intermediate coordination numbers around 6–7. The interplay between covalent bonding ability and metallic bonds was proposed to be at the basis for the existence of a liquid–liquid phase transition in the unaccessible region of the metastable liquid range (no man's land).

Ge has been the subject of investigations since the early days of EXAFS spectroscopy [42]. The temperature parameter was exploited in a number of successive studies to investigate thermal expansion and disorder [43, 44] up to the liquid phase [36]. Evidence for a large undercooling ability of the liquid phase in micrometric droplet samples was early recognized [36] and confirmed by temperature scanning

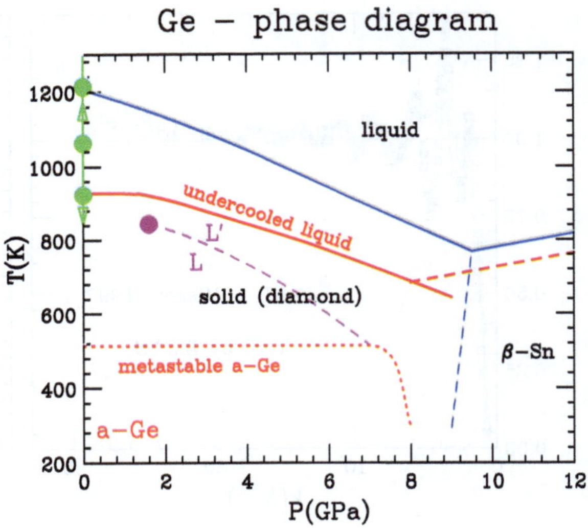

Fig. 9.7 Phase diagram of Ge in the range 0–12 GPa. The region of stability of the diamond c-Ge structure with respect to the metallic β-Sn phase and metallic liquid Ge is reported. Metastable phases include semiconducting a-Ge and undercooled liquid Ge down to about 200–300 K below the melting line. The proposed liquid phase transition line in the no man's land is also reported

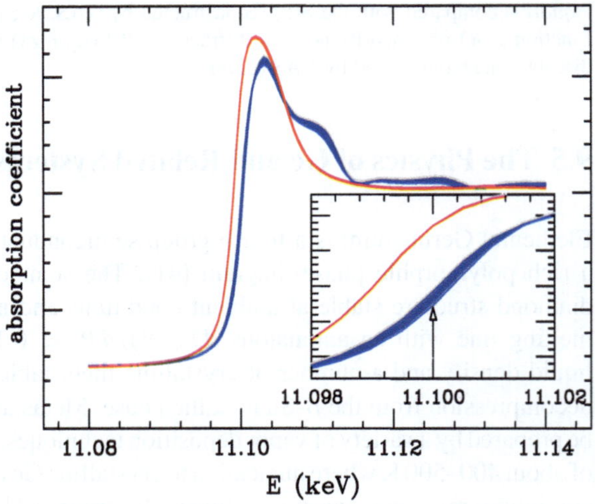

Fig. 9.8 Sequence of high energy resolution XANES spectra of the Ge K-edge of a micrometric c-Ge powder sample collected from ambient temperature up to and above the melting point every about 80 K

techniques [12]. Investigations of the c-Ge phase were able to highlight the phonon contribution to the first shell bond length distribution and to highlight the peculiar EXAFS sensitivity to thermal expansion. High resolution XANES spectra at the Ge K-edge contain important information on the evolution of the unoccupied p-like density of states and transition matrix element. A sequence of spectra collected every ≈80 K from ambient temperature up to the liquid phase at BM29 [10] is shown in Fig. 9.8. The high stability and energy resolution of the double crystal Si (311) monochromator allow to identify the evolution of fine details with T. The evident shift of the absorption edge to lower energies with increasing T, highlighted in the inset of Fig. 9.8, is associated with the progressive renormalization of the semiconductor band-gap by phonons [45]. A discontinuous shift and spectral change occurs upon melting reflecting the metallic character of the liquid state. The quality of the collected spectra is compatible with the application of core-hole lifetimes broadening deconvolution techniques [46] which may provide a further insight into the nature of the electronic states available for the transition.

EXAFS investigations of c-Ge and a-Ge with pressure were further performed using energy scanning and energy dispersive beamlines [47–50]. With these experiments not only the effect of the pressure compression on the first shell bond length distribution was clearly determined, but also evidence for a metallization transition in a-Ge analogous to the one occurring for c-Ge around 8 GPa was observed. The relevance of these findings with the polyamorphism issue [48, 49] was discussed in details. Moreover, the role of sample morphology and defect density in pressure-induced phase transitions for inhomogeneous amorphous samples [49, 50] using a combination of techniques were addressed.

References

1. J.J. Rehr, R.C. Albers, Rev. Mod. Phys. **72**, 621 (2000)
2. A. Filipponi, J. Phys.: Condens. Matter **13**, R23 (2001)
3. A. Filipponi, A. Di Cicco, Nucl. Instrum. Methods in Phys. Res. B **93**, 302 (1994)
4. Y. Nakamoto, M. Sakata, H. Sumiya, K. Shimizu, T. Irifune, T. Matsuoka, Y. Ohishi, Rev. Sci. Instrum. **82**(6), 066104 (2011)
5. X. Hong, M. Newville, V.B. Prakapenka, M.L. Rivers, S.R. Sutton, Rev. Sci. Instrum. **80**(7), 073908 (2009)
6. F. Coppari, A. Polian, N. Menguy, A. Trapananti, A. Congeduti, M. Newville, V.B. Prakapenka, Y. Choi, E. Principi, A. Di Cicco, Phys. Rev. B **85**, 045201 (2012)
7. L.G. Khvostantsev, L.F. Vereshchagin, A.P. Novikov, High Temp. High Pressures **9**, 637 (1977)
8. J.M. Besson, R.J. Nelmes, G. Hamel, J.S. Loveday, G. Weill, S. Hull, Phys. B **180&181**, 907 (1992)
9. Y. Katayama, M. Mezouar, J.P. Itié, J.M. Besson, G. Syfosse, P. Le Fèvre, A. Di Cicco, J. Phys. IV France **7**(C2), C2 (1997)
10. A. Filipponi, M. Borowski, D.T. Bowron, S. Ansell, S. De Panfilis, A. Di Cicco, J.P. Itié, Rev. Sci. Instrum. **71**, 2422 (2000)
11. A. Filipponi, V.M. Giordano, S. De Panfilis, A. Di Cicco, E. Principi, A. Trapananti, M. Borowski, J.P. Itié, Rev. Sci. Instrum. **74**, 2654 (2003)
12. A. Filipponi, M. Borowski, P.W. Loeffen, S. De Panfilis, A. Di Cicco, F. Sperandini, M. Minicucci, M. Giorgetti, J. Phys.: Condens. Matter **10**, 235 (1998)
13. T. Matsushita, R.P. Phizackerley, Jpn. J. Appl. Phys. **20**(11), 2223 (1981)
14. S. Pascarelli, O. Mathon, Phys. Chem. Chem. Phys. **12**(21), 5535 (2010)
15. J.P. Itié, A. Polian, G. Calas, J. Petiau, A. Fontaine, H. Tolentino, Phys. Rev. Lett. **63**(4), 398 (1989)
16. A.V. Sapelkin, S.C. Bayliss, A.G. Lyapin, V.V. Brazhkin, J.P. Itiè, A. Polian, S.M. Clark, A.J. Dent, Phys. Status Solidi B **198**(1), 503 (1996)
17. A. Dadashev, M.P. Pasternak, G.K. Rozenberg, R.D. Taylor, Rev. Sci. Instrum. **72**(6), 2633 (2001)
18. F. Coppari, A. Di Cicco, E. Principi, A. Trapananti, N. Pinto, A. Polian, S. Chagnot, A. Congeduti, High Press. Res. **30**(1), 28 (2010). 47th Meeting of the European-High-Pressure-Research-Group (EHPRG 47), Univ Pierre & Marie Curie, Cordeliers Campus, Paris, France, 06–11 Sep 2009

19. V.V. Brazhkin, A.G. Lyapin, J. Phys.: Condens. Matter **15**(36), 6059 (2003)
20. Y. Katayama, K. Tsuji, O. Shimomura, H. Oyanagi, J. Non-Cryst. Solids **205**(Part 1), 199 (1996). 9th International Conference on Liquid and Amorphous Metals, Chicago, IL, 27 Aug–01 Sep 1995
21. K. Yang, Q. Cui, Y. Hou, B. Liu, Q. Zhou, J. Hu, H.K. Mao, G. Zou, J. Phys.: Condens. Matter **19**(42), 425220 (2007)
22. H. Liu, L. Wang, X. Xiao, F. De Carlo, J. Feng, H.k. Mao, R.J. Hemley, Proc. Nat. Acad. Sci. USA **105**(36), 13229 (2008)
23. K. Tamura, M. Inui, M. Yao, H. Endo, S. Hosokawa, H. Hoshino, Y. Katayama, K. Maruyama, J. Phys.: Condens. Matter **3**, 7495 (1991)
24. M. Inui, Jpn. J. Appl. Phys. Part 1-Regul. Pap. Short Notes Rev. Pap. **32**(32–2), 165 (1993). 7th International Conference on X-ray Absorption Fine Structure (XAFS 7), Kobe, Japan, 23–29 Aug 1992
25. Y. Katayama, O. Shimomura, K. Tsuji, J. Non-Cryst. Solids **250**(Part 2), 537 (1999). 10th International Conference on Liquid and Amorphous Metal (LAM-10), University of Dortmund, Dortmund, Germany, 30 Aug–04 Sep 1998
26. A. Kolobov, H. Oyanagi, K. Tanaka, K. Tanaka, Phys. Rev. B **55**(2), 726 (1997)
27. K. Tamura, M. Inui, S. Hosokawa, Rev. Sci. Instrum. **66**(2), 1382 (1995)
28. Y. Katayama, K. Tsuji, H. Oyanagi, O. Shimomura, J. Non-Cryst. Solids **232**, 93 (1998). 7th International Conference on the Structure of Non-Crystalline Materials (NCM 7), Cagliari, Italy, 15–19 Sep 1997
29. Y. Katayama, J. Synchrotron Radiat. **8**(Part 2), 182 (2001). 11th International Conference on X-ray Absorption Fine Structure (XAFS XI), Ako, Japan, 26–31 July 2000
30. L. Properzi, A. Polian, P. Munsch, A. Di Cicco, High Press. Res. **33**(1), 35 (2013)
31. L. Properzi, Xas and raman scattering experiments on stable and metastable phases of se under high pressure. Master thesis, University of Camerino, Italy, 2012
32. A. Di Cicco, G. Aquilanti, M. Minicucci, E. Principi, N. Novello, A. Cognigni, L. Olivi, J. Phys.: Conf. Ser. **190**(1), 012043 (2009)
33. A. Filipponi, A. Di Cicco, C.R. Natoli, Phys. Rev. B **52**, 15122 (1995)
34. A. Filipponi, A. Di Cicco, Phys. Rev. B **52**, 15135 (1995)
35. A. Di Cicco (ed.), *GNXAS. Extended Suite of Programs for Advanced X-ray Absorption Data-Analysis: Methodology and Practice* (TASK Publishing, Gdansk, 2009)
36. A. Filipponi, A. Di Cicco, Phys. Rev. B **51**, 12322 (1995)
37. A. Di Cicco, M.J. Rosolen, R. Marassi, R. Tossici, A. Filipponi, J. Rybicki, J. Phys.: Condens. Matter **8**, 10779 (1996)
38. A. Di Cicco, M. Taglienti, M. Minicucci, A. Filipponi, Phys. Rev. B **62**, 12001 (2000)
39. M. Misawa, K. Suzuki, Trans. Jpn. Inst. Met. **18**(5), 427 (1997)
40. A. Di Cicco, M. Minicucci, A. Filipponi, Phys. Rev. Lett. **78**, 460 (1997)
41. G.J. Ackland, Rep. Prog. Phys. **64**, 483 (2001)
42. C.E. Bouldin, E.A. Stern, B. von Roedern, J. Azoulay, Phys. Rev. B **30**, 4462 (1984)
43. G. Dalba, P. Fornasini, M. Grazioli, F. Rocca, Phys. Rev. B **52**, 11034 (1995)
44. G. Dalba, P. Fornasini, R. Grisenti, J. Purans, Phys. Rev. Lett. **82**, 4240 (1999)
45. P. Lautenschlager, P.B. Allen, M. Cardona, Phys. Rev. B **31**, 2163 (1985)
46. A. Filipponi, J. Phys. B: At. Mol. Opt. Phys. **33**, 2835 (2000)
47. A. Di Cicco, E. Principi, M. Minicucci, S. De Panfilis, A. Filipponi, F. Decremps, F. Datchi, J. Itie, P. Munsch, A. Polian, High Press. Res. **24**(1), 93 (2004). Meeting on Matter Under Extreme Conditions, Paris, France, 16 May 2003
48. E. Principi, A. Di Cicco, F. Decremps, A. Polian, S. De Panfilis, A. Filipponi, Phys. Rev. B **69**(20), 201201 (2004)
49. A. Di Cicco, A. Congeduti, F. Coppari, J.C. Chervin, F. Baudelet, A. Polian, Phys. Rev. B **78**(3), 033309 (2008)
50. F. Coppari, J.C. Chervin, A. Congeduti, M. Lazzeri, A. Polian, E. Principi, A. Di Cicco, Phys. Rev. B **80**, 11 (2009)

Part III
Semiconductor Nanostructures

Chapter 10
Group IV Quantum Dots and Nanoparticles

Alexander V. Kolobov

Abstract This chapter describes the application of X-ray absorption spectroscopy to group IV—predominantly Germanium—nanostructures. It starts with the description of importance of Ge nanostructures and the experimental methods to study them. This is followed by the results, obtained for Ge quantum dots such as embedded Ge nanoparticles and Ge quantum dots epitaxially grown on Si substrates, by using conventional X-ray absorption spectroscopy as well as more sophisticated approaches such as diffraction anomalous fine structure (DAFS). The chapter is concluded by a description of potential future applications of X-ray spectroscopy, e.g. for pure isotope studies, femtometer-precision X-ray absorption fine structure (XAFS), and time-resolved studies.

10.1 Introduction

At the beginning, a few words about terminology should be said. We start with a definition of a quantum dot (QD). Quantum dots are tiny particles, or nanoparticles, of a semiconductor material, that range from 2 to 10–100 nm in diameter. From the electronic perspective, a quantum dot is a portion of matter whose excitons are confined in all three spatial dimensions. Exciton confinement in nanostructures was discovered in the early 1980s [1]. To reflect the quantum nature of the nanostructures, the term "quantum dot" was suggested [2].

The quantum confinement [3, 4] manifests itself as an increase in the bandgap of quantum dots with respect to the bulk material of the same composition, i.e. one can tune the band gap of a material by varying its dimensions, which opens new degrees of freedom in manufacturing of semiconductor lasers and light emitting diodes. In particular, the ability to precisely control the size of a quantum dot enables one to determine the wavelength of the emission, which in turn determines the colour of light the human eye perceives. Quantum dots can therefore be tuned during production

A.V. Kolobov (✉)
Nanoelectronics Research Institute, National Institute of Advanced Industrial Science and Technology, Tsukuba Central 4, 1-1-1 Higashi, Tsukuba, Ibaraki 305-8562, Japan
e-mail: a.kolobov@aist.go.jp

C.S. Schnohr and M.C. Ridgway (eds.), *X-Ray Absorption Spectroscopy of Semiconductors*, Springer Series in Optical Sciences 190,
DOI 10.1007/978-3-662-44362-0_10

to emit the colour of light desired. Ge QDs have also been considered for other applications, including quantum computing [5].

Confinement in quantum dots can also arise from electrostatic potentials (generated by external electrodes, doping, strain, or impurities). Thus, while bulk Ge is an indirect band gap semiconductor, stressed Ge grown, for example, on Si substrates can posses a direct gap [6], thus opening a possibility to use Ge in light-emiting devices. Recently, there were numerous reports on direct band gap photoluminescence in strained Ge (see e.g. [7]).

Quantum dots can be fabricated using different approaches. One way to produce quantum dots is to confine small portions of a material within a foreign, usually a wide band-gap matrix, e.g. by annealing a material that contains atoms of the needed material, e.g. SiO_2 co-sputtred with Ge. Subsequent annealing results in the formation of embedded nanocrystals [8–11]. Other matrixes have also been used, such as AlN [12]. Confinement of Ge nanocrystals in matrixes of opal and zeolite have also been reported [13–15].

Self-assembled quantum dots can also nucleate under certain conditions during epitaxial growth techniques, when a material is grown on a substrate to which it is not lattice matched. The resulting strain produces coherently strained islands on top of a two-dimensional wetting layer. This growth mode is known as the Stranski-Krastanov growth mode. The islands can be subsequently buried by capping with Si or another material. This approach has been successfully used to grow Ge QDs on Si substrates with different substrate orientations [18–31]

Below in this review we shall use the terms ‘quantum dot’ and ‘nanostructure’ interchangeably. While often the term ’quantum dots’ preferentially refers to epitaxially grown Ge nanostructures, it should be kept in mind that from the electronic perspective embedded nanocrystals are just as good quantum dots and in fact the term “quantum dot” was first introduced to describe the embedded nanocrystals.

An optimal method to investigate the nanostructures depends on the kind of information one is interested in. For example, to study optical properties, photoluminescence is usually used. Phonon confinement may be studied using Raman scattering. For structural investigations, atomic force microscopy (AFM) and its derivatives can be used to study the shape of epitaxially grown quantum dots (Fig. 10.1), while transmission electron microscopy (TEM) is an effective tool to investigate the shape and orientation of nanostructures (Fig. 10.2), be it epitaxially grown layers or embedded nanocrystals. To investigate the local structure of quantum dots, two major techniques used are Raman scattering and X-ray absorption spectroscopy. Application of the latter is the main subject of this chapter.

10.2 Raman Scattering and Its Pitfalls

In the particular case of Ge quantum dots grown on Si substrates, Raman scattering has severe limitations, which makes X-ray absorption spectroscopy especially valuable. Since readers of this chapter may be potential users of Raman scattering

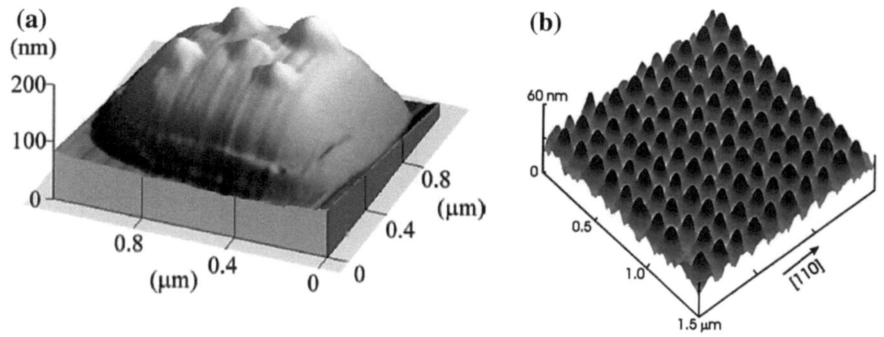

Fig. 10.1 Examples of AFM images of arrays of Ge QDs. **a** A 3D AFM image of 5 Ge *dots* grown on a square Si mesa [16] and **b** An AFM image of an array of Ge islands on prepatterned Si(001) surface [17]. Reproduced with permission from American Institute of Physics

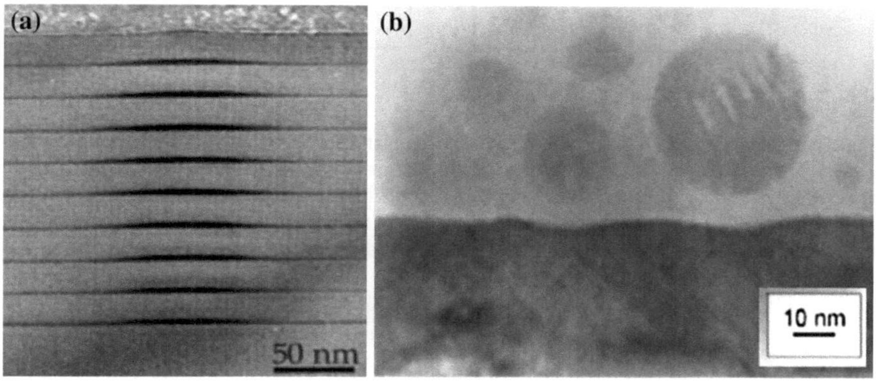

Fig. 10.2 **a** A cross-sectional TEM image taken along the (011) azimuth of a sample with 10 Ge/Si bilayers. The average island size is ca. 95 nm and the average height is 6–7 nm. The image shows that the islands in the stack are located one above the other, have almost equal size and height in all layers [32]. **b** A high-resolution TEM image of annealed Ge nanocrystals inside SiO_2 in the vicinity of the Si(100) interface. The image reveales that the nanocrystals are sperical and their presence modifies the flatness of the SiO_2/Si interface [10]. Reproduced with permission from American Physical Society

or its results, we start this section by describing the limitations of Raman scattering as applied to the study of Ge nanostructures, demonstrating some pitfalls of Raman scattering.

There is a large number of papers where Raman scattering spectroscopy has been used to investigate the structure of Ge nanostructures (see e.g. [33] and references therein). Most authors report a Ge–Ge peak which is located at about 300 cm^{-1} as well as a Ge–Si peak at about 400 cm^{-1} and a peak at about 435 cm^{-1} attributed to the local Si-Si vibrations. It should be noted that silicon, which is usually the substrate material, also possesses peaks in the Raman spectrum located at very similar frequencies. This fact is, unfortunately, ignored in many studies. As a result,

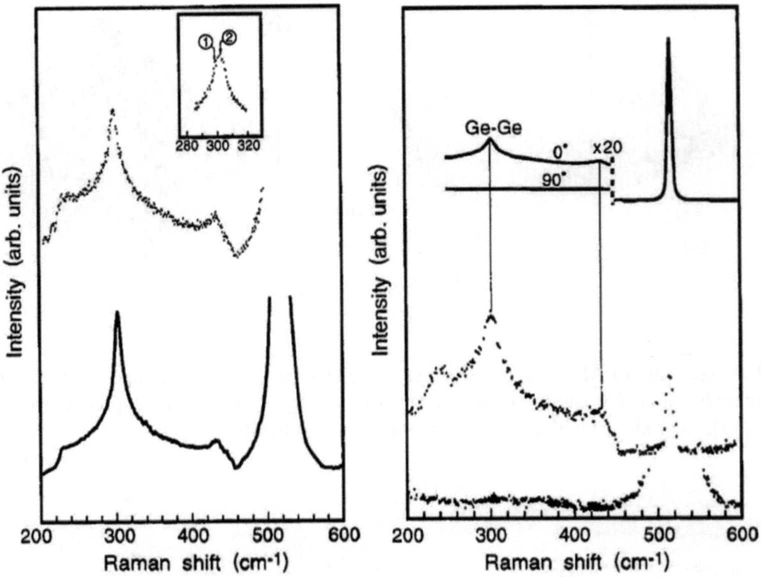

Fig. 10.3 *Left panel* Raman spectra of Si measured using an Ar-ion (*top*) [34] and a He-Ne laser (*bottom*). The *inset* shows the major peak with high-resolution. *Right panel* The *top two curves* are polarised Raman spectra from a sample that contains Ge nanowires grown on a Si substrate [36] and the *bottom curves* are polarised Raman spectra for pure Si. One can see that in both cases there is a pronounced—and very similar—polarisation dependence, generating some doubts as to the correctness of the offered interpretation

Raman spectra obtained from Ge nanostructures (and attributed to Ge quantum dots) are often dominated by a two-phonon peak originating from the silicon substrate, leading to potential misinterpretations of the results.

While the Raman spectrum of silicon is dominated by a strong peak located at 520 cm^{-1}, it also has a peak around 300 cm^{-1}. The latter peak has been interpreted as arising from a peak in the inverse lifetime of the zone-center optical photon. Uchinokura et al. [34] revealed the complete structure which included the peak located at about 300 cm^{-1} and two weaker features at 229 and 435 cm^{-1}. In their experiment, an Ar-ion laser was used and the result is shown in Fig. 10.3 (top) together with a spectrum obtained using a He-Ne laser (bottom). While the ratio of the features located at 229, 300, and 435 cm^{-1} is somewhat different for the two excitation sources, possibly due to different background signals, the overall spectrum is not influenced by the excitation light source. It was concluded that these peaks represent a two-phonon spectrum and reflect the structure of a combined phonon density of states.

This feature is strongly polarized [34, 35] as shown in the left panel of Fig. 10.3. The latter fact must not be ignored when Raman scattering of strongly oriented Ge nanowires is considered [36] since the apparent polarisation dependence attributed to one-dimensional nature of the nanowires may be caused by the selectron rules of Raman scattering from the Si substrate, as seems to be the case for the example shown in the figure.

Self-organized Ge quantum dots are often formed as a result of strain accumulated due to the lattice mismatch. The strain is released due to the formation of islands (the Stranski-Krastanov growth mode). The fact that the Raman peak shifts with strain (to higher wave numbers for compressive strain) enables one to discuss—in principle—the degree of strain relaxation on the basis of the peak position. The shift of the peak position to lower wave numbers due to size effects allows one to check whether quantum confinement is taking place.

At the same time, as to the differences in the reported position of the Ge–Ge peak (300–305 cm^{-1}) attributed to phonon confinement or stresses, we would like to notice that the Si peak located in this energy region is in fact a doublet (see insert to Fig. 10.3 (top)), the critical points being (299 cm^{-1}) and (303 cm^{-1}). The difference between the two values is 4 cm^{-1}, i.e., about the same as the scatter in the reported values of the peak position. Under different experimental conditions either one or the other can be dominant and lead to potential misinterpretation in terms of stresses or quantum confinement.

Various examples of correct and erroneous interpretatations of the Raman signal can be found in [33], titled "Raman scattering from Ge nanostructures grown on Si substrates: power and limitations" and the interested readers are referred to this work. A question arises whether there are ways to tackle this problem. Several solutions can be suggested. Unfortunately, none of them is a complete remedy. One possibility is to make use of resonant Raman scattering. Another possible way to differentiate between the contributions from the Ge nanostructures and the substrate is to make use of the polarization dependence of the silicon two-phonon peak described above. While it may not work for Ge wires, which themselves are strongly oriented, it may be an efficient way to investigate isotropic nanostructures such as QDs and embedded nanocrystals. Finally, use of isotopes may serve to differentiate between the signal coming from the Si substrate and Ge quantum dots but considering the cost of pure isotopes this approach only has an academic interest.

From the above description it follows that use of Raman scattering for cases where Si substrates are used—and such cases represent the majority—has severe limitations. This makes X-ray absorption the most efficient way to investigate the local structure of Ge quantum dots [37–39]. The rest of this chapter is dedicated to a review of X-ray absorption study of epitaxially grown Ge QDs and embedded nanoscrystals.

10.3 X-Ray Absorption Spectroscopy of Ge QDs and Nanocrystals

In this section the results of experimental studies of Ge quantum dots (epitaxially grown and embedded in insulating matrixes) using X-ray absorption spectroscopy are described. The section sets in with the epitaxially grown quantum dots starting with the uncapped QDs.

10.3.1 Epitaxially Grown Uncapped Ge QDs

Uncapped Ge QDs were studies by X-ray absorption by different groups and the main results agree rather well with each other. In what follows, we describe the experimental results in close relationship with the conditions used to fabricate the QDs. Direct comparison of the results is not always straightforward because different growth methods and conditions were used but the trends are rather well reproduced in different experiments.

Boscherini et al. [19, 20, 40] were the first to investigate the structure of epitaxial Ge QDs using XAFS. In a series of publications, they reported on the local structure of both Ge/Si(001) and Ge/Si(111) QDs grown under different conditions. For both cases they found a pronounced Ge–Si intermixing. Below we describe their results in more detail.

For the Si(001) substrate orientation, low pressure chemical vapour deposition at 600 °C was used and samples with equivalent Ge thicknesses from 5.8 to 38 nm were studied. In case of the Si(111) orientation, samples were grown in a temperature range of 450–530 °C using physical vapour deposition in ultra high vacuum. The equivalent Ge layer thickness was varied from 1 to ca. 20 nm.

Figure 10.4 shows the dependence of the partial Ge–Si coordiantion numbers N_{Ge-Si} as a function of the equivalent Ge layer thickness. The partial N_{Ge-Ge} coordination number can be easily calculated as 4—N_{Ge-Si}. For the Si(001) orientation, the Ge–Si coordination number was found to be essentially unchanged. The authors performed measurements for both parallel and perpendicular orientations but no meaningful difference between the two cases could be detected.

For the Si(111) orientation, the authors found (Fig. 10.4) that with increasing equivalent thickness of the Ge layer, the number of Si atoms surrounding Ge decreases from 2 to 1. Therefore, the observed average intermixing is higher in sample in which only a wetting layer is present (50 % Si average content). It was further found (Fig. 10.5) that the Ge–Si coordination number decreased faster for the sample grown at a lower temperature. The authors propose that the Si content in the islands is limited both by the diffusion factor (the height of the islands can reach 50 nm), and by the

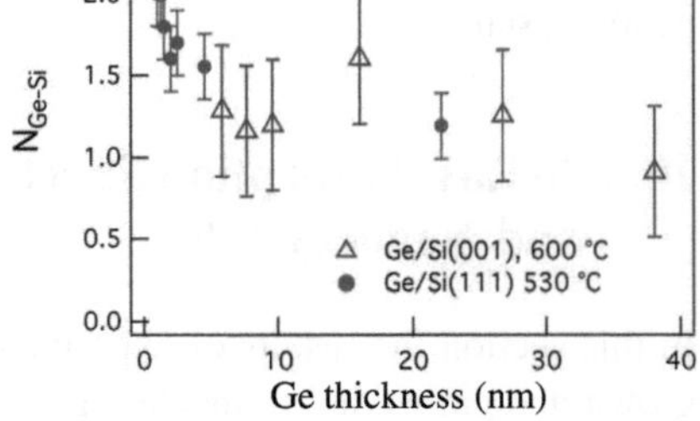

Fig. 10.4 Partial Ge–Si coordination number for Ge quantum dots grown on Si(001) (after [20]) and Si(111) (after [21]). The growth conditions are described in the text

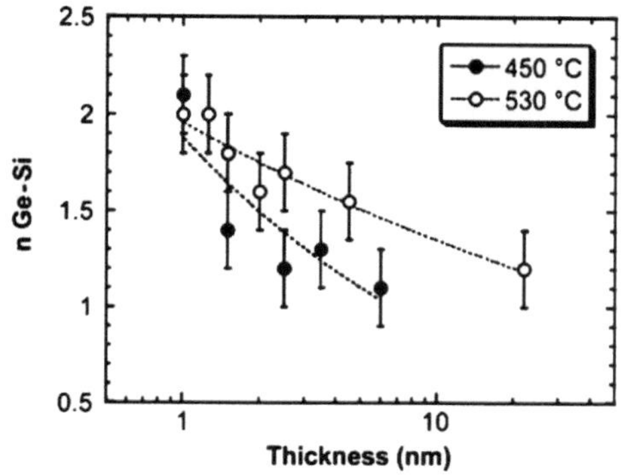

Fig. 10.5 Average number of Si atoms found around a Ge absorber by fitting the EXAFS data of the Ge/Si(111) samples grown at 450 and 530 °C substrate temperature [21]. Reproduced with permission from Elsevier

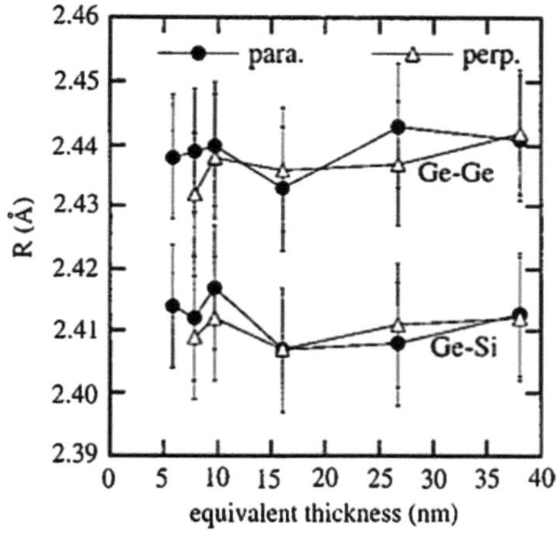

Fig. 10.6 Ge–Ge and Ge–Si bond lengths for Ge/Si(001), after [20]. Reproduced with permission from American Institute of Physics

fact that the lattice in the islands is more relaxed than in the wetting layer, reducing the driving force for the intermixing.

It should be noted that while the observed difference in intermixing could be related to the substrate orientation, it could also be caused by different equivalent Ge layer thicknesses used in the two studies and the smaller thicknesses of Ge for the case of Si(111) substrates may also partially account for the stronger intermixing observed.

In Fig. 10.6 the Ge–Ge and Ge–Si bond lengths as a function of Ge equivalent thickness are shown for the Si(001) substrate orientation. Measurements were taken in two different polarisation geometries but no significant difference was found. The authors conclude that the bond lengths remain essentially the same within the studied range of equivalent thicknesses and correspond to an average composition of $Ge_{0.7}Si_{0.3}$. The obtained results are in agreement with earlier work where the composition dependence of the bond lengths in relaxed GeSi crystalline alloys has been studied by XAFS [41–43] (cf. Fig. 10.7 and Chap. 2). At the same time, it should be noted that as the equivalent thickness increases two parameters change

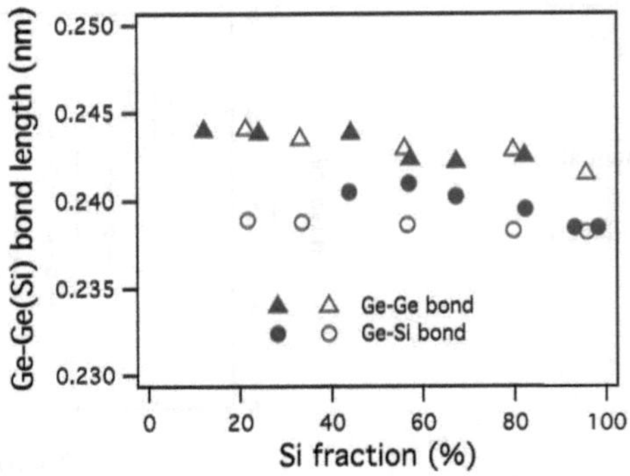

Fig. 10.7 Ge–Ge and Ge–Si bond lengths in bulk relaxed Ge–Si alloys. *Open symbols* represent the result of [42] and *closed symbols* are from [43]

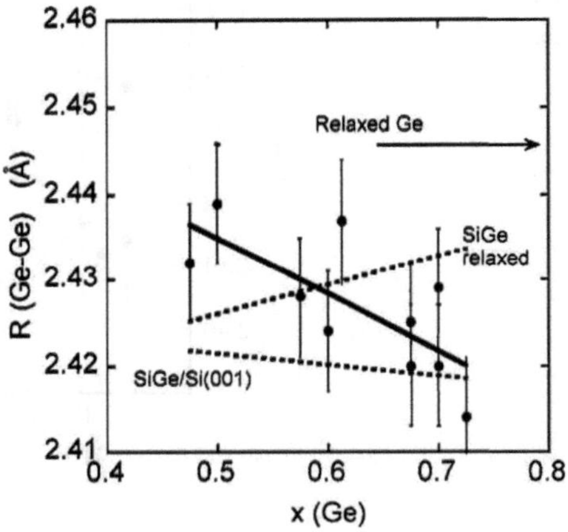

Fig. 10.8 Ge–Ge NN distances measured by XAFS (dots with error bars). *Continuous line* linear fit to the XAFS results obtained for strained mrGe_xSi_{1-x} epilayers [22]. *Dashed lines* model for SiGe/Si(001) [44] and experimental results for relaxed SiGe alloys [41]. The *arrow* indicates the value of the Ge–Ge distance in a relaxed, bulk Ge; reproduced after [22] with permission from American Physical Society

simultaneously, namely, the Ge concentration and the strain, which makes unambiguous conclusions difficult.

While for a relaxed crystal the Ge–Ge bond length increases with an increase in Ge content, for strained layers an opposite dependence has been observed as demonstrated in Fig. 10.8, where Ge–Ge bond length is shown as a function of average Ge concentration for the Si(111) substrate orientation.

In a different study [24, 26], the samples were grown by solid source molecular beam epitaxy on Si(100) substrates at a temperature of about 745 °C. On a 150 nm Si buffer, six monolayers of Ge were deposited at a rate of 0.2 Å/s. On a Ge wetting layer of about five monolayer nominal thickness, Ge islands form in the Stranski-Krastanov growth mode. The islands were about 180 nm in diameter, about 12 nm in

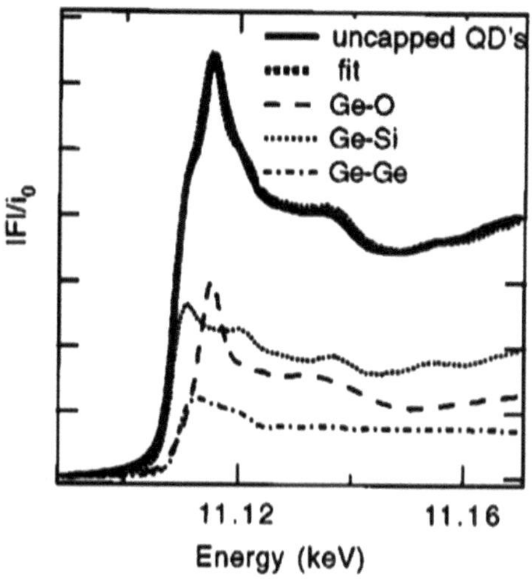

Fig. 10.9 XANES (normalized fluorescence) spectra for uncapped Ge QDs. The fitting is performed with the reference samples of bulk Ge, Ge oxide, and $Ge_{0.006}Si_{0.994}$ [24]

height, and had an areal density of 10^9 cm^2. The sample with uncapped islands was cooled down immediately.

It was found that Ge is partially oxidized with the fraction of Ge-oxide being around 35 %. The formation of Ge-O bonds have also been reported in [45]. The remaining Ge is coordinated to Ge and Si, the partial contribution of Ge–Ge and Ge–Si bonds being, respectively, 25 and 40 %. Because of the large number of fitting parameters, the accuracy in these numbers is on the order of ± 10 %.

The Ge–Ge bond length equals 2.46 Å, i.e., the Ge fraction which exists as a Ge phase, presumably in the QDs core, is unstrained. The result suggests that the QDs are relaxed as a result of intermixing and/or oxidation which are conducive to the more flexible network formation while most of the misfit strain is accommodated by the alloyed region. Most likely this fact accounts for why the remaining Ge is stable within the Ge phase.

In addition to EXAFS, the authors have also analyzed XANES spectra [26]. The spectrum for the uncapped QDs (Fig. 10.9) was fitted by a linear combination of three references with the partial weight of each component being as follows: bulk Ge, 25 %; Ge-oxide, 30 %; $Ge_{0.006}Si_{0.994}$, 45 %. The fit reproduces well all the features of the experimental spectrum and the obtained numerical values agree with the results of the EXAFS data analysis.

10.3.2 Capped Ge QDs

As in the previous section, we describe individual results obtained by different groups in relationship with the growth conditions.

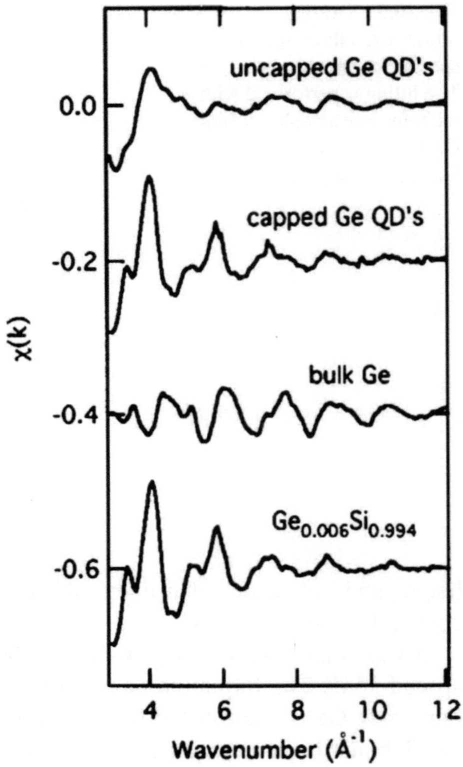

Fig. 10.10 Raw EXAFS oscillations for the uncapped and Si-capped Ge QDs compated with those for bulk Ge and a dilute solid solution of Ge in Si [26]

In one study [26], the effect of the growth temperature on Si(100) was studied. On a 150 nm Si buffer, 6 monolayers (MLs) of Ge were deposited at a rate of 0.2 Å/s at a temperature of 745 °C. On a Ge wetting layer of about 5-monolayer nominal thickness, Ge islands formed in the Stranski-Krastanov growth mode. The islands were about 180 nm in diameter, about 12 nm in height, and with an areal density of $1 \times 10^9\,cm^{-2}$. The sample was subsequently capped by 90 nm Si deposited at a rate linearly increasing from 0.05 to 0.5 Å/s. The capped sample revealed island-related photoluminescence at low temperature at an energy of 0.89 eV which indicated a Ge content of about $x = 0.5$ in the center of islands [26]. In addition to the sample grown at 745 °C, two samples were grown at lower temperatures. One sample was grown at 510 °C and had a nominal thickness of 7 ML Ge, the dots had the hut shape and are ca. 20 nm in diameter and 2 nm in height with $6 \times 10^{10}\,cm^{-2}$ density. The other sample with a nominal thickness of 8.5 ML Ge was grown at 550 °C. The islands were larger (70 nm in diameter and 6 nm in height) and had a density of $5 \times 10^9\,cm^{-2}$.

Figure 10.10 compares raw EXAFS oscillations for the uncapped and Si-capped Ge QDs grown at 745 °C. Also shown in the same figure for comparison are EXAFS oscillations for bulk Ge and for a very dilute solid solution of Ge in silicon ($Ge_{0.006}Si_{0.994}$). Visual comparison of the raw spectra already demonstrates that capping drastically modifies the local structure around Ge species. One can also

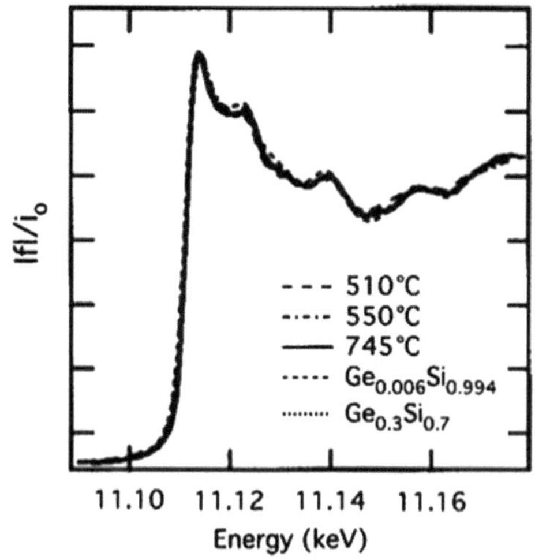

Fig. 10.11 XANES (normalized fluorescence) spectra for the Si-capped Ge QDs. The fitting is performed with the reference samples of bulk Ge, Ge oxide, and $Ge_{0.006}Si_{0.994}$ [24]

see a striking similarity between the spectra of the Si-capped Ge QDs and that of the dilute Ge solid solution which suggests that very strong Ge/Si intermixing occurs. Fitting of the first peak with both Ge–Ge and Ge–Si correlations yielded the Ge–Si fractional coordination number, N_{Ge-Si}, of 3.9 ± 0.3 and the Ge–Si bond length of 2.37 ± 0.01 Å. The uncertainty in the obtained Ge–Si coordination number does not allow one to completely exclude the presence of the Ge phase in the Si-capped dots.

A similar conclusion has been reached from fitting XANES spectra (Fig. 10.11). The samples grown at different temperature have very similar XANES spectra and reasonably good fits were obtained for Ge concentrations in a range of 5–30 %. A very similar result was also obtained in [28].

The authors have also performed a curve fitting for shells higher than the first-nearest neighbors and obtained the following values for the bond lengths: Ge–Si(1) 2.37 ± 0.02 Å, Ge–Si(2) 3.83 ± 0.02 Å, and Ge–Si(3) 4.50 ± 0.02 Å. These values agree very well with the interatomic distances in the silicon crystal which are equal to: 2.35, 3.84, and 4.50 Å, respectively (see also Chap. 2).

For the samples grown at lower temperatures the situation is different. The Ge phase clearly exists. A somewhat larger fractional Ge–Ge coordination number observed for smaller dots grown at the lowest studied temperature of 510 °C is easy to understand since intermixing at lower temperature is much slower. The authors estimate that the amount of Ge (and intermixed Si) in the QDs is about 40 % of the total amount of deposited Ge atoms. Assuming that most of Ge–Si bonding is due to interfaces, they propose that the Ge-rich phase exists as a core of the QDs. Taking the observed average coordination number into account the authors estimate the average Ge concentration in the QDs grown within the 510–550 °C range to be ca. 70 %.

A similar result for the lower growth temperatures was obtained in [29] where the following growth conditions were used. The Si/Ge/Si structures were grown on Si(001) substrates. The epitaxial Si buffer layer with the thickness of 100 nm was produced at the substrate temperature of 600 °C. In order to reduce the size of the quantum dots, the temperature of Si buffer was decreased to 210 °C and then layers of germanium with thickness of 6–25 ML were grown (1 ML of Ge corresponds to ca. 1.4 Å). The structures were subsequently capped by the 20 nm Si layer formed at the temperature of 430 °C. It has been shown that germanium quantum dots start to form when the nominal thickness of the germanium layer exceeds 4 ML. The planar density of the quantum dots in structures grown in similar conditions was approximately $10^{11}\,\mathrm{cm}^{-2}$.

From EXAFS analysis the authors found that in the Si/Ge/Si structure the core of QDs consists of Ge atoms. There is some intermixing of Ge and Si atoms only at the surface of the Ge QDs. For 12 ML sample it was found that the Ge–Ge bonds are shorter and under compressive strain in comparison with the bond length in pure Ge. With increasing of Ge layer thickness, partial stress relaxation occurs in QDs and already for 20 ML sample the Ge–Ge bond length is in agreement with the Ge–Ge bond length in Ge crystal (2.45–2.46 Å) [41–43].

In a similar study [23], the samples were deposited in an ultrahigh vacuum chemical vapor deposition (UHV-CVD) reactor whose base pressure was in the low 10^{-10} Torr range. The Si(100) substrates, after preliminary outgassing, were cleaned at 1,100 °C in H_2 atmosphere. Subsequently, a 500 nm-thick Si buffer layer was grown from high purity silane at a deposition temperature of 800 °C. Then a Ge island layer of 1.3 nm equivalent thickness was deposited at T = 750 °C with a germane pressure of 0.4 mTorr and no carrier gas. In order to investigate the evolution of the Ge islands during the Si overgrowth, the islands were capped with Si layers having thickness ranging from 1 to 33 nm. The growth rate was 1 Å/s, the silane pressure 0.7 mTorr, and the deposition temperature T = 750 °C.

It was found that the average Ge content x decreases from x = 0.55 down to x = 0.25 upon increasing the silicon deposited thickness which is reflected by an increased Ge–Si partial coordination number (Fig. 10.12), i.e. the average Ge content drops at the early stage of the capping (less than 10 nm), and then levels off for larger amounts of deposited silicon. Therefore, once the islands are intermixed and buried, i.e., once a silicon layer thicker than 10 nm has been deposited over the islands, the

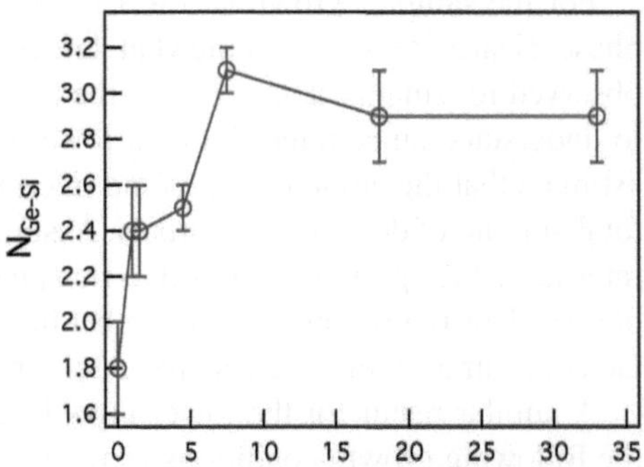

Fig. 10.12 An increase in the partial Ge–Si coordination number as a function of the Si capping layer (plotted after the tabulated data of [23])

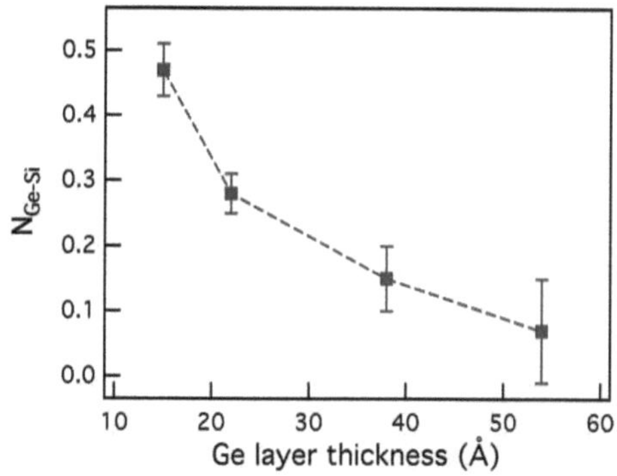

Fig. 10.13 Partial Ge–Si coordination number for the inverted hut structures as a function of the Ge wetting layer, plotted after the tabulated data of [31]

Ge content x does not change anymore. The behaviour was qualitatively the same for different capping temperatures but the saturated value of Ge content decreased with the increased capping temperature. This suggests that, during the capping process, the island shape and composition are stabilized as soon as the critical alloy fraction is reached and the mass transport on the surface is inhibited. A further silicon deposition results in a true Si cap layer [23].

Similar results were reported for the inverted hut nanocrystals [31]. In contrast to the usual nanometer-sized Ge hut clusters commonly grown on top of Si layers using the conventional Stranski-Krastanow self-organized growth mode, SiGe-alloy nanocrystals can be formed beneath the Ge wetting layer and grown into the Si layer in Si/Ge superlattices prepared in a low-temperature molecular beam epitaxy growth mode, and exhibit inverted hut nanocrystal structures regularly spaced along the Si/Ge interface. The EXAFS results obtained with varying Ge wetting layer thickness provide a direct evidence that intermixing of Ge and Si atoms takes place in a zone of about 13 monolayers on each side of the Si/Ge interface. The intermixing of constituent atoms allows a mechanism other than the usual formation of misfit dislocations to release the strain energy resulted from lattice mismatch between Si and Ge at the interface. The average percentage of Ge neighbors being replaced by Si atoms as a result of the presence of the Si/Ge interfaces and Si–Ge intermixing, decreases from 47 to 7 % as the Ge layer thickness increases from 15 to 54 Å (Fig. 10.13).

The conclusion about strong Ge–Si intermixing in both uncapped and especially Si-capped Ge QDs is in agreement with the results obtained using different other techniques [46–49].

10.3.3 Ge Nanoislands Grown on Oxidised Si Surfaces

Ge QDs on oxidized surfaces were studied by XAFS spectroscopy for both SiO_2/Si(001) and SiO_2/Si(111) substrate orientations.

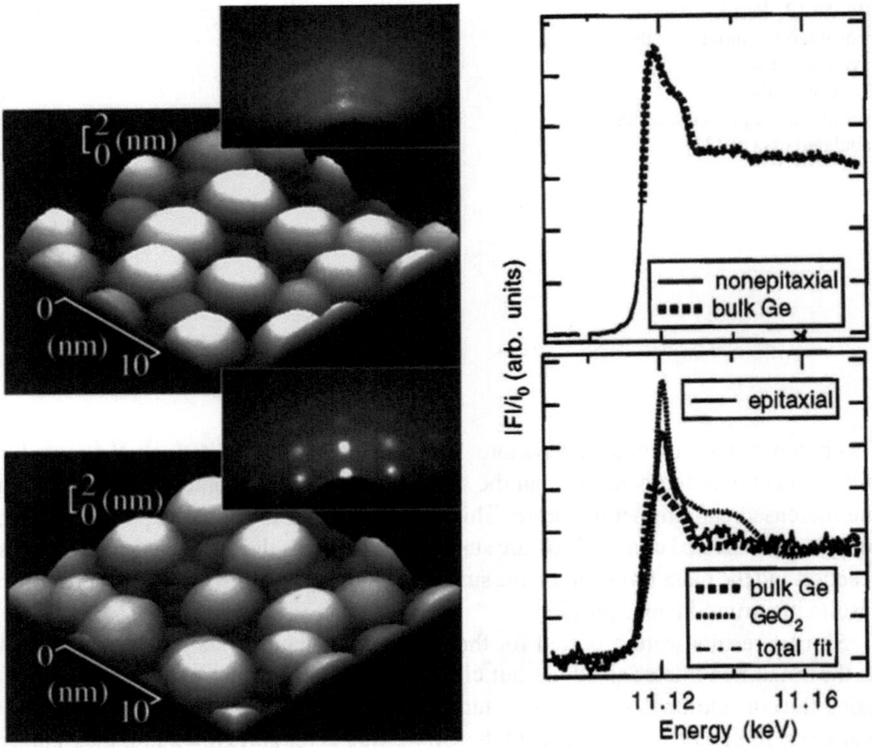

Fig. 10.14 *Left* STM images of Ge islands after 2.6 ML Ge deposition at a deposition rate of 0.5 ML/min on a 0.3 nm thick SiO_2 film on the Si (111) substrate. The substrate temperature during the Ge deposition was (*top*) 390 °C and (*bottom*) 450 °C. Insets show reflected high-energy electron diffraction (RHEED) patterns of the islands, indicating that the islands are (*top*) nonepitaxial and (*bottom*) epitaxial to the Si substrate. *Left* The XANES spectrum (normalized fluorescence) of the nonepitaxial islands (*top*) is identical to that of bulk Ge, while that of the epitaxial islands (*bottom*) has a shoulder at low energies characteristic of unoxidized Ge and a white line typical of Ge-oxide [30]

In the study that used Si(111) substrates, an ultrahigh-vacuum chamber with a base pressure of about 1×10^{-8} Pa was used [30]. Clean Si surfaces were prepared by flash direct-current heating at 1,200 °C. To oxidize the surface, the sample temperature was raised from room temperature to 620 °C for 10 min after oxygen had been introduced into the chamber at a pressure of 2×10^{-6} Torr. The oxide film thickness was estimated to be 0.3 nm and the oxide films were mainly composed of silicon dioxide (SiO_2). A Knudsen cell was used to deposit Ge that formed islands with a base diameter of 7–8 nm and a height of 2–2.5 nm, with the island number density of $\approx 2 \times 10^{12}$ cm^{-2} as shown in Fig. 10.14 (left panel). It is essential that the islands were epitaxial to the Si substrate when the Ge was deposited at temperatures higher than 430 °C, and nonepitaxial at lower temperatures.

In the right panel, XANES spectra for both kinds of Ge islands are shown. One can see that the spectrum for the nonepitaxial islands is identical to that of bulk Ge, with no traces of Ge-oxide being detected. The spectrum for the epitaxial islands, on the other hand, can be fitted as a sum of the metallic Ge and GeO_2 spectra. The best fit is obtained for the fraction of oxidized Ge atoms of about 50 %.

In the study that used Si(100) substrates [27], the samples were grown at 250 °C on substrates that had 0.3 and 1.2 Å SiO_2 coverage. After depositions the films were annealed. After a certain temperature, a RHEED pattern appeared that was indicative of the formation of three-dimensional objects. Ge dots grown though 1.2 nm thick SiO_2 were round shaped with a Ge content of ca. 60 % whereas those grown through 0.3 nm thick SiO_2 were faceted and composed of almost pure Ge.

10.3.4 Embedded Ge Nanoparticles

Ge nanocrystals embedded in SiO_2 were studied for both quartz and Si(100) substrates. While the behaviour was almost similar, slight differences between the substrates were detected (Fig. 10.15). It was found that Ge nanocrystals were predominantly formed from the amorphous Ge phase, pre-existing in samples with higher Ge concentrations. The formed nanocrystals were randomly oriented and had a Ge–Ge bond length of 2.45 ± 0.01 Å, i.e. the crystals were relaxed.

Local structure of GeSi nanocrystals embedded in SiO_2 prepared by co-sputtering of Ge, Si, and SiO_2 targets onto a Si(100) substrate and subsequent annealing was

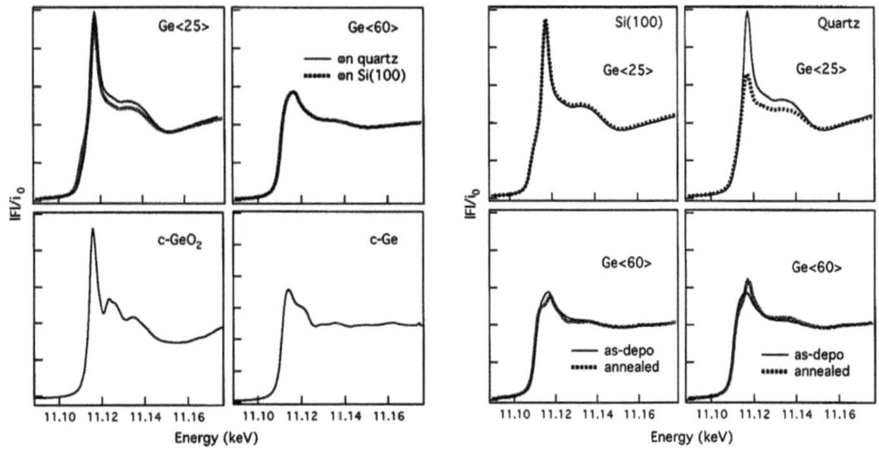

Fig. 10.15 *Left panel* XANES spectra (normalized fluorescence) of as-deposited samples with different low and high Ge content on different substrates (*top pair*) and XANES of reference samples (*bottom pair*). *Right panel* Modification of XANES spectra upon annealing. Ge concentrations and substrate materials are marked in the figure [10]

also studied [9]. The formed nanocrystals, whose size depended on the annealing temperature, were randomly oriented. The nanocrystals were found to consist of a relaxed Ge core with a typical diameter of ca. 4 nm and the Ge–Ge bond length of 2.45 Å and of a GeSi outer shell, the Ge–Si bond length being 2.39 Å. The average composition of the grown nanocrystals was estimated to be $Ge_{0.75}Si_{0.25}$.

10.3.5 Other-Than Ge Quantum Dots

While Ge K-edge energy value makes it easily accessible for X-ray absorption studies, XAFS measurements at other edges, e.g. K-edge of Si, were also done to investigate the local order of nanostructured samples [50, 51]. Sn nanocrystals embedded into a SiO_2 matrix have also been studied [52].

10.4 Beyond Conventional XAFS

10.4.1 Multiple Scattering Analysis of EXAFS

Multiple scattering effects on the analysis of EXAFS spectra of bulk polycrystalline Ge and Ge nanocrystal distributions with mean sizes from 4 to 9 nm were studied in [53]. It was shown that a complete description of the EXAFS signal up to the third shell of nearest neighbours for both c-Ge and Ge nanocrystals is only achieved by including at least two double scattering and one triple scattering path. The authors argued that including only the most prominent double scattering path is insufficient for accurately ascertaining the structural parameters of the second and third shells, leading to unphysically small coordination numbers for the nanocrystals.

Inclusion of multiple scattering into the fitting procedure has also been done for Ge QDs grown on Si(001) where multiple-scattering-EXAFS analysis was performed to obtain the structural parameters from the first to third coordination shells [54]. It is unambiguously demonstrated that in the capping process much stronger Ge/Si intermixing occurs at the temperature of 510 °C than at 300 °C. For the Ge dots capped by Si at 300 °C, from the first shell bond lengths, $R_{Ge\text{–}Ge}$ (2.43 Å) and $R_{Ge\text{–}Si}$ (2.37 Å) as well as the estimated bond angle distortion $\Delta\theta$ (2.0°) and $\Delta\theta'$ (1.0°), the local strain in the nearest neighbor around Ge is mainly accommodated by the bond bending. The observed Ge–Ge interatomic distances of the second (3.93 ± 0.03 Å) and third (4.61 ± 0.04) shells in the islands are 0.07–0.08 Å shorter than the corresponding values in c-Ge, indicating the accommodation of compressive strain by modifying the higher shell Ge–Ge bonds. This implies that the mismatch strain in the Ge core of the QDs appears mainly in the second and higher Ge–Ge shells (See also Chap. 2). For Ge dots capped at 510 °C, Ge is strongly intermixed with Si, but the local structure of the compressively strained pure Ge phase in the core of Ge QDs is hardly changed.

10.4.2 Diffraction Anomalous Fine Structure Experiments

An X-ray spectroscopic, structural, and crystallographic method called the diffraction anomalous fine structure technique (DAFS), which measures the elastic Bragg reflection intensities versus photon energy combines the long-range order and crystallographic sensitivities of X-ray diffraction with the spectroscopic and short-range order sensitivities of X-ray absorption techniques. In the extended fine structure region, DAFS provides the same short-range structural information as EXAFS: the bond lengths, coordination numbers, neighbor types, and bond disorders for the atoms surrounding the resonantly scattering atoms. In the near-edge region, DAFS provides the same structural and spectroscopic sensitivities as XANES: the valence, empty orbital and bonding information for the resonant atoms.

Because DAFS combines the capabilities of diffraction, EXAFS and XANES into a single technique, it has two enhanced sensitivities compared to the separate techniques, namely, (i) non-equivalent crystal site selectivity where DAFS can provide EXAFS- and XANES-like information for the specific subset of atoms selected by the diffraction condition and (ii) site selectivity, where DAFS can provide site-specific absorption-like spectroscopic and structural information for the inequivalent sites of a single atomic species within the unit cell. The major disadvantage of DAFS is a rather complicated data analysis process. DAFS was successfully used to disentangle strain and composition and to detect atomic ordering inside GeSi nanoislands grown on Si(001) [55] as well as to selectively study local order in intermixed nanocrystalline and amorphous phases [56].

10.4.3 Femtometer Precision XAFS

While the accuracy in distance determination is usually considered to be ±0.01 Å, significantly higher femtometer accuracy has been reported for the first, second and third shells of Ge where meaningful differences were detected between ^{70}Ge and ^{76}Ge pure isotope samples [57].

10.4.4 Spectroscopy of Empty States

X-ray absorption spectroscopy has also been successfully applied to investigate empty states in Ge/Si(111) epitaxial structures grown at room temperature. Based on XANES analysis of L_3-edge spectra, the authors conclude that initially the Ge layers are amorphous. Upon annealing the behavior depended on the Ge layer thickness: for thicknesses below 3 ML continuous intermixed films were formed while for larger thicknesses three-dimensional islands were formed [58].

10.4.5 Time-Resolved Studies

XAFS has a potential to investigate the kinetics of structural evolution of Ge with sub-ns time resolution, for example under laser-induced heating, as has been demonstrated by recent studies [59].

10.5 Summary and Outlook

In this chapter, use of X-ray absorption fine structure for investigations of group IV quantum dots and nanocrystals has been described. For QDs grown on Si substrates, both uncapped and capped, strong intermixing between Ge and Si takes place. In most cases, structures often referred to as Ge QDs are in reality strongly intermixed GeSi QDs. Growth and Si capping at lower temperatures tend to generate QDs with a higher Ge content but epitaxial growth of pure Ge QDs remains a challenge. On the other hand, embedded nanocrystals produced by annealing composite samples can more readily generate the Ge-rich phase. Further progress in understanding the structure and properties of Ge nanostructures can be expected from application of more sophisticated X-ray absorption studies beyond conventional EXAFS/XANES, such as use of DAFS or time-resolved experiments.

References

1. A.I. Ekimov, A.A. Onushchenko, JETP Lett. **34**(6), 345 (1981)
2. M.A. Reed, J.N. Randall, R.J. Aggarwal, R.J. Matyi, T.M. Moore, A.E. Wetsel, Phys. Rev. Lett. **60**(6), 535 (1988)
3. E.G. Barbagiovanni, L.V. Goncharova, P.J. Simpson, Phys. Rev. B **83**(3), 035112 (2011)
4. E.G. Barbagiovanni, D.J. Lockwood, P.J. Simpson, L.V. Goncharova, J. Appl. Phys. **111**(3), 034307 (2012)
5. K.L. Wang, J.L. Liu, G. Jin, J. Cryst. Growth **237**(4), 1892 (2002)
6. J. Menendez, J. Kouvetakis, Appl. Phys. Lett. **85**(7), 1175 (2004)
7. Y. Kanemitsu, H. Uto, Y. Masumoto, Y. Maeda, Appl. Phys. Lett. **61**(18), 2187 (1992)
8. A.V. Kolobov, Y. Maeda, K. Tanaka, J. Appl. Phys. **88**(6), 3285 (2000)
9. A.V. Kolobov, H. Oyanagi, Y. Maeda, K. Tanaka, J. Synchrotron Radiat. **8**(2), 511 (2001)
10. A.V. Kolobov, S. Wei, W. Yan, H. Oyanagi, Y. Maeda, K. Tanaka, Phys. Rev. B **67**(19), 195314 (2003)
11. W. Yan, Z. Li, Z. Sun, Z. Pan, S. Wei, A. Kolobov, J. Appl. Phys. **101**(10), 104318 (2007)
12. K. Hassan, A. Sharma, J. Narayan, J. Muth, C. Teng, R. Kolbas, Appl. Phys. Lett. **75**(9), 1222 (1999)
13. F. Meseguer, A. Blanco, H. Mıguez, F. Garcıa-Santamarıa, M. Ibisate, C. Lopez, Colloids Surf. A **202**(2), 281 (2002)
14. H. Miguez, V. Fornes, F. Meseguer, F. Marquez, C. Lopez, Appl. Phys. Lett. **69**(16), 2347 (1996)
15. P. Castrucci, R. Gunnella, N. Pinto, M. De Crescenzi, M. Sacchi, G. Dufour, F. Rochet, Surf. Sci. **416**(3), 466 (1998)

16. G. Jin, J.L. Liu, S.G. Thomas, Y. Luo, K.L. Wang, B.Y. Nguyen, Appl. Phys. Lett. **75**(18), 2752 (1999)
17. T. Kitajima, B. Liu, S.R. Leone, Appl. Phys. Lett. **80**(3), 497 (2002)
18. K. Brunner, Rep. Progr. Phys. **65**(1), 27 (2001)
19. F. Boscherini, G. Capellini, L. Di Gaspare, M. De Seta, F. Rosei, A. Sgarlata, N. Motta, S. Mobilio, Thin Solid Films **380**(1), 173 (2000)
20. F. Boscherini, G. Capellini, L. Di Gaspare, F. Rosei, N. Motta, S. Mobilio, Appl. Phys. Lett. **76**(6), 682 (2000)
21. N. Motta, F. Rosei, A. Sgarlata, G. Capellini, S. Mobilio, F. Boscherini, Mater. Sci. Eng. B **88**(2), 264 (2002)
22. N. Motta, F. Boscherini, A. Sgarlata, A. Balzarotti, G. Capellini, F. Ratto, F. Rosei, Phys. Rev. B **75**(3), 035337 (2007)
23. G. Capellini, M. De Seta, L. Di Gaspare, F. Evangelisti, F. dAcapito, J. Appl. Phys. **98**(12), 124901 (2005)
24. A.V. Kolobov, H. Oyanagi, K. Brunner, P. Schittenhelm, G. Abstreiter, K. Tanaka, Appl. Phys. Lett. **78**(4), 451 (2001)
25. A.V. Kolobov, H. Oyanagi, K. Brunner, G. Abstreiter, Y. Maeda, A. Shklyaev, S. Yamasaki, M. Ichikawa, K. Tanaka, J. Vacuum Sci. Tech. A **20**(3), 1116 (2002)
26. A.V. Kolobov, H. Oyanagi, S. Wei, K. Brunner, G. Abstreiter, K. Tanaka, Phys. Rev. B **66**(7), 075319 (2002)
27. R. Dujardin, V. Poydenot, T. Schulli, G. Renaud, O. Ulrich, A. Barski, M. Derivaz, S. Colonna, T. Metzger, J. Appl. Phys. **99**(6), 063510 (2006)
28. I. Demchenko, K. Lawniczak-Jablonska, S. Kret, A. Novikov, J. Laval, M. Zak, A. Szczepanska, A. Yablonskiy, Z. Krasilnik, Nanotechnology **18**(11), 115711 (2007)
29. I. Demchenko, K. Lawniczak-Jablonska, E. Piskorska, K. Zhuravlev, A. Nikiforov, E. Welter, J. Alloys Comp. **382**(1), 206 (2004)
30. A.V. Kolobov, A.A. Shklyaev, H. Oyanagi, P. Fons, S. Yamasaki, M. Ichikawa, Appl. Phys. Lett. **78**(17), 2563 (2001)
31. Y. Soo, G. Kioseoglou, S. Huang, S. Kim, Y. Kao, Y. Peng, H. Cheng, Appl. Phys. Lett. **78**(23) (2001)
32. V. Le Thanh, V. Yam, P. Boucaud, F. Fortuna, C. Ulysse, D. Bouchier, L. Vervoort, J.M. Lourtioz, Phys. Rev. B **60**(8), 5851 (1999)
33. A.V. Kolobov, J. Appl. Phys. **87**(6), 2926 (2000)
34. K. Uchinokura, T. Sekine, E. Matsuura, J. Phys. Chem. Solids **35**(2), 171 (1974)
35. P.Y. Yu, M. Cardona, *Fundamentals of Semiconductors: Physics and Materials Properties* (Springer, Berlin, 1999)
36. G. Jin, Y.S. Tang, J.L. Liu, K.L. Wang, Appl. Phys. Lett. **74**(17), 2471 (1999)
37. A.V. Kolobov, H. Oyanagi, A. Frenkel, I. Robinson, J. Cross, S. Wei, K. Brunner, G. Abstreiter, Y. Maeda, A. Shklyaev et al., Nucl. Instrum. Methods B **199**(1), 174 (2003)
38. A.V. Kolobov, J. Mater. Sci.: Mater. Electron. **15**(4), 195 (2004)
39. S.B. Erenburg, N.V. Bausk, A.V. Dvurechenskii, Z.V. Smagina, A.V. Nenashev, A.I. Nikiforov, V.G. Mansurov, K.S. Zhuravlev, A.I. Toropov, J. Surf. Invest. **1**(1), 26 (2007)
40. F. Rosei, N. Motta, A. Sgarlata, G. Capellini, F. Boscherini, Thin Solid Films **369**(1), 29 (2000)
41. J. Aubry, T. Tyliszczak, A. Hitchcock, J.M. Baribeau, T. Jackman, Phys. Rev. B **59**(20), 12872 (1999)
42. M.C. Ridgway, K.M. Yu, C.J. Glover, G. Foran, C. Clerc, J.L. Hansen, A.N. Larsen, Phys. Rev. B **60**(15), 10831 (1999)
43. I. Yonenaga, M. Sakurai, Phys. Rev. B **64**(11), 113206 (2001)
44. C. Tzoumanekas, P.C. Kelires, Phys. Rev. B **66**(19), 195209 (2002)
45. M. De Seta, G. Capellini, L. Di Gaspare, F. Evangelisti, F. d'Acapito, J. Appl. Phys. **100**, 093516 (2006)
46. J. Wan, Y.H. Luo, Z.M. Jiang, G. Jin, J.L. Liu, K.L. Wang, X.Z. Liao, J. Zou, J. Appl. Phys. **90**(8), 4290 (2001)
47. M. De Seta, G. Capellini, F. Evangelisti, C. Spinella, J. Appl. Phys. **92**(1), 614 (2002)

48. M. De Seta, G. Capellini, F. Evangelisti, Superlattices Microstruct. **46**(1), 328 (2009)
49. Z.M. Jiang, X.M. Jiang, W.R. Jiang, Q.J. Jia, W.L. Zheng, D.C. Qian, Appl. Phys. Lett. **76**(23), 3397 (2000)
50. R. Sammynaiken, S. Naftel, T.K. Sham, K.W. Cheah, B. Averboukh, R. Huber, Y. Shen, G. Qin, Z. Ma, W. Zong, J. Appl. Phys. **92**(6), 3000 (2002)
51. A.P. Hitchcock, T. Tyliszczak, P. Aebi, J.Z. Xiong, T.K. Sham, K.M. Baines, K.A. Mueller, X.H. Feng, J.M. Chen, B.X. Yang, Z.H. Lu, J.M. Baribeau, T.E. Kackman, Surf. Sci. **291**(3), 349 (1993)
52. S. Spiga, R. Mantovan, M. Fanciulli, N. Ferretti, F. Boscherini, F. d'Acapito, B. Schmidt, R. Grötzschel, A. Mucklich, Phys. Rev. B **68**(20), 205419 (2003)
53. L.L. Araujo, G.J. Foran, M.C. Ridgway, J. Phys.: Condens. Matter **20**(16), 165210 (2008)
54. Z. Sun, S. Wei, A.V. Kolobov, H. Oyanagi, K. Brunner, Phys. Rev. B **71**(24), 245334 (2005)
55. M. Richard, N. Katcho, M. Proietti, H. Renevier, V. Favre-Nicolin, Z. Zhong, G. Chen, M. Stoffel, O. Schmidt, G. Renaud et al., Eur. Phys. J. **167**(1), 3 (2009)
56. A.I. Frenkel, A.V. Kolobov, I.K. Robinson, J.O. Cross, Y. Maeda, C. Bouldin, Phys. Rev. Lett. **89**(28), 285503 (2002)
57. J. Purans, N. Afify, G. Dalba, R. Grisenti, S. De Panfilis, A. Kuzmin, V. Ozhogin, F. Rocca, A. Sanson, S. Tiutiunnikov et al., Phys. Rev. Lett. **100**(5), 55901 (2008)
58. P. Castrucci, R. Gunnella, M. De Crescenzi, M. Sacchi, G. Dufour, F. Rochet, Phys. Rev. B **60**(8), 5759 (1999)
59. D.L. Brewe, E.A. Stern, S.M. Beck, K. L. Heald, Y. Feng, http://www.aps.anl.gov/apsar2002/BREWED1.PDF

Chapter 11
Group IV Nanowires

Xuhui Sun and Tsun-Kong Sham

Abstract X-ray absorption fine structure spectroscopy and related techniques such as X-ray emission spectroscopy and X-ray Excited Optical luminescence play a significant role in understanding the electronic structures of group IV semiconductor nanowires. This chapter reviews how these techniques have been recently used to reveal the unique properties of group IV semiconductor nanowires, especially silicon and germanium, and their oxides. Other group IV nanowires, such as carbon and tin nanowires are also noted.

11.1 Introduction

The emergence of group IV nanowires [1–7] as a class of unique materials that are fundamentally interesting and technologically important has drawn considerable attention in X-ray absorption spectroscopy studies [8]. X-ray absorption fine structure spectroscopy (XAFS) including X-ray near edge absorption fine structure (XANES) [9] and related techniques such as X-ray emission spectroscopy (XES) [10–13] and X-ray Excited Optical luminescence (XEOL) [14–16], play a unique role in tracking the densities of states of the conduction band, the valence band and defect states in group IV semiconductor nanowires [17, 18]. This chapter reviews how these techniques have been recently used to reveal unique properties of group IV semiconductor nanowires, especially silicon, germanium and their oxides. Other group IV nanowires, such as carbon and tin nanowires will be noted. The emphasis is placed on X-ray absorption spectroscopy and related studies, especially in the soft

X. Sun (✉)
Soochow University-Western University Centre for Synchrotron Radiation Research, Institute of Functional Nano & Soft Materials (FUNSOM), Soochow University, Suzhou, Jiangsu 215123, People's Republic of China
e-mail: xhsun@suda.edu.cn

T.-K. Sham (✉)
Department of Chemistry, Soochow University-Western University Joint Centre for Synchrotron Radiation Research, University of Western Ontario, London, ON N6A 5B7, Canada
e-mail: tsham@uwo.ca

C.S. Schnohr and M.C. Ridgway (eds.), *X-Ray Absorption Spectroscopy of Semiconductors*, Springer Series in Optical Sciences 190,
DOI 10.1007/978-3-662-44362-0_11

X-ray region. In addition to XANES, the use of X-ray emission, a photon-in photon-out technique when the core level is just below the valence band is also noted. A special technique, the X-ray Excited Optical Luminescence (XEOL) in both energy and time domain, a unique photon–in photon-out probe in which the X-ray energy absorbed by the specimen is transferred to a radiative de-excitation channel in the optical region (UV, visible and near IR) via core hole cascade and secondary processes, will be discussed in some details [19]. This capability is most suitable for the investigation of light emitting nanomaterials [8].

11.2 Si and Ge Nanowires: Morphology and Structure Via Top-down and Bottom up Strategies

Group IV nanowires can be prepared by "bottom-up" and "top-down" strategies. The "bottom-up" strategy can be used to synthesize large super molecules of predetermined sizes and shapes by chemical means (e.g. via self-assembly) [20–23]. The "top-down" approach fabricates nanomaterials from bulk materials using novel techniques such as electron beam, electrochemistry, X-ray lithography, nanoimprint, nanomachining, etc. [24–27].

Many bottom-up techniques have been developed to produce bulk quantities of Si nanowire (SiNW). A widely used one is the vapor-liquid-solid (VLS) method [28], in which silicon from a gaseous source is "dissolved" in the liquid phase of the catalyst, in the form of a nanoparticle, forming a eutectic. Oversaturated silicon subsequently crystallizes at the liquid-solid interface, promoting one-dimensional growth [1, 29]. A variant of the VLS technique is the solution-liquid-solid (SLS) process wherein the gaseous source is replaced by a solution source [3].

Another commonly used technique is the oxide-assisted growth (OAG) via thermal evaporation of silicon suboxide [2, 30]. In the OAG method, oxides play an important role in inducing the nucleation and growth of nanowires. Its major advantage is that it does not require a metal catalyst which can affect the performance of nanowires in a device.

Recently, several other top-down methods have been developed to prepare SiNWs from bulk Si such as lithography and etching, [31–33] and metal-catalyzed electroless etching (MCEE) [34–36]. Ordered arrays of specifically oriented SiNWs of controlled size, density, and electronic properties can be easily produced on a Si wafer by a combination of reactive ion etching (RIE) and lithography. For example, Peng et al. reported that a wafer-scale SiNW array could be readily produced via electroless etching at room temperature by simply immersing Si wafers into a HF–$AgNO_3$ solution [34].

Compared to Si, renewed interest in Ge as a material of choice for future electronics is more recent and is due to its significantly higher electron and hole mobility than Si [37]; this is especially desirable as electronic devices are scaled down to the sub-100 nm regime. Therefore, Ge nanowires are expected to offer potential performance gain over Si-based planar and nanowire devices. Moreover, the exciton radius of Ge (24.3 nm) is larger than that of Si (4.9 nm), which consequently leads to more prominent quantum confinement effects [38]. However, strategies for the synthesis of germanium nanowires are not as extensive as for Si. Several growth methods such as laser ablation [1], vapor transport [39, 40], low-temperature CVD [41, 42], and supercritical fluid-liquid-solid synthesis [43], have been developed. In most cases, GeNW growth using Au nanoparticle as the catalyst benefits from the low eutectic temperature of Ge-Au alloy (360 °C), and follows the VLS [28] process. Low temperature synthesis helps reduce growth defects, prevents nanowire from outgassing or deformation, and minimizes dopant diffusion where local doping is strictly demanded in device design.

A systematic study of the effect of temperature on GeNW synthesis shows that Ge nanowires can be produced within the temperature range of 410–850 °C using Ge powder as the feeding source, following the metal-catalytic VLS mechanism [7]. The morphology of GeNW thus produced exhibits significant dependence on the growth temperature. Nanowires grown from 650 to 850 °C show low density on the substrates with non-uniform diameter. The best growth of nanowires that are straight, uniform in size and with high density (yield) is observed from 450 to 600 °C. Below 450 °C, in addition to nanowires, Ge particles appear with increasing size as temperature decreases. Only Ge thin-film growth was observed from 380 to 340 °C. Two alternative industrial benign catalysts are indium (In) and antimony (Sb). Similar to Au, both In and Sb exhibit eutectic behavior with Ge (In-Ge: 158 °C, Sb-Ge: 587 °C), and have been used as both the catalyst and the dopant sources for GeNW synthesis [44].

11.3 Soft X-Ray Spectroscopy: Yield Measurements, XANES, XES and XEOL

11.3.1 X-Ray Absorption Fine Structure Spectroscopy

X-ray absorption fine structure spectroscopy (XAFS) is concerned with the measurement and interpretation of the absorption coefficient above an absorption edge [45]. In a free atom, as the incident photon energy is approaching the threshold of a core level, it will excite the core electron to Rydberg states via dipole selection rules ($\Delta l = \pm 1$, where l is the angular momentum). Further increase in photon energy excites the core electron to the continuum (ionization). Thus XAFS for free atom will appear as a series of sharp peaks approaching the ionization threshold followed by the edge jump and a featureless region beyond where the absorption coefficient decreases monotonically with increasing photon energy.

When an atom is in a chemical environment, the Rydberg states are quenched, the molecular and crystal potential sets up molecular orbitals and band states, respectively; these states can be bound (such as the lowest unoccupied molecular orbital (LUMO)), or quasi-bound (states trapped by the potential barrier, also known as

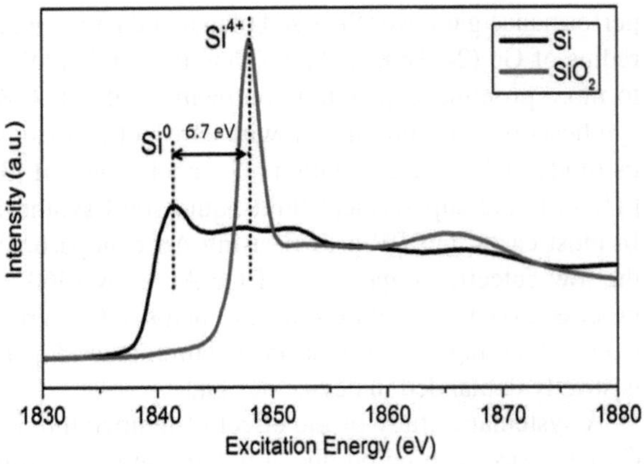

Fig. 11.1 Si K-edge XANES of Si and SiO_2 nanowires; the *dotted line* shows the first resonance (known as *whiteline*) of Si in Si^o and Si^{4+} oxidation state with a separation of ~6 eV

multiple scattering states). The absorption coefficient above the threshold shows resonances ranging from several eV below the threshold up to ~50 eV above the threshold. This region is referred to as XANES or NEXAFS (Near Edge X-ray Absorption Fine Structures) where one can observe energy shifts of the threshold due to different oxidation state and resonances characteristic of coordination number, neighboring atoms and local symmetry.

Figure 11.1 shows the Si K-edge XANES of Si and SiO_2 nanowires illustrating the spectral feature (band-like states in Si and sharp resonance in SiO_2) and the sensitivity of the probe (a ~6 eV whiteline shift from element Si(0) to Si(IV)) [14, 15]. Beyond the XANES is the extended region known as EXAFS (Extended X-ray Absorption Fine Structures) which exhibits sinusoidal oscillations arising from the interference of the outgoing and backscattered electron waves of the outgoing electrons with sufficient kinetic energy such that single scattering process dominates [45]. Since the absorption edge is element specific, XANES will be element and chemical specific, a most desirable capability for speciation and chemical analysis.

11.3.2 Soft X-Ray Absorption Measurements: Yield and De-excitation Spectroscopy

XAFS in the hard X-ray energy region (e.g. >5 keV) is often recorded in transmission in which, the incident photon flux I_o and the transmitted photon flux I through a homogeneous sample with thickness t are measured. The absorption coefficient μ can be obtained according to the Beer's law,

$$I = I_o e^{-\mu t} \tag{11.1}$$

Table 11.1 Attenuation length ($t = 1/\mu$) of photon energy at relevant edges of group IV elements [46]

Element	Edge (eV)	Just below (μm)	Just above (μm)
C	K (284)	2.5	0.088
Si	$L_{3,2}$ (100)	0.55	0.056
	K (1840)	14	1.27
Ge	L_3 (1216)	1.57	0.24
	K (11103)	67	9.2
Sn	M_5(484)	0.30	0.29
	L_3 (4156)	1.6	1.2
	K (29205)	193	32

where $\mu = \sigma \times \rho$, μ is the absorption coefficient (cm^{-1}), σ is mass absorption cross-section in cm^2/g and ρ is density in g/cm^3; μ contains all information about the local structure and bonding. In soft X-ray spectroscopy, we deal with very strong absorption such that the X-ray only penetrates a shallow region of the specimen and is totally absorbed; it is therefore undesirable or impossible to prepare such a thin sample for transmission measurements. A rule of thumb for good statistics is to use a sample thickness of one absorption length, also known as attenuation length (the incident photon intensity falls off to 1/e of its initial value). The attenuation lengths for group IV elements with edges of interest are shown in Table 11.1.

It is apparent from Table 11.1 that the most interesting edges fall below 4 keV with attenuation length in the micron to submicron range. Thus measurements at these edges are often made with yield spectroscopy; that is that we use the products such as electrons and photons and sometimes ions and even neutrals resulting from the absorption to monitor μ provided that the yield is proportional to $I_o(1-e^{-\mu t})$, where the yield is often total electron yield (TEY) and total fluorescent X-ray yield (FLY). More recent studies utilizing Si solid state detector facilitates partial fluorescence yield as well as inverse partial fluorescence yield [47]. TEY contains photoelectrons, Auger electrons and to the largest extent secondary electrons from the thermalization of energetic electrons in the specimen. Since electrons have short escape depths, thus TEY is surface sensitive while FLY, which has an attenuation length typically two orders of magnitude higher than that of electron, is bulk sensitive. Hence Soft X-ray spectroscopy adds a new dimension in probing the depth profile of a specimen. In addition, nanomaterials provide inadvertently other advantages: the reduction of saturation effect in FLY and the truncation (confinement) of the thermalization track. The latter plays a significant role in the site specificity of XEOL measurements. This will be discussed further below.

It should be noted that yield measurement is generally valid in TEY and FLY for thin samples (e.g. nanostructures) although FLY often suffers from saturation effect if the sample is too thick. This can be amended by inverse fluorescence yield [47]. Expansion of the exponent term, $I_o(1 - e^{-\mu t})$ gives a yield proportional to μt. More details can be found in the literature [48]. For light emitting nanostructures, we can track the absorption at an edge using XEOL in both the energy (selected wavelength) and the time domain (using the pulse structure of the synchrotron) [19, 49–51] providing another perspective for the study of energy transfer dynamics in nanostructures.

When a shallow core electron is excited, the core hole decays immediately via Auger and XES. We can track the decay using de-excitation spectroscopy. XES is particularly useful when a shallow core level is directly below the valence band, for example, the C 1s (K-edge) and the Si $2p_{3/2,1/2}$ ($L_{3,2}$-edge) core levels [10, 12]. The core hole in C 1s and Si 2p are filled directly by valence electrons of p and s character, respectively, emitting fluorescence X-rays, tracking the density of states of the valence band. Thus XES yields the valence band density of states similar to that of photoemission with two added advantages: first, it is core specific; that is that for a compound composed of two or more elements, it can probe the contribution of a specific element to the valence band; second, it is bulk sensitive. It has a very low cross-section however and requires a high brightness source [11, 13].

Figure 11.2 shows the schematic of XANES, XEOL, XES and XANES with PLY. XES records the energy distribution of the fluorescence X-rays from the filling of the core hole by electrons in the valence band via dipole transition. XEOL tracks the energy transfer to the optical channel via a process in which the energetic electrons (holes) are thermalised in the solid resulting in electrons at the bottom of the conduction band and holes at the top of the valence band. The electrons and holes can recombine to form an exciton which then decays radiatively emitting an optical

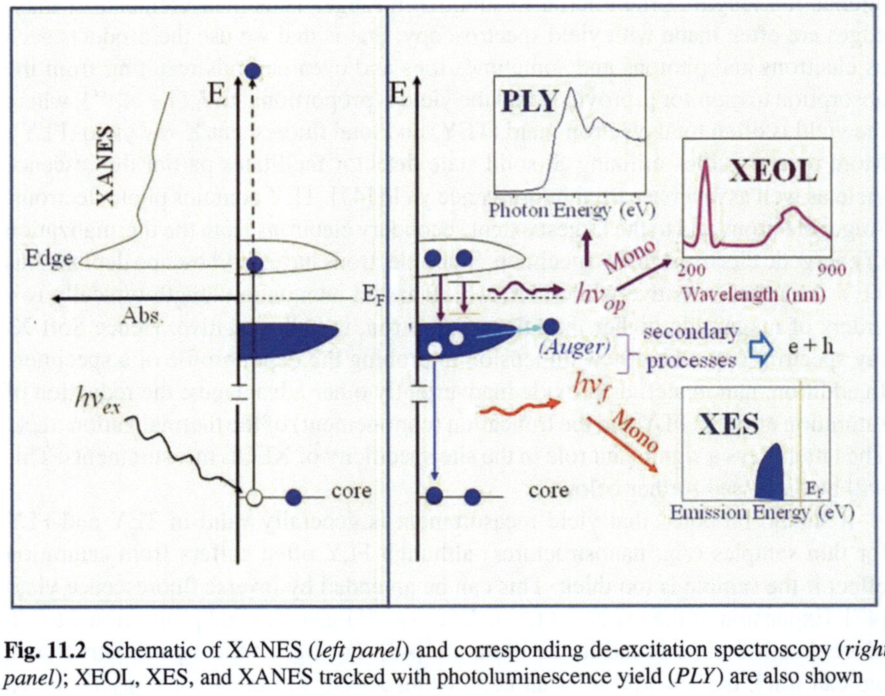

Fig. 11.2 Schematic of XANES (*left panel*) and corresponding de-excitation spectroscopy (*right panel*); XEOL, XES, and XANES tracked with photoluminescence yield (*PLY*) are also shown

photon with energy near that of the band gap energy. This emission is often called near band gap (NBG) emission for its energy is the difference between the band gap energy and the binding energy of the exciton ($E_B - E_{ex}$) and it is a fast process. When defects are present, energy transfer to various optical channels within the band gap can take place, emitting at longer wavelengths with a much longer life time. Finally, XANES can also be monitored by PLY as shown in the right panel of Fig. 11.2. In this case, the PLY can be the yield of all optical photons (zero order) or wavelength-selected and the PLY-XANES either resembles that of the absorption in TEY or exhibits variations such as distortion or even inversion depending on the origin of the luminescence, which in turn depends on the absorbing atom and its environment, the sample thickness and more importantly the size, morphology and crystallinity of the specimen [48]. These situations are discussed below.

11.3.3 XEOL in the Time Domain

The perhaps most unique capability for the investigation of group IV nanowires is XEOL in both the energy and time domain; the latter is often referred to as TR-XEOL (time-resolved XEOL) or time-gated XEOL spectroscopy [19, 49–51]. A schematic is shown in Fig. 11.3.

From Fig. 11.3, we see that the electron bunch produces a light pulse (green arrow) which is used to trigger the start/stop, and the dark gap between the pulses is used to track the luminescence (red arrow) excited by the X-ray pulse within a selected time window; the top right graph shows the decay of all optical photons (zero order) and a selected time window marked as horizontal bars (red for fast and blue for slow), which can be used to track the optical yield, as shown on the far right.

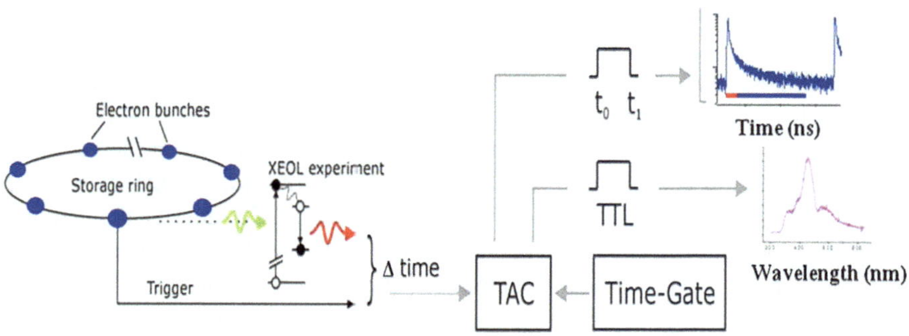

Fig. 11.3 Schematic for TRXEOL using the time structure of the synchrotron: *Left panel* an electron bunch emits a short pulse (35 ps at the CLS) with a dark gap between bunches (570 ns at the CLS for single bunch). *Right panel* Time-gated XEOL by selecting a time window (e.g. *red* and *blue* bar for fast and slow window) of the decay curve within the dark gap

By exciting the system of interest with varying photon energy across absorption edges, we can preferentially inject the photon energy into a particular site of interest and watch how it is transferred to the optical channel in the time domain. This process in turn reveals many properties of the nanostructures as is illustrated below in group IV nanowires.

11.4 Si and Ge Nanowires and Related Materials: X-Ray Spectroscopy Studies

11.4.1 Si Nanowires

We present below representative XANES and de-excitation (XES and XEOL) spectroscopy studies of Si NW. Cases include (i) SiNW from OAG bottom-up method, (ii) SiNW from MCEE top-down method, and (iii) nanowire composites.

11.4.1.1 Bottom-up VLS Si Nanowires

SiNW prepared from SiO is typically coated with a thin layer of oxide which in turn helps guide its growth direction [2, 30]. This material is light emitting under the excitation of UV, electrons and X-rays. Figure 11.4a summarizes the Si K-edge results of such a SiNW, more appropriately a Si/SiO_2 core/shell NW. Figure 11.4b shows the site specific emission with excitation energy that falls on the whiteline (WL), a sharp resonance at the Si K-edge of Si and SiO_2 [17].

It is apparent from Fig. 11.4a that the Si K-edge XANES recorded in TEY and FLY exhibit characteristics of both Si and SiO_2. The edge jump for elemental Si increases markedly in the FLY compared to the TEY, confirming a core/shell structure. The XEOL in Fig. 11.4b shows that the specimen emits a wide range of optical photons (blue-green-orange). Three bands at 460, 530 and 630 nm are clearly discernible. It is well known that Si has an indirect band gap of ~1.1 eV (1,127 nm) and emission in the visible is forbidden. It is also recognized however that in nanostructures, quantum confinement takes effect and the presence of surface oxide or defects will also contribute to the luminescence [14, 15]. To identify the origin of the luminescence, XEOL was excited at the WL of Si and SiO_2. The result shows that the branching ratio of 460 nm emission increases markedly at the SiO_2 WL (inset, Fig. 11.4b). This observation immediately points this band to a surface oxide origin. A series of partial PLY (wavelength selected) XANES at the three emission bands is shown in Fig. 11.4a. The PLY at 460 nm exhibits the most intense SiO_2 WL and is attributed to surface SiO_2, confirming the XEOL results. The 530 nm PLY shows little SiO_2 WL and the largest Si/SiO_2 WL intensity ratio, and is attributed to the quantum confined silicon nanocrystals in the NW. The 630 nm PLY exhibits a Si/SiO_2 WL ratio between that of 460 and 530 nm yield and is attributed to the

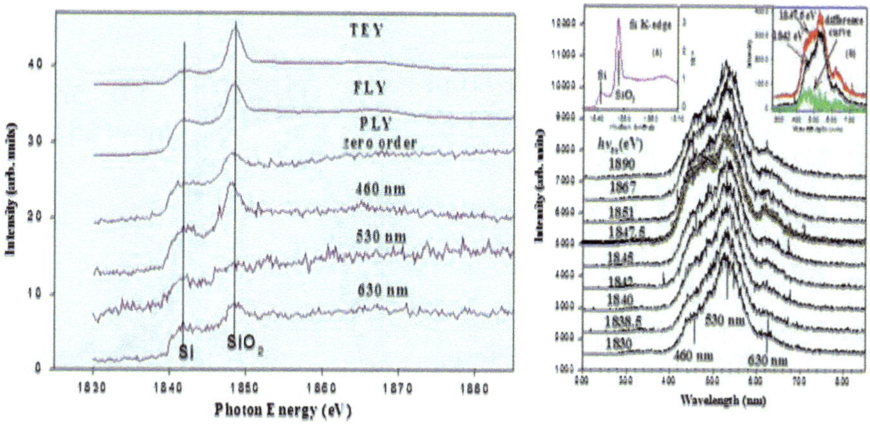

Fig. 11.4 **a** Si K-edge XANES of a core/shell Si/SiO_2 NW recorded in TEY, FLY, and wavelength-selected PLY; *vertical line* marks the WL of Si and SiO_2. **b** XEOL excited with photon energy from below to above the Si K-edge and at the WL corresponding to the Si core and the SiO_2 shell (*top right inset*); the TEY XANES is also shown (*top left inset*) [17]

interfacial states between the Si nanocrystal and SiO_2. A similar specimen, which exhibits three similar XEOL bands at 477, 564 and 689 nm, has been investigated by TRXEOL [52]. The results are shown in Fig. 11.5.

From Fig. 11.5, we see that TRXEOL using a fast and a slow window yields different emission patterns with different branching ratio (Fig. 11.5a, b). A step by step tracking of the yield of the emission branches across the Si K-edge XANES in 0.5 eV steps shows that the Si and SiO_2 whiteline intensity thus obtained varies significantly: The 2.6 eV and the 3.4 eV emission within the slow window is respectively, more and mostly oxide like, while the 2.2 and 1.8 eV within the fast window, is respectively, more and mostly Si like. In addition, the 1.8 eV (689 nm) emission yield shows a weak oxide whiteline at energy slightly lower than that of SiO_2; this is consistent with the presence of sub-oxide or Si/Si oxide interface confirming the XEOL results observed in Fig. 11.4.

The Si/SiO_2 core shell NW has been probed with XES. The left panel of Fig. 11.6 shows the Si L_l XES excited above the Si $L_{3,2}$-edge. These data are to be distinguished from pre-edge excitations where Resonant Inelastic X-ray Scattering (RIXS) takes place; RIXS is beyond the scope of this article [11, 13]. Thus the XES resembles the valence band from X-ray Photoelectron Spectroscopy (XPS) as shown in the right panel of Fig. 11.6 [17].

Let us look at the left panel of Fig. 11.6 and compare the "SiNW as prepared", which has a Si/SiO_2 core/shell structure, with the "SiNW HF" (oxide shell removed). We see that the intensity of the band closest to the Fermi level is drastically reduced indicating that this band in the as-prepared SiNW is mainly from SiO_2. A comparison of the XES of the HF treated SiNW with that of the clean Si (100) confirms that the oxide on the surface of the as-prepared SiNW has been removed by HF.

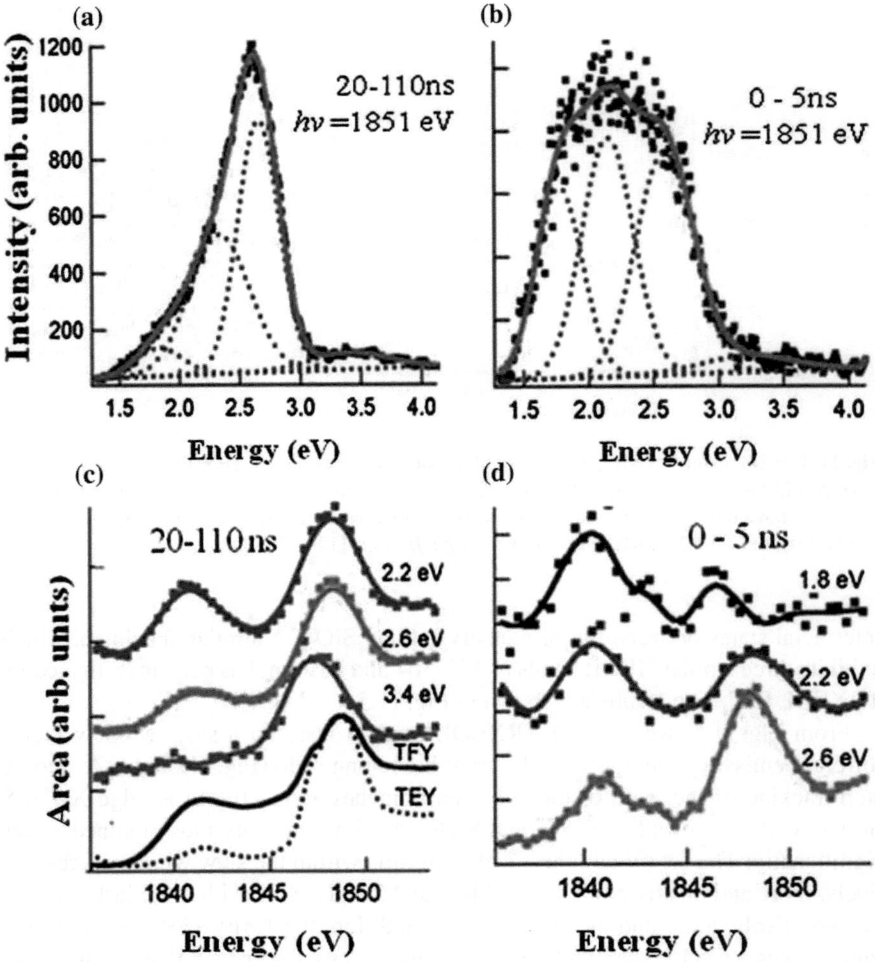

Fig. 11.5 TRXEOL and Time-gated XANES of a core/shell Si/SiO_2 NW: **a** With a slow window (20–110 ns), **b** With a fast window (0–5 ns), the dark gap is 153 ns (APS top-up mode). **c** Area under the *curve* within the slow window is plotted as a function of excitation energy from 1835 to 1855 eV at 0.5 eV steps. **d** Area under the *curve* within the fast window is plotted as a function of excitation energy as in **d** [52]

We now compare the XES with the XPS shown in Fig. 11.6b. At first glance, the spectra look very similar. Closer examination reveals subtle differences. First, the XPS is more blurry than the XES (e.g. as prepared Si NW and PS), this is because XPS is surface sensitive while the XES is bulk sensitive revealing the sharper features of the more crystalline bulk. Second, the XES selection rule is dipole, and it is only sensitive to Si 3s character involved in the bonding while XPS looks at all valence electrons, i.e. the s and p of all elements (e.g. O and Si in Si oxide) with photon energy dependent cross-sections. Thus XES complements XPS in revealing the electronic structure of these materials.

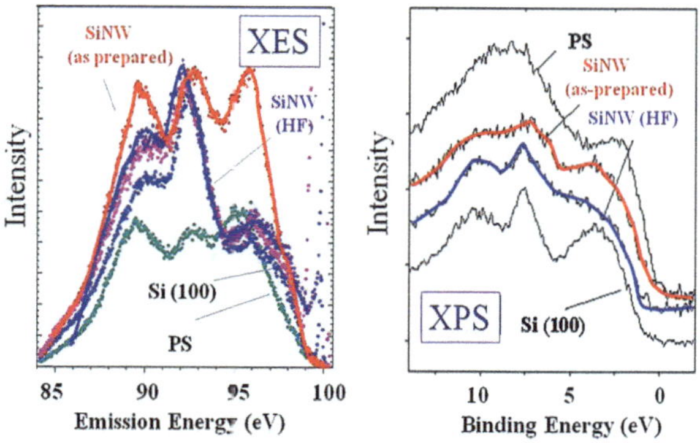

Fig. 11.6 *Left panel* XES excited with photon energy above the $L_{3,2}$-edge. XES of interest are "SiNW (as prepared)" (Si/SiO_2 core/shell), SiNW (HF) with oxide shell removed, clean Si(100) wafer and porous silicon, PS. *Right panel* XPS valence band of same sample of SiNW (as prepared and HF treated) as well as references Si(100) and PS [17]. The smooth *curves* are drawn to guide the eye

11.4.1.2 Top-down, Electroless and Chemically Etched Si Nanowires

The Si NWs discussed here were prepared by MCEE using a lightly and heavily B doped p-type Si(100) wafer with the resistivity of 3–10 Ω cm (lightly doped), and 0.005–0.025 Ω cm (heavily doped), respectively, henceforth denoted LDNW and HDNW [53]. It is found that silicon nanowires with a porous surface can also be obtained [54–57].

Liu et al has studied the XANES and XEOL of the afore-mentioned system of LDNW and HDNW and tracked their behaviour as the system underwent oxidation in air at elevated temperature [53]. The results are summarized in Fig. 11.7. From Fig. 11.7a, b, we observe (1) the as-prepared LDNW and HDNW are nearly oxide free as little SiO_2 WL is observed in both TEY and FLY; (2) HDNW, which has a porous surface, is more susceptible to oxidation but to a lesser extent than PS as expected and (3) the oxidation of all the nanostructure of Si is incomplete at 800 °C as we can still see the Si resonance in the FLY of all the heated samples even though the Si WL signal disappears in the TEY. It is not until the LDNW was oxidized at 1,000 °C for 6 h, that the Si WL disappears from the FLY XANES and the sample is fully oxidized and henceforth denoted SiO_2 NW.

The most intriguing observation is the optical behaviour displayed in Fig. 11.7c. It shows that the XEOL of all the oxidized samples exhibit spectroscopic features which fall into three regions: (i) blue light at ~460 nm, (ii) orange light at 600–700 nm, and (iii) a weak UV peak at ~300 nm (inset). The branching ratio also varies. For LDNW-800, HDNW-800 and PS-800, the blue and orange emissions have comparable intensity. For LDNW-1000, as shown in Fig. 11.7c, the main emission is at 460 nm and the orange luminescence only appears as a weak broad peak, and SiO_2 NW only shows a single blue emission. The disappearance of the orange

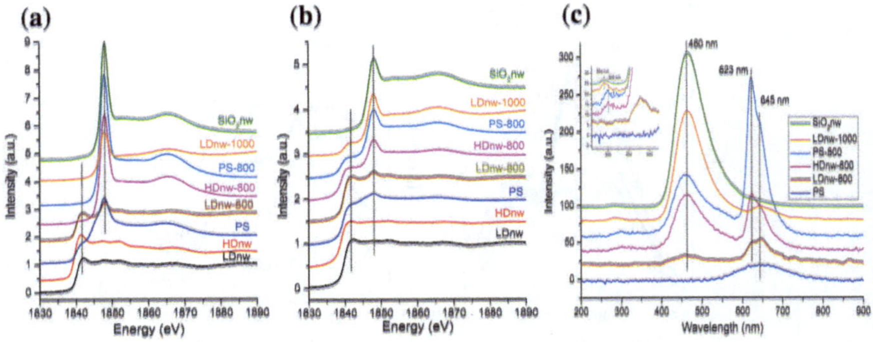

Fig. 11.7 **a** and **b** Si K-edge XANES recorded in TEY and FLY, respectively for LDNW and HDNW at each stage of the oxidation together with porous silicon (PS); the *vertical line* marks the WL of Si and SiO_2. **c** XEOL excited at 1,900 eV. The number after the *dashed line* denotes the temperature (°C). The sample labeled SiO_2nw is LDNW oxidized at 1,000 °C for 6 h whence the element Si signal is absent at both TEY and FLY of the Si K-edge

emission in the completely oxidized nanowire confirms that these bands originate from nanocrystallite Si and SiO_2 interface which disappear upon complete oxidation [17].

More details of the three luminescence regions have been investigated by focusing on the HDNW-800 sample. Figure 11.8 shows the Si K, O K and Si $L_{3,2}$-edge data for the HDNW oxidized at 800 °C, which is essentially a Si/SiO_2 core shell system with a porous surface.

Several interesting observations can be made from Fig. 11.8. First, HDNW-800 is a Si/SiO_2 core/shell structure with a thick SiO_2 layer and the core is only detectable using FLY at the Si K-edge but not at the Si $L_{3,2}$-edge. Second, excitation at all edges exhibit similar XEOL with 3 emission bands but different branching ratio. At the Si K-edge, the ~460 nm emission intensity increases relative to the ~640 nm as the excitation varies from the Si WL resonance to that of the SiO_2. This is consistent with the observation in the SiNW discussed above (Fig. 11.1). At the O K-edge, the ~640 nm band is more intense and changes little below and at resonance. Third, the PLY exhibits the most noticeable changes due to the variable penetration depth of the excitation photon and the truncation of the thermalization track. For example the PLY at the Si $L_{3,2}$-edge are all inverted, this is due to a sudden change to a very shallow penetration depth at resonance and the Auger electrons will more likely escape without being completely thermalized, hence reducing the yield at resonance. The 2-D display gives the overall emission properties across the entire region of excitation, each horizontal slice gives a XEOL spectrum at a given excitation energy and each vertical slice gives a wavelength select PLY XANES [58].

Finally, the most important observation is the comparison between XEOL from bottom up SiNW and top-down SiNW. From Fig. 11.4 we observe that the Si/SiO_2 core/shell SiNW exhibits three noticeable bands at 460, 530 and 630 nm and a weak UV band at ~300 nm. In Fig. 11.8, we see that the XEOL from HDNW-800, a

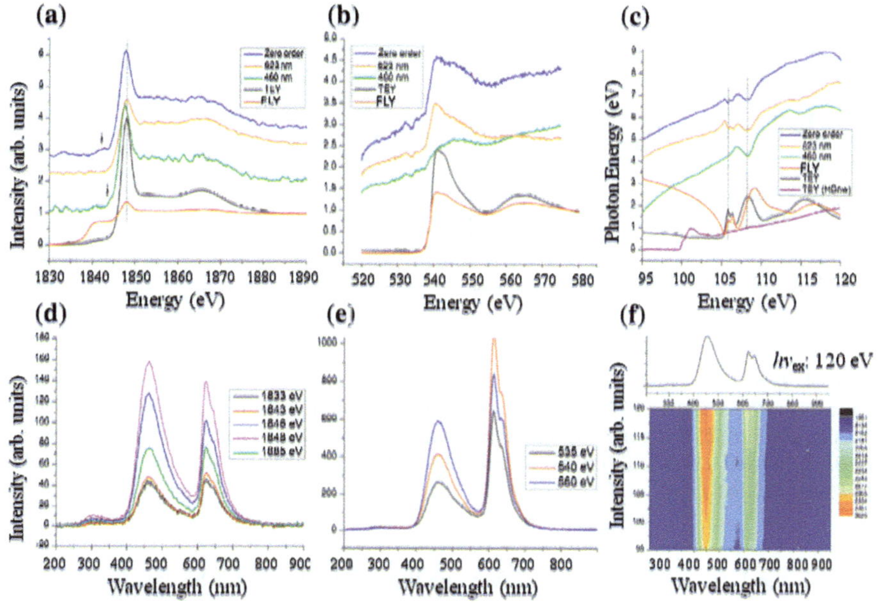

Fig. 11.8 **a** Si K-edge, **b** O K-edge and **c** Si$L_{3,2}$-edge XANES of HDNW oxidized at 800 °C and **e**, **f** and **g** the corresponding XEOL, respectively. All XANES are in TEY, FLY and PLY (total and wavelength selected). The 2-D display in () shows the entire data set of XEOL as a function of excitation energy across the Si $L_{3,2}$-edge, The XEOL of a horizontal cut at 120 eV excitation energy is shown at the *top* [53]

highly doped SiNW oxidized at 800 °C and a Si/SiO_2 core/shell system, the 530 nm emission is conspicuously missing or at best very weak. This emission band is attributed to small and quantum confined Si nanocrystallites embedded in the nanowire. This observation shows that such nanocrystallites are negligible if the silicon nanowire is prepared from a top-down method, even in the case of Si NW with a porous surface. It should be noted that for both LDNW and HDNW, their as-prepared Si NW exhibits no luminescence prior to oxidation. The 300 and 460 nm bands are associated with defects in SiO_2 and the 630 nm band is largely associated with the Si core and the Si-SiO_2 interface [11, 53].

11.4.1.3 SiNW Chains and Heterostructures

We conclude the discussion on SiNW with two more cases in which SiNW is a component in a nanocomposite. In one case, alternate SiNW and SiO_2 sections of a chain-like nanowire made by alternating the carrier gas pressure in the synthesis is studied [59]. In the other, Si participates as a nanowire in a variety of CdSe-Si nanocomposite [60, 61]. These systems can be conveniently tracked with XANES and XEOL. Figure 11.9 shows the most interesting XEOL features of these studies.

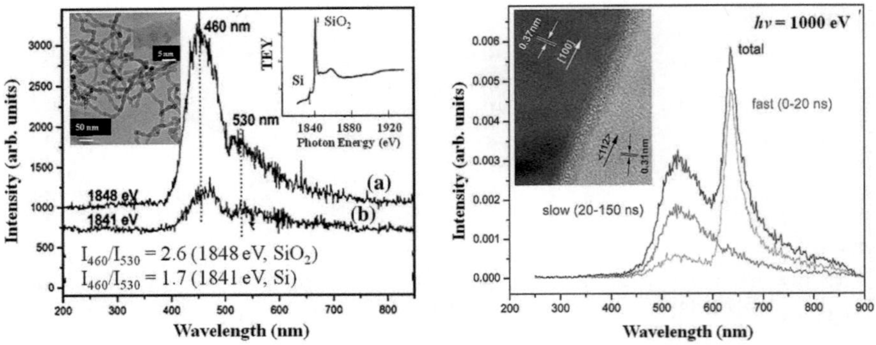

Fig. 11.9 *Left panel* XEOL from an alternate Si, SiO_2 chain-like nanowire (*left inset*) with excitation energy at the Si and the SiO_2 WL resonance (*right inset*); the relative intensity of the 450–530 nm emission is noted [59]. *Right panel* TRXEOL from a biaxial CdSe-Si nanowire composite with a fast (0–20 ns) and a slow (20–150 ns) time window. The inset shows the interface between CdSe (*dark*) and Si (*light*) [60]

From the left panel of Fig. 11.9, we see that both elemental Si and Si oxide are present. The XEOL displays the 460 nm and the 530 nm emission, attributable to Si oxide and Si nanocrystallites in the nanowire, respectively. The 460 nm intensity increases markedly when excited at the SiO_2 WL resonance confirming that this emission originates from the oxide. The right panel of Fig. 11.9 shows the un-gated and the TRXEOL from CdS-Si biaxial nanostructure excited at 1,000 eV. It is interesting that this system shows two very different and somewhat unusual emission features with the band at the longer wavelength showing a narrower linewidth. The fast window shows a dominant emission at longer wavelength which is the near band gap emission of CdSe. The slow window shows only the 530 nm band from the quantum confined emission of Si nanocrystallites, trapped in the interface. This is consistent with the observation shown in Fig. 11.4.

11.4.2 Ge Nanowires and GeO_2 Nanowires

11.4.2.1 Ge Nanowires

The GeNW XAFS exhibits similar oscillatory behavior as bulk Ge albeit with noticeable broadening of the oscillations and reduction in amplitude relative to the sharp features observed in bulk Ge attributable to disorder and degradation of long range order due to the truncation of the lattice at ~26 nm (Fig.11.10) [18]. The Ge K-edge XANES shows an interesting mismatch despite the similarity in the EXAFS. We see a broad peak at the threshold (11,103 eV) in bulk Ge, which narrows considerably in GeNW. In addition, a sharp peak is seen in the GeNW XANES at 11,110 eV. We attribute the narrowing of the white line to confinement and the presence of the sharp

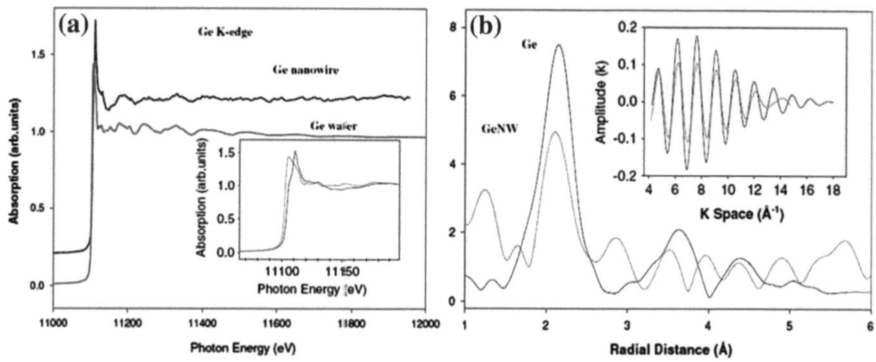

Fig. 11.10 **a** Normalized XAFS of Ge nanowire and a Ge wafer; the inset shows the XANES region, **b** FT of Ge EXAFS, the inset shows backtransform of the first shell [18]

peak to surface Ge oxide on the GeNW. The FT of the Ge K-edge EXAFS clearly shows the first, second and third shell Ge-Ge distance of the bulk Ge. The GeNW sample only shows one clearly identifiable peak at the first shell with nearly the same position and a noticeable reduction in amplitude. The phase analysis shows that the Ge-Ge interatomic distance in GeNW is slightly shorter than that in bulk Ge but cannot be quantified with certainty.

The amplitude reduction is attributed to disorder and the presence of an oxide-Ge interface in the GeNW. The Debye-Waller factor, exp $[-k^2\sigma^2]$, where σ is the root mean square displacement, will damp the amplitude noticeably. The reduction in amplitude of the filtered Fourier backtransform of the first shell for GeNW relative to Ge is clearly displayed. It should be noted that amplitude reduction can be seen even in perfect nano single crystals; this is because the coordination number decreases at the surface and interface of the crystal, which also introduces chemical inhomogeniety.

11.4.2.2 GeO_2 Nanowires

Germanium dioxide (GeO_2) has attracted much attention due to its unique electronic and optical properties with potential applications in optoelectronics [62, 63], especially with one-dimensional nanowires [64, 65]. GeO_2 also has a larger band bap (5 eV) than other transparent conductive oxides, making it attractive as a host for optical impurities for luminescent devices, from the ultraviolet-blue to the near-infrared range [66]. Typically, GeO_2 nanocrystals emit violet and blue photoluminescence (PL) due to the oxygen vacancy or other surface defects [67, 68]. The nature of the oxygen defects affects the luminescent property [69].

High quality single crystalline germanium oxide nanowires (GeONWs) have been synthesized via a simple thermal evaporation method following a vapor-liquid-solid (VLS) mechanism [65]. SEM image of the as-synthesized GeONWs on a Si substrate and a representative TEM image of a single GeONW with uniform diameter of about

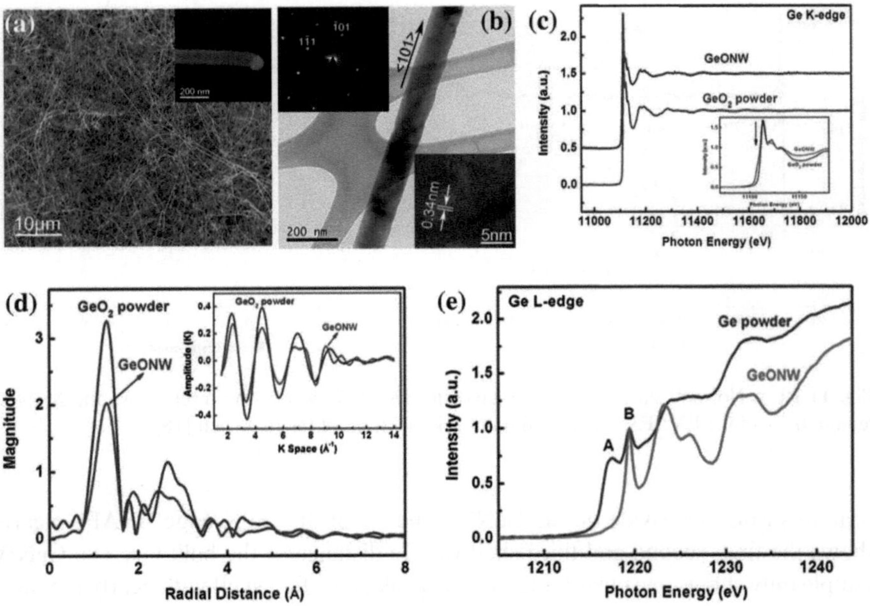

Fig. 11.11 **a** SEM of GeONWs, **b** HRTEM and SAED of an individual GeONW. **c** Ge K-edge XAFS of GeONW and GeO_2 powder, the inset shows the XANES, **d** FT of the EXAFS with backtransform shown in the inset, **e** Ge $L_{3,2}$-edge of GeONW and Ge powder [65]

100 nm along its entire length are shown in Fig. 11.11a, b, respectively. The EDS analysis of an individual nanowire reveals that the nanowire is composed of only germanium and oxygen with an atomic ratio close to 1:1.8. Thus the as-prepared nanowires is $GeO_x (x = 1.8)$.

Figure 11.11c, d shows the Ge K-edge XAFS with the XANES in the inset and the FT of the EXAFS (data range: k = 2–12 Å for GeO_2 and GeONWs) with the back transform shown in the inset, respectively. We can see from the XAFS that the oscillation pattern for GeONWs resembles that of bulk GeO_2 indicating that they have essentially the same crystal structure. For GeONWs, the broadening of the oscillations of the sharp features observed in bulk GeO_2 is attributed to disorder. The Ge K-edge XANES shows an interesting mismatch despite the similarity in the EXAFS. We see a shoulder before the edge jump (11,110 eV) in GeONWs, which is absent in the GeO_2. We attribute this shoulder to Ge with low oxidation states. The FT of the Ge K-edge EXAFS in Fig. 11.11d clearly shows the first and second shell Ge-O distance of the bulk GeO_2. The GeONW only shows one clearly identifiable peak of the first shell at nearly the same position with a noticeable reduction in amplitude. Table 11.2 shows the coordination number and interatomic distance of Ge-O in GeONWs by fitting the phase of the Ge-O first shell of GeONWs to that of bulk GeO_2. There is an obvious deficiency in coordination number in GeONWs (2.4 $\pm$ 0.2) comparing to that of bulk GeO_2 (4.0). The amplitude reduction is attributed to the lack of long-range order due to the oxygen vacancies and a large surface area

Table 11.2 The parameters for GeONWs using FEFF and IFEFFIT*

Samples	Shell	CN[a]	R(Å)[b]	$\Delta\sigma^2$(x10^{-3}Å^2)[c]	ΔE_0 (eV)
GeO_2 (quartz)	Ge-O	4.0	1.73	–	–
GeONWs	Ge-O	2.4 ± 0.2	1.74 ± 0.02	1.07 ± 0.21	5.21 ± 0.52
GeONWs (annealed)	Ge-O	3.9 ± 0.4	1.74 ± 0.02	0.05 ± 0.10	7.75 ± 1.08

[a]Coordination number (CN)
[b]Coordination distance
[c]The Debye-Waller factor accounting thermal vibrations and static disorder
*The many body reduction factor (S_0^2) was fixed to 0.98. R-space fit: $\Delta k = 2.0$ to 12.0 Å^{-1}, $\Delta r = 0.8$ to 2.0 Å, 8 statistically justified free parameters [65]

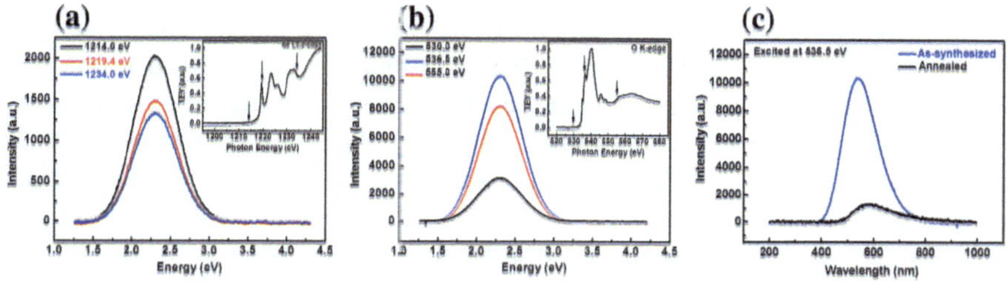

Fig. 11.12 **a** XEOL excited across the Ge L_3-edge, **b** XEOL excited across the O K-edge [65]. **c** XEOL spectra of the GeONWs before and after annealing excited at O K-edge (536.5 eV) [65]

(reduced coordination number) in the nanosize GeONW. The Ge $L_{3,2}$-edge XANES of GeONWs compared to Ge powder is shown in Fig. 11.11e. The first resonance (marked A) at the Ge $L_{3,2}$-edge in Ge powder corresponds to the 2p to 4s transition from the un-oxidized Ge and the second resonance (marked B) is characteristic of Ge in an oxygen environment (GeO_2) from surface oxidation of Ge. In GeONWs, Ge $L_{3,2}$-edge exhibits a sharp peak with a small edge at the A position due to the presence of Ge with low oxidation states consistent with the Ge K-edge results. Peak A is less intense at the Ge L-edge compared to the shoulder at the Ge K-edge due to shallower penetration depth indicating that the low oxidation state is in the bulk of the nanowire.

Figure 11.12a, b show the XEOL of the GeONWs excited at selected energies across the Ge L-edge and O K-edge, respectively. The inset of the Fig. 11.12a, b are the corresponding XANES. XEOL of GeONWs excited at both the Ge L-edge and the O K-edge exhibit same broad PL emission band centered at 2.3 eV (540 nm), corresponding to the PL spectrum of individual nanowires excited by laser [65]. When excited at the pre-edge (1,214.0 eV), the edge (1,219.4 eV), and the post-edge (1,234.0 eV) of Ge, GeONW shows a weak PL, and the emission intensity decreases with increasing excitation energy. However, when GeONWs were excited at the O K-edge, (pre-edge (530 eV), edge (536.5 eV) and post-edge (555 eV)), the emission intensity is much stronger than that excited at Ge L-edge. We also see the strongest emission at O K-edge (536.5 eV), which is about 5 time stronger than that excited at O K pre-edge, then PL weakens at O K post-edge excitation. From the above observation, we can immediately conclude that the strong green light emission at 2.3 eV peak is related to oxygen in GeONWs.

In order to confirm that the XEOL of GeONWs comes from the oxygen deficiency, a thermal annealing of the as-synthesized GeONWs was performed to remove the oxygen vacancies in the GeONWs. The as-synthesized GeONWs sample was annealed in air at 450 ~ 500 °C for 1 h. The XEOL of annealed GeONWs exhibits a much more attenuated XEOL emission compared with that of as-synthesized GeONWs. Ge K-edge XAFS of the annealed GeONWs compared with the as-synthesized GeONWs shows that the shoulder in the XANES of as-synthesized GeONWs arisen from the Ge with low oxidation state (mainly Ge^{3+}) disappears. The XANES of the annealed GeONWs shows similar features as that of the bulk crystalline GeO_2. The Fourier transform of the Ge K-edge EXAFS of annealed GeONWs clearly shows the first and second shells Ge-O distance as observed in the bulk GeO_2 and the amplitude restored significantly compared to that of as-synthesized GeONWs. These observations confirm that the as-synthesized GeONWs were further oxidized to GeO_2 NW with much of the oxygen vacancies replenished by thermal annealing in air.

Finally, we use a study of GeONW deposited on a silica substrate to illustrate that XEOL, in combination with probing depths either achieved by varying the excitation photon energy or the angle of incidence, can be used to obtain the depth profile of luminescence sites [70]. Relevant data are displayed in Fig. 11.13.

It is apparent from Fig. 11.13a that photons with energy at the O K-edge and the Ge L_{32} – edge have different attenuation lengths. The average diameter of the GeONW

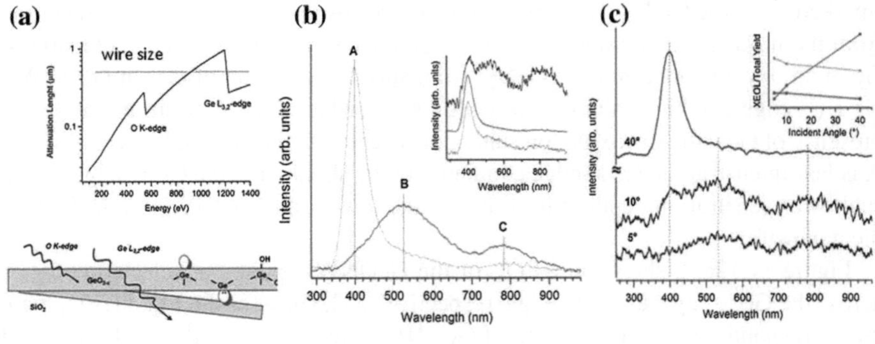

Fig. 11.13 **a** A schematic and attenuation length (*top*) of GeONW deposited on a silica substrate with surface, bulk, and interface defects. The *horizontal line* denotes the average size of the nanowire and the wavy arrow indicates the penetration depth of the photons, **b** PL of GeONW at different excitation energies: XEOL in the vicinity of the O K-edge (530 eV, *full line*) and the Ge L_3-edge (1220 eV, *dotted line*). Three emission bands, marked as A, B, and C, are centered at ~410 nm (~3.02 eV, *violet*), ~525 nm (~2.36 eV, *green*), and ~780 nm (~1.59 eV, *red*), respectively. Inset: XEOL at 1,110, 1,220 eV (Ge L-edge) and 1,230 eV (*bottom* to *top*). **c** XEOL from the GeONW as a function of the angle of incidence at a fixed photon energy (1,220 eV); relative intensity of the *violet* (~400 nm), *green* (~520 nm) and *red* (~780 nm) components as a function of angle is shown in the inset [70]

is ~ 500 nm which is larger than the attenuation length of photon at the O K-edge but smaller than the attenuation length at the Ge $L_{3,2}$-edge. Thus O K-edge photons will not be able to probe the GeONW-silica interface whereas the Ge $L_{3,2}$-edge photons will. If the luminescence sites have a depth profile, as in this case where the defects are on the surface, in the bulk of the NW and at the GeONW-SiO_2 substrate interface, photons with different attenuation length will probe different regions of the sample. From Fig. 11.13b we see a drastic difference in the XEOL between excitation at the O K-edge and the Ge $L_{3,2}$-edges (increasing penetration depth). The commonly observed O defect peak at ~540 nm and the surface defect at ~780 nm are observed at the O K-edge while at the Ge $L_{3,2}$-edge, an intense peak at ~400 nm originated from the GeO_2 NW- SiO_2 substrate interface emerges and becomes dominant. Above the Ge L-edge, all three emission bands are clearly observed.

The variation of probing depth can also be tuned by changing the angle of incidence to glancing, as shown in Fig. 11.13c. It is clearly seen that as the angle of incidence (90° for normal incidence) is tuned towards glancing e.g. from 40 to 15° then 5°, the branching ratio of the surface defect increases and the branching ratio of the interface defect emission decreases drastically.

11.4.3 Other Group IV Nanowires (C and SnO_2 Nanowire)

We present here two examples of nanowires of other group IV elements, C and Sn. Although carbon nanotube (CNT) is perhaps the most fascinating pseudo 1-D nanostructure, the discussion here is confined to nanowires. We will present the XANES study of an amorphous carbon nanowire (CNW), which is a precursor of CNT [71], and a 1-D structure of the nanoribbon (NR) [72] and nanowire (NW) of SnO_2 [73], an ideal candidate for sensor and Li battery applications.

Amorphous CNW is another variation of carbon based materials that can be synthesized in large quantities [71]. Figure 11.14 shows the C K-edge XANES of CNW in comparison with those of a multiwall carbon nanotube (MWNT).

It is clear from Fig. 11.14 that CNW contains both sp^2 and sp^3 bonding and the surface contains C-H and others, most likely carbonyl and carboxylic functionalities (~287–289 eV) as well. The splitting in the π^* resonance at ~286 eV in the TEY is due to chemical inhomogeneity arising from distortion due to the presence of surface species, which broadens in the FLY due to self-absorption; the σ^* is graphitic but shows no fine structures, indicating considerable disorder. The polarization dependent study shows that there is noticeable texture in which the π^* like orbital are graphitic, thus the CNW can be a precursor of carbon nanotube or multilayer graphene.

Figures 11.15 and 11.16 show the morphology, Sn $M_{5,4}$ and O K-edge XANES, and XEOL, from SnO_2 NR [72] and SnO_2 NW [73] respectively.

The most interesting feature from Figs. 11.15 and 11.16 is that the Sn $M_{5,4}$-edge can be clearly tracked. Its pre-edge resonance at ~485 eV (Fig. 11.16) arises from defects which are present mainly on the surface of the NR and in the bulk of SnO_2

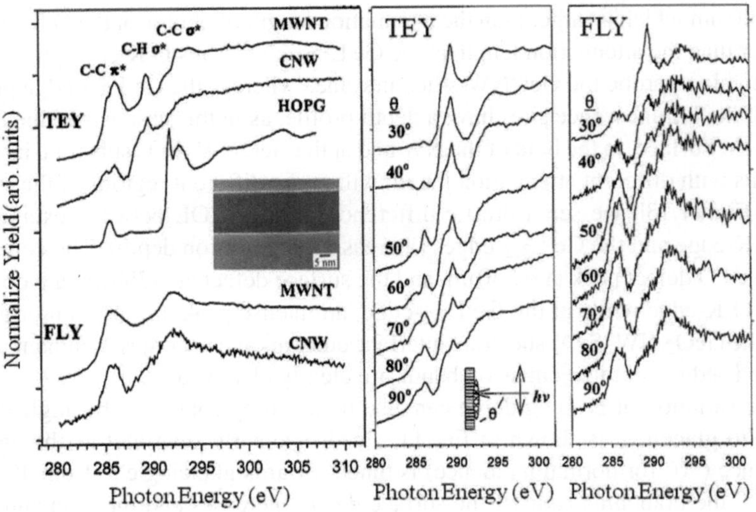

Fig. 11.14 C K-edge XANES of CNW together with MWNT and HOPG in TEY and FLY (*left panel*), and polarization dependent studies (*right two panels*); the inset in the *left panel* shows the HRTEM of an amorphous CNW, the inset in the *middle panel* depicts the orientation of the incident beam with respect to the surface of the specimen [71]

NW grown on SS (Stainless Steel) but absent if the substrate is Cu. The TRXEOL in Fig. 11.15 clearly reveals a fast decay channel at ~450 nm, which is not apparent in the un-gated XEOL. Thus TRXEOL can be used to reveal the near band gap emission of various nanomaterials. The substrate dependent growth, and electronic and optical properties of SnO_2 NW, can also be revealed with XANES and XEOL showing that SS substrate induces bulk defect and near IR emission which again diminishes when Cu substrate is used [73].

11.5 Summary and Outlook

We have illustrated in this chapter the interplay of group IV nanowires and X-ray absorption spectroscopy, especially in the soft X-ray region where de-excitation spectroscopy like XES, XEOL and TRXEOL can play an important role in their characterization. We have noted that SiNW, depending on the method of preparation, exhibits subtle difference in its electronic and optical behaviour in that the bottom up method like VLS introduces nanocrystallites into the 1-D structure, which emits light because of quantum confinement and interfacial effects while such effect are absent from SiNW prepared from the top-down, electrode-less chemical etching method. Similarly, substrate and preparation condition can be used to tune the properties of nanowires as in the case of GeO_2 NW on Si vs. GeO_2 NW on silica, SnO_2 NR versus SnO_2NW and SnO_2 NW on SS vs. SnO_2NW on Cu. Clearly XEOL in the energy

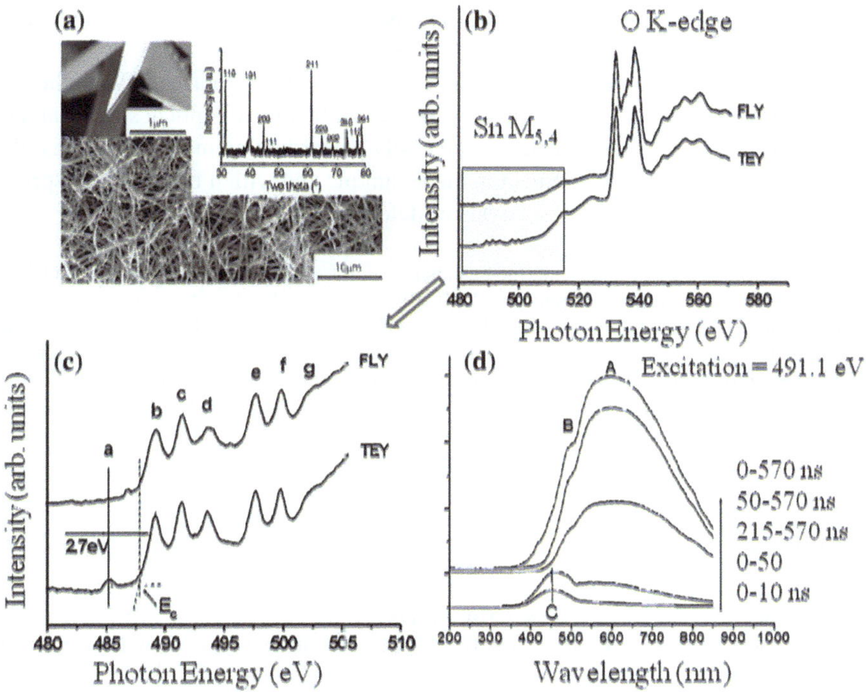

Fig. 11.15 **a** SEM of SnO_2 NR, inset show the XRD, **b** $SnM_{5,4}$ and O K-edge, **c** $SnM_{5,4}$ edge enlarged, **d** TRXEOL from SnO_2 NR with a set of time windows, (0–570 ns is un-gated, *dark gap* at the CLS in single bunch is 570 ns) [72]

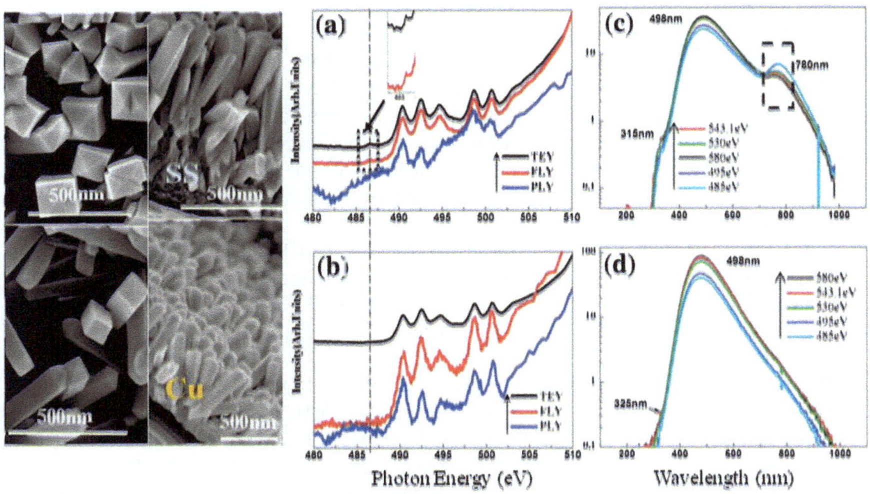

Fig. 11.16 SnO_2 NW grown on stainless steel (*SS*) and Cu substrate (*Left panel*) and corresponding XANES at the Sn $M_{5,4}$ edge, **a** and **b** and XEOL(in log plot), **c** and **d** respectively. The near band gap and maximum emission wavelengths are noted [73]

and time domain, together with XANES and XES, is a powerful tool to reveal many properties of these and many other nanomaterials.

XEOL and TRXEOL experiments can now be conducted routinely. With the use of a two dimensional XEOL scheme [74] and the application of optical streak camera [75], this type of experiments can now be easily performed on many beamlines. We fully anticipate further development in the technique and with it more widely spread applications will emerge in the analysis of defects in light emitting materials.

Acknowledgments Xuhui Sun thanks the financial support from the National Basic Research Program of China (973 Program) (Grant No. 2010CB934500) and the Priority Academic Program Development of Jiangsu Higher Education Institutions. TKS wishes to thank all his students, postdoctoral fellows and visiting scientists as well as colleagues at the Canadian Synchrotron Radiation Facility, Synchrotron Radiation Center, University of Wisconsin-Madison, Canadian Light Source, and Advanced Photon Source for their participation and collaboration in some of the work described herein.

References

1. A.M. Morales, C.M. Lieber, Science **279**, 208 (1998)
2. Y.F. Zhang, Y.H. Tang, N. Wang, D.P. Yu, C.S. Lee, I. Bello, S.T. Lee, Appl. Phys. Lett. **72**, 1835 (1998)
3. J.D. Holmes, K.P. Johnston, R.C. Doty, B.A. Korgel, Science **287**, 1471 (2000)
4. Y.H. Tang, N. Wang, Y.F. Zhang, C.S. Lee, I. Bello, S.T. Lee, Appl. Phys. Lett. **75**, 2921 (1999)
5. B.K. Teo, X.H. Sun, Chem. Rev. **107**, 1454 (2007)
6. D.W. Wang, Pure Appl. Chem. **79**, 55 (2007)
7. B. Yu, X.H. Sun, G.A. Calebotta, G.R. Dholakia, M. Meyyappan, J. Cluster Sci. **17**, 579 (2006)
8. T.K. Sham, Int. J. Nanotech. **5**, 1194 (2008)
9. J. Stohr, *NEXAFS Spectroscopy* (Springer, Berlin, 1992)
10. Y. Ma, N. Wassdahl, J. Guo, J. Nordgren, P.D. Johnson, J.-E. Rubensson, T. Boske, W. Eberthardt, S.D. Kevan, Phys. Rev. Lett. **66**, 2598 (1992)
11. E.J. Nordgren, S.M. Butorin, L.C. Duda, J.H. Guo, J. Rubensson, Soft X-ray fluorescence spectroscopy for materials science and chemical physics, in *Chemical Applications of Synchrotron Radiation Part I*, ed. by T.K. Sham (World Scientific, Singapore, 2002)
12. J.-E. Rubensson, D. Mueller, R. Shuker, D.L. Ederer, C.H. Zhang, J. Jia, T.A. Callcott. Phys. Rev. Lett. **64**, 1047 (1990)
13. F. de Groot, A. Kotani, *Core Level Spectroscopy of Solids* (CRC Press, Boca Raton, 2008)
14. T.K. Sham, D.-T. Jiang, I. Coulthard, J.W. Lorimer, S.P. Frigo, X.H. Feng, K.H. Tan, D.C. Houghton, B. Bryskiewicz, R.A. Rosenberg, Nature **363**, 331 (1993)
15. D.T. Jiang, I. Coulthard, T.K. Sham, J.W. Lorimer, S.P. Frigo, X.H. Feng, R.A. Rosenberg, J. Appl. Phys. **74**, 6335 (1993)
16. T.K. Sham, R.A. Rosenberg, ChemPhysChem **8**, 2557 (2007)
17. T.K. Sham, S.J. Naftel, P.-S.G. Kim, R. Sammynaiken, Y.H. Tang, I. Coulthard, A. Moewes, J.W. Freeland, Y.-F. Hu, S.T. Lee, Phys. Rev. B **70**, 045313 (2004)
18. X.H. Sun, C. Didychuk, T.K. Sham, N.B. Wong, Nanotechnology **17**, 2925 (2006)
19. F. Heigl, A. Jurgensen, X.-T. Zhou, M. Murphy, J.Y.P. Ko, T.K. Sham, R. Rosenberg, R. Gordon, D. Brewe, T. Regier, L. Armelao, in *AIP Proceedings, 9th International Conference on Synchrotron Radiation Instrumentation* (AIP CP879, 2007), p. 1202
20. J.-M. Lehn, *Supramolecular Chemistry: Concepts and Perspectives* (VCH, Weinheim, 1995)
21. F. Vogtle, *Supramolecular Chemistry: An Introduction* (Wiley, Chichester, 1991)

22. D.B. Amabilino, J.F. Stoddart, Chem. Rev. **95**, 2725 (1995)
23. G. Ungar, Y.S. Liu, X.B. Zeng, V. Percec, W.D. Cho. Science **299**, 1208 (2003)
24. I. Amato, Science **282**, 402 (1998)
25. J.W. Judy, Smart Mater. Struct. **10**, 1115 (2001)
26. H. Heinrich, G. Bauer, F.E. Kuchar, *Physics and Technology of Submicron Structures* (Springer, Berlin, 1988)
27. R.F. Service. Science **293**, 785 (2001)
28. R.S. Wagner, W.C. Ellis, Appl. Phys. Lett. **4**, 89 (1964)
29. Y. Cui, L.J. Lauhon, M.S. Gudiksen, J.F. Wang, C.M. Lieber, Appl. Phys. Lett. **78**, 2214 (2001)
30. R.Q. Zhang, Y. Lifshitz, S.T. Lee, Adv. Mater. **15**, 635 (2003)
31. R. Juhasz, N. Elfstrom, J. Linnros, Nano Lett. **5**, 275 (2005)
32. C.M. Hsu, S.T. Connor, M.X. Tang, Y. Cui, Appl. Phys. Lett. **93**, 133109 (2008)
33. E. Garnett, P.D. Yang, Nano Letters **10**, 1082 (2010)
34. K.Q. Peng, Y.J. Yan, S.P. Gao, J. Zhu, Adv. Mater. **14**, 1164 (2002)
35. K.Q. Peng, Y. Wu, H. Fang, X.Y. Zhong, Y. Xu, J. Zhu, Angew. Chem. Int. Ed. **44**, 2737 (2005)
36. K.Q. Peng, A.J. Lu, R.Q. Zhang, S.T. Lee, Adv. Funct. Mater. **18**, 3026 (2008)
37. S.M. Sze, *Physics of Semiconductor Devices* (Wiley, New York, 1981)
38. Y. Maeda, N. Tsukamoto, Y. Yazawa, Y. Kanemitsu, Y. Masumoto, Appl. Phys. Lett. **59**, 3168 (1991)
39. Y.Y. Wu, P.D. Yang, Chem. Mater. **12**, 605 (2000)
40. P. Nguyen, H.T. Ng, M. Meyyappan, Adv. Mater. **17**, 549 (2005)
41. D.W. Wang, H.J. Dai, Angew. Chem. Int. Ed. **41**, 4783 (2002)
42. S. Mathur, H. Shen, V. Sivakov, U. Werner, Chem. Mater. **16**, 2449 (2004)
43. B.A. Korgel, T. Hanrath, Adv. Mater. **15**, 437 (2003)
44. X.H. Sun, G. Calebotta, B. Yu, M. Meyyappan, J. Vac. Sci. Technol. B **25**, 415 (2007)
45. G. Bunker, *Introduction to XAFS* (Cambridge University Press, Cambridge, 2012)
46. X-ray calculator, http://www-cxro.lbl.gov/
47. A.J. Achkar, T.Z. Regier, H. Wadati, Y.-J. Kim, H. Zhang, D.G. Hawthorn, Phys. Rev. B: Condens. Matter Mater. Phys. **83**, 081106(R) (2011)
48. X.H. Sun, S. Lam, T.K. Sham, F. Heigl, A. Jürgensen, N.B. Wong, J. Phys. Chem. B **109**, 3120 (2005)
49. F. Heigl, S. Lam, T. Regier, I. Coulthard, T.K. Sham, J. Am. Chem. Soc. **128**, 3906 (2006)
50. R.A. Rosenberg, G.K. Shenoy, L.C. Tien, D. Norton, S. Pearson, X.H. Sun, T.K. Sham, Appl. Phys. Lett. **89**, 093118 (2006)
51. X.T. Zhou, F. Heigl, J.Y.P. Ko, M.W. Murphy, J.G. Zhou, T. Regier, R. Blyth, T.K. Sham, Phys. Rev. B **75**, 125303 (2007)
52. R.A. Rosenberg, G.K. Shenoy, P.-S.G. Kim, T.K. Sham, J. Phys. Chem. **112**, 13943 (2008)
53. L.J. Liu, T.K. Sham, Small **8**, 2371 (2012)
54. K.Q. Peng, Y. Xu, Y. Wu, Y. Yan, S.T. Lee, J. Zhu, Small **1**, 1062 (2005)
55. M.L. Zhang, K.Q. Peng, X. Fan, J.S. Jie, R.Q. Zhang, S.T. Lee, N.B. Wong, J. Phys. Chem. C **112**, 4444 (2008)
56. Y. Qu, L. Liao, Y. Li, H. Zhang, Y. Huang, X. Duan, Nano Letters **9**, 4539 (2009)
57. L. Lin, S. Guo, X. Sun, J. Feng, Y. Wang, Nanoscale Res. Lett. **5**, 1822 (2010)
58. M.J. Ward, J.G. Smith, T.Z. Regier, W.Q. Han, T K. Sham, SRI 2012 IOP Proc (2012) (accepted)
59. X.H. Sun, N.B. Wong, C.P. Li, S.T. Lee, T.K. Sham, J. Appl. Phys. **96**, 3447 (2004)
60. R.A. Rosenberg, G.K. Shenoy, X.H. Sun, T.K. Sham, Appl. Phys. Lett. **89**, 243102 (2006)
61. X.H. Sun, T.K. Sham, R.A. Rosenberg, G.K. Shenoy, J. Phys. Chem. C **111**, 8475 (2007)
62. C.Y. Yan, M.Y. Chan, Z. Zhang, P.S. Lee, J. Phys. Chem. C **113**, 1705 (2009)
63. Z.Y. Lin, B.K. Garside, Appl. Opt. **21**, 4324 (1982)
64. Z.G. Bai, D.P. Yu, H.Z. Zhang, Y. Ding, Y.P. Wang, X.Z. Cai, Q.L. Hang, C.C. Xiong, S.Q. Feng, Chem. Phys. Lett. **303**, 311 (1999)
65. M.F. Peng, Y. Li, J. Gao, D. Zhang, Z. Jiang, X.H. Sun, J. Phys. Chem. C **115**, 11420 (2011)
66. P. Hidalgo, E. Liberti, Y. Rodríguez-Lazcano, B. Méndez, J. Piqueras, J. Phys. Chem. **C113**, 17200 (2009)

67. M. Zacharias, P.M. Fauchet, J. Non-Cryst. Solids **227**, 1058 (1998)
68. X.C. Wu, W.H. Song, B. Zhao, Y.P. Sun, J.J. Du, Chem. Phys. Lett. **349**, 210 (2001)
69. A.S. Zyubin, A.M. Mebel, S.H. Lin, J. Phys. Chem. A **111**, 9479 (2007)
70. L. Armelao, F. Heigl, P.-S.G. Kim, R.A. Rosenberg, T. Regier, T.K. Sham, J. Phys. Chem. C **116**, 14163 (2012)
71. Y.H. Tang, P. Zhang, Y.F. Hu, X.H. Sun, M.K. Fung, Y.F. Zheng, C.S. Lee, S.T. Lee, T.K. Sham, Appl. Phys. Lett. **79**, 3773 (2001)
72. X.T. Zhou, F. Heigl, T. Regier, I. Coulthard, T.K. Sham, Appl. Phys. Lett. **89**, 213109 (2006)
73. D. Wang, J. Yang, X. Li, R. Li, T.K. Sham, X.-L. Sun, Cryst. Growth Des. **12**, 397 (2012)
74. M.J. Ward, J.G. Smith, T.Z. Regier, T.K. Sham, SRI2012 Proc. J. Phys. Conf. Ser. **425**, 132009 (2013)
75. M.J. Ward, T.Z. Regier, J.M. Vogt, R.A. Gordon, W.-Q. Han, T.K. Sham, SRI 2012 Proc. J. Phys. Conf. Ser. **425**, 092006 (2013)

Chapter 12
Group III–V and II–VI Quantum Dots and Nanoparticles

Alexander A. Guda, Mikhail A. Soldatov and Alexander V. Soldatov

By decreasing the size of semiconducting material the novel properties appear from the increased surface/bulk ratio and quantum confinement effects. X-ray absorption spectroscopy (XAS) can be applied to quantum dots and nanoparticles in-situ in the colloid, on the substrate or inside a solid matrix. The structural information about average size, phase composition, and growth orientation can be extracted from XAS along with information about electronic and magnetic properties of pure and doped nanostructures. In the chapter we shall describe the general properties of quantum dots and their applications that will help to understand the tasks for spectroscopy. Case studies will provide the information that is obtained from XAS complementary to other methods.

12.1 Properties and Applications of Quantum Dots

Nanoparticle (NP) and quantum dot (QD) are regarded as zero-dimensional nanostructures. The term "QD" is usually used for semiconductor nanoparticles and islands where quantum confinement of electrons and excitons determines their properties. Decades passed after the dependence of the spectral position of the exciton absorption lines of the nanosized CuCl crystals was attributed to the quantum size effect [1] and three-dimensionally confined semiconductor quantum wells or zero-dimensional semiconductor nanostructures showing discrete electronic states were named "QDs"

A.A. Guda (✉) · M.A. Soldatov · A.V. Soldatov
Southern Federal University, Sorge str. 5, Rostov-on-Don 344090, Russia
e-mail: guda@sfedu.ru

M.A. Soldatov
e-mail: mikhailsoldatov@sfedu.ru

A.V. Soldatov
e-mail: soldatov@sfedu.ru

C.S. Schnohr and M.C. Ridgway (eds.), *X-Ray Absorption Spectroscopy of Semiconductors*, Springer Series in Optical Sciences 190,
DOI 10.1007/978-3-662-44362-0_12

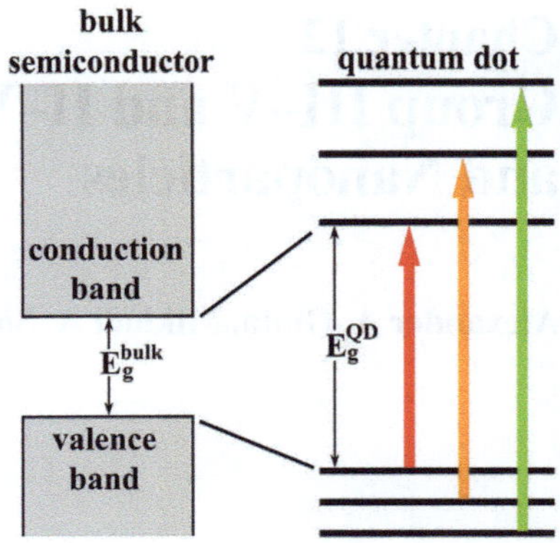

Fig. 12.1 The changes in the density of electronic states are shown schematically for different sizes of semiconductors

[2]. However it is still a hot topic due to the recored-breaking pace of developments in nanotechnology.

A QD can be thought of as an artificial atom, as its electron density of states consists of a series of very sharp peaks, and its physical properties resemble, in many respects, those of an atom in a cage. The discrete structure of energy states shown in Fig. 12.1 leads to a discrete absorption spectrum of QDs, which is in contrast to the continuous absorption spectrum of a bulk semiconductor. However these artificial atoms are also expected to have some significantly different properties compared with real atoms, as they can be filled with both electrons and holes [3].

The charge carriers inside QDs are confined in all three dimensions by the surface and thus the size and shape of QDs determine the energy spectrum of electronic levels [4]. It is possible to estimate the change of the QDs band gap energy with size R by applying the simple model of spherical potential well:

$$E_g^{QD} \approx E_g + \frac{\pi^2 \hbar^2}{2m_{eh} R^2} \tag{12.1}$$

$$m_{eh} = \frac{m_e m_h}{m_e + m_h}$$

where m_e and m_h are the effective electron and hole masses correspondingly. The ability to precisely control the size of a QD enables the manufacturer to determine the desired emission wavelength.

The main advantages of QD include high quantum yields, broad absorption spectra, size-tunable emission spectra, and good resistance to chemical and photochemical degradation. QDs have the large surface which is highly reactive. Thus in colloids they are stabilized with surface reactants. Usually the appropriate shell covers the core of the QD in order to improve the photo-luminescence yield and prevent from degradation. Unterminated dangling bonds on the QD's surface can affect the emission efficiency because they lead to a loss mechanism wherein electrons are rapidly trapped at the surface before they have a chance to emit a photon.

Cadmium selenide has been used intensively as a cellular marker because CdSe emits in the visible domain and is an excellent contrast agent. ZnS layer can be used as a shell to protect the core from oxidation and also the leeching of CdSe into the surrounding solution. QDs are more resistant to degradation than other optical imaging probes such as organic dyes. Figure 12.2 shows pseudocoloured images depicting five-colour QD staining of fixed human epithelial cells. Each color was used for labeling the nucleus, Ki-67 protein, mitochondria, microtubules and actin filaments correspondingly. Right panel shows confocal fluorescence images of fixed HepG2 cells stained with glutathione-capped CdTe QDs. Nucleoli and cytoplasm were stained by QDs emitting green and red light correspondingly.

The possibility to tune the energy gap between electronic levels by changing the size of QD explains the great interest to such objects from nanoelectronics, quantum computing, light emitters and makes them desirable for solar cell use [7, 8].

Schematic explanation of the operation principle of the electron spin memory device based on Ga(In)As QDs is shown Fig. 12.3 [9]. Resonant excitation produces excitons in the QD. Axial electric field provided by the applied bias V_{store} ionizes

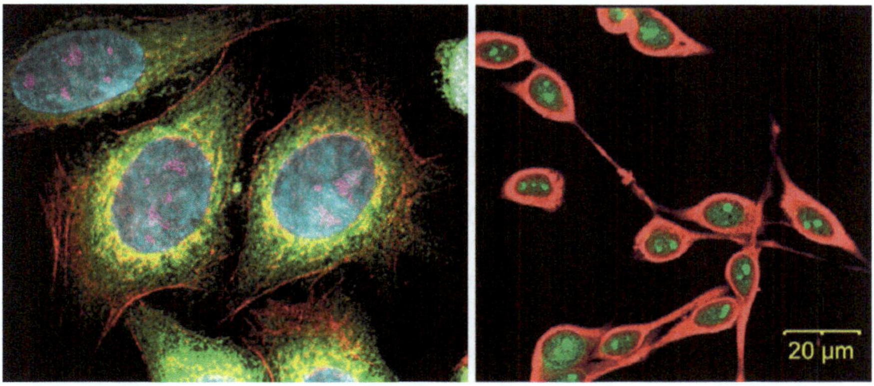

Fig. 12.2 Fluorescence images of epithelial cells (*left*, courtesy of Quantum Dot Corp) and HepG2 cells (*right*) stained with QDs QDs of specified size. Adapted from [5, 6]

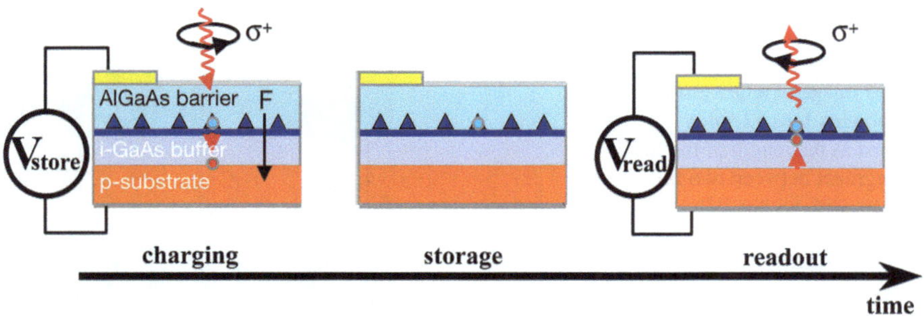

Fig. 12.3 Optically programmable electron spin memory using Ga(In)As QDs. Adapted from [9]

excitons. The AlGaAs barrier above the QD layer inhibits electron escape from the QDs and after charge separation, they are stored directly in the QDs. Readout is provided by applying the forward biasing to the device. When holes drift back into the QDs, they neutralize the stored charge. The polarization of generated photon probes the electron spin orientation.

12.2 Synthesis

First QDs were CuCl nanocrystals grown during high-temperature heating in the SiO_2 matrix [1]. Self-assembled QDs could be fabricated during the epitaxial growth, e.g. molecular beam epitaxy. QDs in material systems InAs/GaAs and InAs/InP are formed in Stranski–Krastanow regime as shown in Fig. 12.4. This means that InAs, is depositing layer-by-layer on the substrate with significantly mismatched lattice parameter [10]. Due to the large strains in the structure part of this material forms array of islands with diamond-like shape after the thickness of planar layer of InAs exceeds the critical value (in case of InAs on top of GaAs it is 1, 7 monolayer).

Using the chemical beam epitaxy the QDs can be realized as layers inside nanowires as shown in Fig. 12.5. Longitudinal quantization exceeding 10 meV has been observed in transport measurements on 50–70 nm diameter InAs nanowires with two InP barriers [11]. Changing the dot size L allows designing QDs along a nanowire with different specific spin splittings in a constant magnetic field. In future this is expected to allow individual gate tunable spin splittings in a series of dots along a nanowire which makes nanowire QDs containing a single electron spin interesting

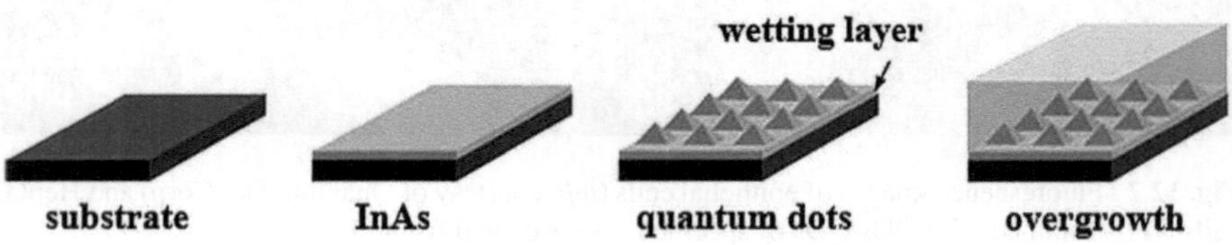

Fig. 12.4 Stranski-Krastanow regime of InAs/GaAs formation process

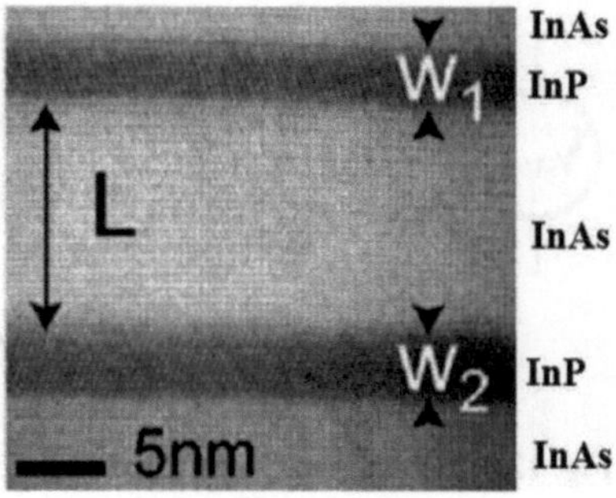

Fig. 12.5 High-resolution STEM image of a QD inside a nanowire. L = 12 nm is defined as the length of the In As segment between two InP barriers with widths W1 and W2. Figure adapted from [11]

systems for the realization of cubits. Authors in [12] have shown the possibility to produce QDs inside a homogeneous nanowire. By changing the InP crystal structure from zinc-blende to wurtzite a system similar to one in Fig. 12.5 was formed.

A metal-organic chemical vapor deposition technique was used to for controllable growth of ZnO/MgO quasi core-shell QDs [13]. Silicon ⟨111⟩ wafers were used as the substrates. Diethylzinc (DEZn), bis(methylcyclopentadienyl) magnesium ($(MeCp)_2Mg$), and O_2 were used as the zinc source, magnesium source, and oxygen source, respectively. N_2 was used as the carrier gas. The growth temperature was 400 °C. The chamber pressure was maintained at 3 and 6 Pa during the growth of ZnO and MgO, respectively. An exciton emission from ZnO QDs was greatly enhanced after the growth of MgO layer. TEM image of ZnO/MgO core shell QD is shown in Fig. 12.6.

The easiest way to produce macroscopic quantities of QDs is the hydrothermal synthesis. A unified approach through liquid–solid–solution (LSS) process can be used to synthesize a broad range of nanocrystals—from noble metals to the semiconductors [14]. Figure 12.7 shows cadmium selenide QDs and their size distribution after hydrothermal synthesis.

The QDs in colloid solution are covered by the layer of organic molecules which prevents uncontrolled growth and agglomeration of the nanoparticles. When coated with a suitable, chemically active surface layer, QDs can be coupled to each other or to different inorganic or organic entities and thus serve as useful optical tags.

XAS can probe the chemical bonding between surface atoms and surfactants. The ligand for the biological applications should possess a strong affinity for the QD surface. Ligand exchange is the method used to replace the original hydrophobic surface ligands on the QD surface with hydrophilic ligands or to attach to the QD a cell-penetrating peptide in order to allow QDs penetration into the living cells [15, 16]. Thiol groups as in Fig. 12.8 have been the most frequently utilized as anchors on the QD surface since they have strong affinity with Cd and Zn. In order to avoid surface reactants the in-water laser ablation can be used [17].

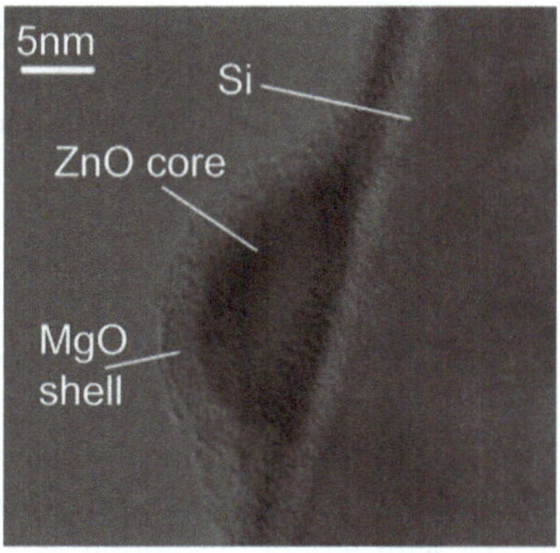

Fig. 12.6 Cross-sectional TEM image of a selected ZnO/MgO core-shell QD. Adapted from [13]

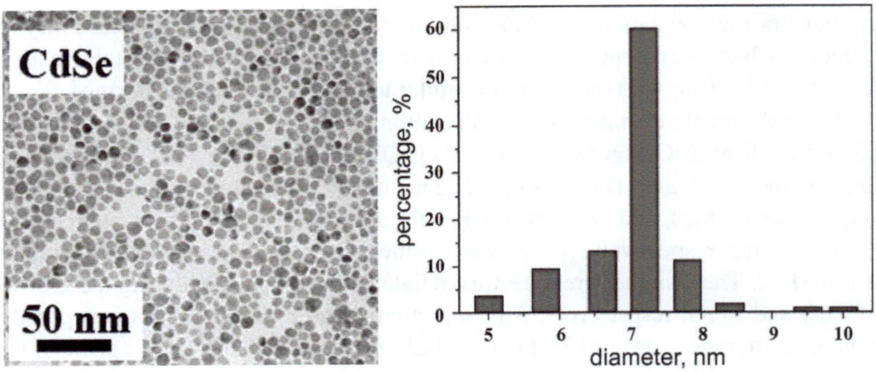

Fig. 12.7 CdSe nanocrystals and their size distribution. 300 particles were measured to get the size distribution. Adapted from supplementary data of [14]

Fig. 12.8 Ligands attached to the QD in colloid. **a** trioctylphosphine (TOP/TOPO) **b** monodentate thiol; **c** bidentate thiol; **d** histidine-appended peptides. Adapted from [15]

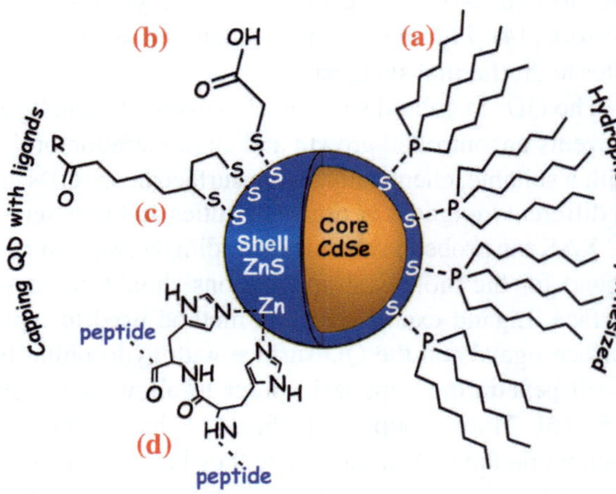

Nanoparticles can be effectively produced by means of high-energy ball milling. For example ZnS QDs with cubic phase have been prepared by mechanical alloying the stoichiometric mixture of Zn and S powders at room temperature in a planetary ball mill under Ar [18].

12.3 Methods to Study the QDs

The size distribution of QDs can be studied by direct methods such as electron microscopy or atom probe microscopy. These methods allow studying the crystal planes with atomic resolution as well as single dopants in semiconductors [19]. Indirect methods including X-ray diffraction, small-angle X-ray scattering, optical scattering, optical absorption and emission can be used as well.

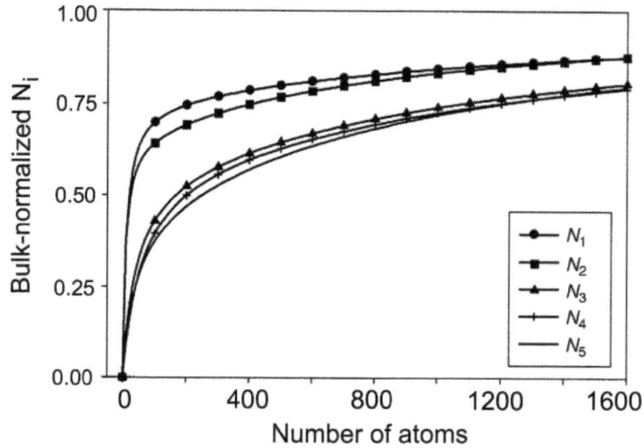

Fig. 12.9 Correlation between N_i coordination number normalized to the bulk value and total number atoms in the nanoparticle. Adapted from [20]

The X-ray absorption fine structure (XAFS) can be used to analyze the oxidation state, crystal structure and size of QDs. This method is favorable for the local defects inside QDs and when QDs of a small size are analyzed. The extended X-ray absorption fine structure (EXAFS) contains information about coordination numbers of absorbing atoms i.e. one can determine the first, second, third etc. coordination numbers for a given sort of atoms in QD. Figure 12.9 shows how first, second, third etc. coordination numbers depend on the total number of atoms in QD [20]. These curves represent a fit over wide range of densely packed nanoclusters with different shape.

Coordination numbers decrease when the cluster size decreases and thus the values obtained from EXAFS-analysis can be used to estimate the cluster size. Functions for the first N_1 and second N_2 coordination shell approach much faster the values from bulk compared with the higher coordination shells. Consequently, the accuracy of the clusters size determination strongly depends on the number of atoms as well as on the order of the coordination shell analyzed.

The near-edge X-ray absorption fine structure (NEXAFS) is a useful detection method for crystal structure modifications in nano-scale systems. Gilbert et al. [21] found that low energy L-edge absorption is sensitive to the structural transition from wurtzite into zinc blende for ZnS due to high energy resolution obtained in the experiment. Figure 12.10 shows the corresponding experimental spectra above S L-edge and Zn L-edge.

One can use different X-ray polarizations to study the crystal phases as well as the growth direction of nanostructures. For the wurtzite structure with a preferential direction, *c-axis,* one can observe X-ray linear dichroism (XLD) in contrast to the case of zinc blende where X-ray absorption is isotropic [22]. In case of wurtzite nanostructures the XLD signal will be present for the oriented nanowires while nanoparticles and QDs shall not exhibit dichroism, as shown in Fig. 12.11.

The Zn K-edge was less sensitive to the W-ZB phase transformation. However in case of high pressure transition into the NaCl phase the Zn K-edge X-ray absorption

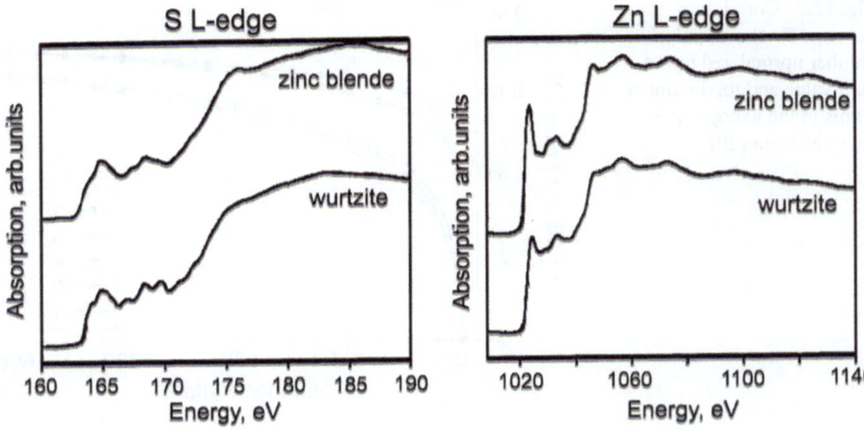

Fig. 12.10 Experimental NEXAFS of ZnS zinc blende and wurtzite polytypes. Adapted from [21]

Fig. 12.11 Polarization-dependent X-ray absorption spectra of zincite ZnO isotropic spherical nanoparticles and anisotropic rod-like morphology. Adapted from [23]

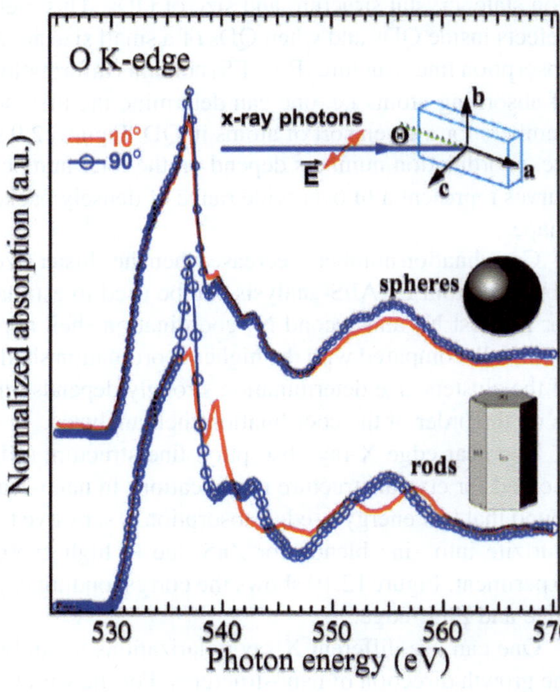

near-edge structure (XANES) spectra reveal striking differences [24]. Figure 12.12 shows how Zn K-edge fine structure changes for the ZnO under different pressure.

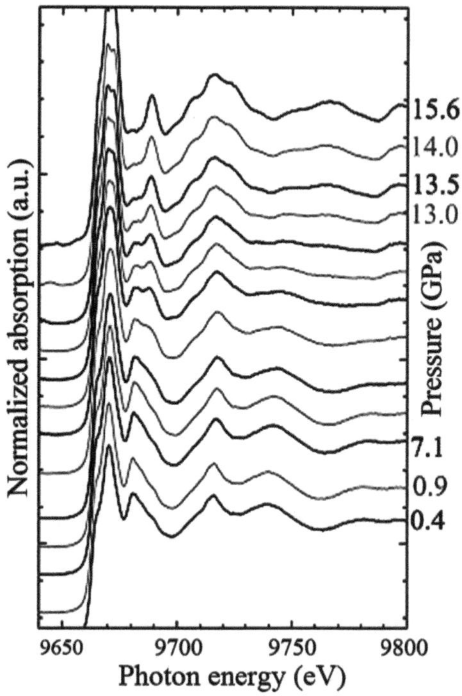

Fig. 12.12 Zn K-edge XANES spectrum of bulk ZnO as a function of applied pressure. Adapted from [24]

12.4 Case Studies

The determination of the magnitudes of elastic deformations in thin layers and nanocrystals by conventional methods of X-ray diffraction analysis or electron diffraction techniques is limited by their sensitivity and by the absence of long-range order in such systems. EXAFS and XANES spectroscopy methods, in which powerful continuous-spectrum X-ray sources are used, provide a unique opportunity to solve such problems. Using powerful synchrotron radiation, EXAFS and XANES spectroscopies allow one to determine the local environment characteristics of atoms (interatomic distances, coordination numbers, environment symmetry, types of neighboring atoms, and Debye-Waller factors). Electron level shifts and charge state of atoms can be analyzed as well..

12.4.1 Group III–V QDs and Nanoparticles

Variations in the microscopic structural parameters of Ge/Si, GaN/AlN, and InAs/AlAs QDs such as interatomic distances, coordination numbers, and types of neighboring atoms are determined by the EXAFS and XANES spectroscopy methods

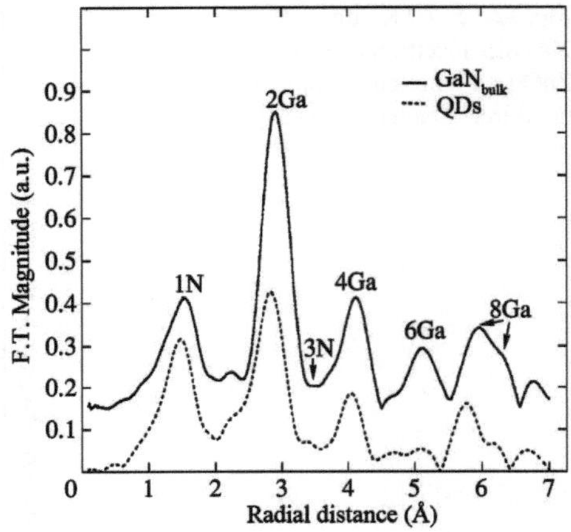

Fig. 12.13 The magnitude of FT for the bulk GaN and the GaN/AlN QDs at the Ga K-edge. Adapted from [25]

[25]. A detailed description is given to GaN/AlN Ga K-edge EXAFS (Figs. 12.13 and 12.14 and 12.15) and InAs/AlAs heterosystems (Fig. 12.16).

The authors fit Fourier-filtered data with k- and k^2 -weighting in the range of photoelectron wave vectors from 3 to $14\,\text{Å}^{-1}$. The Fourier-filtered experimental data in the ranges $1.0\,\text{Å} < R < 1.8\,\text{Å}$ and $2.5\,\text{Å} < R < 3.3\,\text{Å}$ for the first (N) and the second (Ga) coordination spheres of Ga, respectively, were used to perform the fitting procedure.

It is found that the Ga-N interatomic distance in the heterostructure with GaN QDs decreases by $\sim$0.02 Å in comparison with bulk GaN and is equal to $\sim$1.93 Å. The Ga–Ga interatomic distance in the heterostructure with GaN QDs decreases by $\sim$0.05 Å in comparison with pure GaN and is equal to $\sim$3.13 Å. Further XAS studies of GaN are presented in Chaps. 3 and 13.

Eu-doped GaN QDs embedded in AlN matrix have been studied by means of EXAFS technique [26]. Each QD sample consisted of 10 stacked planes of GaN QDs separated from one another by a 12 nm-thick AlN spacer. Due to the large difference in band gap energy between AlN and GaN, strong confinement of carriers in GaN QDs is expected, which leads to a remarkable persistence of photoluminescence up to room temperature.

To investigate Eu incorporation Eu L_3 EXAFS experiments were performed at the FAME beamline (BM30B) at the European Synchrotron Radiation Facility in Grenoble. Spectra were recorded in fluorescence mode and samples were cooled with liquid nitrogen. All recorded spectra clearly show two contributions in the L_3 edge, corresponding to the Eu^{2+} edge and the Eu^{3+} edge. Eu^{2+} can be attributed to the europium oxide monolayer that typically accumulates on the surface. Thus EXAFS oscillations were extracted relative to the Eu^{3+} edge.

EXAFS results obtained for a sample with 2.7 % Eu-doped QDs are presented in Fig. 12.14. It is important to determine whether Eu has been substitutionally incorporated inside GaN or undesirable AlN doping occurred. The first coordination shell of Eu both in AlN and GaN consists of nitrogen and thus one has to assess the chemical

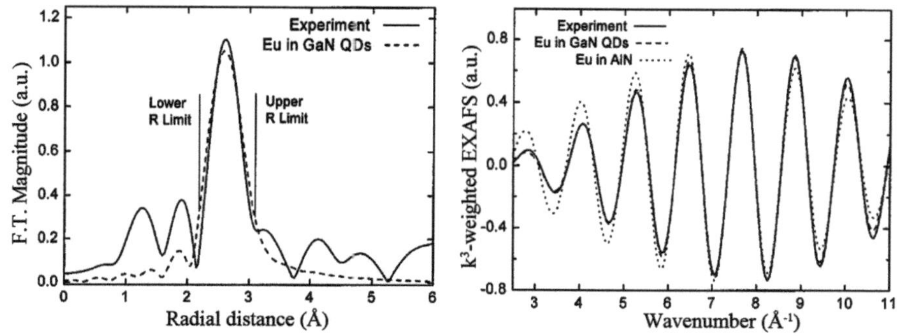

Fig. 12.14 Experimental and calculated Eu L_3-edge EXAFS in R- and q-space for Eu doped GaN QDs embedded in AlN matrix. *Dashed line* correspond to the simulation for Eu in GaN. *Dotted line* is a simulation for Eu in AlN. Adapted from [26]

nature of the second nearest neighbor shell of Eu. Second shell consists of 12 Ga for Eu inside GaN QDs and 12 Al inside the AlN matrix. These backscatterers are easily distinguishable because of the large discrepancy in mass between Al and Ga. The pseudoradial distribution in R (left part in Fig. 12.14) was obtained by taking the norm of the Fourier transform of $k^3\chi(k)$, with k ranging from 2.6 to 11.1 Å^{-1}. Nitrogen and oxygen in the first coordination shell of Eu give rise to the small two peaks between 1.0 and 2.1 Å. Second shell gives rise to the peak at 2.6 Å. EXAFS analysis was conducted on this second peak by Fourier backtransform (right panel in Fig. 12.14) with R ranging from 2.2 to 3.1 Å. Backtransform spectra clearly indicate that Eu is embedded into GaN QDs while poor agreement with experiment is found for the model when Eu incorporates in AlN. Doping of GaN is also discussed in Chaps. 4 and 13.

It was shown that EXAFS in total reflection geometry (ReflEXAFS) can be effective in determining the structure of $In_xGa_{(1-x)}As$ QDs grown on GaAs [27]. ReflEXAFS data from thin surface systems are not contaminated by anomalous dispersion effects so they can be directly treated as conventional EXAFS spectra. In particular, authors claim that it is possible to recognize the strained or relaxed state of the dots as well as their composition by comparing theoretical models to experimental data on first shell bond lengths. The Fourier transforms with the superimposed best fitting curves are shown in Fig. 12.15 whereas the inset shows the raw EXAFS spectra. Well defined first shell coordination is evident at approximately 2.25 Å (value without phase correction). XAS studies of unstrained $In_xGa_{1-x}As$ are also discussed in detail in Chap 2.

Specific variation of XAFS method—capacitance XAFS was applied to study the local structure of an As atom in the self-organized InAs QD arrays inside GaAs matrix [28]. The X-ray-induced photoemission of confined electrons in a QD via inner-shell absorption was detected by a capacitor and the photon energy dependence of the capacitance provided the XAFS spectrum of the atom in the QD. It was found that when the bias voltage applied to the system aligned the Fermi energy with the

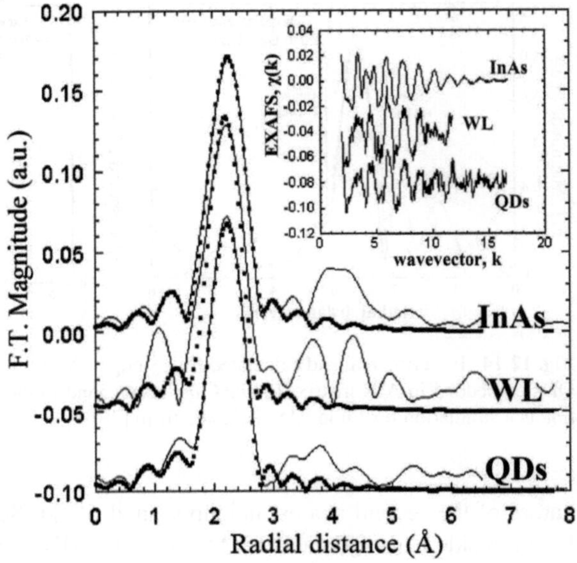

Fig. 12.15 Fourier transforms of the k^2-weighted EXAFS spectra calculated in the range $k = 3 \ldots 10\,Å^{-1}$. Inset show $\chi(k)$ data. Continuous *lines* are the experimental data, the *dotted lines* are the best fits done in R space in the range $R = 1.5 \ldots 2.7\,Å$. InAs—*bulk*, WL-wetting layer, *Dots* -QDs. Adapted from [27]

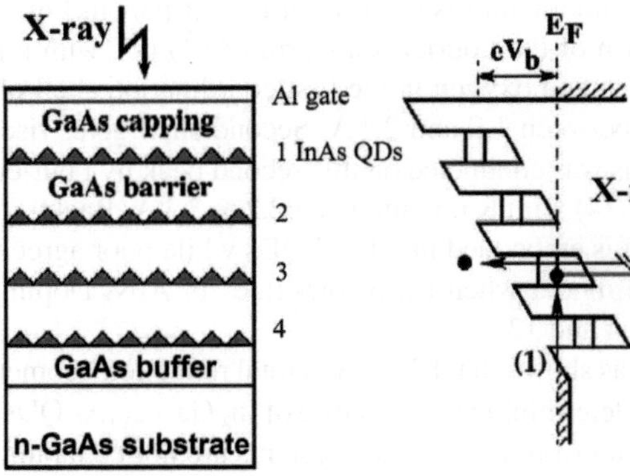

Fig. 12.16 Proposed concept for selective observation of QD by the capacitance XAFS method. *Left* typical sample structure. *Right* expected electron transition process in the sample. Adapted from [28]

quantum electronic level, the site-selectivity of XAFS could be enhanced. The peak energy shift of about 1.5 eV was observed in the site-selective XAFS spectrum of an As atom in the system of QD. Theoretical simulations indicated that this energy shift originated from the atoms in the point defect position at the boundary between an InAs QD and a GaAs barrier layer [28].

The interface effect of InSb QDs embedded in SiO_2 matrix was investigated by XAFS (both of EXAFS and XANES regions) [29]. The EXAFS showed in Fig. 12.17 suggested the bond length of the Sb-In first shell of the InSb QD to be contracted by about 0.02 Å compared with that of the bulk InSb. Theoretical analysis of the Sb K-XANES of InSb QDs embedded in SiO_2 matrix made it possible to conclude that the intensity increase and broadening of the white line peak were mainly due to the increase of Sb p-hole occupation and the change of Sb intra-atomic potential affected

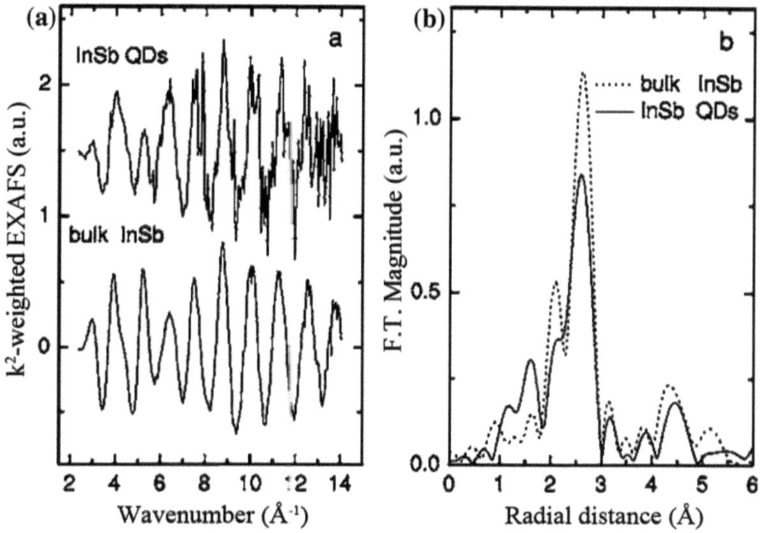

Fig. 12.17 **a** Sb K-edge EXAFS functions $k^2\chi(k)$ of the bulk InSb and the InSb QD. **b** Fourier transforms of the bulk InSb and the InSb QD. Adapted from [29]

by the SiO_2 matrix. Thus, the interaction between the InSb QDs and SiO_2 matrix resulted not only in slight lattice contraction of InSb QDs and structural distortion of the interface, but also leaded to the significant change of the Sb intra-atomic potential and the charge redistribution of Sb atoms.

12.4.2 Group II–VI QDs and Nanoparticles

Seehra et al. [30] reported size-dependent ferromagnetism in TOPO-capped (tri-n-octylphosphine oxide) CdSe QDs observed by means of combined XANES and EXAFS study. The article covers such points as measuring the number of unoccupied states or d-holes on the Cd site for 2.8, 4.1, and 5.6 nm CdSe QDs. It is shown that this ferromagnetism is due to d-holes on the Cd site bonded to TOPO created by experimentally observed charge transfer from Cd to TOPO.

The authors measured charge transfer from Cd to TOPO invoked in their interpretation of ferromagnetism using XANES spectra of TOPO-capped CdSe QDs. Figure 12.18a shows an increase in the Cd K-edge white line intensity with QDs size reduction indicating an increase in the number of unoccupied states above the Fermi level. An increasing charge loss in the occupied states of Cd ion with decreasing QDs size is also stated. The additional charge transfer to Se in CdSe QDs induced by capping with TOPO lead to decrease in the lattice constant with decreasing QDs size. Figure 12.18b shows Fourier transform (FT) of the EXAFS $k^2\chi$ data at the Cd K-edge. It is claimed that weaker intensity of the oscillation for the smaller sizes

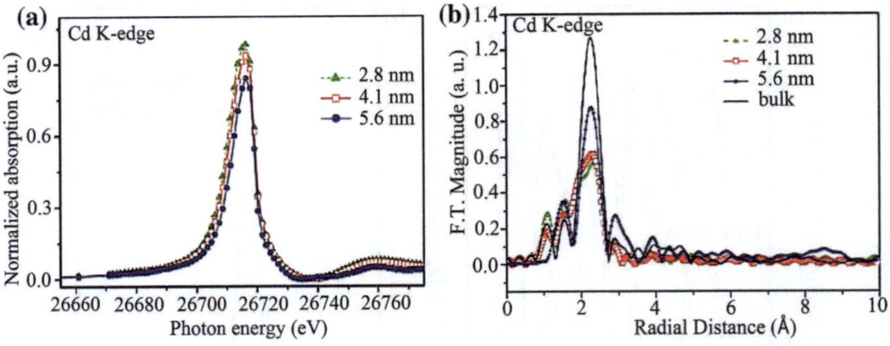

Fig. 12.18 **a** Normalized and background corrected Cd K-edge XANES spectra for CdSe QDs. **b** Fourier transforms amplitudes of the EXAFS $k^2\chi$ data at the Cd K-edge for 2.8, 4.1, 5.6 nm, and bulk CdSe. Adapted from [30]

indicates more disordered local structure. It is also pointed out that the decrease in the amplitude of the peak near 2.27 Å in the FTs due to Cd–Se bond length with decreasing size indicates reduction of the coordination number of Cd and increase of the local disorder. Also the additional feature that is shown on the low energy side is due to Cd–O bond for the smaller QDs. The paper claims that these observations strongly support the assumption of charge transfer between Cd and oxygen that would deplete the otherwise full $4d^{10}$ band of Cd, creating holes and producing a net magnetic moment for Cd atoms bonded to TOPO on the surface of the CdSe QDs.

EXAFS has been applied to investigate a local structure for the CdSe/ZnSe QDs grown by molecular beam epitaxy (MBE) and migration-enhanced epitaxy (MEE) [31]. The article covers such points as the intermixing of Cd and Zn atoms, chemical compositions and strain induced by cap-layer studied by means of EXAFS at the Cd K-edge. The paper claims that from the qualitative analysis the number of oscillations for QDs and the bulk CdSe is similar, however the second coordination shell between 3.5 and 4.5 Å depicted in Fig. 12.19 is broader for the QDs in comparison to the bulk one, suggesting the existence of more than one species' of atoms in this shell.

Structural parameters for investigated and reference CdSe result from the fitting analysis. It is reported that the first coordination shell in the QDs is composed by 4 Se near-neighbors, while the second one is composed by a mixture of Cd and Zn and the coordination number is kept 12 as follows from the nominal zinc-blende lattice. It is also reported that the Cd–Se bond length was determined as 2.61 Å and the next-neighbor distance Cd–Cd in the second coordination shell is larger by about 0.06 Å from that in the pure CdSe compound (4.31 Å), whereas the bond length Cd–Zn was found to be 4.18 Å. It is also stated that the value of the Debye–Waller factor for distances Cd–Se and Cd–Cd is similar to this found for reference CdSe compound.

The array of ZnO nanowires decorated by different-sized InP QDs were proposed for splitting water with a substantially enhanced photocurrent. The conduction band properties of ZnO nanowires with InP QDs was investigated by the Zn K-edge XAFS

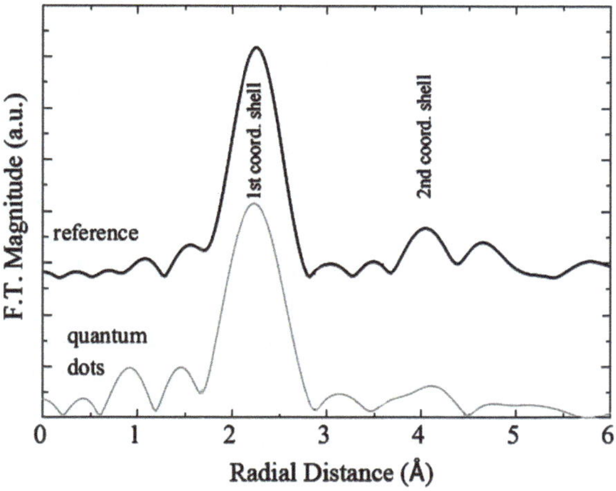

Fig. 12.19 The magnitude of FT for the bulk CdSe and the CdSe/ZnSe QDs at the Cd K-edge. Adapted from [31]

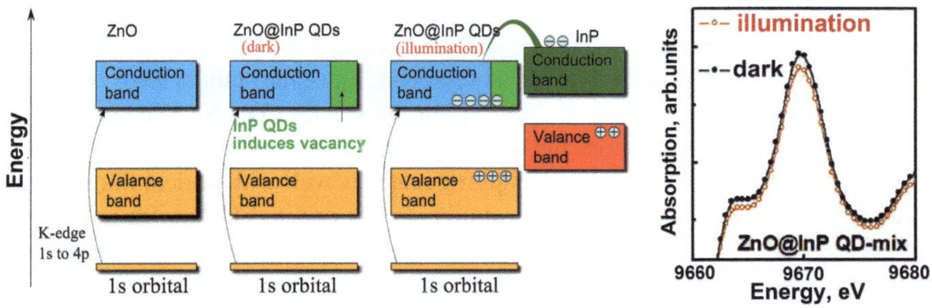

Fig. 12.20 Band structure evolution of the ZnO nanowires system covered with InP QD and changes in the Zn K-edge X-ray absorption spectrum under illumination. Adapted from [32]

analysis for both solar illuminated and not illuminated cases [32]. It was found that the orbital coupling occurred in the c axis direction of the ZnO nanowire. Because the white line intensity of XANES is proportional to the density of unoccupied states of the absorbing atoms these unoccupied states of the conduction band could enhance the transition probability from valence band to conduction band as shown in Fig. 12.20. Solar illumination of the system under the study made it possible to investigate the contribution from photoelectrochemical response. ZnO decorated with InP QDs had a significant increase in photogenerating electrons in the 4p orbital. One could attribute this increase of photogenerating electrons to the absorption of InP QDs in the visible region and the photogenerating electrons transfer from the conduction band of InP to that of ZnO. The contribution of photoresponse from ZnO nanowires or InP QDs could be distinguished by comparing the XAFS spectra collected under illuminated and not illuminated conditions. Further XAS studies of ZnO nanowires, both pure and doped, are reviewed in Chap. 13.

Cr-doped ZnSe QDs were studied by the Cr $L_{2,3}$-edge XAFS analysis [33]. Because it is well established that the Cr ion can be incorporated into ZnSe with variable oxidation states one should control the oxidation state of Cr ion inside the ZnSe QD. The energy positions of the spectral features and the shape of the Cr $L_{2,3}$-edge of the $x = 2.5\,\%$ $Zn_{1-x}Cr_x$ Se QD in Fig. 12.21 allowed an assignment of the Cr oxidation state as Cr^{3+} on the basis of a comparison to simulated $L_{2,3}$-edge XANES spectra from [34].

Chromium ion was found to be in the tetragonally distorted O_h site with a crystal field of $10\,Dq = 2.07$ eV due to pair formation of Cr octahedral by removal of three Zn ions for each pair of Cr ions in the lattice. The analysis of the data obtained made it possible to conclude that the sharing of the O_h edge by the Cr(III) ions in the ZnSe lattice produces a minimally distorted site in the lattice analogous to spinel inclusions in bulk metal chalcogenide structures [33]. XAS studies of Cr doped ZnTe are discussed in Chap. 16.

Microbeam X-ray absorption near edge structure (μ-XANES) technique was used for the analysis of the physicochemical changes of CdSe-ZnS core-shell QD in vivo [35]. A specific kind of worms—namely Caenorhabditis elegans have been studied to investigate the biological effects of QDs. Because in CdSe-ZnS core-shell QDs, selenium atoms are located in the core encapsulated by a ZnS shell, its chemical state changes only if the ZnS shell structure is degraded. Thus, the authors analyzed in situ Se K-edge μ-XANES spectra of QDs within the worms. The Se K-edge XANES of CdSe QDs mainly monitors the unoccupied p states of the Se element. It was found that at positions where XRF mapping matches well with optical fluorescence imaging, all the XANES spectra are similar to the original QDs with only slight differences as shown in Fig. 12.22. Interestingly, XANES of QDs after digestion where QDs fluorescence quenched showed obvious differences in intensity of some peaks and the energy position of the peaks (mostly a shift to higher energy by about 2 eV) and intensity of a shoulder gradually decreased. All these changes could be together assigned to a transition to oxidized Se like in Na_2SeO_3 compound. Thus, it was suggested that Se ion in the CdSe core is oxidized to SeO_3^{2-} after digestion in a worm. The changes of Se XANES spectra may be attributed to the collapse of core/sell structure after digestion, and the efficient protection of the CdSe core by ZnS shell is destroyed after 24 h of digestion in a worm.

XAFS spectroscopy at Cd L_3 edge and ab initio modeling of the experimental spectra were used to investigate the effects of surface passivation on the unoccupied electronic states of CdSe QDs covered with trioctylphosphine oxide (TOPO) or hexadecylamine (HDA) ligands [36]. Authors indicate that nitrile impurities present in the HDA bind to the surface of CdSe QDs during synthesis and lead to a mixing HDA and nitrile surfactant coating. Thus QDs synthesized in HDA are referred to as CdSeHDA/CN. It was found that there are considerable variations in the XAS spectral features as a function of size. The spectra for bulk CdSe and the 25 Å CdSe-HDA/CN QDs are closely comparable but increasingly pronounced changes in the absorbance are evident beyond the absorption edge as the QD radius decreases. The dependence of the XAS data on size provides a strong indication that these spectral variations are surface-related because the reduction in QD radius is accompanied by

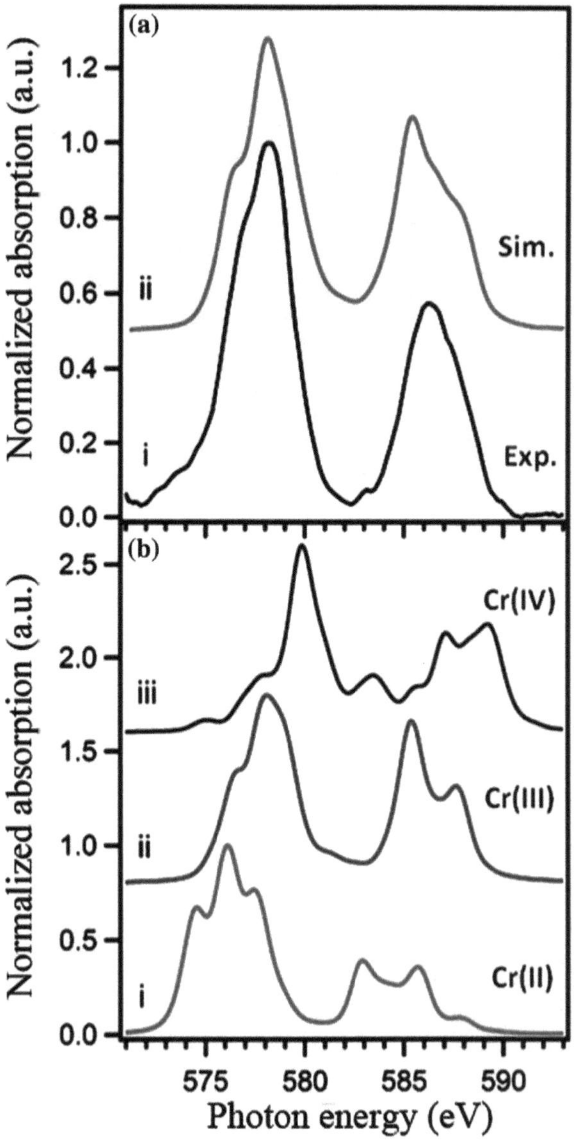

Fig. 12.21 Cr $L_{2,3}$-edge XAFS spectra [33]. **a** experiment on Cr-doped ZnSe QD and theory for tetragonally distorted O_h crystal field. **b** XAFS simulations of Cr ions in a perfect O_h field

a substantial increase in the fraction of surface Cd atoms and by a greater extension of the surface contribution to the overall signal. The conclusion was done that the surface ligands are primarily responsible for the differences in the shape of the XAS spectra.

The ligand exchange results in profound changes to the XAS data as shown in Fig. 12.23. XAS spectra demonstrate that the associated changes in unoccupied electronic structure arise due to interactions between the nitrile moieties and surface

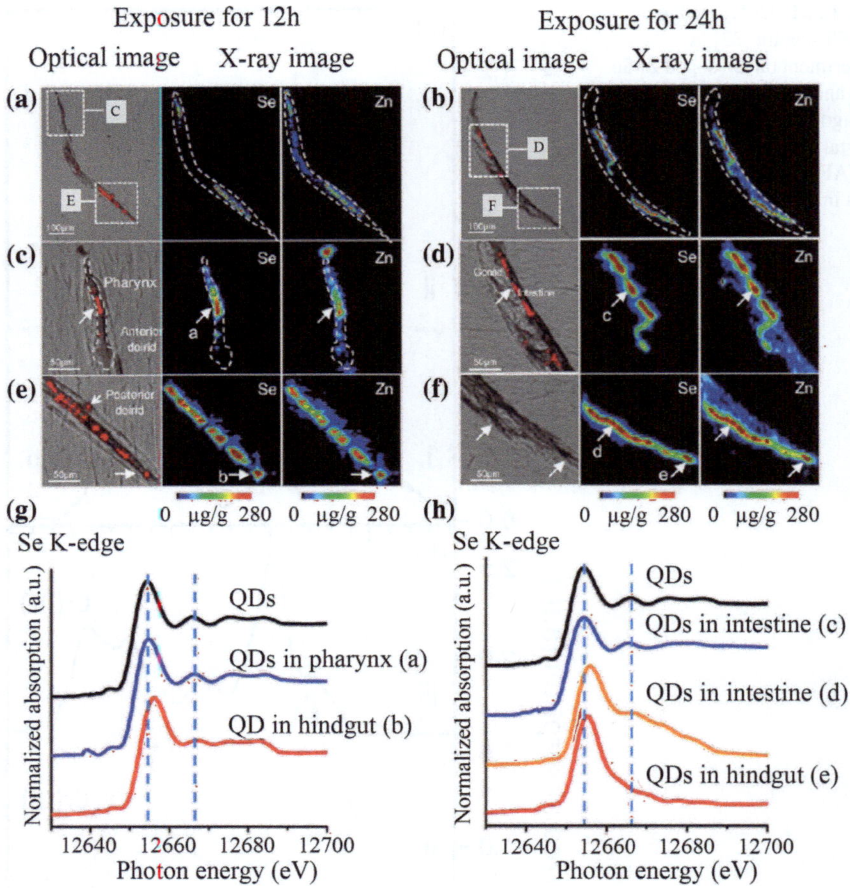

Fig. 12.22 Elemental analysis and QD degradation in a worm hermaphrodite *C. elegans*. (*A, B*)—optical fluorescence image (*gray*) versus X-ray fluorescence mapping of the whole worm exposed in CdSe-ZnS QDs for 12 and 24 h correspondingly. (*C–F*) similar images for the enlarged parts of the worm body shown by the *dashed lines* in *A* and *B*. (*G, I*) - in-situ Se K-edge μ-XANES at selected points *a–e* shown by *arrows*. X-ray spot size 5 × 5 μm. Adapted from [35]

Cd atoms. A particularly interesting feature of CdSe QDs is that the studied ligands primarily affect unoccupied electronic states several eV above the absorption edge rather than in the gap, as one might have predicted for surface states. This is consistent with the fact that the optical properties are not affected yet there is a ligand dependent change in the electronic structure, which could be responsible for the novel magnetic properties of the QDs [36].

Electronic structure and atomic arrangement of CdSe QD doped with Cu were studied by the Cu L_3 and Se L_3XAFS [37]. It was found that the L_3-edge stays at nearly a constant energy through the doping range of 2–15 molar percent. The observation of a constant L_3-edge energy (equal to the energy for CuI spectrum)

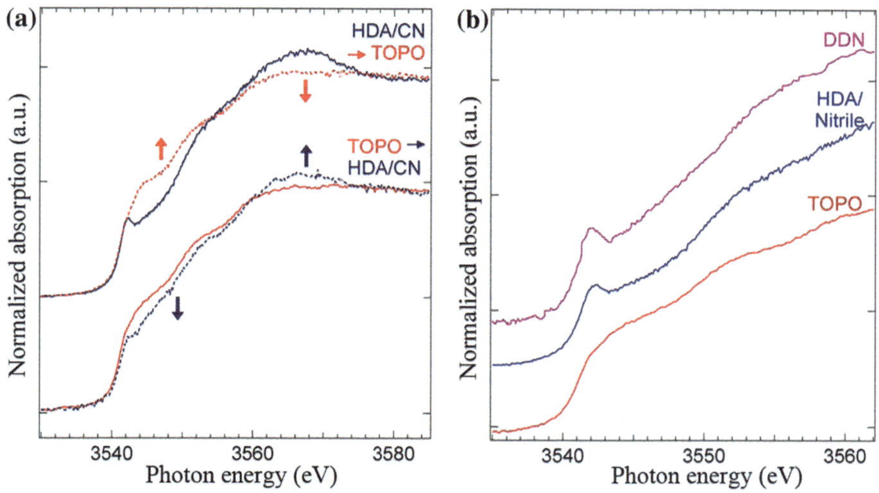

Fig. 12.23 The Cd L_3 -edge XANES **a** as prepared CdSe-HDA/CN QDs following ligand exchange with TOPO and as prepared CdSe-TOPO QDs following ligand exchange with 90 % HDA (including nitrile contaminants). **b** 15 Å CdSe-TOPO QDs following ligand exchange with DDN and reference 15 Å CdSe-TOPO and CdSe-HDA/CN QD spectra. Adapted from [36]

indicates that the oxidation state of the Cu ions in the CdSe QD is Cu(I) through all doping concentrations studied. This constant oxidation state is also an indication for the fact that Cu ion in the CdSe QD is a highly localized defect with very little charge transfer to the host lattice. The Se L_3 absorption edge shows the shifts to lower energy by 0.6–0.7 eV when going from an undoped CdSe particle to a Cu-doped CdSe particle. This is consistent with the Se environment effectively becoming "ionized" as the more electronegative Cu ion is substituted for a Cd site in the CdSe lattice. It was also found that there is an evidence for statistical ion doping, as opposed to ion clustering. The introduction of Cu^{+1} led to charge imbalances in the nanoparticle, which produced deep trap levels due to Se vacancies [37]. Doping of II-VI semiconductors is also discussed in Chaps. 4, 13, and 16.

QDs are sensitive to the optical radiation and can degrade under visible light in the oxygen-rich atmosphere. A photoetching technique is commonly used for controlled reduction of QD size. In order to understand a long-term photostability of colloidal CdSe QDs they were suspended in toluene and stored in combinations of light/dark and N_2/O_2[38]. By subjecting the CdSe suspension to air and light oxidative transformations of the surface are accelerated. On the contrary, one can minimize surface oxidation through storage in a dark environment under an inert atmosphere. The protecting TOPO surface layer of QDs was decreased by washing in a toluene suspension with methanol. Figure 12.24 shows the Se K-edge XANES spectra for the QDs with reduced protective layer before and after light exposition.

XANES spectrum can be recorded by measuring X-ray fluorescence or UV-visible luminescence. Figure 12.25 shows the optically X-ray excited optical luminescence

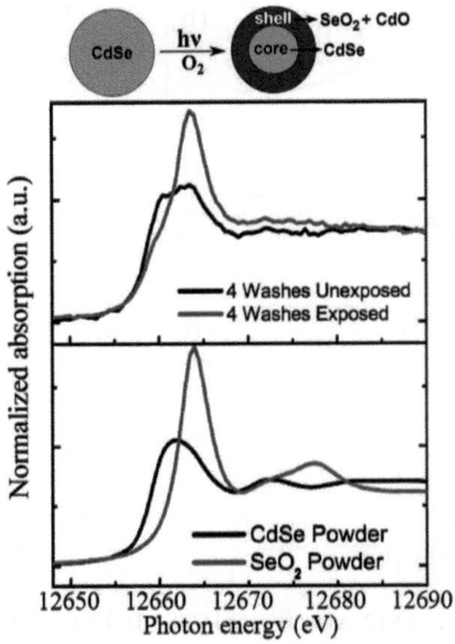

Fig. 12.24 Se K-edge XANES spectra for the CdSe colloidal nanoparticles after four washes. Comparison to spectra of CdSe and SeO_2 samples suggests that photoetched samples are partially oxidized. Adapted from [38]

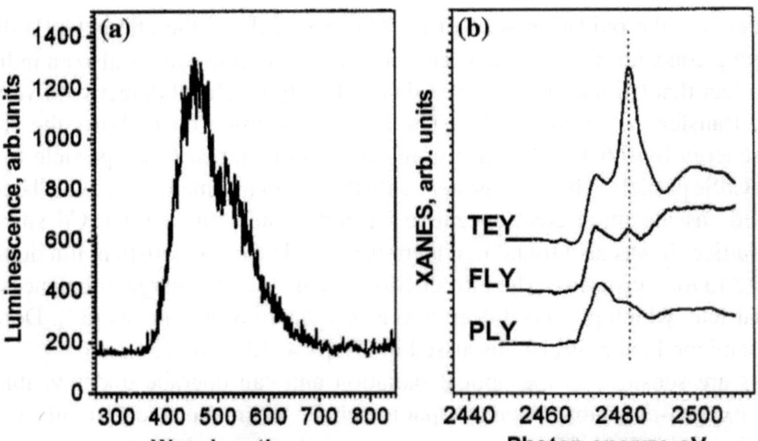

Fig. 12.25 **a** XEOL spectrum of CdS QDs excited at 2474 eV. **b** S K-edge XANES of CdS QDs measured in total electron yield (*TEY*), X-ray fluorescence yield (*FLY*) and total photoluminescence yield (*PLY*) modes. Adapted from [39]

(XEOL) spectrum of CdS QDs [39]. It displays a blue-light maximum emission at 465 nm followed by a shoulder at 520 nm. XEOL is very similar to the UV-excited spectrum because they involve similar recombination of holes in the valence band and electrons in the conduction band while only excitation process is different.

Authors have found that blue band in XEOL is very stable for more than 2 months in the atmosphere, indicating that each QD in the dendrimer aggregates does not degrade. When total UV-visible luminescence is counted as a function of an energy incident X-ray beam, a photoluminescence yield (PLY) XANES can be measured. Figure 12.25b shows the S K-edge XANES spectra recorded in bulk-sensitive X-ray fluorescence yield mode (TEY), surface sensitive total electron yield mode and PLY mode. Since only sulfur sites that produce photoluminescence efficiently will be responsible for the optical yield, the PLY spectrum provides information about the local structure of sulfur species contributing to the luminescence of the QDs.

A high resonance at 2482 eV in the TEY recorded XANES can be attributed to the absorption of sulfur in the form of sulfate SO_4^{2-} species it the nanometer-thick sulfate layer around QDs. Intensity of this resonance is reduced in the bulk sensitive FLY recorded XANES which reflects the CdS structure. Further reduction of the intensity of the SO_4^{2-} related resonance at 2482 eV is observed in PLY recorded XANES. And an important conclusion can be made that it is the S^{2-} species in CdS, but not SO_4^{2-} species that is responsible for the luminescence behavior of the QDs.

References

1. A.I. Ekimov, A.A. Onushchenko, JETP Lett. **34**, 345 (1981)
2. M.A. Reed, J.N. Randall, R.J. Aggarwal, R.J. Matyi, T.M. Moore, A.E. Wetsel, Phys. Rev. Lett. **60**, 535 (1988)
3. P. Harrison, *Quantum Wells, Wires and Dots: Theoretical and Computational Physics of Semiconductor Nanostructures* (Wiley, New York, 2005)
4. V.I. Klimov, *Nanocrystal Quantum Dots* (CRC Press Inc, 2010)
5. I.L. Medintz, H.T. Uyeda, E.R. Goldman, H. Mattoussi, Nat. Mater. **4**, 435 (2005)
6. Y. Zheng, S. Gao, J.Y. Ying, Adv. Mater. **19**, 376 (2007)
7. N. Fuke, L.B. Hoch, A.Y. Koposov, V.W. Manner, D.J. Werder, A. Fukui, N. Koide, H. Katayama, M. Sykora, ACS Nano. **4**, 6377 (2010)
8. K.S. Leschkies, A.G. Jacobs, D.J. Norris, E.S. Aydil, Appl. Phys. Lett. **95**, 193103 (2009)
9. M. Kroutvar, Y. Ducommun, D. Heiss, M. Bichler, D. Schuh, G. Abstreiter, J.J. Finley, Nature **432**, 81 (2004)
10. L. Goldstein, F. Glas, J.Y. Marzin, M.N. Charasse, G. Le Roux, Appl. Phys. Lett. **47**, 1099 (1985)
11. M.T. Björk, A. Fuhrer, A.E. Hansen, M.W. Larsson, L.E. Fröberg, L. Samuelson, Phys. Rev. B **72**, 201307 (2005)
12. N. Akopian, G. Patriarche, L. Liu, J.C. Harmand, V. Zwiller, Nano Letters **10**, 1198 (2010)
13. Y.J. Zeng, Z.Z. Ye, F. Liu, D.Y. Li, Y.F. Lu, W. Jaeger, H.P. He, L.P. Zhu, J.Y. Huang, B.H. Zhao, Cryst. Growth Des. **9**, 263 (2008)
14. X. Wang, J. Zhuang, Q. Peng, Y. Li, Nature **437**, 121 (2005)
15. J.B. Delehanty, K. Susumu, R.L. Manthe, W.R. Algar, I.L. Medintz, Anal. Chim. Acta **750**, 63 (2012)
16. W.R. Algar, K. Susumu, J.B. Delehanty, I.L. Medintz, Anal. Chem. **83**, 8826 (2011)
17. S. Horoz, L. Lu, Q. Dai, J. Chen, B. Yakami, J.M. Pikal, W. Wang, J. Tang, Appl. Phys. Lett. **101**, 223902 (2012)
18. S. Patra, B. Satpati, S.K. Pradhan, J. Appl. Phys. **106**, 034313 (2009)
19. P.M. Koenraad, M.E. Flatte, Nat. Mater. **10**, 91 (2011)
20. A. Jentys, Phys. Chem. Chem. Phys. **1**, 4059 (1999)

21. B. Gilbert, B.H. Frazer, H. Zhang, F. Huang, J.F. Banfield, D. Haskel, J.C. Lang, G. Srajer, G.D. Stasio, Phys. Rev. B **66**, 245205 (2002)
22. C. Brouder, J. Phys.: Condens. Matter **2**, 701 (1990)
23. J.H. Guo, L. Vayssieres, C. Persson, R. Ahuja, B. Johansson, J. Nordgren, J. Phys.: Condens. Matter **14**, 6969 (2002)
24. F. Decremps, F. Datchi, A.M. Saitta, A. Polian, S. Pascarelli, A. Di Cicco, J.P. Itié, F. Baudelet, Phys. Rev. B **68**, 104101 (2003)
25. S.B. Erenburg, N.V. Bausk, A.V. Dvurechenskii, Z.V. Smagina, A.V. Nenashev, A.I. Nikiforov, V.G. Mansurov, K.S. Zhuravlev, A.I. Toropov, J. Sur. Invest. X-ray, Synchrotron Neutron Tech. **1**, 26 (2007)
26. Y. Hori, X. Biquard, E. Monroy, D. Jalabert, F. Enjalbert, L.S. Dang, M. Tanaka, O. Oda, B. Daudin, Appl. Phys. Lett. **84**, 206 (2004)
27. F. d'Acapito, S. Colonna, F. Arciprete, A. Balzarotti, I. Davoli, F. Patella, S. Mobilio, Nucl. Instrum. Methods Phys. Res. Sect. B **200**, 85 (2003)
28. M. Ishii, K. Ozasa, Y. Aoyagi, Microelectron. Eng. **67–68**, 955 (2003)
29. D. Chen, C. Li, Z. Zhu, J. Fan, S. Wei, Phys. Rev. B **72** (2005)
30. M.S. Seehra, P. Dutta, S. Neeleshwar, Y.-Y. Chen, C.L. Chen, S.W. Chou, C.C. Chen, C.-L. Dong, C.-L. Chang, Adv. Mater. **20**, 1656 (2008)
31. E. Piskorska-Hommel, V. Holý, O. Caha, A. Wolska, A. Gust, C. Kruse, H. Kröncke, J. Falta, D. Hommel, J. Alloy Comp. **523**, 155 (2012)
32. H.M. Chen, C.K. Chen, C.C. Lin, R.-S. Liu, H. Yang, W.-S. Chang, K.-H. Chen, T.-S. Chan, J.-F. Lee, D.P. Tsai, J. Phys. Chem. C **115**, 21971 (2011)
33. W. Zheng, K. Singh, Z. Wang, J.T. Wright, J. van Tol, N.S. Dalal, R.W. Meulenberg, G.F. Strouse, J. Am. Chem. Soc. **134**, 5577 (2012)
34. E. Stavitski, F.M.F. de Groot, Micron **41**, 687 (2010)
35. Y. Qu, W. Li, Y. Zhou, X. Liu, L. Zhang, L. Wang, Y.-F. Li, A. Iida, Z. Tang, Y. Zhao, Z. Chai, C. Chen, Nano Letters **11**, 3174 (2011)
36. J.R.I. Lee, H.D. Whitley, R.W. Meulenberg, A. Wolcott, J.Z. Zhang, D. Prendergast, D.D. Lovingood, G.F. Strouse, T. Ogitsu, E. Schwegler, L.J. Terminello, T. van Buuren, Nano Letters **12**, 2763 (2012)
37. R.W. Meulenberg, T. van Buuren, K.M. Hanif, T.M. Willey, G.F. Strouse, L.J. Terminello, Nano Letters **4**, 2277 (2004)
38. D.A. Hines, M.A. Becker, P.V. Kamat, J. Phys. Chem. C **116**, 13452 (2012)
39. P. Zhang, S.J. Naftel, T.K. Sham, J. Appl. Phys. **90**, 2755 (2001)

Chapter 13
Group III–V and II–VI Nanowires

Francesco d'Acapito

Abstract In this contribution a review on the studies based on X-ray absorption spectroscopy (XAS) on III–V and II–VI nanowires in the recent years is presented. Examples of structural characterization of several systems like pure semiconductors (mainly ZnO, ZnS, GaN) and doped materials (with transition metals Co, Mn, Fe and rare earth ions Eu, Er) are presented. XAS in the various cases has played a major role in explaining the observed physical properties or to validate new production routes. Advanced data collection and analysis methods like micro-XAS, site selective XAS via optical emission, comparison between XAS and results of structural modeling have demonstrated to be valuable tools for a more complete understanding of the XAS data.

13.1 Introduction

X-ray absorption spectroscopy (XAS) has revealed to be a powerful tool for the structural characterization of semiconductor nanowires (NW) of the III–V and II–VI families. A considerable effort has been devoted to ZnO and GaN NWs doped with magnetic species (Co, Mn) for their attractiveness, as good candidates for room temperature ferromagnetic semiconductors [1] whereas ZnO was also studied for its interesting properties in chemistry and optoelectronics [2]. Further examples of II–VI systems studied by XAS include ZnS and ZnSe binary alloys whereas among III–Vs GaAs and InAs have been investigated.

Several peculiar aspects of XAS have been exploited for the various studies: the chemical sensitivity has permitted the characterization of the incorporation sites of dopants in the different host matrices [3] whereas the crystal-orientation dependence of the absorption coefficient (linear dichroism) has evidenced structural details in well oriented wires assemblies [4]. Advanced data collection methods recently developed have permitted a more complete description of the nanowire material like the

F. d'Acapito (✉)
CNR-IOM-OGG c/o European Synchrotron Radiation Facility-GILDA CRG,
ESRF 71, avenue des Martyrs - CS 40220, 38043 Grenoble Cedex 9, France
e-mail: dacapito@esrf.fr

C.S. Schnohr and M.C. Ridgway (eds.), *X-Ray Absorption Spectroscopy of Semiconductors*, Springer Series in Optical Sciences 190,
DOI 10.1007/978-3-662-44362-0_13

microscopic studies on single wires based on X-ray nanobeams [5] or the use of X-ray excited optical luminescence (XEOL) which demonstrated to allow site-sensitive studies [59]. In situ studies coupled to rapid data collection (Quick-XAS) have permitted a time resolved descrition of the formation of NWs [6].
In the following these investigations will be presented in detail in different sections dedicated to III–V and II–VI systems with this last part largely dedicated to ZnO.

13.2 III–V Wires

13.2.1 GaAs and InAs

In the field of III–V diluted magnetic semiconductors (see also Chap. 15) a series of papers have been devoted to the use of Manganese at the same time as wire-growth catalyst and magnetic dopant for GaAs and InAs wires. The possibility for Mn being incorporated in GaAs could lead to the realization of 1-Dimensional spintronic devices. Indeed, a widely adopted method to produce NWs is by evaporating the NW components on a substrate containing small particles of a metal. This one is called VLS (Vapor Liquid Solid) process [7] and it has been used to describe the formation of Si NWs using Au (see also Chap. 11). First, a layer of metallic nanoparticles (NP) is deposited on a substrate, successively Si is added by evaporation and it is absorbed by the NPs forming an alloy. When the concentration of Si exceeds the equilibrium value for the growth temperature the NP will release this component that in turn segregates in a crystalline wire stemming from the NP. The same principles can be applied when using manganese as catalyzer for the formation of GaAs and InAs wires and indeed their formation has been evidenced in literature [8] in a temperature range 540–620 °C for GaAs or 370–410 °C for InAs.

In order to find if Mn enters the wire and which is its incorporation site, XAS investigations were carried out on these systems by Martelli et al. [8] and d'Acapito et al. [9]. The XAS studies have revealed that Mn is actually incorporated in the GaAs structure with a Mn-As bond length of 2.56–2.60 Å. This value is considerably longer than the values observed in substitutional Mn_{Ga} in GaAs (2.50 Å [10]) and practically coincident with the distance observed in hexagonal MnAs (2.57 Å). It was then interpreted as the formation of MnAs nuclei (as no further coordination shells were visible) in agreement with previous studies on Mn:GaAs thin films where a long Mn-As distance appeared for annealing temperature above 550 °C [11] (see also Chap. 16).

An even clearer picture arose in a study by Jabeen et al. on InAs wires catalyzed by Mn [12] where the formation of extended (i.e. multiple coordination shells) crystals of hexagonal MnAs was evidenced by XAS. These investigations have thus shown the possibility of doping the wires with the catalyzer metal but the temperature range for wire formation is sufficiently high to induce the formation of the unwanted MnAs phase (in crystals or small nuclei) rather then the substitutional incorporation of Mn in the Ga sites.

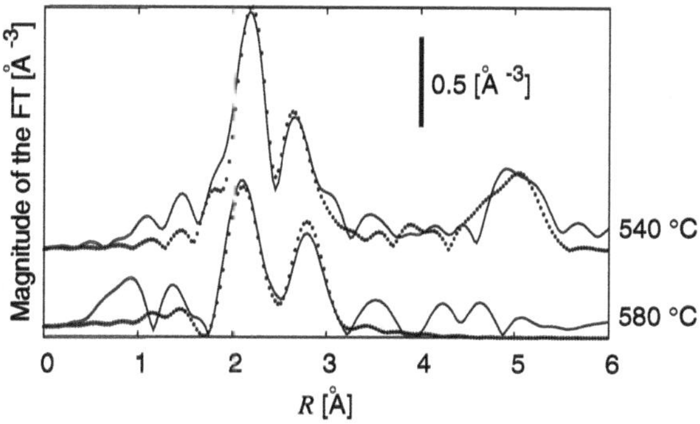

Fig. 13.1 Fourier transforms of EXAFS data of Mn-doped GaAs wires grown using Au as catalyzer, [9]. Both *curves* show a double peak in the first coordination shell testifying the presence of a heavy backscatterer around Mn: Au. In the *upper curve* (sample grown at 540 °C) also a second peak at higher *R* values is visible due to collinear chains Mn-Au-Mn typical of the MnAu structure that is absent in the *lower curve* relative to the sample grown a 580 °C

Attempts to form Mn-doped GaAs wires via the more conventional Au-catalyzed way have been reported [9] in which Mn was added to the wires during the evaporation stage. XAS has shown that in this case Mn forms an alloy with Au (cubic MnAu) that, for lower growth temperatures (540 °C), can also exhibit a measurable degree of chemical ordering (Fig. 13.1). In conclusion, Mn doping of GaAs or InAs is possible but Mn forms predominantly a MnAs phase or intermetallic compound with Au (when this metal is used as catalyzer) that prevents the observation of the (potentially possible) Mn_{Ga} sites in the wires.

13.2.2 GaN and AlGaN

A more clear situation is found in the case of Mn-doped nitrides, in particular GaN. In [13] by Farvid et al. Mn:GaN wires produced by Chemical Vapor Deposition (CVD), were analyzed using X-ray Absorption Near Edge Structure (XANES). In this study, it was possible first of all to determine the valence state of Mn in the wires looking at the position of the half-absorption point of the normalized spectra compared with those from model oxide compounds and retrieve an average state slightly more than 2+. This means that Mn in the wires is a mixture of Mn^{2+} and Mn^{3+} states, at difference with what was observed in the case of Mn:GaN thin films [14] where a dominant Mn^{3+} phase was found. The XANES spectrum of Mn exhibited also a pre-edge resonance at 6540.5 eV which is due to 1s-3d transitions that are partially allowed in sites lacking centro-symmetry like the tetrahedrons [15]. From this observation it was derived that Mn was substitutional for Ga in the structure.

In [16] by Baik et al. Mn:GaN wires were obtained by Au-catalyzed growth and successively some specimens were exposed to nitrogen plasma in a radio-frequency

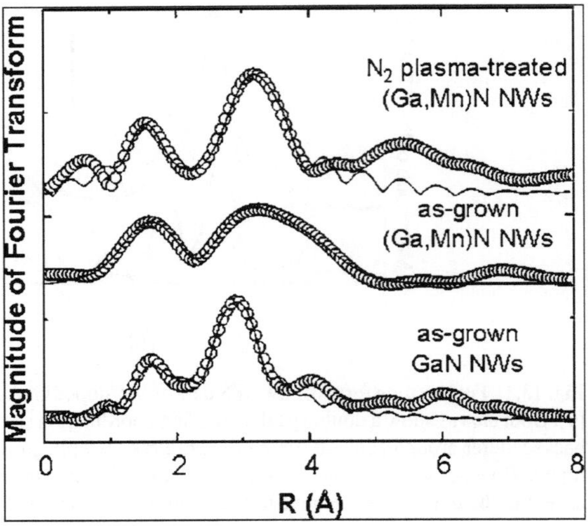

Fig. 13.2 Effect of N_2 plasma treatment on Mn-doped GaAs wires from [16]. The fourier transforms of EXAFS spectra at the Ga-K edge (*bottom curve*) and Mn-K edge (*middle* and *upper curve*) show that prior the plasma treatment the Mn spectrum exhibits broad features whereas after the treatment the Mn spectrum strongly resembles that of Ga. Reprinted with permission from [16] © 2006 American Institute of Physics

reactor. The exposure to the N plasma resulted in an increase of the ferromagnetic (FM) response of the material and EXAFS provided an explanation for that. Indeed, by comparing the XAS spectra of Ga-K edge of the wires, and the Mn-K edge of the samples prior and after the exposure to the plasma, it was evident that in the first case the Fourier Transforms of the spectra were markedly different and this was interpreted as caused to the presence of secondary phases induced by a deficiency of N in the wires. The exposure the the plasma re-established the N content to the stoichiometric value and the FT took the same shape as the Ga-K edge demonstrating in this way the incorporation of Mn as substitutional of Ga (Fig. 13.2). A further study based on XANES and anomalous X-ray Diffraction, demonstrated the substitutional location of Cu in GaN NWs [17] and its 2+ valence state so permitting to explain the FM response of the material. The substitutional site is not always successfully achieved as shown in a further study by Guda et al. [18] where tiny (50–70 atoms) metallic Cu clusters were detected in AlN nanorods. Doping of GaN is also discussed in Chaps. 4 and 12.

The potentialities of the linear dichroic response of NWs have been exploited in [19] by Han et al. about GaN/ZnO core-shell rods studied at the Ga-K edge. Indeed the fact of having assemblies of well-aligned wires of noncubic structure (wurtzite in the presented case) permits the distinct study of bonds parallel or perpendicular to the preferred growth direction (usually the crystallographic *c* axis). Namely for the wurtzite structure the first shell of the cation is composed of 4 anions in a tetrahedral arrangement one of which lies along the c axis. The contribution from this bond can be significantly enhanced adopting a data collection with the beam polarization vector parallel to the c axis and completely suppressed when setting the beam polarization parallel to the surface. A similar situation happens for the second coordination shell which consists of 12 cations, 6 lying on the growth plane (i.e. perpendicular to the c axis) and 3+3 atoms above and below this plane. The contribution of the 6 plane cations can be suppressed in the case of beam polarization perpendicular to the surface. Using a double data collection (i.e. with the X-ray beam polarization vector parallel or perpendicular to the wires c axis) the authors have revealed a marked dichroism in the XANES part of the Ga-K edge spectrum similar to what was observed in epitaxial thin firms [20] so demonstrating the high orientation degree of the wires. Moreover, the study of the bonds revealed that the structure of GaN was in that case a slightly distorted wurtzite with a distortion more evident on the Ga-Ga bonds rather than on the Ga-N bonds.

Site selectivity can be achieved by using a combined approach between X-ray absorption and diffraction, a technique that is known as DAFS (Diffraction Anomalous Fine Structure) [21]. This method is based on the analysis of the intensity of a given diffraction peak as a function of the X-ray energy when crossing the absorption edge of an element of the crystal under study. If the planes responsible for the diffraction peak contain this element the peak will exhibit modulations similar to EXAFS. The modulation are predominantly linked to the real part of the scattering coefficient (at difference of the standard EXAFS that is linked to the imaginary part) and the quantitative analysis of these data is not straightforward. Nonetheless, this technique has been applied in the analysis of the structure of AlGaN NW grown on AlN via Molecular Beam Epitaxy (MBE) [22]. This technique is preferable to VLS in the production of some semiconductor materials because it guarantees that no pollution from external sources (i.e. the catalyst metal in VLS) goes into the wires. In this case the problem to solve was, as Ga diffuses much faster than Al, to find the real composition of the AlGaN wires. The analysis of the diffraction peaks around the AlN and GaN (105) reflections reported the presence of two peaks at (h,k,l) = (1.02, 0, 5.05) [A region] and (1.03, 0, 5.12) [B region] corresponding to Ga-rich and and Al-rich zones. Successively, the intensity of these two peaks was collected as a function of the X-ray energy around the Ga-K edge. The oscillating signal was then analyzed with a formalism similar to that used for EXAFS and the local geometrical parameters (relative to Ga atoms in each of the two zones) were collected. Apart from the technical difficulty, this procedure permits to distinguish the Ga sites in two crystallographic different regions that would instead be averaged in a conventional XAS analysis. The result was that the B region contained 45 % of Al (estimated through the analysis of the number of second neighbors of Ga) and corresponded to the wires whereas the A region was considerably depleted in Al ($\approx$10 %).

Information on the electronic structure can be retrieved by the analysis of the XANES spectra at the cation L_{III} edges i.e. in the soft X-ray regime. A study by Chiou et al. [23] reported on the comparison of the Ga-K and N-K edges XANES in GaN wires and thin films. The modulations of the absorption coefficient were less marked for the wires at the N-K edge whereas they were larger at the Ga-K edge. This observation was interpreted as an increased amount of unoccupied Ga-4p states and occupied N-2p states that led the authors to suggest an increased electron transfer from Ga to N in wires with respect to thin films. A later paper by Pao et al. [24] on a wider variety of wires (InN, GaN, AlN) confirmed the previous observations at the cation-K edge whereas the N-K XANES of wires appeared to be more similar to those of thin-films. This was explained by as an increased localization of the cation (free) p orbitals in the wires with respect to the thin films. XAS studies of III-N semiconductors are also presented in Chaps. 3 and 12.

13.3 II–VI Wires

13.3.1 ZnO

13.3.1.1 ZnO and MgZnO

Zinc Oxide is a particularly interesting material due to peculiar properties that make it interesting in optoelectronics [25] and sensors [26] among others. This is the reason why a considerable number of papers have appeared on this material and in particular on its nanowire form. By exploiting its luminescent properties it has been possible to derive interesting results using the X-ray excited optical luminescence (XEOL) technique. XEOL is a particular data collection technique [27, 28] which derives the absorption coefficient μ from the optical luminescence yield that follows the absorption of an X-ray photon in particular materials. The photo- and Auger- electrons emitted in each absorption event produce a shower of lower energy electrons (that extends far from the X-ray absorbing atom depending on the emission energy) in the conduction band that eventually decay in the valence band by emitting a photon either directly or via impurity or defect intra-gap states. As the intensity of this signal is proportional to the number of the initial (photo-, Auger-) electrons, this is in turn proportional to the absorption coefficient and will contain the XAS signal related to the neighboring absorbing centers. The site selectivity between luminescent and non-luminescent species under investigation can be achieved provided that the two families are sufficiently ($\approx$10 nm [27] at low X-ray energies) spatially separated. In general to guarantee site selectivity analyses in the soft-Xray regime are to be preferred as the lower photon energy generate more localized secondary excitation showers. In practical terms, the data collection is realized by taking the optical emission from the sample with a lens focusing on an optic fiber or directly to the entrance slit of a wavelength dispersive monochromator. Different portions of the emission spectrum can be thus used for the data collection. When using sub-micrometric beams XEOL has been demonstrated to be a useful tool also for imaging luminescent species as shown in a recent study of Martinez-Criado et al. on nano-LEDs based on InGaN wires [29].

Site-selective experiments have been successfully carried out as that reported in [30] by Rosemberg et al. In this study the authors considered $Mg_xZn_{1-x}O$ core shell wires of two types: type I were wurtzite wires with growing Mn content from the center to the boundary, in type II wires a wurtzite pure ZnO core was surrounded by a Rocksalt shell with a high Mg content. A first part of the study concerned the

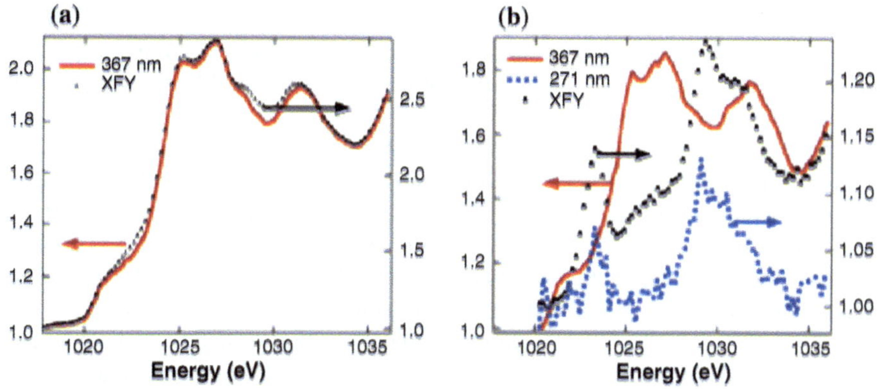

Fig. 13.3 Zn-L_{III} edge XANES spectra collected in X-ray fluorescence yield (XFY) or XEOL modes from [30]. Here two cases are compared: panel (**a**) ZnO wires (type-I) ; panel (**b**) ZnO wires cladded by MgO. While XFY is equally sensitive to any Zn site (so providing averaged information on all the Zn present in the material) different optical emission bands can be relative to different environments for the metal. Indeed, collecting the XEOL data at two different emission bands (367 and 271 nm), permitted in the type-II case to evidence the presence of two distinct sites for Zn. The emission at 367 nm is associated to the wurtzite whereas that at 271 nm comes from the Rocksalt ZnO phase. Reprinted with permission from [30] © 2007 American Chemical Society

time properties of the optical emission. In all samples the analysis of the optical luminescence at low temperature evidenced a line at 367 nm with a very fast decay time (<1 ns) originating from band-edge excitons. In type II samples a considerable emission from other bands (250–350 and 450–700 nm) was also found with a much longer decay time attributed to defect states. Collecting the data with a fast (0–3ns) gating[1] permitted to completely eliminate the defect states signal from the type-II samples. This shows how, in addition to the wavelength selection, also the separation of the signal in the time domain can contribute to the selection of the emitting sites. The analysis of the Zn-L_{III} edge XANES (Fig. 13.3) provided striking structural information: the type-I samples measured using the 367 nm emission presented a spectrum identical to that of the X-ray fluorescence Yield and coinciding to that of wurtzite ZnO. On the contrary, the analysis of type-II samples revealed the wurtzite structure collecting the 367 nm emission and a spectrum typical of Rocksalt ZnO when collecting the signal at 271 nm. The band previously attributed to defect states (due to the fact that it has a longer decay time) was then due to the Zn atoms in the Mg rich region in the shell portion of these wires. Mg-K edge gave an interesting complement to this analysis, exhibiting a spectrum typical of cubic MgO in the X-ray fluorescence of 270 nm optical luminescence spectra and no structure at all in the faint 370 nm band. This further demonstrates that, within the dimensions of the wire, i.e. about 30 nm, the excitation created on the Mg sites was not capable to reach the Zn luminescent sites of the core part thus assuring the local character

[1] This supposes the availability of a pulsed source as frequently encountered in 3rd generation synchrotrons: in this case the X-ray pulses were 80 ps wide with a repetition time of 153 ns.

of the information. The summary of this investigation is that, by using XEOL on luminescent materials, the combined use of wavelength- and time- domains permits the separation of signals related to different parts (and in turn different sites) of the sample under investigation.

XAS can provide not only static properties of the system under study but also their kinematical behavior. This is achieved by using time-resolved data collection either using fast scanning methods (Quick-EXAFS, [31]) or parallel acquisition (Dispersive XAS, [32]). In the field of ZnO nanowires the fast scanning technique was used by McPeak et al. to elucidate the process of ZnO NW formation during Chemical Bath Deposition (CBD) [6]. In the particular process taken into account, the deposition of ZnO wires originates from an aqueous solution of zinc nitrate and hexamethylenetetramine (HTMA). The nitrate provides Zn^{2+} ions to the solution whereas HTMA decomposes in formaldeide and ammonia. The latter compound reacts with water giving rise to ammonium plus hydroxide ions. Eventually the reaction between Zn^{2+} and OH^- leads to ZnO + water. If this is a coarse description of the process, it could be supposed that intermediate complexes, namely Zn-hydrates, Zn-HTMA or Zn-ammine, could be formed during the ZnO deposition. A time-resolved XAS study at the Zn-K edge could in principle evidence the formation of these intermediate phases. In this study the authors have realized a reactor where ZnO could grow by CBD on the walls of an X-ray transparent (Kapton) tube. The tube was internally coated with CdO in order to act as seed for the ZnO growth and was heated in order to study the reaction at different temperatures (in this case 60 and 90 °C were used). A fast scanning setup permitted to collect 100 eV wide XANES spectra in about 1.5 min over a total reaction period of 2 hours. In this way, the collected spectra changed in a continuous way from Zn^{2+} in water (actually, $[Zn(H_2O)_6]^{2+}$ complex) to ZnO (Fig. 13.4). Successively, the data were analyzed via the Principal Component Analysis (PCA) in order to reveal the phases that enter into play in the formation of the various XANES spectra. PCA [33] is a particular data analysis procedure that, based on the Single Value Decomposition algorithm, permits to determine how many distinct components are present in the spectrum of a mixed phase sample. Successively, by testing several model compounds likely to be those forming the material under study, the actual components participating to the reaction can be determined. By applying PCA to the time resolved spectra and testing a wide variety of Zn^{2+} compounds the authors came to the conclusion that only $[Zn(H_2O)_6]^{2+}$ and ZnO were present in the spectra with a relative ratio depending on the reaction time. This gave a first answer to the problem investigated, i.e. the ZnO deposition happens without any detectable intermediate phase. By reproducing the individual spectra as linear combination of these two phases it has been possible to follow their evolution with time: namely it has been possible to evidence an incubation time of about 8 min during which Zn remains as ion in the water. Successively ZnO starts growing first at a high rate corresponding to a nucleation-and-growth process and then, once the nucleation centers were saturated, as linear growth until all the metal ions in the solution have reacted. The regimes of nucleation and linear growth change with the concentration of Zn^{2+} in the solution.

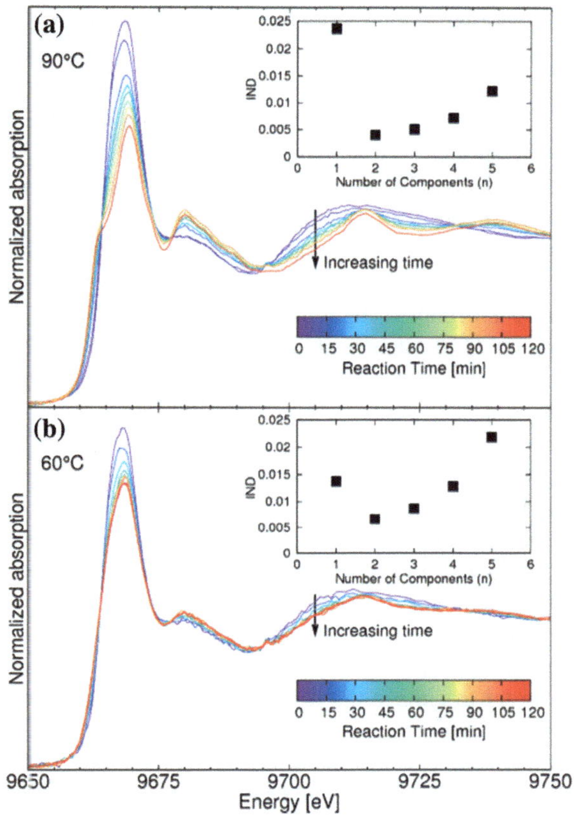

Fig. 13.4 In situ time resolved Zn-K edge XANES spectra of ZnO nanowire growth by the chemical bath deposition process ([6]). Data relative at the process carried out at 90 and 60 °C are shown. The inset presents the indicator IND function of the PCA analysis: having a minimum at 2 components this means that only these two phases contribute to the spectra in the time evolution. These two phases are identified as being hydrated Zn^{2+} ions and wurtzite ZnO with no further contribution. Reprinted with permission from [6] © 2010 American Chemical Society

The effects of preparation routes on the optical properties of ZnO wires were also studied by Chen et al. [34] in the case of pure hydrothermal growth or with the addition of ammonia and polyethileneimine (PEI). The analysis of the EXAFS data revealed to the authors the presence of interstitial Zn in the first case and interstitial O in the second case and this had a considerable effect on the optical properties of the wires.

Also in the case of ZnO the possibility of growing NW assemblies with a defined orientation with respect to the surface (i.e. with the growth direction coinciding with the c axis and perpendicular to the surface) has permitted polarized studies. In a first investigation [35] the data collection with the incoming beam polarization vector parallel and perpendicular to the surface permitted a detailed analysis on NWs of different length, revealing an increased disorder in the shorter ZnO bond (parallel to the c axis) for the shorter wires. A technically similar study [36] has permitted to investigate ZnO wires heavily implanted with N^+ ions and revealing a partial substitution of O neighbors of Zn with N.

The effect of the morphology of the wires on their electronic structure has been studied by Chiou et al. [37] in a series of ZnO wires of different sizes (d = 150 nm ×

$l = 600\,nm, d = 80\,nm \times l = 540\,nm, d = 45\,nm \times l = 260\,nm$) studied by XANES at the Zn-K and O-K edges. The study, based on the analysis of the features of the XANES spectra, revealed an increased number of O-2p and Zn-4p free states upon NW dimension reduction that was interpreted as due to the increase of electron states at the surface. A following paper by Sipr et al. [38] showed, based on theoretical XANES calculation of the surface and bulk Zn and O atoms, that surface effects should rather lead to an opposite result, leaving then open the interpretation of the origin for this phenomenon. The same procedure based on differences on the XANES spectra was used by Chiou et al. [39] to correlate the occupation of O-2p and Zn 4s/3d unoccupied states to the dimension of the Au particle used to catalyze the wire growth. H.M. Chen et al. used the difference of XANES under dark and illumination condition to explain the improved behavior of ZnO nanowires sensitized by InP nanodots [40] (see also Fig. 12.19 in Chap. 12) or Au nanodots [41] in materials for the photolysis of water.

A final example will be presented here based on the use of submicron sized X-ray beams (see also Sect. 13.3.1.2). Scanning Transmission X-ray Microscopy has permitted the study of the interface in complex structures consisting in ZnS ribbons with ZnO rods stemming from them [42]. This study was conducted in the soft X-ray region at the Zn-L_{III}, S-K and O-K edges using a 30 nm wide beam. In this energy region the focusing was achieved via a zone plate. The difference in edge position for the Zn-L_{III} edge in ZnS and ZnO permitted to realize chemical sensitive images in order to choose the regions of interest (ROI) where to collect XANES spectra. These, were collected along the axis joining one of the rods with the ZnS substrate in order to detect the possible variation of the local order around the components of the system. An abrupt change from ZnS-like to ZnO-like pattern was detected at the Zn edge meaning that the interface between the two phases was sharper than the beam size. But it was possible to reconstruct the origin of the rods as initiating from the (001) Zn-terminated faces of the ZnS ribbon. Also the position of the edge, recorded along the passage from the sulphide to the oxide, exhibited a sharp increase with no hint of an extended inter-diffused region.

13.3.1.2 Dopants in ZnO

The possibility of producing 1-dimensional semiconductors has brought the idea of adding dopants for the realization of real devices. Apart from conventional electric doping (i.e. done with non-isoelectronic dopants in order to add carriers in the conduction band, see also Chap. 4) also magnetic doping has been explored in the last years as already mentioned in Sect. 13.2 and as discussed in detail in Chaps. 14–16. A widely studied magnetic dopant for ZnO is Co (see also Chaps. 5 and 16) and the possibility of introducing it in wires opens several possibilities in 1-D devices for spintronics and optoelectronics. With the advent of new achromatic focusing devices for hard X-rays [43] a new era opened in the field of micro-XAS. Indeed, achromaticity is a key issue for XAS as this technique requires a wide (at least 500 eV, corresponding to a $\Delta E/E$ variation of $\approx 5\,\%$ at 10 keV) energy range that

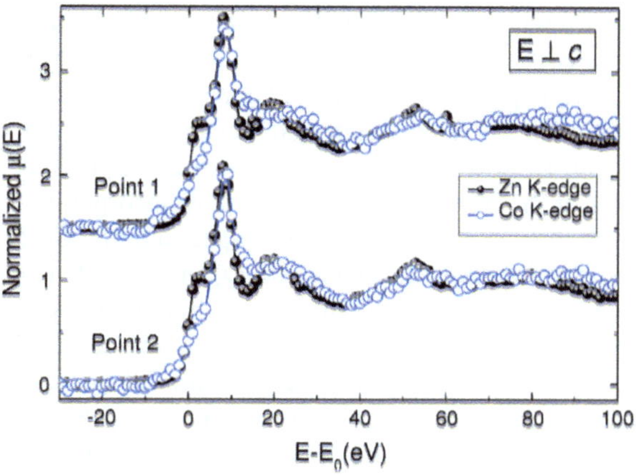

Fig. 13.5 XANES spectra collected in two different points of a single nanowire of ZnO doped with Co [5]. The similarity between the two datasets (Zn-K and Co-K edge) demonstrate that Co substitutes for Zn in the matrix. Reprinted with permission from [5] © 2011 American Chemical Society

cannot be covered with chromatic devices like Fresnel Lenses. The improvement of the technology of mirrors and in particular of the Kirckpatrick-Baez geometry has led to the demonstration of the possibility of collection of extended XAS data in sub-μm regions [44] of semiconductor devices. These concepts have been applied in a recent study on Co-doped ZnO NW [5] by Segura-Ruiz et al. In this investigation, Co:ZnO wires (length $\approx$7 μm, diameter $\approx$170 nm) grown on Si by the VLS method were doped by ion implantation of Co and analyzed with an X-ray sub-μm beam of dimensions $0.1 \times 0.1\,\mu m^2$. A suitable sample orientation stage permitted to place a single wire in the beam and to choose its orientation relative to the beam polarization. A first step of the investigation consisted in collecting X-ray fluorescence maps of single wires that are capable to provide a quantitative determination of the metal content in the material. Based on these maps 3 different spots in the wire were chosen to carry out the data collection at the Co-K and Zn-K edges in the XANES (see Fig. 13.5) and EXAFS regions. XANES data at the Zn-K edge were collected with the polarization vector parallel and perpendicular to the wire axis and showed a markedly dichroic signal. This, as already mentioned in the previous sections, comes from the fact that the wires have a noncubic crystal lattice and in particular the observed signals were typical of the wurtzite structure. Also The XANES data at the Co-K edge exhibited linear dichroism and revealed a marked analogy with that of Co:ZnO in thin film form [45] demonstrating in this way that Co was substitutional for Zn in the wire. The edge position showed that the dopant had a 2+ valence state. A detailed analysis of the EXAFS data at the Zn-K edge revealed structural parameters typical of wurtzite, i.e. with no major distortions as it could be expected after Co ion implantation. This shows that the annealing process that followed the implant was really effective in recovering the crystal damage.

An analogous study was presented some years before by Yuhas et al. on Co and Mn doped ZnO wires produced by solution-based synthesis [3]. In that case a part of the investigation was carried out in the soft X-ray regime where smaller beams (about 40 nm sized) were available for microanalysis. Transmission images taken in difference mode (i.e. taken at two different energies above and below the dopant L_{III} edge) permitted to visualize the wires and to choose points where to collect the XANES spectra. These spectra demonstrated that the dopant (both for Co and Mn) was in the 2+ valence state and that a non octahedral O cage was present around the metal. These findings are consistent with the presence of the dopant in the tetrahedral sites of Zn. A more stringent test on the incorporation site of the doping metals came from the analysis of the EXAFS spectra at the Co-K and Mn-K edges. In this case information about the first (metal-O) and second (metal-Zn) coordination shells could be retrieved in a quantitative form although on an extended spatial region because the beam spot was in that case considerably larger (5*5 μm^2) compared to the soft-X ray measurement. The analysis showed that Co produces a small distortion of the first shell with respect to the Zn site (+0.6 %) whereas Mn alters considerably this bonds (+3.5 %): this was interpreted by the authors as the origin of the simpler incorporation of Co in ZnO (well above 10 at %) compared to Mn (no more than a few at %). The dilution of the dopants and their occupation of the substitutional site was put in relation with the observed magnetic properties i.e. the spins of the dopants placed in crystallographic positions align with those of the carriers giving rise to weak ferromagnetism in addition to paramagnetism as frequently reported on those samples by magnetic measurements.

These experiments are a demonstration that with the brilliant synchrotron radiation sources available nowadays and with the use of efficient focusing devices it is possible to push XAS spectroscopy (both in the near edge and extended regions) below the 1 μm limit opening the possibility of investigation of single microstructures like quantum dots, wires, biological cells [46].

The study presented in [47] by Guda et al. presented a different approach to determine the site of Mn in ZnO. The Mn-K edge XANES (rather than EXAFS as commonly carried out) was fitted to a structural model following the original procedure presented in [48]. The geometrical results were then compared to a structural modeling simulation (Density Functional Theory) in order to correctly interpret the experimental data. The comparison between the experiment and the theory confirmed the substitutional occupation of Mn in the ZnO lattice and its 2+ valence state. Substitutional occupation of Transition Metals (TM) in ZnO was found in further structural results on Co ([49]), Fe ([50]), Mn ([51]) and Ni [52].

The behavior of Cu as a magnetic impurity for ZnO NWs has been investigated by Kataoka et al. in [53]. Here NWs obtained by vapor phase transport and of diameter 120–400 nm and length 4 μm were investigated via XANES at the Cu-L_{III} edge. Two distinct data collection schemes were used: Total Electron Yield (TEY) for surface (estimated by the authors to be ≈5 nm) analysis and Total Fluorescence Yield (TFY) for the analysis of the bulk, with a depth sensitivity of about 100 nm. A comparison between X-ray Photoemission Spectroscopy data and XANES permitted to identify a majority of Cu^{3+} ions in the surface of the NWs (XANES measured by TEY) whereas the bulk (measured by TFY) contained a mixture of Cu^{2+}/Cu^{3+} ions. X-ray Magnetic Circular Dichroism data, taken in TEY and TFY modes, yielded different results with a faint dichroic signal from surface Cu ions and a well detectable signal from the TFY data. This permitted the authors to state that the ferromagnetic response of Cu:ZnO NW comes from Cu ions in the bulk of the nanostructures where the FM coupling is favored by the Cu^{2+}-Cu^{3+} interaction. Doping of ZnO with Mn, Cu, and Fe is also discussed in Chap. 16.

The addition of dopants Co and Mg in ZnO was shown to increase the negative charge on O as shown by XANES at the O-K edge by Chiou et al. in [54, 55]. In addition to magnetic issues arising from doping with Co, Mn, or Cu several studies were devoted to attempts to change also the optical behavior of ZnO wires. In particular, a series of works aimed to luminescent Rare-Earth doped ZnO in order to study the incorporation site of the RE ion. However it has been recognized by Huang et al. [56] that a number of critical issues dominate this field like the considerable difference in ionic radius between Zn and some REs (Eu, Er) and the fact that the ZnO exciton decays too rapidly to efficiently transfer its energy to luminescent RE centers. A work on ZnO wires nominally doped with Eu [56] made at the Zn-K and Eu-L_{III} edges revealed that the structure of the wires is not changed from the pure ones upon doping when observing the Zn edge. Looking to the Eu edge it was evident that the spectrum was identical to that of the oxide Eu_2O_3 suggesting that this phase can be formed as a surrounding shell of the wires. In reality the Photoluminescence data exhibited a different line pattern in the Eu-ZnO wires with respect to pure Eu_2O_3 thus supporting the idea of a tight interaction of these two phases. X-ray diffraction carried out on the same samples presented no sign of Europium oxide and this shows the strength of a local technique like XAS that permits to obtain information on a minority phase whose diffraction signal lies buried below that of the dominating ZnO. A further study on Er_2O_3-coated ZnO [57] presented detailed High Resolution Transmission Electron Microscopy study demonstrating the presence of a RE oxide around the wires. Also in this case the XAS spectrum at the Er-L_{III} edge resulted to be identical to powder Er_2O_3 supporting the model of core(ZnO)-shell(Er_2O_3) for these systems and the difficulty of insertion of REs in the crystalline lattice of ZnO NWs.

13.3.2 Other II–VI

In the field of II–VI further examples on the use of site selective XAS using XEOL can be found in [58] by Rosemberg et al. Here, ZnS nanoribbons exhibiting the wurtzite structure (note: in the bulk form the stable phase of ZnS is zincblende) have been analyzed at the Zn-L_{III} edge by collecting the optical emission signals. These NWs, obtained by the VLS method using Au as growth catalyst, exhibit a luminescence spectrum characterized by 3 emission lines [59], one at 338 nm originating from the wurtzite structure, one at 430 nm due to defects and one at 520 nm related to Au_{Zn} substitutional sites near the metal tip terminating the wire. Transmission Electron Microscopy showed that the zone below the metal tip was of Zincblende structure. XEOL spectra were collected by recording separately the emission from the three lines and different spectra were obtained. In the first case (emission at 338 nm) the XEOL spectrum closely reproduced that collected in conventional X-ray Fluorescence mode on the same sample as well as that of bulk wurtzite ZnS demonstrating that this emission is related to this part of the wire (by far the spatially dominating one even if this emission is minoritary in the total spectrum). More interesting are the XEOL spectra collected using the 430 and 520 nm emission lines: here the two spectra are similar and different from the previous one. The pattern strongly resembles that of Zincblende ZnS demonstrating that this emission comes from the highly defective or Au-containing zones near the gold tip i.e. those identified by TEM as Zincblende ZnS. Note that this phase has a negligible volume fraction with respect to wurtzite but its strong luminescence and energy-resolved XEOL permits its detection and characterization. This is a further example where XEOL, in presence of two spatially separated phases with different luminescence properties has made possible a site selective analysis of the material. Again, it is worth to stress that spatial separation (as already noted in the case of core-shell Zn-Mg-O wires in Sect. 13.3.1.1) is a crucial issue in XEOL as the electronic excitations extend by several nm around the absorbing atom. The comparison between XEOL (full spectrum) and fluorescence XAS at the S-K edge on the same samples [59] permitted to evidence the effect of oxygen removal from the S atoms on the wire walls. Indeed, the Sulphur-K edge in fluorescence mode presented a feature at 2482 eV (interpreted as due to the white line in a zinc sulfate phase) that disappeared after annealing at 600 °C and the spectrum obtained in this way resulted to be identical to that collected with the optical emission.

Zhou et al. [60] have exploited XEOL for the characterization of ZnS and CdS nanoribbons grown via VLS method. The analysis of the optical luminescence excited with X-rays above the S-K edge in ZnS permitted to evidence a faint emission at 329 nm coming from the band gap photoluminescence plus two lines at 494 and 523 nm due to defects. CdS presented a main emission at 528 nm due to the band gap and a weaker emission at 445 nm also due to defects. XANES spectra in Total Electron Yield (TEY) mode were collected on the wires and on the bulk sulphides whereas XEOL was collected only on the wires as no XEOL signal was detected in the bulk material. For ZnS TEY and XEOL XANES of wires were similar to the spectrum of the bulk with the exception of a shoulder at 2484 eV in the TEY spectrum of ribbons. As shown in the previous example this peak is due to S-O interactions and disappears upon annealing. The difference between TEY (showing the S-O interactions) and XEOL (more similar to the bulk powder) further confirms the possibility of site selectivity of XEOL with respect to luminescent species. The analysis of CdS presented instad deceiving results with a XEOL signal presenting no edge step upon crossing the S-K edge. This is a phenomenon sometimes encountered in XEOL and in this case it was interpreted by the authors as the fact that the luminescence is originated by Cd rather than by S sites.

The possibility of producing well oriented arrays of NWs has permitted Aruguete et al. to obtain a detailed description of the structure of colloidal CdSe wires [4]. In this investigation Se-K edge XAS was collected on arrays of oriented wires with varying incidence angles between the wire main axis and the beam polarization. As commonly encountered in NWs, also CdSe exhibits the wurtzite crystal arrangement with the *c* axis coinciding with the growth direction of the structure. The use of a small (16*7 μm^2) beam spot permitted the authors to carry out the investigation on a well oriented domain of wires. Spectra were collected at several orientation angles and the variations of the derived bond length value for the first (Se-Cd) and second (Se-Se) shell were analyzed as a function of the angle. It was found that the first shell bond was less anisotropic with respect to what was expected from the bulk structure and that the second shell distance had a marked contraction of the distances along the c axis (growth direction) with respect to those present on the a/b plane (the growth plane). In order to understand this behavior a structural modeling based on Density Functional Theory was carried out by relaxing the atomic positions of a model NW of 2.4 nm in diameter and then analyzing the angular behavior of the interatomic distances averaged on the whole wire or in selected zones (core and surface). The results of the DFT modelization followed qualitatively the EXAFS results with a null dependence of the first shell distance on the angle and a contraction of the second shell distance along the growth direction. The NW core vs surface analysis of the DFT calculated structure permitted to evidence how the principal source of the observed anisotropy in the bond distances came from surface atoms. This example also shows how the use of theoretical structural modeling can provide a fundamental support for the interpretation of the XAS data. Further XAS studies of II–VI semiconductors are also presented in Chaps. 3, 12, and 16.

Mn-doped ZnS is a widely used material for its electroluminescence properties. For this material a great attention is devoted to the production in wire form. [61] by Brieler et al. presented a study on the production of Mn:ZnS wires within the pores of ordered mesoporous silica. This method permits the realization of arrays of wires with a well defined order (hexagonal or cubic depending on the host matrix) in the growth plane. XAS at the Mn-K edge was used in order to distinguish whether Mn remained in a MnS phase or it was really incorporated in the ZnS matrix. The analysis of the XANES and the bond parameters in the first and second shell permitted to confirm the incorporation of the doping metal in the matrix. Substitutional sites were also detected by XAS in Mn:ZnS particles grown by solvothermal methods in [62] by Cao et al. that, depending on the solvent used in the sample preparation, could either produce cubic nanoparticles or wurtzite nanowires. Also ternary $Zn_{0.99-x}Mn_{0.01}Cu_xS$ produced by hydrothermal method as described in [63] reported substitutional occupancy for Mn and Cu in the matrix. Doping of II–VI semiconductors is also discussed in Chaps. 4, 12, and 16.

A detailed analysis of the XANES part of the X-ray Absorption spectrum was used by Wu et al. [64] for the structural characterization of NiO nanowires grown by a low temperature method. XANES collected at the Ni-K and O-K edges were compared to the corresponding bulk forms of NiO and to the theoretical calculation of increasingly bigger clusters. The comparison yielded that the nanowire form retains only a few features of the extended crystal demonstrating that the wires lack a long range ordering and thus contain a considerable degree of disorder.

13.4 Conclusion

In this section it has been shown how X-ray Absorption Spectroscopy can contribute to the structural characterization of semiconductor nanowires. The local view of this technique permits to easily determine the environment of dopants whereas additional site selectivity can be obtained in certain cases by collecting XAS spectra using optical luminescence. The availability of polarized light and highly ordered NW arrays has permitted in several cases a detailed structural characterization by exploiting the linear dichroism of the absorption coefficient.

Focusing of X-rays in submicron spots represents at present a major frontier for the analysis of materials and the advent of advanced instruments dedicated to this issue [65] will boost research in this field. Also new techniques like time resolved data collection in the ns regime based on the time structure of the synchrotron [28, 66] will contribute improving the possibility of site selection. Further opportunities for time resolved studies will arise as soon as ultrafast (10^2 fs) X-ray Free Electron Laser (X-FEL) sources will become available to the users community.

References

1. T. Dietl, H. Ohno, F. Matsukura, J. Cibert, D. Ferrand, Science **287**, 1019 (2000)
2. S.J. Pearton, D.P. Norton, K. Ip, Y.W. Heo, T. Steiner, J. Vac. Sci. Technol. B Microelectron. Nanometer Struct. **22**(3), 932 (2004)
3. B.D. Yuhas, S. Fakra, M.A. Marcus, P. Yang, Nano Letters **7**(4), 905 (2007)
4. D.M. Aruguete, M.A. Marcus, L.s. Li, A. Williamson, S. Fakra, F. Gygi, G.A. Galli, A.P. Alivisatos. J. Phys. Chem. C **111**(1), 75 (2007)
5. J. Segura-Ruiz, G. Martinez-Criado, M.H. Chu, S. Geburt, C. Ronning, Nano Letters **11**(12), 5322 (2011)
6. K.M. McPeak, M.A. Becker, N.G. Britton, H. Majidi, B.A. Bunker, J.B. Baxter, Chem. Mater. **22**(22), 6162 (2010)
7. M. Law, J. Goldberger, P. Yang, Ann. Rev. Mater. Res. **34**(1), 83 (2004)
8. F. Martelli, S. Rubini, M. Piccin, G. Bais, F. Jabeen, S.D. Franceschi, V. Grillo, E. Carlino, F. D'Acapito, F. Boscherini, S. Cabrini, M. Lazzarino, L. Businaro, F. Romanato, A. Franciosi, Nano Letters **6**, 2130 (2006)
9. F. d'Acapito, M. Rovezzi, F. Boscherini, F. Jabeen, G. Bais, M. Piccin, S. Rubini, F. Martelli, Semicond. Sci. Technol. **27**(8), 085001 (2012)
10. F. d'Acapito, G. Smolentsev, F. Boscherini, M. Piccin, G. Bais, S. Rubini, F. Martelli, A. Franciosi, Phys. Rev. B **73**(3), 035314 (2006)
11. Y.L. Soo, S. Wang, S. Kim, G. Kim, M. Cheon, X. Chen, H. Luo, Y.H. Kao, Y. Sasaki, X. Liu, J.K. Furdyna, Appl. Phys. Lett. **83**(12), 2354 (2003)
12. F. Jabeen, M. Piccin, L. Felisari, V. Grillo, G. Bais, S. Rubini, F. Martelli, F. d'Acapito, M. Rovezzi, F. Boscherini, J. Vac. Sci. Technol. B Microelectron. Nanometer Struct. **28**(3), 478 (2010)
13. S.S. Farvid, M. Hegde, I.D. Hosein, P.V. Radovanovic, Appl. Phys. Lett. **99**(22), 222504 (2011)
14. W. Stefanowicz, D. Sztenkiel, B. Faina, A. Grois, M. Rovezzi, T. Devillers, F. d'Acapito, A. Navarro-Quezada, T. Li, R. Jakieła, M. Sawicki, T. Dietl, A. Bonanni, Phys. Rev. B **81**, 235210 (2010)
15. F. Farges, Phys. Rev. B **71**, 155109 (2005)

16. J.M. Baik, Y. Shon, T.W. Kang, J.L. Lee, Appl. Phys. Lett. **89**(15), 152113 (2006)
17. H.K. Seong, J.Y. Kim, J.J. Kim, S.C. Lee, S.R. Kim, U. Kim, T.E. Park, H.J. Choi, Nano Letters **7**(11), 3366 (2007)
18. A.A. Guda, S.P. Lau, M.A. Soldatov, N.Y. Smolentsev, V.L. Mazalova, X.H. Ji, A.V. Soldatov, J. Phys. Conf. Ser. **190**(1), 012136 (2009)
19. S.W. Han, H.J. Yoo, S.J. An, J. Yoo, G.C. Yi, Appl. Phys. Lett. **88**(11), 111910 (2006)
20. F. d'Acapito, F. Boscherini, S. Mobilio, A. Rizzi, R. Lantier, Phys. Rev. B **66**, 205411 (2002)
21. H. Stragier, J.O. Cross, J.J. Rehr, L.B. Sorensen, C.E. Bouldin, J.C. Woicik, Phys. Rev. Lett. **69**, 3064 (1992)
22. N.A. Katcho, M.I. Richard, O. Landr, G. Tourbot, M.G. Proietti, H. Renevier, V. Favre-Nicolin, B. Daudin, G. Chen, J.J. Zhang, G. Bauer, J. Phys. Conf. Ser. **190**(1), 012129 (2009)
23. J.W. Chiou, J.C. Jan, H.M. Tsai, W.F. Pong, M.H. Tsai, I.H. Hong, R. Klauser, J.F. Lee, C.W. Hsu, H.M. Lin, C.C. Chen, C.H. Shen, L.C. Chen, K.H. Chen, Appl. Phys. Lett. **82**(22), 3949 (2003)
24. C.W. Pao, P.D. Babu, H.M. Tsai, J.W. Chiou, S.C. Ray, S.C. Yang, F.Z. Chien, W.F. Pong, M.H. Tsai, C.W. Hsu, L.C. Chen, C.C. Chen, K.H. Chen, H.J. Lin, J.F. Lee, J.H. Guo, Appl. Phys. Lett. **88**(22), 223113 (2006)
25. M.H. Huang, S. Mao, H. Feick, H. Yan, Y. Wu, H. Kind, E. Weber, R. Russo, P. Yang, Science **292**(5523), 1897 (2001)
26. J. Mller, S. Weienrieder, Fresenius' J. Anal. Chem. **349**, 380 (1994)
27. A. Rogalev, J. Goulon, in *Chemical Applications of Synchrotron Radiation: X-ray Applications*, vol. 2, Chap. 15. Advanced Series in Physical Chemistry (World Scientific Publishing Co., Singapore, 2002)
28. T.K. Sham, R.A. Rosenberg, Chem. Phys. Chem. **8**(18), 2557 (2007)
29. G. Martinez-Criado, J.A. Sans, J. Segura-Ruiz, R. Tucoulou, A.V. Sol, A. Homs, J. Yoo, J.G.C. Yi, B. Aln, Phys. Status Solidi (c) **9**(3–4), 628 (2012)
30. R.A. Rosenberg, G.K. Shenoy, M.F. Chisholm, L.C. Tien, D. Norton, S. Pearton, Nano Letters **7**(6), 1521 (2007)
31. R. Frahm, Rev. Sci. Instrum. **60**(7), 2515 (1989)
32. T. Matsushita, R.P. Phizackerley, Jpn. J. Appl. Phys. **20**(11), 2223 (1981)
33. T. Ressler, J. Wong, J. Roos, I.L. Smith, Environ. Sci. Technol. **34**(6), 950 (2000)
34. L.Y. Chen, Y.T. Yin, C.H. Chen, J.W. Chiou, J. Phys. Chem. C **115**(43), 20913 (2011)
35. H.J. Yu, E.S. Jeong, S.H. Park, S.Y. Seo, S.H. Kim, S.W. Han, AIP Conf. Proc. **879**(1), 1865 (2007)
36. C.H. Kwak, Y.B. Lee, S.Y. Seo, S.H. Kim, C.I. Park, B.H. Kim, D. Jeong, J. Kim, Z. Jin, S.W. Han, Curr. Appl. Phys. **11**(3 Suppl.), S328 (2011)
37. J.W. Chiou, K.P.K. Kumar, J.C. Jan, H.M. Tsai, C.W. Bao, W.F. Pong, F.Z. Chien, M.H. Tsai, I.H. Hong, R. Klauser, J.F. Lee, J.J. Wu, S.C. Liu, Appl. Phys. Lett. **85**(15), 3220 (2004)
38. O. Sipr, F. Rocca, J. Phys.: Condens. Matter **23**(31), 315501
39. J.W. Chiou, S.C. Ray, H.M. Tsai, C.W. Pao, F.Z. Chien, W.F. Pong, M.H. Tsai, J.J. Wu, C.H. Tseng, C.H. Chen, J.F. Lee, J.H. Guo, Appl. Phys. Lett. **90**(19), 192112 (2007)
40. H.M. Chen, C.K. Chen, C.C. Lin, R.S. Liu, H. Yang, W.S. Chang, K.H. Chen, T.S. Chan, J.F. Lee, D.P. Tsai, J. Phys. Chem. C **115**(44), 21971 (2011)
41. H.M. Chen, C.K. Chen, C.J. Chen, L.C. Cheng, P.C. Wu, B.H. Cheng, Y.Z. Ho, M.L. Tseng, Y.Y. Hsu, T.S. Chan, J.F. Lee, R.S. Liu, D.P. Tsai, ACS Nano **6**(8), 7362 (2012)
42. Z. Wang, J. Wang, T.K. Sham, S. Yang, J. Phys. Chem. C **116**(18), 10375 (2012)
43. R. Barrett, R. Baker, P. Cloetens, Y. Dabin, C. Morawe, H. Suhonen, R. Tucoulou, A. Vivo, L. Zhang, in *Proceedings of the SPIE 8139*, 813904 (2011). doi:10.1117/12.894735
44. L. Mino, D. Gianolio, G. Agostini, A. Piovano, M. Truccato, A. Agostino, S. Cagliero, G. Martinez-Criado, F. dAcapito, S. Codato, C. Lamberti, Small **7**(7), 930 (2011)
45. G. Ciatto, A. Di Trolio, E. Fonda, P. Alippi, A.M. Testa, A.A. Bonapasta, Phys. Rev. Lett. **107**, 127206 (2011)
46. G.E. Ice, J.D. Budai, J.W.L. Pang, Science **334**(6060), 1234 (2011)

47. A. Guda, N. Smolentsev, J. Verbeeck, E. Kaidashev, Y. Zubavichus, A. Kravtsova, O. Polozhentsev, A. Soldatov, Solid State Commun. **151**(19), 1314 (2011)
48. N. Smolentsev, A. Soldatov, G. Smolentsev, S. Wei, Solid State Commun. **149**(4142), 1803 (2009)
49. T. Yao, W. Yan, Z. Sun, Z. Pan, Y. Xie, Y. Jiang, J. Ye, F. Hu, S. Wei, J. Phys. Chem. C **113**(32), 14114 (2009)
50. S. Gautam, S. Kumar, P. Thakur, K.H. Chae, R. Kumar, B.H. Koo, C.G. Lee, J. Phys. D. Appl. Phys. **42**(17), 175406 (2009)
51. X. Zhang, Y. Zhang, Z.L. Wang, W. Mai, Y. Gu, W. Chu, Z. Wu, Appl. Phys. Lett. **92**(16), 162102 (2008)
52. S. Chiu, J. Huang, J. Non-Cryst. Solids (2012). doi:10.1016/j.jnoncrysol.2011.12.053 (in press)
53. T. Kataoka, Y. Yamazaki, V.R. Singh, A. Fujimori, F.H. Chang, H.J. Lin, D.J. Huang, C.T. Chen, G.Z. Xing, J.W. Seo, C. Panagopoulos, T. Wu, Phys. Rev. B **84**, 153203 (2011)
54. J.W. Chiou, H.M. Tsai, C.W. Pao, K.P.K. Kumar, S.C. Ray, F.Z. Chien, W.F. Pong, M.H. Tsai, C.H. Chen, H.J. Lin, J.J. Wu, M.H. Yang, S.C. Liu, H.H. Chiang, C.W. Chen, Appl. Phys. Lett. **89**(4), 043121 (2006)
55. J.W. Chiou, H.M. Tsai, C.W. Pao, F.Z. Chien, W.F. Pong, C.W. Chen, M.H. Tsai, J.J. Wu, C.H. Ko, H.H. Chiang, H.J. Lin, J.F. Lee, J.H. Guo, J. Appl. Phys. **104**(1), 013709 (2008)
56. W.L. Huang, J. Labis, S.C. Ray, Y.R. Liang, C.W. Pao, H.M. Tsai, C.H. Du, W.F. Pong, J.W. Chiou, M.H. Tsai, H.J. Lin, J.F. Lee, Y.T. Chou, J.L. Shen, C.W. Chen, G.C. Chi, Appl. Phys. Lett. **96**(6), 062112 (2010)
57. D. Mustafa, D. Biggemann, J. Wu, J.L. Coffer, L.R. Tessler, Superlattices Microstruct. **42**(16), 403 (2007)
58. R.A. Rosenberg, G.K. Shenoy, F. Heigl, S.T. Lee, P.S.G. Kim, X.T. Zhou, T.K. Sham, Appl. Phys. Lett. **87**(25), 253105 (2005)
59. R.A. Rosenberg, G.K. Shenoy, F. Heigl, S.T. Lee, P.S.G. Kim, X.T. Zhou, T.K. Sham, Appl. Phys. Lett. **86**(26), 263115 (2005)
60. X.T. Zhou, P.S.G. Kim, T.K. Sham, S.T. Lee, J. Appl. Phys. **98**(2), 024312 (2005)
61. F.J. Brieler, P. Grundmann, M. Frba, L. Chen, P.J. Klar, W. Heimbrodt, H.A. Krug von Nidda, T. Kurz, A. Loidl, J. Am. Chem. Soc. **126**(3), 797 (2004)
62. J. Cao, J. Yang, Y. Zhang, Y. Wang, L. Yang, D. Wang, Y. Liu, X. Liu, Z. Xie, Opt. Mater. **32**(5), 643 (2010)
63. J. Cao, J. Yang, Y. Zhang, L. Yang, Y. Wang, D. Wang, M. Gao, Y. Liu, X. Liu, Z. Xie, Mater. Res. Bull. **45**(6), 705 (2010)
64. Z.Y. Wu, C.M. Liu, L. Guo, R. Hu, M.I. Abbas, T.D. Hu, H.B. Xu, J. Phys. Chem. B **109**(7), 2512 (2005)
65. G. Martinez-Criado, R. Tucoulou, P. Cloetens, P. Bleuet, S. Bohic, J. Cauzid, I. Kieffer, E. Kosior, S. Labouré, S. Petitgirard, A. Rack, J.A. Sans, J. Segura-Ruiz, H. Suhonen, J. Susini, J. Villanova, J. Synchrotron Radiat. **19**(1), 10 (2012)
66. T. Regier, R. Sammynaiken, J. Vogt, M. Ward, T. Sham, in *Canadian Light Source 2009 Activity Report* (2010)

Part IV
Magnetic Semiconductors

Chapter 14
Magnetic Ions in Group IV Semiconductors

Roberto Gunnella

Abstract XAS experiments at the TM K-edge in semiconductors, will be employed as a crucial technique to spread light not only on the mechanisms of the atomic substition, but also to investigate how the semiconductor surrounding the metal is perturbed by the insertion of an extrinsic atomic species. Wherever nominal dilution of Mn is obtained, such investigations put in light the physical constraints that must be considered to describe the electronic problem of the ferromagnetism on-set. Contrarily to the case of the III-V group doped semiconductors the research field is still incomplete and many systems have not been dealt with yet in any details, possibly because of more difficult realization and of a more controversial interpretation.

14.1 Introduction

One open problem in condensed matter physics is represented by the possibility of mutual control of transport (by spin and charge) and ferromagnetic (FM) ordering in electronic devices or spintronics devices [1]. In such devices, like spin-diodes and spin field effect transistors, achievements would be non-volatility, increased processing speed, decreased power consumption. Ferromagnetic semiconductors are known since the 60s, like europium and chromium chalcogenides in rock salt (like EuO) and spinel structures(like $CdCr_2S_4$) [2, 3]. However, difficulties in material preparation discouraged their use from the application point of view. Manganites (like perovskite(La,Sr)MnO3), which show colossal magnetoresistance (CMR),are magnetic semiconductors, whose studies have been particularly rich over the recent years [4] but there is still concern about their combination with semiconductor technology used in nowadays devices. On a different basis, ferromagnetic ordering can be achieved by the inclusion of suitable atomic doping (less than 10 %) of transition metal (TM) in semiconductors (SCs) providing transport and ferromagnetism to the pristine insulating and non-magnetic solid with ferromagnetism persisting up to RT.

R. Gunnella (✉)
Scienze e Tecnologie, Università di Camerino, Via Madonna delle Carceri 9,
62032 Camerino, MC, Italy
e-mail: roberto.gunnella@unicam.it

C.S. Schnohr and M.C. Ridgway (eds.), *X-Ray Absorption Spectroscopy of Semiconductors*, Springer Series in Optical Sciences 190,
DOI 10.1007/978-3-662-44362-0_14

In such a way many problems related to spin injection at the interfaces would also be minimized [5]. However recently most of the attention has been conveyed on Mn doped III-V SCs [6, 7], with GaMnAs representing the undisputed prototype example reaching a Curie Temperature (T_C) of 110 K [6]. Transition metal doping of group III-V semiconductors is discussed in detail in Chap. 15. Another option is represented by II-VI SCs [8] as reviewed in Chap. 16. With respect to the former ones, II-VI group semiconductors (like CdMnTe), show high solubility but are quite difficult to be doped (by both n and p type shallow impurities), while relatively weak bonds make the practical applications more cumbersome. On the contrary difficulties in dissolution of dopants in the III-V SCs could be at the origin of presence of defects like anti-sites occupation responsible for lower performances. For instance curing such defects led recently to rise the Curie temperature of GaMnAs from 110 to 170 K [9]. So far, compared to III-V group SCs, much less attention has been devoted to TM doping of group-IV SCs, despite their leading technological role in the nowadays electronic industry and their reduced complexity on the side of the chemical environment (as the absence of As_{Ga} anti-site defects to name one). The main reason of this poor attention was the problem in obtaining a good quality of the material. Renewed interest in IV-group SCs followed the refinement of non equilibrium growth techniques like low temperature molecular beam epitaxy (LTMBE) [10], laser ablation and ion implantation [11]. Park and coworkers discovered a Curie temperature up to 120 K increasing with concentration in the case of Mn_xGe_{1-x} alloy [10]. However, due to the very low solubility limit of Mn in Ge ($10^{15}\,\mathrm{cm}^{-3}$), corresponding to an atomic percentage of about 0.2×10^{-6}, i.e., several orders of magnitude below the typical Mn concentration employed in the fabrication of MnGe alloys, growth of homogeneous phases of diluted alloys is challenging. The magnetic response of most of the $\mathrm{Mn}_x\mathrm{Ge}_{1-x}$ alloys fabricated so far, is then probably due to Mn-rich separated phases [12]. And only in few cases, a prove of a diluted magnetic semiconductor (DMS), i.e., an efficient and homogeneous dilution of magnetic acceptor impurities in the semiconducting matrix, has been obtained [10, 13–15]. In this review we shall report how XAS contributed to support many studies on the local structures of Mn and other TM dopants. Many of these systems have been only seldom studied by XAS spectroscopy because of difficult realization or because of their small or zero magnetic response. On the other hand, it is beyond the scope of this review, to deal with secondary phases of the DMS, sometimes deliberately prepared sometimes obtained accidentally. Such secondary nanostructures could be important to stabilize, otherwise highly metastable states showing higher Curie temperatures and magnetoresistance [16].

14.2 Theoretical Background

When an isolated impurity is introduced in a crystalline host a variety of factors should be considered as the atom's valence state, the lattice relaxation, the many electron effects and localization vs hybridization of electronic states. On these details will depend the possible acceptor/donor behavior and the resulting magnetic moment. Isolated TM impurities can occupy a substitutional or interstitial site. In both cases d states are splitted by the crystal field in three-fold (T2) and two-fold (E) states, which interact differently with the ligand s-p orbitals. In the first case a bonding (merged in the valence band) and an anti-bonding state (in the middle of the gap) are formed. In the second case a non-bonding E state is also forming within the valence band. This leads to a high spin ground state with exchange splitting much larger than the crystal field. Differently, in case of an interstitial site occupation , E state is strongly involved in the hybridization as the second nearest neighbors are octahedrally coordinated. From a band structure point of view a TM dopant in a IV-group semiconductor is hybridized with valence host electrons. For instance, Cr, Mn, Fe and Co have respectively six, seven , eight and nine valence electrons to hybridize. When considering TM atoms from Cr to Fe the following electronic configurations are expected: $E^2_{\uparrow}T2^0_{\uparrow}$,$E^2_{\uparrow}T2^1_{\uparrow}$,$E^2_{\uparrow}T2^2_{\uparrow}$ and $E^2_{\uparrow}T2^3_{\uparrow}$ with the result of a number of Bohr magneton ranging from 2 to 5. About the semimetallic character, as it is clear from Fig. 14.1 this is obtained only for Mn and Fe, with respectively 2 and 1 holes per TM atom [17]. In this case an integer atomic magnetic moment should be obtained by the TM atoms, while in case of insulating system non integer magnetic moment should be observed. In this latter case the hybridization merges the states with valence band. For instance, changing from C to Si and finally to Ge the hybridization will change because of the different s-p ligand states energy with respect to the almost pinned TM levels (internal reference rule [18]) with hybridization slightly dispersing them into the gap. Very often extension of the unit cell is cumbersome from the calculation point of view to discriminate localization properties like activation energies [19]. In particular starting from the calculation done for Cr, Mn and Fe in Si and Ge by Weng and Dong [20]. We can see in Fig. 14.1 the occupation of the majority and minority electron bands in Ge (upper panel) and Si (lower panel) respectively from top to bottom, for 3 % of Cr, Mn and Fe. We can deduce that with respect to Mn one T2g up electron is missing in Cr while in Fe one additional down state is present, which can also be seen from the opening of the gap in spin-up channel in the case of Cr and closure of the spin-down gap compared to that of the Mn in the case of Fe. These changes of total DOS around the Fermi Edge in the spin-channels will lead to different spin polarized conductivity in these systems according to their interplay between carrier mobility and FM ordering of the local magnetic moment. As shown in Fig. 14.1 the Mn in Ge leads to a semi-metal while Cr and Fe to semiconducting and metallic bands respectively. Better stabilization of the half-metalic phase can be achieved in smaller lattice parameters solid (Si, SiC) using Cr instead of Mn [21]. Though the electronic problem seems to be easily assessed, there is nevertheless little knowledge about the basic atomic structure. To avoid such a problem model Hamiltonian theories are likely to remain indispensable because, when applicable, they provide more transparent physical pictures of ferromagnetism and often enable predictions of thermodynamic, transport, and other properties that are sometimes depending on material complexity, beyond the reach of ab initio theory techniques. Nevertheless, the most popular Zener p-d mean-field theory [22] used so far, especially in the III-V case, has been shown to strongly overestimate the magnetic

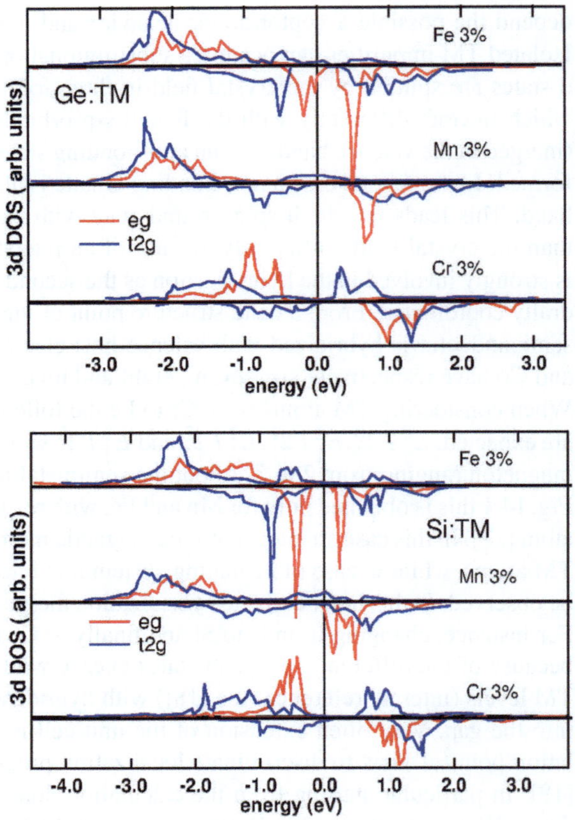

Fig. 14.1 3d-DOS of spin up and down electronic states of Si (*lower panel*) and Ge (*upper panel*) for 3 % doping of (from *top* to *bottom*) Fe, Mn and Cr. From [20]

ordering temperatures in case neither the thermal fluctuation nor the percolation effects are included [23]. Precise atomic structure are therefore fundamental as basis of the ab initio total energy calculations. Recently ab initio calculations have been performed by Stroppa et al. [24]: it was shown that Mn in both Si and Ge is at the same time a source of holes as well as of localized magnetic moment equal to about 3 μ_B. The electronic structure of MnGe alloys showed a half-metallic behavior and an overall ferromagnetic (FM) coupling. A comparison among different TM's as dopants for Ge, revealed that Mn is indeed the best candidate, since it grants holes that can mediate FM coupling as well as relatively high moments [25]. Surprisingly enough a strong environmental effect occurs in practical situations: along specific crystallographic directions, the values of the exchange constant depends on the relative distribution of the Mn atoms. Environmental effects were also recently reported by Sato et al. in DMS via a cluster-embedding method [23]. Of course, these findings imply that a careful prediction of the Curie ordering temperature requires an accurate knowledge of the real distribution of the Mn atoms in the host matrix [17], and the complete solution of the structural problem, especially in case of high concentration DMS.

14.3 Experimental Growth Techniques

Solubility limit of TM in IV-group (10^{15} cm^{-3}) [26], corresponding to an atomic percentage of about 0.2×10^{-6}, is several orders of magnitude below the typical Mn concentrations employed in the fabrication of MnGe alloys. Therefore growth of homogeneous phases of diluted alloys is challenging, strongly enhanced by surface effects like reconstruction, passivation and others giving rise to supersaturation. The magnetic response of most of the alloys fabricated so far, is then probably due to TM-rich separated phases. Formation of both metal-rich precipitates and the unintentional donor-like doping by occupation of interstitial sites [27], represent two main limitations to overcome in the fabrication of a successful DMS. To limit the importance of the above processes and at the same time to improve the solubility of the metal impurities in the matrix, one can reduce even further the TM concentration to about 1 % [28, 29]. However, if the main task is to increment the number of magnetically active atoms, two ways can be pursued: (1) reduction of the growth temperature possibly followed by postgrowth treatment; (2) introduction of energy barriers against undesired reaction paths. There is no doubt that only the use of structural probes with chemical sensitivity, like XAS, can safely reveal the structure of Mn diluted in the matrix even for very small percentage (less than 1 %) [30]. Growth techniques for creating a slightly disordered but uniform doping of a IV group SC by TM atoms must have the following three characteristics: (1) to work in out-of-equilibrium thermodynamic conditions to hinder formation of spurious compounds; (2) to be sufficiently precise to dose the exact amount of dopant resulting in the highest Curie temperature achievable, as the increase the number of holes moves the Fermi level, depleting the impurity bands and reducing the double exchange (p–d) mechanism; (3) to avoid possible spurious and contamination contribution to the weak FM signal. Among growth techniques Bridgman method [31] takes advantage of physical vapor deposition from evaporated polycrystalline material on a seed crystal or amorphous substrates. The elemental vapor is transferred by careful control of the thermal gradient. Contamination could not be completely hindered from the point of view of the crystal quality obtained. For instance, oxygen acts as getter of TMs and should be avoided in DMS preparation. Molecular Beam Epitaxy (MBE) [32] evaporates pure materials using ultra high vacuum (UHV) and liquid nitrogen (LN) shrouded chambers [33, 34]. It allows a precise growth front control useful in the fabrication of multilayered films or in the growth of codoped system. The intrinsic advantage of MBE is the possibility to separate material growth rate from the substrate temperatures, used to tune kinetics during out-of-equilibrium growth. In such a way a more detailed investigation of metastability and spinodal decomposition is possibly leading to the formation of nanostructured doped films [16]. Related to MBE is the solid phase epitaxy (SPE) technique where metal diffusion after growth is controlled by thermal annealing of pre-deposited films. Similar results are achievable by metal-organic chemical vapor deposition [32], in this case fast rates are obtained by the high substrate temperature to activate molecular pyrolysis. Nevertheless isotropic growth rate enhancement is also obtained by means of radio frequencies fields or electric discharges. One interesting way to improve Mn dilution in semiconductors is the growth of δ doped layers of TM [15], i.e., a non-equilibrium growth of a fraction of monolayer of metal alternated with a semiconductor layer. This method is used to fabricate digital layer (DL) alloys heterostructures, when alloys with an high TM concentration are alternated with undoped SC layers. In such DL heterostructures of III-V DMS grown at low temperatures a sizeable increase in the Curie temperature (T_C) has been measured [35]. Effective magnetic doping has been shown to increase if a codopant [36] is added to the main magnetic impurity. Two different explanations are currently considered for this effect: (1) a resulting increase in the hole number mediating the magnetism with higher T_c; (2) in [37] Mn and n-type codopant (As, Sb and P) experienced electrostatic formation of dimers stabilizing Mn atom in the substitutional site and codopant in interstitial site; in addition to a better substitutional incorporation the anti-ferromagnetic interaction is minimized increasing the Mn–Mn first neighbour distance. Another technique able to rise substantially the solubility limit and to avoid the precipitation of intermetallic alloys, is the ion implantation, which in the case of the Ge DMS's has been initially investigated by Liu and coworkers [38] and more extensively in more recent works [11, 30, 39–41] in Ge and in Si [42]. From these studies the coexistence of different phases of alloy with intermetallic character, surrounded by extended regions with Mn dilution was detected for implantation temperatures above 450 K. In order to reduce intermetallic compound formation the implantation temperature was reduced to RT or below [43]. Ion beam implantation using atomic beams of 100 keV is an ideal way to prepare diluted TMs in semiconductors because of the possibility to keep the lattice in conditions far from equilibrium. In this way homogeneous films are obtained, because of the reduced kinetic of the host and doping species. In fact, for Mn implantation in Ge at temperatures ranging from RT to 450 K [39, 44], an onset of strong swelling and amorphization of the film with transformation to a spongelike structure was observed, while a less defected structure was obtained below RT. The main drawback of this method is the little deposition precision in the perpendicular direction with a broad gaussian profile of the concentration of the implanted species. Typical ion energy is in the range of 100 keV for Mn and fluences in the range of 10^{16} at/cm^2. In order to reduce the inhomogeneity of the TM deposition effects like channelling and damaged surface, orientation and postgrowth annealing are carefully chosen to improve the crystal quality. Laser ablation techniques can also lead to DMS by using high purity targets and high vacuum [32]. In this case post-growth annealing is necessary to achieve reasonable crystalline phases. Often laser flash annealing [45] can give some advantages after growth/implantation. In fact short pulses laser can locate the energy in very close range hindering the TM migration on a large volume and the formation of spurious phases. By the simple knowledge of the ion implantation energy a simple model can give the new dopant distribution after laser annealing [46].

14.4 Samples Characterization

The XAS experiments on DMS are performed at synchrotron radiation beamlines within a photon range from 5,989 to 7,112 eV from Cr to Fe K-edge. $L_{2,3}$ edges (ranging around 650 eV) would also deserve to be studied but they are of more difficult assessment, with more electronic states effects, and with little energy extension, amounting to few eV's. On one side such low kinetic energy electrons probe a large portion of the sample around the Mn absorber, loosing sensitivity to the zone perturbed by the dopant, the short range order, and bringing mainly the information on the unoccupied localized d-states. About the experimental data collection, the transmission collection mode during the XAS experiment cannot be used because of the nature of diluted samples, as the absorbed radiation is far from the 10 % of the available incident radiation. In this case the fluorescence signal or the total electron yield is taken. Third generation sources are giving enough number of photons (10^{11} ph/s) from bending magnet. In general the I_O is measured by a gas ion chamber, while the fluorescence is measured at liquid nitrogen temperature in order to minimize thermal disorder, by hyperpure Ge detectors. Contrarily to K-edge, in the case of $L_{2,3}$ edge sputtering of the superficial layer is necessary because of the lower penetration of X-rays and the more surface sensitive detection mode. In case of implanted samples, it could be necessary to remove defects produced by implantation, especially due to porous and swelled surfaces, when the implantation is done between RT and 473 K. A careful examination of the composition of the samples and their phases must precede any kind of refined structural investigation.

For instance in Fig. 14.2a, reporting implantation of 100 kV Mn ions in Ge at 513 K, it is possible to determine, by sputtering controlled XPS, a region where low concentration of dark Mn rich precipitates is accompanied by a large homogeneous concentration of Mn. Such a condition could be even better observed in Fig. 14.2b after annealing at 673 K. A much more inhomogeneous region laying below 60 nm from the surface was due to the presence of Ge atoms freed after the impact with Mn ions. In such a region, after bombardment mobility is risen by several orders of magnitude and the formation of intermetallic crystallites is favored after thermal annealing by the enhanced kinetics of interstitial atoms.

The $L_{2,3}$ Mn edge of a sample implanted at 513 K and subsequently annealed at 673 K in UHV is shown in Fig. 14.3, after sputtering by a 2 kV Ar^+ beam for 30 min. In this way it was possible to remove a region of about 30 nm, corresponding, according to Fig. 14.2 to expose a surface with the maximum Mn concentration value as depicted in panel b) of Fig 14.2 [11] and a minimized number of localized precipitates Mn_5Ge_3. Due to the high surface sensitivity of the $L_{2,3}$ edge, in Fig. 14.3 there is a clear transition from a very disordered configuration to a more ordered one, after the postgrowth treatment, due to the depletion of Ge atoms from interstitial sites in the sub-surface region, and their migration to deeper lattice sites. It has been proved, that such features are not due to oxidation effects, as it can be seen from the oxygen signal in the inset of Fig. 14.3. Changes of the sample composition according

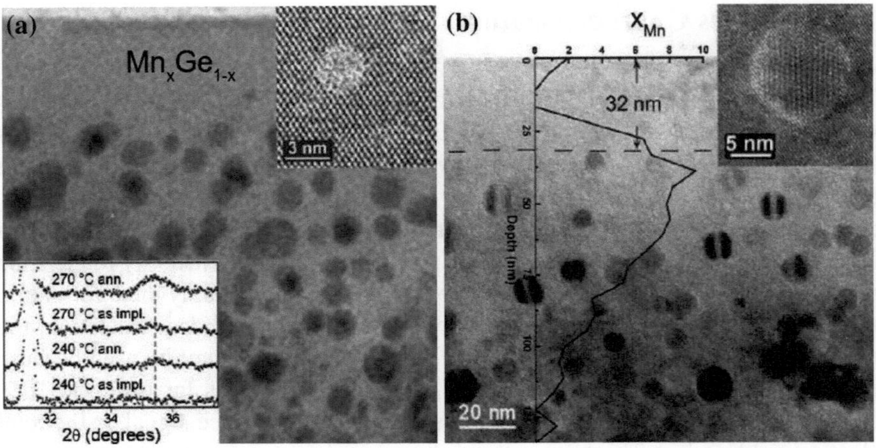

Fig. 14.2 Cross sectional TEM images and close-up images of nanoclusters of MnGe ion implanted alloy at temperature of 513 K. *Panel a* as-implanted sample. Lower left inset in *panel a* XRD Bragg-Brentano plots of the ion implanted samples. *Panel b* sample after 673 K UHV annealing; the corresponding XPS depth Mn concentration profile is superimposed to the TEM image. From [40]

Fig. 14.3 Mn $L_{2,3}$ XAS spectra of ion implanted MnGe as implanted (*dashed line*) and after annealing at 400 °C (continuous). From [47]

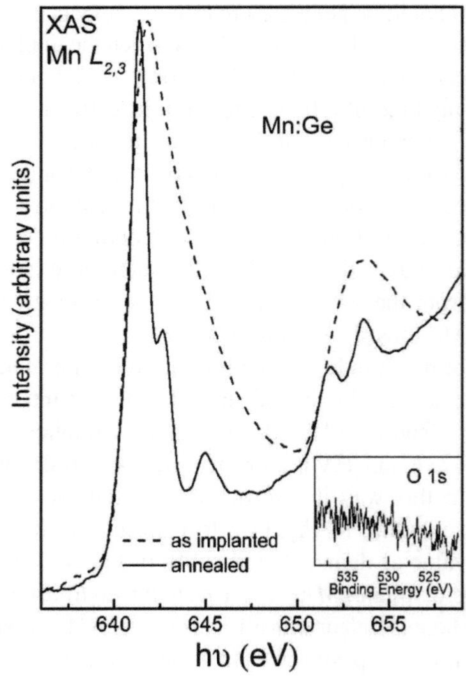

to Wu and coworkers [49], were due to the relevant presence of Mn dimers before annealing. Same evidences were found in other EXAFS analysis [40].

14.5 XANES and EXAFS of TM in IV-Group SCs

14.5.1 XANES

For a quantitative structural determination we limit this review to XAS at the TM K edge, giving information on local geometry surrounding the doping metal. The strategy to keep the concentration to the lowest levels as possible in order to avoid interaction between two metal atoms, is not completely safe in case secondary intermetallic phases are formed. Low concentration alone (less than 1 %) is not a sufficient condition to hinder intermetallic precipitation [50] and XAS can be more sensitive than TEM in detecting such secondary phases. XAS studies at the near edge region (XANES) provides sensitivity to the coordination geometry and bond lengths through a trial and error procedure of many possible geometries. As pointed out in [48] and [40] for MnGe alloys (fabricated either by ion implantation or MBE), TM K-edge XAS experiments allow to monitor: (i) occurrence of phase precipitation in the DMS alloy of Mn into Mn-rich phases (like Mn_5Ge_3 crystalline or amorphous clusters); (ii) formation of Mn–Mn dimers; (iii) discrimination between interstitial and substitutional Mn in a Ge host; (iv) quantitative determination of the local order character; (v) the dilution in the amorphous phase; (vi) the degree of elastic deformation of the first coordination shells around the dopant. The above information is of utmost importance in determining the degree of dilution in the IV-group SC's matrix. As it can be seen by a theoretical calculation in Fig. 14.4 the Mn-K edge of a Mn substitutional impurity in Ge shows several characteristic features. First, a pre-edge peak at 6540 eV that can be assigned to a quadrupole s-d transition. Such a transition is less prominent in Ge with respect to Si [42], where it occurs in the proximity of the Fermi edge (see 3d DOS in Si:Mn Fig. 14.1). Other shape resonances at 6550, 6590 and 6620 eV are already present after the on-set of second shell coordination of purely substitutional impurity as it can be observed in Fig. 14.4. Unfortunately, a clearcut agreement to such a simple model is difficult to achieve experimentally in most systems where a TM dopant is diluted in the host semiconductor matrix. If we consider the experimental Mn-K edge in Ge from [51], we observe in Fig. 14.5 that the samples grown by solid phase epitaxy at 473K (a) and with a nominal concentration $x = 3\,\%$ by codeposition at high temperature (623 K) (b), are quite different from those at the same concentration grown at lower T, respectively 523 and 433 K (Fig. 14.5c and d). It is clearly shown that the transition of the alloy towards a Mn rich phase occurs around 473 K, i.e. the solid phase epitaxy temperature. On the contrary at 433 K (Fig. 14.5d) a very different spectrum is observed. In Fig. 14.6 relationship between MBE and ion implanted samples grown at low temperatures can be established. MBE samples grown at 343 K (spectra (i) and (ii)) at 3 and 6 % Mn concentrations respectively were reported along with samples grown by implantation at room and liquid nitrogen (LN) temperatures. In particular samples (ii) and (iii) looked almost identical, prepared by RT ion implantation (followed by sputtering and annealing at 673 K) and by MBE (T = 343 K) respectively.

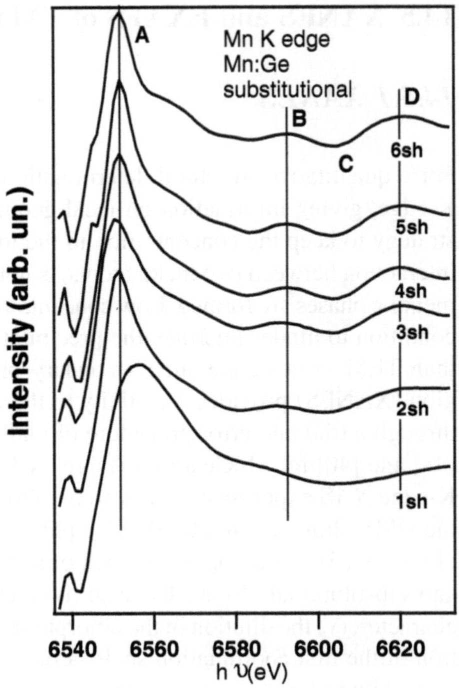

Fig. 14.4 Shell by shell multiple scattering calculations of the Mn K-edge in MnGe alloys within a purely substitutional site occupation model. From [48]

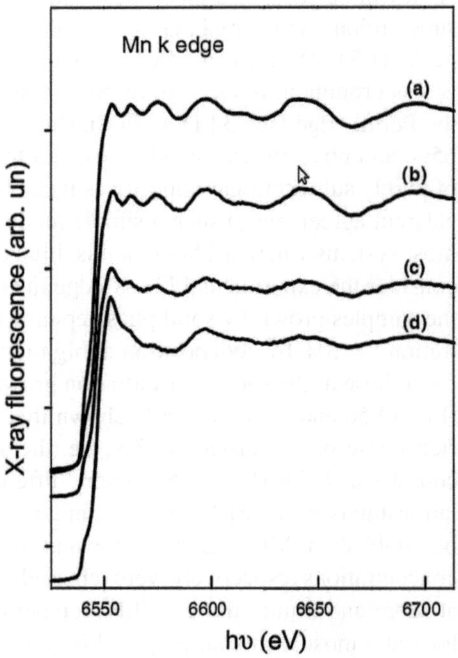

Fig. 14.5 XAS spectra of Mn 1s fluorescence (normalized) yield for: **a** Mn_5Ge_3 SPE alloy; **b** $Mn_{0.03}Ge_{0.97}$ DMS grown at 623 K; **c** $Mn_{0.03}Ge_{0.97}$ DMS grown at 523 K; **d** $Mn_{0.03}Ge_{0.97}$ DMS grown at 433 K. From [51]

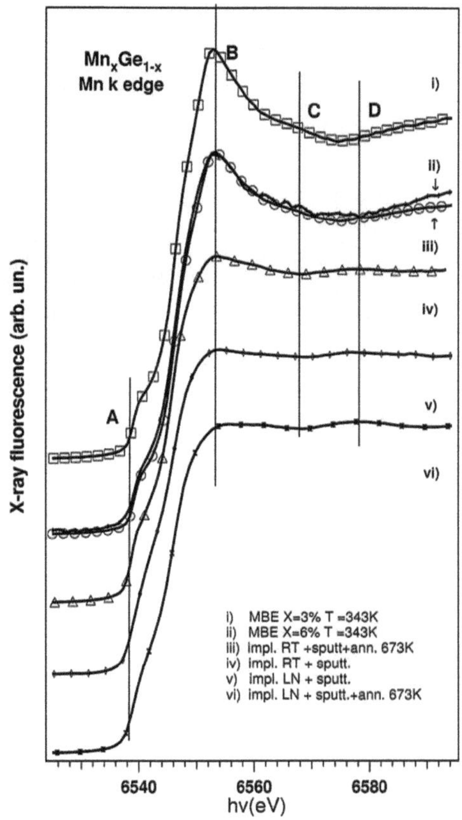

Fig. 14.6 Mn-K edge XANES of MBE films grown at 343 K at 3 % (*i*) and 6 % (*ii*) concentration compared with RT implanted samples after sputtering and annealing (*iii*) or sputtering without annealing at 673 K (*iv*). Spectrum (*v*) reports the case of LN implanted sample and (*vi*) the same sample after annealing at 673 K. Continuous lines represent guides for the eye. From [48]

Thermal effects can be followed and singled out in spectra from Fig.14.6 (iv–vi). A clear change in the spectra trend is visible showing that the RT implanted samples if not annealed are equivalent to LN implanted samples. Moreover the LN implantated samples are stable against annealing.

Additional information on the MnGe alloys structure could be obtained by means of multiple scattering calculations of the XAS of selected models to be compared to the experiments. Calculations were performed by means of a muffin-tin multiple scattering approach, recently reviewed in [52]. Such calculations are based on a suitable choice of a spherical cluster surrounding the absorbing atom and proper convolution due to experimental and lifetime broadening amounting to 0.7 eV in this case. Little influence of the Mn concentration on the XAS calculations was verified for values ranging between 6 and 25 %. The calculations were performed for the substitutional (S) and interstitial tetrahedral (T) sites occupation (see Fig. 14.7) of the dopant Mn, while the hexagonal site occupation resulted to have structureless absorption.

In Fig. 14.8 we plotted by symbols experimental data for spectra from (i) to (v) and by continuous lines the theoretical XAS spectra according to the relevant

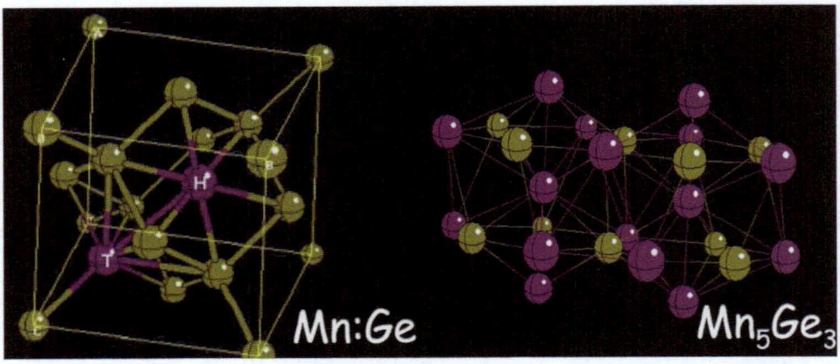

Fig. 14.7 *Left panel* Stick and ball model of interstitial T and H occupation sites within the Ge unit cell (*left panel*). The c axis is pointing towards the *bottom*. *Right panel* Stick and ball sketch of Mn_5Ge_3 alloy

Fig. 14.8 Comparison of Mn K-edge of selected samples (from i to v) compared with calculations of models related to the following systems: MnGe alloy, interstitial T and substitutional Mn. From [48]

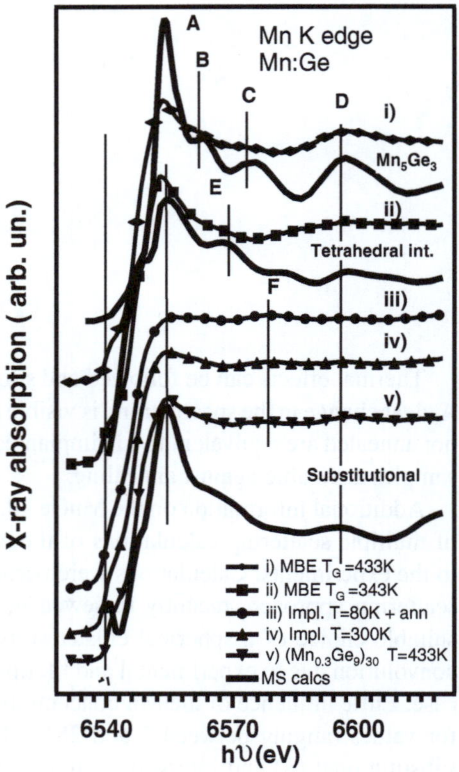

Table 14.1 Coordination numbers and distances of different coordination shells for the three occupation sites of interest

Site occupation	N1, $\Re 1$	N2, $\Re 2$	N3, $\Re 3$	N4, $\Re 4$	N5, $\Re 5$	N6, $\Re 6$
Substitutional (S)	4, 2.44	12, 3.99	12, 4.68	6, 5.66	12, 6.16	24, 6.91
Int. hexagonal (H)	6, 2.34	8, 3.66	6, 4.63	9, 5.42	13, 6.11	6, 6.72
Int. tetrahedral (T)	4, 2.44	6, 2.82	12, 4.68	8, 4.89	12, 6.16	24, 6.31

Table 14.2 Coordination numbers and distances of different coordination shells for two inequivalent Mn atoms of the Mn_5Ge_3 alloy

Mn type	N1, $\Re 1$	N2, $\Re 2$	N3, $\Re 3$	N4, $\Re 4$	N5, $\Re 5$
Mn1	Mn(2) , 2.53	Ge(6) , 2.54	Mn (6), 3.06	Mn(3), 4.14	Ge(3), 4.38
Mn2	Ge(2), 2.4	Ge(1), 2.6	Ge(2), 2.77	Mn(10), 3.02	Mn(2), 4.27

models. The calculations included the Mn_5Ge_3 alloy, the tetrahedral interstitial (T) and the substitutional (S) occupation sites according to the shell parameters reported in Tables 14.1 and 14.2.

As it can be seen from Fig. 14.8, the Mn_5Ge_3 alloy showed spectral features at well defined energies: 6553 (A) , 6563 (B), 6575 (C) and 6596 (D) eV. In the tetrahedral interstitial site case, intensity maxima at 6566 (E) and at 6596 (D) eV were found. MBE samples grown at low temperatures or samples implanted at low temperatures showed the main features at 6579 (F) and 6596 (D) eV.

From the comparison of calculations with spectrum (i) of Fig. 14.8 we found that samples grown by MBE at 433 K or higher temperatures underwent a transition to an intermetallic compound as shown by the agreement with the Mn_5Ge_3 alloy. In a similar manner to what found for the III-V DMS [53], the low temperature MBE grown MnGe samples (T = 343 K) (Fig. 14.8 spectrum (ii)) showed a peculiar difference from the samples grown at 433K at the same concentration of x = 3 % (spectrum (i)). For instance, similarities with the theoretical calculations of Mn occupation in interstitial T site in Fig. 14.8 were visible. In particular the resonance at 6566 eV (E) is a quite peculiar fingerprint of the absorption for this latter occupation site. Within the sensitivity of this qualitative analysis, a rather good agreement between the pure substitutional model in Fig. 14.8 and the three spectra grown at lower temperatures, namely, the LN implanted samples (spectrum (iii)), the after sputtering RT implanted samples (spectrum (iv)), and the spectrum of the DL samples $(Mn_{0.3}Ge_9)_{30}$ grown at 433 K (v) was found. The only notable exception was the prominent peak at 6579 eV (F) which was not properly reproduced by the calculations. An attempt to recover such a spectral feature (F) at 6579 eV with Mn–Mn scattering of near-neighbor impurities in the substitutional model, or in general by increasing the number of Mn absorbers per volume, was not successful. The reason was again the short range order sensitivity of the XANES spectroscopy. A more realistic possibility was represented by the in-plane uniaxial compressive strain induced by the growth on the oriented substrate Ge(001). In this latter case, we reported in left panel of Fig. 14.9a) the

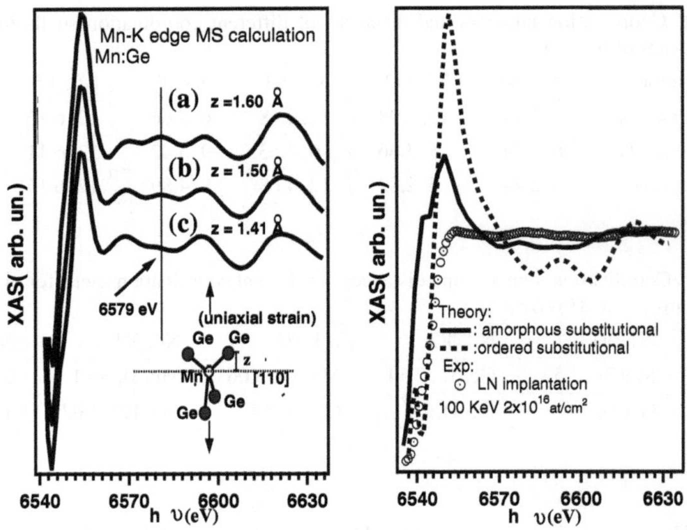

Fig. 14.9 *Left panel* MS XAS calculations of substitutional Mn model at the Mn-K edge. Here the first coordination shell is uniaxially stressed compressively as shown in the inset. Calculations are reported for different strain values: 0, 4 and 12 % for (*a*), (*b*) and (*c*) respectively. From [48]. *Right panel* Mn K-edge XAS absorption data for Mn-Ge LN inplanted. The ordered (*dashes*) and the disordered (*solid*) MS theory of the XAS are compared with the LN implanted experimental spectrum (*symbols*). From [43]

calculation of the unstrained substitutional Mn XAS case, while in (b) and (c) the spectra including an in-plane compressive strain giving rise to an elastic elongation along the normal to the sample. Such an elongation amounted respectively to 4 and 12 % and corresponded to a [100] interplanar distance z change from 1.415 to 1.50 Å and to 1.60 Å (see the sketch in left panel of Fig. 14.9). As a result the first Ge neighbor distance reached the value of 2.56 Å while the coordination angle $Ge-\hat{Mn}-Ge$ was reduced to 102° from the nominal value of 109°. One main consequence in the XAS spectra was that the peak at 6,579 eV was enhanced because of this large local strain.

Our intent here was to give a first order estimate of the local strain as supported by the EXAFS determination of the first Mn-Ge distance close to 2.50 Å (see next section). Such a parameter change would be difficult to be detected by XRD because of the very local character of the deformation. For sake of completeness, the degree structural disorder as the result of the implantation at low temperatures should be considered. The Mn K-edge XAS spectrum of the Mn-Ge system under investigation is reported in right panel of Fig. 14.9 by empty circles. This XAS spectrum is very similar to the amorphous MnGe evidenced by TEM [39]. This occurrence is verified once the swelled amorphous Ge layer of the RT implanted alloy is removed (Fig. 14.8iv) and the subsurface disordered Mn-implanted amorphous Ge layer is measured with XAS. In this case XAS experiments allowed to rule out with high confidence the presence of interstitial occupation of Mn. Interstitials are very prone to diffusion and cluster formation upon annealing of the samples [48] producing notable changes in the XAS spectra under annealing, due to the formation of precipitated phases. Noteworthily, instead, the XAS spectrum of right panel of Fig. 14.9 remains totally unchanged upon annealing the sample even up to 400 C. Any detectable trace of Mn-rich precipitates can be ruled out by analyzing this spectrum either in the amorphous or crystalline phase; this occurrence would lead to very specific peaks in the XAS spectrum at the energies of 6,553, 6,563, 6,575, and 6,596 eV, which are definitely absent in the spectrum of right panel of Fig. 14.9. Direct comparison of the two MS calculated spectra with the experimental one, clearly shows that a better agreement is observed when passing from the crystalline diamond MS spectrum to the disordered model named ST12 [54], where local distortion of the tetrahedra is taken into account (solid curve in right panel of Fig. 14.9). The calculated XAS spectra have been adjusted to the experimental one by normalization of the jump of the cross section between 6,535 and 6,625 eV. The ST12 model is still a crystalline phase, but in terms of short range nearest-neighbor pair distribution function, it very nicely mimics germanium in the amorphous phase [54]. It has 12 atoms per unit cell and a space group with few symmetry operations. Unlike diamond, the ST12 crystalline network exhibits seven- and five-fold atom rings and a fairly broad distribution of configurations of bonds on neighbors atoms, similar to a-Ge [54]. Simulations of a-Ge with such a distorted crystal phase model allow to grasp most of the important properties of the real a-Ge system (see also Chap. 7). As we can see the rise of a scattering resonance around 6,580 eV is consistent with the experimental data, and certainly connected with our previous finding of a strain component in the ordered structure seen in left panel of Fig. 14.9. Furthermore this is in agreement with the 2 % positive strain of the lattice parameter in the very similarly prepared a-Ge system of Ridgeway and coworkers [55] (Ge implantation). It is far from obvious that the Ge amorphous structure obtained by ion implantation is consistent with the one obtained in MD simulations. Though, this can be tempered by right agreement with XANES spectra of the calculated ST12 model, the heat of formation of substitutional Mn at concentration close to experiment (a few %) is quite lower in amorphous than in diamond Ge [43]. This finding, together with the results obtained for the magnetic interactions between Mn's, suggests that an amorphous environment might favor incorporation of substitutional Mn impurity and possibly results in a stronger ferromagnetic interaction with respect to a perfect diamond-crystalline coordination.

14.5.2 EXAFS

The present alloys with sizeable structural and compositional disorder, showed negligible EXAFS signals from long distance shells and only few contributions from closer shells are needed for a successful signal fitting. In particular, the occurrence of precipitation of Mn_5Ge_3 crystallites was investigated by monitoring the bond

Table 14.3 Fitting parameters (Average atomic distances and variances for the three nearest neighbor Ge atoms in different structural configuration) (S, T and A) for differently grown samples

		Substitution (S)	Tetrahedral (T)	Alloy (A)
Impl. LN +	R(Å)	2.51(3)		
ann 673K	σ^2(Å^2)	0.008(3)		
Impl. RT +	R(Å)	2.53(3)	2.75(3)	
ann 673K	σ^2(Å^2)	0.011(3)	0.016(3)	
Impl. 523K +	R(Å)	2.48(3)	2.82(3)	
ann 673K	σ^2(Å^2)	0.011(3)	0.011(3)	
MBE	R(Å)	2.52(3)	2.82(3)	
343K	σ^2(Å^2)	0.011(3)	0.029(5)	
MBE	R(Å)	2.50(3)	2.87(3)	3.06(3)
433K	σ^2(Å^2)	0.009(3)	0.011(3)z	0.003(1)
MBE	R(Å)	2.53(3)	2.81(3)	3.02(4)
523K	σ^2(Å^2)	0.006(3)	0.011(3)	0.003(1)
Mn_5Ge_3	R(Å)	2.50(3)	2.83(3)	3.02(4)
	σ^2(Å^2)	0.006(3)	0.013(3)	0.003(1)
$(Mn_{0.3}Ge_9)_{30}$	R(Å)	2.52(3)		
	σ^2(Å^2)	0.010(3)		

length region between 2.0 and 3.8 Å around the Mn impurity. In such a range the possibilities of substitutional (S), interstitial (both in the H or T coordination) and alloy (A) occupation sites were represented by peculiar shell distances. In Table 14.1 the nearest and next nearest neighbors distances of the Mn atomic site for the three cases of Mn site occupation (S, H and T) were reported, while in Table 14.2 the mean distances and occupation numbers for the case of two independent Mn sites in the Mn_5Ge_3 can be found. In fact the interstitial T short range region was different from that of the pure substitutional because of a strong scattering shell (6 Ge atoms) at 2.82 Å. Furthermore, a full occupation of the T sites would correspond to a 4:6 ratio of the first to second shell occupation numbers. An interstitial Mn in H site would have respectively 6 and 8 Ge atoms at 2.35 and 3.66 Å respectively. Finally an atomic shell at 4.7 Å with 12 Ge atoms was present in both cases of S and T interstitial occupation. The EXAFS analysis was performed by means of the multiple scattering ab initio GnXAS package [56]. Theoretical signals from atomic shells surrounding the Mn absorber were considered to describe the experimental modulation absorption coefficient. In the case of Mn a strong background component was given by additional edges $\alpha_{exc}(E)$ due to many-electron excitations [57]. A smooth and monotonic background by two third-order spline functions was obtained by adding a double electron excitation related to a KM shake-off ($1s- > \epsilon p$, $3s- > \epsilon s$ at about 6637 eV). Such a double electron excitation was located about 95 eV above the edge, close to the Fe 3s ionization energy (92 eV), as expected on the basis of the Z+1 rule in X-ray absorption. From the above considerations, seven combinations

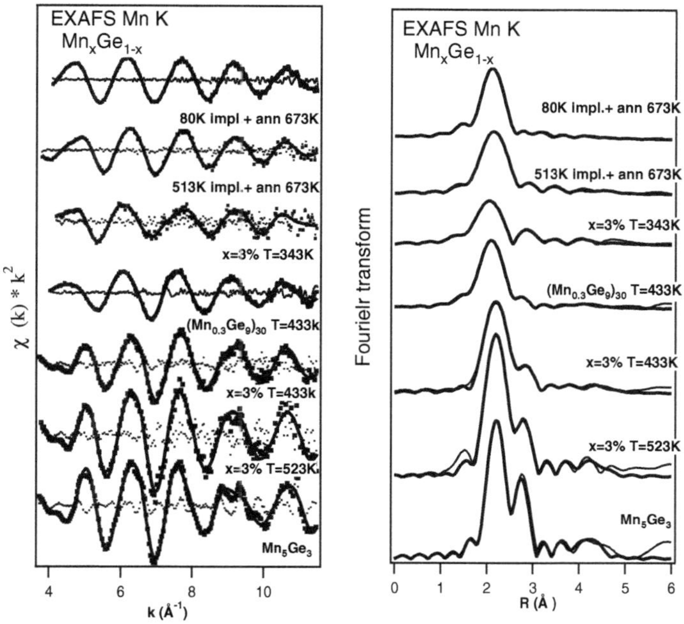

Fig. 14.10 *Left panel* Comparison between theoretical and experimental EXAFS $\chi(k)$ functions for the selected set of samples which best represents the relevant effects observed in Mn_xGe_{1-x} MBE and implanted alloys of the present study. *Right panel* Comparison between theoretical and experimental Fourier Transform of EXAFS $\chi(k)$ functions. From [48]

of EXAFS signals were used to represent the data, namely six two-body γ and one three-body η signals to describe the short range order around a single Mn atom [56]. A single three-body signal (η_1) was used to describe the triangle Mn-Ge-Ge with the two Ge atoms at 2.45 Å and an angle of about 109° between them. The latter signal included the most important among the multiple scattering contributions and a long bond scattering from the shell of 12 Ge atoms at 4.0 Å. In the interstitial T case it was difficult to resolve the scattering from the shell at 2.82 Å from that of the Mn–Mn distance in the Mn_5Ge_3 alloy (an average of 8 Mn atoms at 3.04 Å) [25]. It was nevertheless observed that splitting of the Mn–Mn signal in two close shells at such a distance, turned out to be crucial to obtain the right value of 8.4 Mn scatterers in second shell of Mn_5Ge_3. Such a value was usually underestimated when only a single shell distance was used in the multiparametric fit [40]. Using the above procedure we obtained the results reported in Table 14.3 for the main parameters of the signals considered (average atomic distances and their variances). A controversial aspect of the whole analysis was the value of the nearest neighbor (NN) distance in the substitutional and in the interstitial case which should also be in relation with the possible strain effects induced by the substrates, as outlined in the previous section. In local density approximation (LDA) Mn NN distance in interstitial T occupation was found to be larger than the Ge-Ge NN distance (2.45 Å) in bulk Ge. On the

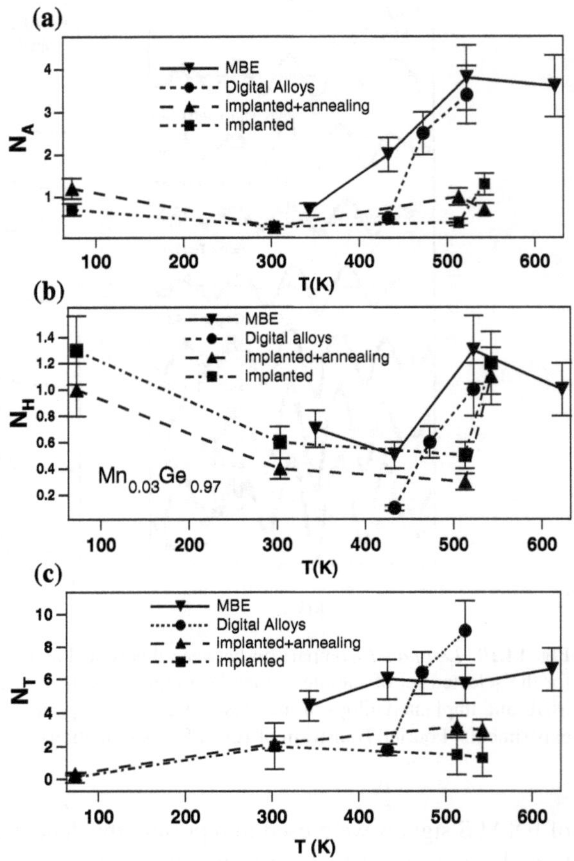

Fig. 14.11 Summary of shell occupation numbers obtained by EXAFS. From **a** to **c** we observe the occupation numbers of the alloy (N_A) the interstitial hexagonal (N_H) Mn and the interstitial tetrahedral (N_T) Mn. From [48]

contrary, it was found to be smaller in the case of substitutional Mn atoms [25]. In the present EXAFS analysis we observed an extended Ge–Mn bond length with respect to bulk Ge, in agreement with a similar study about III-V semiconductors [58]. Such discrepancies were not completely unexpected in LDA of small band gap semiconductors and a further improvement in the theoretical approach, possibly including correlation due to the d-state localization of Mn valence electrons has been proposed [59]. In left panel of Fig. 14.10, the theoretical EXAFS curves compared to the normalized experimental absorption were reported. Such a set of data showed how, in the case of MBE samples, the amplitude of the $\chi(k)$ oscillations increased with the growth temperature indicating that the number and/or the configurational order of scatterers around the Mn absorber increased with the temperature. Moreover a sizeable interference between two main signals was observed, while only a single main signal was dominant in samples grown at low temperatures. Different is the case for the implanted samples where only limited degrees of freedom are left to atoms migration even after thermal annealings. In right panel of Fig. 14.10 the corresponding Fourier transforms of the EXAFS spectra of Fig. 14.10 were reported, clearly indicating that the low temperature implanted and the digital alloy samples were intrinsically less affected by strong signals from shells around 3.0 Å when the high temperature formation of the alloy occurred. A complete comparison between the two different classes of samples (implanted versus MBE) must take into consideration possible differences in the Mn concentration. We believe that this difference cannot affect our analysis as the Mn distribution in the implanted samples is very well centered around the value of x = $4 \pm 1\,\%$. To support this view we also found that EXAFS data were more influenced by the temperature and by the growth process rather than by the Mn concentration, showing also at very low concentrations the same EXAFS signatures of precipitation of more concentrated samples at similar growth temperatures. The results of the EXAFS data analysis were reported in Table 14.3 and for the occupation numbers in form of diagram in Fig. 14.11. In panel a) of Fig. 14.11 the occupation of the alloy (A) site (N_A) was reported to rapidly increase in the case of MBE samples, while the hexagonal contribution (N_H) (panel b) to the interstitial site appeared to be negligible . A clear combination of A and T sites was seen to occur at higher temperature growth by looking at the panel c). The latter observation barred out the occurrence of the simultaneous occupation of the interstitial T and H sites, with their characteristic distance of 3.06 Å able to explain the frequent EXAFS observation of the main alloy Mn–Mn distance (see Fig. 14.7) detected also at low Mn (less than 3 %) concentration. Indeed, in the MBE growth samples and to a lower extent in implanted samples, the content of alloy component (N_A in panel a) was accompanied by a correspondent presence of interstitial tetrahedral site occupation (N_T) (panel c) independently of the growth temperature. By reducing the growth temperature in MBE samples only the N_A is reduced with respect to N_T. In the case of implanted samples both the occupations numbers N_A and N_T remained well below the 30 % of the nominal value, indicating the formation of an almost pure substitutional occupation, but with some degree of disorder, as observed from the EXAFS signal. In particular the implanted sample at 80 K showed remarkable single phase and thermal stability. A possible interpretation of these data is that tetrahedral interstitial occupation is the precursor phase of the alloy and the failure in occupation of such interstitial sites hindered the alloy formation. MBE samples grown at 343 K did not turn completely to alloy and showed mostly Mn in tetrahedral interstitial site. A second relevant case is given by the MBE DL alloy samples $(Mn_{0.3}Ge_9)_{30}$ grown at 433 K. In this case the EXAFS analysis showed the least amount of alloy ($N_A = 1.2$) amounting to about one fourth of the contribution found in the MBE samples at the same temperature, and a negligible tetrahedral interstitial site occupation. From this evidence we can estimate that the DL alloys did not simply rise the temperature threshold for the alloy transformation but rise a potential barrier against the interstitial tetrahedral occupation. DL alloys are clearly metastable structures as it can be observed from the abrupt change in phase composition as the growth temperature is slightly increased to $T > 433$ K [50]. It must be noted that clusters of neighboring substitutional and interstitial sites have been demonstrated to be more stable than clusters of pure substitutional sites in the case of Mn in GaAs [60] though a correspondent experimental confirmation is lacking.

Moreover, the experimental data presented here, and in particular the fact that the alloy formation began from the occupation of interstitial sites, were supported by the observations presented in [61] for MnGe. In that study one of the tetrahedra composing the Ge_3Mn_5 structure was identified as being the building block of the GeMn nanocolumnar structures. It can be shown that the base of this tetrahedron, containing distances Ge–Mn = 2.50 and Mn–Mn = 2.80, can be reproduced in the Ge structure by occupying a T site (center of the cube) and a S site (center of a face) with Mn atoms. Finally it is worth to comment about the amorphous structure on the determined value of the Mn–Ge nearest-neighbor distance 2.51 Å. This is the typical Mn–Ge coordination distance for Mn diluted in a host Ge matrix [48] This value is about 2 % larger than the Ge-Ge coordination distance in crystalline Ge (c-Ge), and this mismatch is a sign of the difficulty of Mn dilution in the Ge matrix. On the other hand, once Ge is amorphized (by Ge ion implantation), as reported by Ridgeway et al. [55], there is a clear trend (with the ion fluence) to the increase of the Ge-Ge nearest- neighbor distance toward 2.47 Å. This indicates the presence in the Ge matrix of point defects, i.e., threefold (T3) and fivefold (T5) coordinated atoms. Noteworthily, such defects are characterized by a nearest-neighbor distance of 2.52 and 2.57 Å, respectively [62]. Accordingly we suggest that Mn atoms are most likely accommodated in T3 and T5 sites. The fact that the average number of nearest neighbors as determined by the EXAFS analysis of right panel of Fig. 14.9 is about 4.0 is because the five-fold coordinated Mn likely occurs with the same probability as the three-fold one, as in [55]. Finally we note that the direct observation in our case of a 2 % strain of the lattice parameter, together with a similar finding for the Ge–Ge distance in the very similarly prepared a-Ge system of Ridgeway et al [55] are perfectly in line with the determination of a 2.0 % density decrease of Ge implanted Ge at LN. To further support this point we also note that the Fourier transform (FT) of sample grown at the LN temperature (right panel of Fig. 14.10) is very similar to the FT of amorphized Ge (EXAFS experiments) obtained by LN implantation of Ge into Ge by Ridgeway et al. [55]. Thus this evidence also indicates, from a microscopic point of view, that the host Ge matrix in our Mn–Ge system is amorphous and because of this character a RT ferromagnetic signal was clearly detected. Furthermore codoping by As in such structures seems to be an efficient way to increase the number of Mn active magnetic moments [63]. Scarce examples of similar studies for systems different from Mn in Ge are present in literature. Soo et al. [64] performed investigation on magnetically doped (Mn and Cr) amorphous silicon prepared by deposition on glass and hydrogenation. In this way single shell of Si around Mn and Cr were formed with a surprising high concentration (about 20 %) keeping the sample free of metallic inclusions. Quite surprisingly with respect to the elecronic band structure of the Si showing metallicity in case of Mn, here the latter dopant reach good ferromagnetic properties because of the lack of antiferromagnetic contributions, indicating that electronic calculation of bulk system are inadequate to explain ferromagnetism in disordered systems [43]. In hydrogenated a-Si:Mn, Cr the number of nearest neighbors is clearly larger than 4 because of the larger volume available to include TM atoms and one or two additional Si atoms around the dopants should be reasonable. Here a single shell analysis did not detect any interstitial contribution. This is reasonable as tetrahedral and hexagonal interstitials in the crystalline Si that can complicate the EXAFS analysis simply do not exist in the amorphous Si structure which is basically comprised of continuous random network of tetrahedrally bonded Si atoms in the short-range order regime. In another study of implanted Si with Mn in fact [42], similar result to those of implanted Ge reported here were found, pointing to a natural hindering of occupation of interstitial sites also in the case of less amorphous structures. In that case the XANES region is, nevertheless a clearcut indicator that Mn-Si phases are formed, as the authors describe by using several models of alloy. The inclusion of TM in amorphous carbon, and diamond like carbon is a natural extension of the present work. If we consider the possibility that p-d hybridization can increase with the reduced host semiconductor bond length, rising the Curie Temperature of the system. It is in any case crucial the amount of space to be left to the TM dopant in order to change its properties from the magnetic or electric point of view. In recent works, Hsu and coworkers [65] and Colón- Santana and coworkers [66] reported quite different results (dilution without cluster formation of intermetallic compounds in the former case) probably due to the characteristic of the disordered alloy, in particular the amount of sp3/sp2 bond ratio, which deserves right attention especially from the point of view of theoretical simulations.

14.6 Conclusions

In conclusion the path to an efficient magnetic dopant incorporation in Ge and other IV-group SCs is still far from being considered well established due to the difficulties to obtain a unique way of operation. Among many proposed systems and techniques, uncertainty between real effectiveness and spurious effects is large. Until the best choice, overcoming by far all the other proposed systems, will not be achieved, we believe that only a comparative method on a large set of samples, prepared in different ways to single out the peculiarites of any experiment, and analysed under the same methodology could help to draw the line of reseach towards the best conditions for a succesful comprehension of the magnetic doping of semiconductors. Such experimental procedure could be of founding importance for dedicated total energy calculations based on a specific set of structural parameters, to identify stable systems and ferromagnetic response. For what is specifically related to the present study, possible future studies with the use of codoping and refinement of flash annealing technique of implanted samples [63] would represent a nice completion of the data reported here, also at the light of their apparent substantial improvement of the magnetic response obtained, will be fundamental to give a unifying picture of the complex physical behavior of these new functionalities.

References

1. A. Fert, Nobel lecture: origin, development and future of spintronics. Rev. Mod. Phys. **80**, 1517 (2008)
2. T. Kasuya, A. Yanase, Rev. Mod. Phys. **40**, 684 (1968)
3. A. Mauger, C. Gotard, Phys. Rep. **141**, 51 (1986)
4. J.M.D. Coey, M. Viret, S. von Molnar, Adv. Phys. **48**, 167 (1999)
5. A. Fert, H. Jaffres, Phys. Rev. B **64**, 184420 (2001)
6. H. Ohno, H. Munekata, T. Penney, S. von Molnar, L.L. Chang, Phys. Rev. Lett. **68**, 2664 (1992)
7. H. Ohno, A. Shen, F. Matsukura, A. Oiwa, A. Endo, S. Katsumoto, Y. Iye, Appl. Phys. Lett. **69**, 363 (1996)
8. A. Haury, A. Wasiela, A. Arnoult, J. Cibert, S. Tatarenko, T. Dietl, Y. Merle dAubigny, Phys. Rev. Lett. **79**, 511 (1997)
9. T. Jungwirth et al., Phys. Rev. B **72**, 165204 (2005)
10. Y.D. Park, A.T. Hanbicki, S.C. Erwin, C.S. Hellberg, J.M. Sullivan, J.E. Mattson, T.F. Ambrose, A. Wilson, G. Spanos, B.T. Jonker, Science **295**, 651 (2002)
11. L. Ottaviano, M. Passacantando, S. Picozzi, A. Continenza, R. Gunnella, A. Verna, G. Bihlmayer, G. Impellizzeri, F. Priolo, Appl. Phys. Lett. **88**, 061907 (2006)
12. J.-S. Kang et al., Phys. Rev. Lett. **94**, 147202 (2005)
13. S. Cho, S. Choi, S.C. Hong, Y. Kim, J.B. Ketterson, B.-J. Kim, Y. C. Kim, J.-H Jung, Phys. Rev. B **66**, 033303 (2002)
14. A.P. Li, J. Shen, J.R. Thompson, H.H. Weitering, Appl. Phys. Lett. **86**, 152507 (2005)
15. C. Zeng, Z. Zhang, K. van Benthem, M.F. Chrisholm, H.H. Weitering, Phys. Rev. Lett. **100**, 066101 (2008)
16. M. Jamet et al., Nat. Mater. **5**, 653 (2006)
17. S. Picozzi, M. Lezaic, New J. Phys. **10**, 055017 (2008)
18. V.V. Singh, A.S. Vengurlekar, Transition-metal impurities in semiconductors and the Haldane-Anderson model. Phys. Rev. B **30**, 3527 (1984)
19. A. Zunger, U. Lindfelt, Phys. Rev. B **27**, 1197 (1983)
20. H. Weng, J. Dong, Phys. Rev. B **71**, 035201 (2005)
21. Y.-S. Kim, H. Kim, Y.-S. Chung, Nanotech (2003), vol 3, http://www.nsti.org, ISBN 0-9728422-2-5
22. T. Dietl, Ferromagnetic semiconductors. Semicond. Sci. Technol. **17**, 377 (2002)
23. K. Sato et al., Rev. Mod. Phys. **82**, 1634 (2010)
24. A. Stroppa, S. Picozzi, A. Continenza, A.J. Freeman, Phys. Rev. B **68**, 155203 (2003)
25. A. Continenza, G. Profeta, S. Picozzi, Phys. Rev. B **73**, 035212 (2006)
26. X. Luo, S.B. Zhang, S.-H. Wei, Phys. Rev. B **70**, 033208 (2004)
27. R. Goswami, G. Kioseoglou, A.T. Hanbicki, O.M.J. van 't Erve, B.T. Jonker, G. Spanos, Appl. Phys. Lett. **86**, 032509 (2005)
28. J.-P. Ayoub, L. Favre, I. Berbezier, A. Ronda, L. Morresi, N. Pinto, Appl. Phys. Lett. **91**, 141920 (2007)
29. N. Pinto, L. Morresi, M. Ficcadenti, R. Murri, F. D'Orazio, F. Lucari, L. Boarino, G. Amato, Phys. Rev. B **72**, 165203 (2005)
30. M. Passacantando, L. Ottaviano, F. D'Orazio, F. Lucari, M. De Biase, G. Impellizzeri, F. Priolo, Phys. Rev. B **73**, 195207 (2006)
31. V.K. Dixit, B.V. Rodrigues, H.L. Bhat, J. Cryst. Growth **217**, 40 (2000)
32. K. Seshan, Handbook of thin-film deposition processes and techniques. Intel Corporation
33. N. Pinto, L. Morresi, R. Gunnella, R. Murri, F. D'Orazio, F. Lucari, S. Santucci, P. Picozzi, M. Passacantando, J. Mater. Sci.: Mater. Electron. **14**, 337 (2003)
34. F. D'orazio, F. Lucari, S. Santucci, P. Picozzi, A. Verna, M. Passacantando, N. Pinto. L. Morresi, R. Gunnella, R. Murri, J. Magn. Magn. Mater. **262**, 158 (2003)
35. A.M. Nazmul, T. Amemiya, Y. Shuto, S. Sugahara, M. Tanaka, Phys. Rev. Lett. **95**, 017201 (2005)

36. F. Tsui, L. He, L. Ma, A. Tkachuk, Y. Chu, K. Nakajima, T. Chikyow, Phys. Rev. Lett. **91**, 177203 (2003)
37. W. Zhu, Z. Zhang, E. Kaxiras, Phys. Rev. Lett. **100**, 027205 (2008)
38. L. Liu, N. Chen, Z. Yin, F. Yang, J. Zhou, F. Zhang, J. Cryst. Growth **265**, 106 (2004)
39. A. Verna et al., Phys. Rev. B **74**, 085204 (2006)
40. L. Ottaviano, M. Passacantando, A. Verna, R. Gunnella, E. Principi, A. Di Cicco, G. Impellizzeri, F. Priolo, J. Appl. Phys. **100**, 063528 (2006)
41. S. Picozzi, L. Ottaviano, M. Passacantando, G. Profeta, A. Continenza, F. Priolo, M. Kim, A.J. Freeman, Appl. Phys. Lett. **86**, 062501 (2005)
42. A. Wolska, K. Lawniczak-Jablonska, M. Klepka, M.S. Walczak, A. Misiuk, Phys. Rev. B **75**, 113201 (2007)
43. L. Ottaviano, A. Continenza, G. Profeta, G. Impellizzeri, A. Irrera, R. Gunnella, O. Kazakova, Phys. Rev. B **83**, 134426 (2011)
44. L. Ottaviano, A. Verna, V. Grossi, P. Parisse, S. Piperno, M. Passacantando, G. Impellizzeri, F. Priolo, Surf. Sci. **601**, 2623 (2007)
45. S. Zhou, D. Bürger, A. Mücklich, C. Baumgart, W. Skorupa, C. Timm, P. Osterlin, M. Helm, H. Schmidt, Phys. Rev. B **81**, 165204 (2010)
46. T. Kim et al., J. Appl. Phys. **108**, 013508 (2010)
47. L. Ottaviano, M. Passacantando, A. Verna, F. D'Amico, R. Gunnella, Appl. Phys. Lett. **90**, 242105 (2007)
48. R. Gunnella et al., J. Phys.: Condens. Matter **22**, 216006 (2010)
49. D. Wu, D.J. Keavney, R. Wu, E. Johnston-Halperin, D.D. Awschalom, J. Shi, Phys. Rev. B **71**, 153310 (2005)
50. R. Gunnella, N. Pinto, L. Morresi, M. Abbas, A. Di Cicco, J. Non-Cryst. Solids **354**, 4193 (2008)
51. R. Gunnella, L. Morresi, N. Pinto, R. Murri, L. Ottaviano, M. Passacantando, F. D'Orazio, F. Lucari, Surf. Sci. **577**, 22 (2005)
52. D. Sebilleau, R. Gunnella, S. Di Matteo, Z.-Y. Wu, C.R. Natoli, J. Phys.: Condens. Matter **18**, R175 (2006) and references therein
53. K.M. Yu, W. Walukiewicz, T. Wojtowics, I. Kuryliszyn, X. Liu, Y. Sasaki, J.K. Furdyna, Phys. Rev. B **65**, 201303(R) (2002)
54. A. Mujica, R.J. Needs, Phys. Rev. B **48**, 17010 (1993)
55. M.C. Ridgeway, C.J. Glover, K.M. Yu, G.J. Foran, C. Clerc, J.L. Hansen, A. Nylandsted Larsen, Phys. Rev. B **61**, 12586 (2000)
56. A. Filipponi, A. Di Cicco, C. R. Natoli, Phys. Rev. B **52**, 15122 (1995) (1996).
57. A. Di Cicco, Phys. Rev. B **53**, 6174 (1996)
58. R. Bacewicz, A. Twarog, A. Malinowska, T. Wojtowicz, X. Liu, J.K. Furdyna, J. Phys. Chem. Solids **66**, 2004 (2005)
59. M. Rovezzi, F. D'Acapito, A. Navarro-Quezada, B. Faina, T. Li, A. Bonanni, F. Filippone, A. Amore Bonapasta, T. Dietl, Phys. Rev. B **79**, 195209 (2009)
60. P. Mahadevan, A. Zunger, Phys. Rev. B **68**, 075202 (2003)
61. M. Rovezzi, T. Devillers, E. Arras, F. D'Acapito, A. Barski, M. Jamet, P. Pochet, Appl. Phys. Lett. **92**, 242510 (2008)
62. G. Kresse, J. Hafner, Phys. Rev. B **49**, 14251 (1994)
63. L. Ottaviano, Private Communication (2012)
64. Y.L. Soo, J.H. Yao, C.S. Wang, S.L. Chang, C.A. Hsieh, J.F. Lee, T.S. Chin, Phys. Rev. B **81**, 104104 (2010)
65. H.S. Hsu, P.Y. Chung, J.H. Zhang, S.J. Sun, H. Chou, H.C. Su, C.H. Lee, J. Chen, J.C.A. Huang, Appl. Phys. Lett. **97**, 032503 (2010)
66. J.A. Colón Santana, R. Skomski, V. Singh, V. Palshin, A. Petukhov, Ya. B. Losovyj, A. Sokolov, P.A. Dowben, I. Ketsman, J. Appl. Phys. **105**, 07A930 (2009)

36. F. Tsui, L. He, L. Ma, A. Tkachuk, Y.S. Chu, K. Nakajima, T. Chikyow, Phys. Rev. Lett. 91, 177203 (2003)
37. W. Zhu, Z. Zhang, E. Kaxiras, Phys. Rev. Lett. 100, 027205 (2008)
38. L. Liu, N. Chen, Z. Yin, F. Yang, J. Zhou, F. Zhang, J. Cryst. Growth 265, 466 (2004)
39. A. Verna et al., Phys. Rev. B 74, 085204 (2006)
40. L. Ottaviano, M. Passacantando, A. Verna, R. Gunnella, E. Principi, A. Di Cicco, G. Impellizzeri, F. Priolo, J. Appl. Phys. 100, 063528 (2006)
41. S. Picozzi, L. Ottaviano, M. Passacantando, G. Profeta, A. Continenza, F. Priolo, M. Kim, A.J. Freeman, Appl. Phys. Lett. 86, 062501 (2005)
42. A. Wolska, K. Lawniczak-Jablonska, M. Klepka, M.S. Walczak, A. Misiuk, Phys. Rev. B 75, 113201 (2007)
43. L. Ottaviano, A. Continenza, G. Profeta, G. Impellizzeri, A. Irrera, R. Gunnella, O. Kazakova, Phys. Rev. B 83, 134426 (2011)
44. L. Ottaviano, A. Verna, V. Grossi, P. Parisse, S. Piperno, M. Passacantando, G. Impellizzeri, F. Priolo, Surf. Sci. 601, 2623 (2007)
45. S. Zhou, D. Bürger, A. Mücklich, C. Baumgart, W. Skorupa, C. Timm, P. Oesterlin, M. Helm, H. Schmidt, Phys. Rev. B 81, 165204 (2010)
46. J. Kim et al., J. Appl. Phys. 108, 013508 (2010)
47. L. Ottaviano, M. Passacantando, A. Verna, F. D'Amico, R. Gunnella, Appl. Phys. Lett. 90, 242105 (2007)
48. R. Gunnella et al., J. Phys.: Condens. Matter 22, 216006 (2010)
49. D. Wu, D.J. Keavney, R. Wu, E. Johnston-Halperin, D.D. Awschalom, J. Shi, Phys. Rev. B 71, 153310 (2005)
50. R. Gunnella, N. Pinto, L. Morresi, M. Abbas, A. Di Cicco, J. Non-Cryst. Solids 354, 4193 (2008)
51. R. Gunnella, L. Morresi, N. Pinto, R. Murri, L. Ottaviano, M. Passacantando, F. D'Orazio, F. Lucari, Surf. Sci. 577, 22 (2005)
52. D. Sébilleau, R. Gunnella, Z.-Y. Wu, S. Di Matteo, C.R. Natoli, J. Phys.: Condens. Matter 18, R175 (2006) and references therein
53. K.M. Yu, W. Walukiewicz, T. Wojtowicz, I. Kuryliszyn, X. Liu, Y. Sasaki, J.K. Furdyna, Phys. Rev. B 65, 201303(R) (2002)
54. A. Mujica, R.J. Needs, Phys. Rev. B 48, 17010 (1993)
55. M.C. Ridgway, C.J. Glover, K.M. Yu, G.J. Foran, C. Clerc, J.L. Hansen, A. Nylandsted Larsen, Phys. Rev. B 61, 12586 (2000)
56. A. Filipponi, A. Di Cicco, C.R. Natoli, Phys. Rev. B 52, 15122; 15135 (1995)
57. A. Di Cicco, Phys. Rev. B 53, 6174 (1996)
58. R. Bacewicz, A. Twaróg, A. Malinowska, T. Wojtowicz, X. Liu, J.K. Furdyna, J. Phys. Chem. Solids 66, 2004 (2005)
59. M. Rovezzi, F. D'Acapito, A. Navarro-Quezada, B. Faina, T. Li, A. Bonanni, F. Filippone, A. Amore Bonapasta, T. Dietl, Phys. Rev. B 79, 195209 (2009)
60. P. Mahadevan, A. Zunger, Phys. Rev. B 68, 075202 (2003)
61. M. Rovezzi, T. Devillers, E. Arras, F. D'Acapito, A. Barski, M. Jamet, P. Pochet, Appl. Phys. Lett. 92, 242510 (2008)
62. G. Kresse, J. Hafner, Phys. Rev. B 49, 14251 (1994)
63. L. Ottaviano, Private Communication (2012)
64. Y.L. Soo, J.H. Yao, C.S. Wang, S.L. Chang, C.A. Hsieh, J.F. Lee, T.S. Chin, Phys. Rev. B 81, 104104 (2010)
65. H.S. Hsu, P.Y. Chung, J.H. Zhang, S.J. Sun, H. Chou, H.C. Su, C.H. Lee, J. Chen, J.C.A. Huang, Appl. Phys. Lett. 97, 032503 (2010)
66. J.A. Colón Santana, R. Skomski, V. Singh, V. Palshin, A. Petukhov, Ya.B. Losovyj, A. Sokolov, P.A. Dowben, I. Ketsman, J. Appl. Phys. 105, 07A930 (2009)

Chapter 15
Magnetic Ions in Group III–V Semiconductors

Krystyna Lawniczak-Jablonska

The hope that the exploitation of the electron spin in the conventional semiconductor technology will improve electronic devices stimulates the interest in physics of magnetic ions immersed in semiconducting matrix. Short introduction will be given to present the concept of diluted magnetic semiconductors (DMS) and the overview of the recent achievements in theory as well as in growing and characterization of III–V DMS with the highest Curie temperature. The particular attention was given to underline the unique information provided by XAS and X ray magnetic circular dichroism.

15.1 Introduction

The magnetic semiconductors have attracted considerable attention due to the expectation that manipulation of the electron spin, as an alternative to manipulation of the electron charge, can be used for the storage of information in semiconducting devices. This expectation initiated the so called spin electronics or spintronics. The pioneering work on the spintronics in semiconductors was performed already in the early 1970s in the Institute of Physics, Polish Academy of Sciences, in Warsaw, Poland. The technological and experimental work by Galazka [1] showed the giant energetic splitting of the spin electron band in the presence of magnetic field in Mn doped crystal of semiconductors. The effect of ferromagnetism induced by carriers, so important for spintronics, was also first reported by scientists from that institute in [2]. Nevertheless, the world wide interest in magnetic semiconductors began at the end of 1990s when Ohno from Tohoku University in Sendai reported the ferromagnetic properties of III–V classical semiconductors doped by Mn (InMnAs and

K. Lawniczak-Jablonska (✉)
Institute of Physics, Polish Academy of Science, Al. Lotnikow 32/46, 02-668 Warsaw, Poland
e-mail: jablo@ifpan.edu.pl

C.S. Schnohr and M.C. Ridgway (eds.), *X-Ray Absorption Spectroscopy of Semiconductors*, Springer Series in Optical Sciences 190,
DOI 10.1007/978-3-662-44362-0_15

GaMnAs) in [3, 4]. The search for the ferromagnetic semiconductors with the Curie temperature (T_C) close to room temperature dominates the spintronics studies until now. Very important theoretical work by Dietl et al. [5] from the Warsaw Institute of Physics predicting the Curie temperature for many semiconductors doped by transition metals (TM) is still an inspiration for the technological work to achieve the room temperature magnetic semiconductor. In spite of the effort of many technologists and scientists, till now the magnetic semiconductors with uniformly distributed magnetic ions do not exhibit ferromagnetic properties at room temperature, therefore the granular materials are now under increasing interest. In these materials the nanoinclusions of ferromagnetic phases are incorporated into semiconductors matrix. It was recently shown that such materials can be used in constructions of new electronic devices. In all cases the location of TM ions in the host matrix is crucial for the magnetic properties.

15.2 Origin of the Magnetism in Semiconductors

Several different approaches are known for describing the possible origin of the observed magnetism in general and particularly magnetism in semiconductors (see also Chap. 14).

Magnetism in (III, TM)V and other semiconductors matrices originates from magnetic ions local moments. However, the problem of the mechanism of the magnetic ordering is more complex and still subject of discussion. There is no useful theory of magnetism which can be applied to all materials. Particular magnetic properties are always the interplay between the electronic spin degree of freedom, the repulsive Coulomb interaction between electrons and fermionic quantum statistic of electrons. To describe the dependence of the energy of the system on the relative orientation of magnetic ions moments, **an *exchange* interaction** was introduced. Several types of qualitative effects that lead to exchange interactions has been identified when considering magnetic order in (III, TM)V DMS. The applicability of each and the relative importance of different effects depends on the TM kind and doping regime, and on the host semiconductor. In the literature describing the magnetic properties in semiconductors, the following exchange interactions are considered.

Stoner's *itinerant exchange* where the carriers of the magnetism are electrons with unsaturated spins in the d-band defined in [6]. This model emphasizes the "free" nature of electrons in the solid and includes the periodicity of the lattice in the wave function. Electron bands can spontaneously split into up and down spins if the relative gain of the energy in exchange interaction is larger than the loss in kinetic energy. According to fermionic quantum statistic of electrons, two electrons cannot be in the same location at the same time and have spins in the same direction. In the case of transition metals with the increasing temperature electrons behave like they have local moments rather than an itinerant character.

For describing the exchange interaction between local moments several models have been introduced:

Heisenberg's *direct exchange* model described in [6] is stressing the "bound" or highly localized nature of the electrons by treating the solid as a collection of atoms and introducing the "direct exchange" parameter. The simple Heisenberg approach considers each pair of atoms as similar to a hydrogen molecule, i.e., each atom is thought to have a single 3d electron which interacts with its neighbor and the dominant interaction for producing ferromagnetism in the solid is considered to arise from a superposition of this two-electron interactions.

Kramers's *superexchange* (*or **Kramers–Anderson superexchange***) interaction defined in [7, 8] applies to local moments that are separated by a nonmagnetic atom. In (III, Mn)V materials, superexchange gives an antiferromagnetic contribution to the interaction between Mn moments located on neighboring cation sites. An electron from the nonmagnetic atom can interact with an empty shell of the magnetic atom and with electrons forming its local moment. As a result, the nonmagnetic atom is polarized and coupled through a direct exchange with all its magnetic neighbors. However, there is no charge transfer.

Zener's *double-exchange* mechanism introduced in [9] also assumes the involvement of an intermediate nonmagnetic atom. In superexchange, the electrons do not actually move between two magnetic ions. In double exchange, the electrons actually move between the positive ions via the intermediate ligand what results in the material displaying magnetic exchange coupling as well as metallic conductivity. To meet the Hund rule, and to increase the probability of hopping, electrons in magnetic shells of both ions need to have the same spin. Then, the wave function of the hopping electron extends in the space between the two ions, and thus the kinetic energy of the system decrease. Exchange couples magnetic moments ferromagnetically. This model is valid only in the weak coupling limit but it should be kept in mind that going from antimonides to nitrides the p-d hybridization increases. Therefore, double exchange mechanism dominates e.g. for transition metal impurity in a wide band gap semiconductors (Mn or Cr in GaN). The stabilization of the ferromagnetic state arises from the width of the impurity band. **Zener's *kinetic-exchange*** (or **indirect-exchange** interaction) described in [10] assumes that coupling between moments of *d*- or *f*-shell, is mediated by *s*- or *p*-band itinerant carriers. Interaction occurs with the band electrons on the same site (ferromagnetic interaction) or is due to hybridization between the local moments and band electrons on neighboring sites (antiferromagnetic interaction).

Double exchange and kinetic p–d exchange are the extreme cases of a more general type of interaction, where both mechanisms can occur simultaneously.

The double exchange is considered as the dominating mechanism for ferromagnetism in wide-band-gap semiconductors such as GaN. In this semiconductor the atomic *p* levels of N are located at very low energy, leading to deep-lying *p* valence bands and a large band gap. When this material is doped with the TM, the *d* states form impurity bands in this wide gap. Consequently, double exchange is very strong and short ranged. The opposite behavior occurs for relatively narrow-band DMSs such as GaMnAs and GaMnSb. For these semiconductors the atomic *p* levels are located at higher energy, leading to narrower band gaps and, when doped with TM, the majority *d* state is centered in the lower region of the valence *p* band. This leads to increased importance of Zener's *p-d* exchange, which is relatively weak but longer ranged.

Superexchange is claimed to be the dominant antiferromagnetic interaction, leading to the disordered local moment phase of DMSs. This interaction is also rather strong and essentially restricted to the nearest neighbors. It is the largest when the Fermi level falls in the gap between the majority and minority states of impurities. If E_F lies in an impurity band, it competes with the ferromagnetic double exchange.

The considerable theoretical progress has been made in the understanding of the properties of ferromagnetic semiconductor materials both bulk and low dimensional. The presented main results of the many theoretical studies are taken mostly from the excellent overviews published recently by Sato et al. [12] and Jungwirth et al. [13]. As the main interest in these materials is to get the Curie temperature ($\mathrm{T_C}$) above room temperature, we will concentrate on the methods providing the $\mathrm{T_C}$ for this class of materials.

Over the years, the model of Zener's theory of ferromagnetism with kinetics exchange interaction (p-d exchange) was established (e.g. in [5]) and is the most frequently referred when describing the nature of magnetism in semiconductors. This theory considers a coupling between carrier spins and local atomic moments of magnetic impurities and provides an estimate of the ordering temperatures of several III–V DMS materials. Dietl et al. [5] motivated many experimentalists for growing the appropriate materials and, after several years of experimental work, it turned out that Curie temperature predicted by this theory was strongly overestimated since it did not consider thermal fluctuations and disorder (percolations).

Calculations based on density-functional theory (DFT) are more precise and a number of results were reported in the literature (see review by Zunger [14]). The implementation of the local force theory for magnetic exchange in [15, 16] provided a lot of useful information about interatomic exchange interaction. Moreover, the existence of coupling between magnetism and defect concentration was revealed. In the Zener's kinetic exchange approach for the DMS materials in [5] interaction is long ranged and has no angular component. The exchange interaction suggested by DFT is much shorter ranged and reflects the orbital character of the atoms in the host lattice. This implies that the magnetic properties of DMS materials are influenced by magnetic percolation (disorder effects).

Calculation of $\mathrm{T_C}$ applying DFT to the system under consideration usually involves two steps. First, the DFT and related methods are used to calculate the ground-state properties and in addition the exchange coupling constants (exchange integrals). In the second step, the thermal properties are evaluated using statistical methods for the Heisenberg model. To estimate the $\mathrm{T_C}$ and other thermal properties for most of applications, the mean-field approximation (MFA) is used. An important advantage of the MFA is that it provides simple understanding of the principal exchange mechanisms. An essential disadvantage of the MFA is its limited accuracy. The very informative example of the practical results of mean-field calculations presented in [17] is the dependence of the $\mathrm{T_C}$ on the magnetic ions concentration c (Fig. 15.1) of GaMnN, GaMnP, GaMnAs, and GaMnSb. It can be noticed that the increase of MFA T_C with c becomes less pronounced when going from GaN to GaSb.

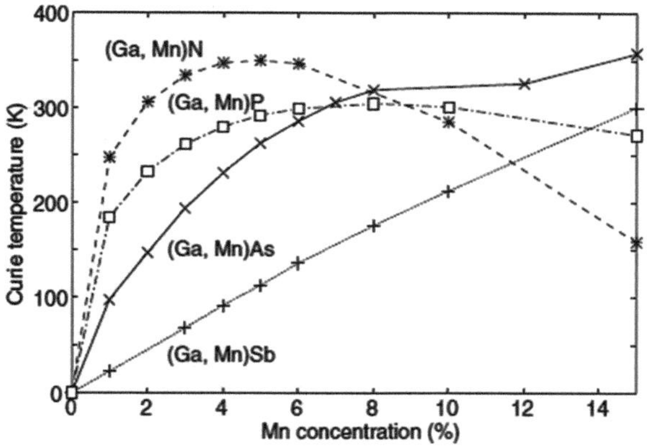

Fig. 15.1 MFA calculations of T_C for GaMnN, GaMnP, GaMnAs, and GaMnSb as a function of Mn concentration. From Sato et al. [17]

This can be explained by a systematic change in the dominant exchange mechanism, with double exchange dominating for GaMnN, GaMnP and *p*-*d* exchange determining the behavior of GaMnSb, while for GaMnAs both mechanisms should be considered.

The carrier doping effects on magnetic properties of the above mentioned materials have been investigated in [18] applying also MFA. The MFA does not give an accurate prediction of T_C, but the qualitative relation can be obtained. The MFA T_C of GaMnN is increased significantly by additional hole doping and has a maximum value at 4 %. In the case of GaMnAs, the hole doping does not affect significantly values of MFA T_C.

The discussed MFA treats the disorder in an approximate *ad hoc* fashion. Discarding the MFA resulted in the decrease of the calculated Curie temperature, and good agreement between theory and observations was obtained (e.g. in [19]). Nevertheless, results obtained from DFT depend on the quality of the effective potential. An extensive review of the results from different approximations for this potential can be found, e.g. [12]. Several examples, how the choice of effective potential influences the calculated properties, are also discussed there.

The main difference between theoretical methods used to handle the problem of magnetism in semiconductors is the choice of solution for two problems, namely, the spin fluctuation problem and the treatment of the substitutional disorder. For the spin fluctuation problem, the discussed above MFA and for more precise calculation the self-consistent local random phases approximation (SC-LRPA) and Monte Carlo (MC) simulation are frequently used. For the disorder treatment, two different methods have been applied: the virtual crystal approximation (VCA) or explicit configurational averaging over different disorder configurations. The latter method is exact, assuming that a sufficiently large number of configurations is included in the averaging.

The values of the Curie temperature calculated from SC-LRPA taken from [12] are shown in Fig. 15.2. In these calculations, perfect ordering in the Mn location in substitutional Ga position was assumed. This corresponds to the optimally annealed

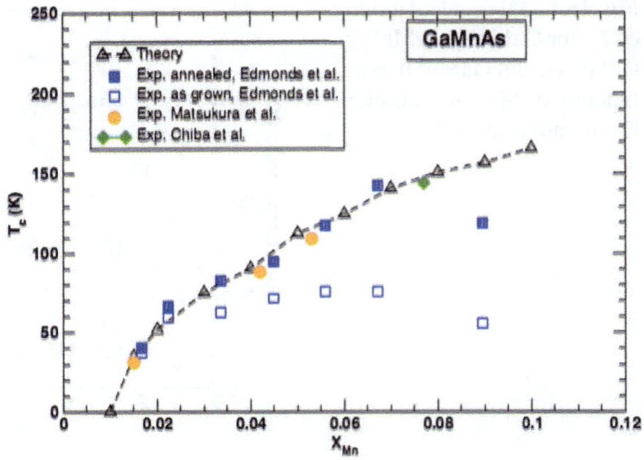

Fig. 15.2 Curie temperature in $Ga_{1-x}Mn_xAs$ as a function of Mn concentration *x*. The *triangles* denote the calculated values within the SC-LRPA. *Other symbols* are the measured values for as-grown reported in [20] and for annealed samples reported in [21, 22]. Experimental values observed by Matsukura et al. [23] and by Chiba et al. [24] are also indicated (taken from [12])

and neutral samples. Consequently, very good agreement with the perfectly ordered samples can be noticed. As grown samples (open squares) have considerably lower T_C due to the presence of defects (Mn in interstitial or As antisites).

In Fig. 15.2, only the values calculated using the LRPA approximation are shown but a good agreement with experimental values was obtained also in [25] from the MC calculations. Deviation observed at high concentration may indicate that, despite of annealing, the sample still contains compensating defects or inclusions of a secondary phase, and not all Mn atoms are magnetically active. The problem of calculating the T_C for a sample that contains both Mn atoms substituting the Ga atoms, denoted usually as Mn_{Ga}, and Mn at interstitial positions, Mn_I, needs first of all information about abundance of these positions of Mn in the sample. In the following chapters it will be demonstrated that, for this purpose, the XAS is a unique experimental tool.

Majority of theoretical approaches assume the ideal ordering in the crystal lattice of the considered DMSs. Recently, it was demonstrated that the origin of ferromagnetic response at high temperatures are, in most of the cases, precipitations of magnetic phase (e.g. in [26]). Due to limited solubility of magnetic atoms in semiconductors, and the existence of many magnetic compounds involving considered elements, nanoclusters containing a large concentration of magnetic ions can precipitate during the growth or processing. In many cases a spinodal decomposition into regions with a low and a high concentration of particular constituents is also observed leading to a large disorder at the cation site and formation of coherent nanoclusters. According to the work by van Schilfgaarde and Mryasov [27] and other authors [28], the evaluated gain in energy by bringing two Ga-substitutional Mn atoms together is 120 meV in GaAs and 300 meV in GaN. Moreover, magnetic nanoclusters embedded by the semiconductor may assume a novel, unknown in bulk, crystallographic form and/or chemical composition. Since spinodal decomposition does not usually involve a precipitation of another crystallographic phase, it is not easily detectable experimentally. It will be demonstrated that EXAFS together with the electron transmission microscopy (TEM) can unambiguously detect such nanoclusters. The nanoclusters that form in this way may only be stable inside the semiconductor matrix. For this reason, is a priori unknown whether they will exhibit ferromagnetic, ferrimagnetic or antiferromagnetic order.

These systems usually show superparamagnetic behavior and should be considered rather as an ensemble of non-interacting ferromagnetic particles than a uniform magnetic alloy. In such a case, the temperature dependences of magnetization and magnetic susceptibility are described by four temperatures: the ordering temperature T_m; the blocking temperature T_B, that corresponds to a maximum of zero-field cooled magnetization; the apparent Curie temperature $T_C^{(app)}$ of the composite material; and the Curie–Weiss temperature Θ, characterizing a weighted magnitude of the exchange interactions between the magnetic ion spins within the nanocluster.

The model of coherent nanoclusters with a large concentration of magnetic constituent explains the origin of ferromagnetic response in DMSs in which an average concentration of magnetic ions is below the percolation limit for the nearest-neighbor coupling and, at the same time, the free carrier density is too low to mediate an efficient long-range exchange interaction. A useful overview of this kind of materials can be found [29]. Surprisingly, it has been reported in [30] that the presence of magnetically active metallic nanoclusters in GaMnAs leads to enhanced magneto-transport and in [31] magneto-optical properties over a wide spectral range. This phenomenon opens an opportunity for various applications of such hybrid systems provided that the methods for controlling the nanoclusters characteristics and the size distribution would be elaborated. To apply these materials for construction of electronic devices, one should control the crystal structure and the size distribution of nanoclusters since the magnetic properties depend on these parameters. An important method to achieve these goals is the XAS since it provides full characterization of nanoclusters. Considerable advantage of XAS is the element specificity, sensitivity to the location of element in the semiconductor matrix, direct information regarding unoccupied density of states distribution and magnetic moment located at the given atoms.

15.3 Location of Transition Metals in the Semiconductor Matrices—EXAFS Studies

15.3.1 Substitutional and Interstitial Positions of the Magnetic Ions

The magnetic ordering in dilute magnetic semiconductors (DMSs) discussed in Sect. 15.2 is interpreted in terms of the p-d Zener model. This model assumes that DMSs are random alloys, where a fraction of the host cations is substitutionally replaced by magnetic ions and the indirect magnetic coupling is provided by delocalized or weakly localized carriers (sp-d exchange interactions). Consequently, location of magnetic atoms in DMSs is correlated with all important physical properties, and it is crucial to known this location before interpretation of magnetic properties of any investigated DMSs. Information about atom location in crystal lattice is provided by only few techniques. Among them, EXAFS is a well established tool and has been proven to be useful as a chemically sensitive local probe for the identification of the site and valence state of magnetic ions in DMSs. Therefore, in many papers the EXAFS technique is used to check the possible presence of secondary phases, precipitates or nanoclusters, as well of the chemical phase separation. The calculated magnetic ion spectra for different location of magnetic atoms show significant differences (e.g. in [32–36]). In majority of the papers, the quantitative comparison of the calculated model spectra with measured spectra, was applied as the homogeneity checking procedure. Due to limited space only selected papers presenting EXAFS studies for Ga nitrides, arsenides and antimonides doped by Mn, Fe or Cr ions will be overviewed. For all matrices the Mn dopant most frequently was studied due to highest solid solubility limit.

In [36] the EXAFS studies confirmed the exclusively substitutional position of Mn atom in the GaMnN. The presence of additional phases Mn_3GaN clusters and octahedral or tetrahedral Mn interstitials has been additionally checked by carrying out fits with a two phases model. The fraction of the Mn in other then Mn_{Ga} position was found to be less than 5 % therefore, in the limit of EXAFS sensitivity.

In the case of Fe dopant, the solubility limit is much lower than for Mn and typically is difficult to introduce more than 1 % of Fe into randomly distributed substitutional sites. Results of XAFS studies carried out on GaFeN and GaFeN:Si samples presented in the [37] allowed to identify the epitaxial parameters controlling the formation of Fe_3N precipitates in GaFeN. Additionally, it was proved that presence of bonds specific to Fe_3N has not been detected in the GaFeN films co-doped with Si. Consequently, it was stated that co-doping is the way to overcome the solubility limit in MBE growth.

The limit of solid solubility in the wurtzite GaCrN was found in [38] to be close to 4 % by comparing the radial distribution EXAFS function simulated by FEFF7 code assuming substitutional Cr position, with the measured for the set of samples. For concentration of Cr higher than this the EXAFS oscillations were very similar to these measured for CrN compound. No Cr atoms at the interstitial were detected. The solubility limit estimated from EXAFS studies in the zinc-blende structure was found in [39] to be much lower (0.2 %). The full EXAFS analysis of wurtzite GaCrN layers in the wide range of concentration was presented in [40].

There are only few experimental reports, which compare the properties of the AlN-, GaN- and InN-based DMSs. In [41], the growth characteristics of the AlCrN, GaCrN and InCrN and their structural properties estimated by XRD and EXAFS were discussed. Radial distances from Cr atoms to the first nearest neighbor N atoms were found differ only slightly, while distances to the second nearest neighbor cation atoms were changed following each lattice constants in AlCrN, GaCrN and InCrN. Crystal field splitting of Cr d level in AlCrN and GaCrN was also observed in the XANES spectra.

The highest solubility limit close to 10 % was reported in [42] for GaMnAs. It is known that in the as-grown MBE samples in the GaAs matrix, the Mn atoms may occupy substitutional and interstitial positions but the abundance of each position

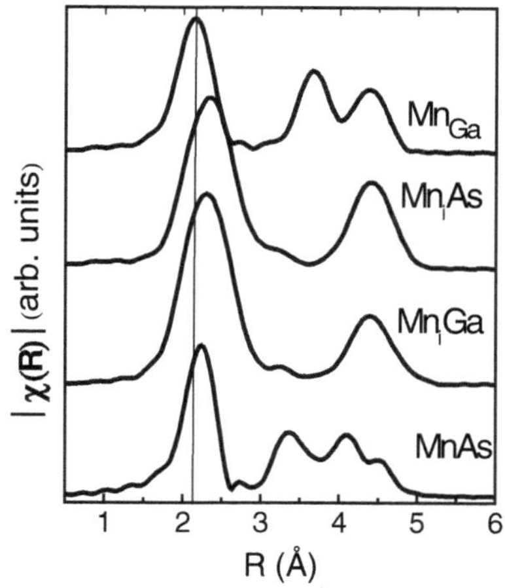

Fig. 15.3 Modulus of the Fourier-transformed calculated Mn spectra using the parameters of the model GaAs structure and Mn located in different position and for hexagonal MnAs. The *line* indicated the position of first shell in the case of Mn substituting Ga

is not easy to find. The number of Mn atoms in the interstitial is increasing with the Mn concentration. It will be demonstrate how the proportion between these two locations can be estimated applying EXAFS.

The Mn EXAFS spectra for different location of Mn in GaAs matrix calculated under assumption that all Mn atoms are located at given position, show significant differences (e.g. in [32–35]). In Fig. 15.3, the calculated moduli of the Fourier-transformed EXAFS spectra for different location of Mn and for hexagonal MnAs compound are presented. One can see that spectra for two possible interstitials with As or Ga atoms as the first neighbours are practically identical, therefore cannot be distinguished in EXAFS analysis. The assumption that atoms may occupy more than one position in the crystal lattice, involves consideration of a superposition of several models and offers the opportunity to estimate the distribution of Mn atoms between all feasible lattice locations. In the [35], the EXAFS data analysis was performed considering superposition of two possible Mn locations in GaAs matrix, substitutional Mn_{Ga}, and interstitial Mn_i.

Due to the fact that electron scattering on Ga and As atoms is very similar, the differences between Mn positions in the EXAFS spectra were not noticed within the first coordination sphere and analysis of the further neighbourhood was necessary. Taking into consideration model with substitutional Mn position only, it was impossible to fit the experimental spectra around 4.5 Å (Fig. 15.4a). The quality of fitting for wide concentration range of samples studied in [35] decreased drastically with increase of Mn content. Adding to the model the possibility of Mn location in the interstitial resulted in much better fit (Fig. 15.4b).

Percentage of atoms located in the substitutional positions decreases with the increase of Mn content. For the sample with $x = 0.06$, only the 53 % of Mn atoms were found in the substitutional position. The distance between Mn and As atoms in first coordination shell for the substitutional position, was found to be the same for all sets of samples (2.48(2) Å), and the same as reported in other papers e.g. in [32–34]. This confirms that this distance is characteristic for Mn atoms in the

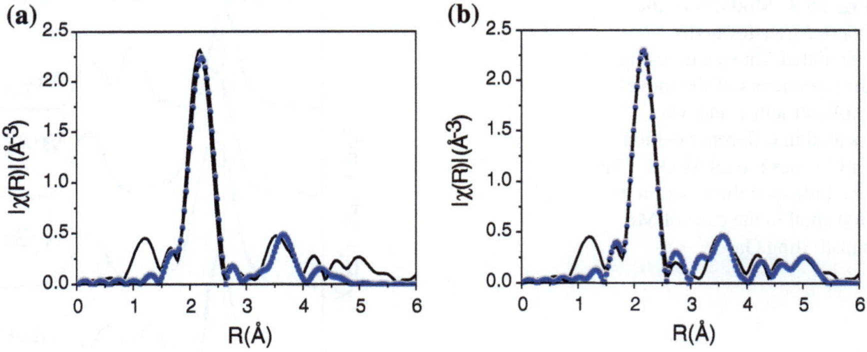

Fig. 15.4 Moduli of the Fourier-transformed $\chi(k)$ functions of samples $Ga_{1-x}Mn_xAs$ with $x = 0.06$ (*solid line*) and the best fit (*circle*) for **a** the substitutional Mn position only, **b** the substitutional and the interstitial position

site of Ga and does not depend on the Mn concentration. Remaining atoms were located in the interstitial positions at a distance of 2.41 Å with increasing distortion of distances in subsequent shells as compared to these resulted from perfect GaAs crystal structure. Impurities lead to a local deformation of the crystal lattice that takes place as an expansion of the interatomic distances compared with those present in the pure crystal. Consequently, with increase of the number of Mn atoms in interstitial position, more stress is imposed on the structure what is leading to the well known effect of changes in the layer lattice constant and defects formation demonstrated e.g. in [34].

The EXAFS approach was also used to provide a description of the geometry of Mn sites in δ-doped GaAs layers grown by MBE as a function of concentration, growth temperature, and Fermi level position. Neglecting of Mn located in the interstitial position or inclusions in the fitted model leads to unphysical coordination of Mn in the first shell. Instead of atoms, the number varying from 1.8 up to 4.9 was reported in [43, 44].

Knowing that there are two different interstitial positions (with Ga or As near-neighbors), can one examine if they are populated with the same probabilities. In the analysis of EXAFS spectra, these two interstitial positions cannot be distinguished (Fig. 15.3). Nevertheless, as will be show in Sect. 15.4, the XANES calculated for these different interstitial position significantly differs since XANES is more sensitive for long range order and chemical bonding. The effect of hydrogenation on $Ga_{1-x}Mn_xAs$ leading to the formation of manganese–hydrogen complexes is discussed in Chap. 5.

The GaAs doped with Fe and Cr has not attracted much attention due to low solubility and paramagnetic properties.

Producing the solid solution of Mn in GaSb was found to be much more difficult than in GaAs, and only very limited amount of Mn was introduced in the matrix providing the Curie temperature below 10 K for GaMnSb solid solution as reported in [45]. Surprisingly, it was found in [46] that MnSb phase segregation during the

MBE growth of GaMnSb, does not inhibit the two-dimensional MBE growth mode. This leads to the producing of single phases, ferromagnetic at room temperature, inclusions distributed in semiconductor matrix, which can be of interest for new devices design. In the case of GaMnAs MBE growth such process leads to nanowires formation.

15.3.2 Formation of Nanoinclusions

The failure in producing DMSs which are ferromagnetic at RT resulted in an increasing interest to study the granular materials with RT ferromagnetic properties. Moreover, an interest in composite materials is presently also increasing in numerous branches of technology. As a result, new questions appeared: how to control the nanoclusters' crystal structure when more than one compound or phase is formed. Electron microscopy commonly used to control technological processes provided information from a very limited volume, which is not representative for entire sample. The XRD is collecting information from representative volume but is not sensitive to presence of small amount of clusters or clusters with similar lattice parameters as a host matrix. We demonstrate that XAS as element selective and sensitive to local atomic order, is a very useful tool for detecting the nanoclusters with different structure and composition. This technique can be used for any composite materials to determine the partitioning of element between the phases.

It is well documented that during the annealing the MBE grown GaMnAs layers above 400 °C the nanoclusters of cubic zinc blende (ZB) and hexagonal MnAs phases are formed. Presence of these nanoclusters change drastically the magnetic properties of layers increasing the T_C above RT. Creation of small and exclusively cubic MnAs clusters was assumed for annealing at 500 °C and larger hexagonal for annealing at 600 °C and above. Recently performed in papers [25, 47, 48] detailed studies of the set of samples with different concentration unambiguously demonstrated that such assumption is not justified and the situation is more complicated. Applying XAS in [25], it was demonstrated that the ZB MnAs clusters do not exist. Instead, small cubic GaMnAs clusters are formed with much higher content of Mn than ever produced in GaMnAs layers with randomly distributed Mn atoms. Clusters with size larger than 8 nm have already the MnAs hexagonal structure. The proportion of Mn in each structure estimated by EXAFS was in perfect agreement with the size distribution of cluster found from TEM analysis (Fig. 15.5). In TEM it was demonstrated that cluster with size 8 nm crystallized already in hexagonal structure.

According to the atomic order in hypothetical ZB MnAs, Mn atoms should have 4 As atoms in the first shell and 12 Mn and 12 As atoms in the next shells. Surprisingly, for a wide range of Mn concentration of samples investigated in [47] in none of these samples the 12 Mn atoms were found in the second coordination sphere assuming the cubic order. Only Ga atoms in the second sphere provided physical parameters of the fits. The amplitude and phase of scattered photoelectrons remarkably differ for Mn and Ga atoms and can be easily distinguished during EXAFS analysis.

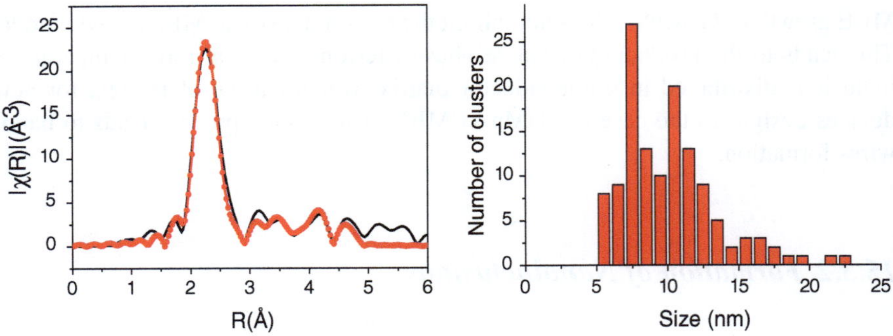

Fig. 15.5 (*Left*) Modulus of the Fourier-transformed $\chi(k)$ functions of sample $Ga_{1-x}Mn_xAs$ with x = 0.06 annealed at 600 °C (*solid line*) and fit with 50 % of hexagonal MnAs clusters (*circles*) and (*right*) histograms of clusters size distribution

Existence of cubic inclusions with 12 Mn in the second shell was never reported in the papers presenting EXAFS studies of the cubic clusters in GaMnAs layers, (see e.g. [34] and references therein). Presence of one Mn atom in the second shell is at the level of the error in the EXAFS analysis and cannot be excluded. This provides close to 20 % of Mn concentration when considering the 6 nm cluster. Details of the performed EXAFS analysis together with SIMS, XRD, electron transmission microscopy (TEM) and magnetic studies for the discussed set of samples were presented in papers [25, 47, 48].

It was shown that for the sample with the smallest content of the Mn x = 0.025 annealed at 500 and 600 °C, only cubic GaMnAs nanoclusters were detected. Existence of only small inclusions has been confirmed by TEM. The characteristic feature of this sample was that 90 % of Mn atoms occupied the substitutional position in the as-grown layer. It implies that substitutional position of Mn atoms when formed is stable. To produce hexagonal MnAs clusters in such a layer, the annealing temperature higher than 60 °C is necessary.

To get successful fit to the EXAFS data in samples with higher content of Mn it was necessary to consider both kinds of nano-clusters: the cubic with GaMnAs and hexagonal MnAs. For these clusters, the atomic order was assumed according to crystallographic data. In the case of hexagonal clusters, which have dimension larger than 8 nm, the influence of surface is not important and the atomic distances were set to be constant. To keep the number of fitting parameters as small as possible, the weighting parameter for the number of atoms in the coordination spheres according to the fraction of Mn atoms in each kind of nanoclusters was introduced. Such approach allowed to not exceed the number of unknown parameters over independent data points. It is important to underline that in the NiAs-type hexagonal MnAs compound, the order of atoms in the subsequent shells is different from that of cubic structure. In the first shell 6 As atoms are present followed by 2 and 6 Mn atoms, therefore, existence of such cluster at the level higher than 5 % is easy discovered in EXAFS analysis. Assumption of only cubic nanocluster in such a sample leads to problems with the fitting parameters (the number of atoms and distances) as was discussed in [34], and it is necessary to introduce cumulants to describe deviation of the pair distribution from the Gaussian shape. In this way, the presence of second phase (or high disorder) in the sample is manifested during the EXAFS analysis when fitting with one model only. While cumulants are not introduced an unphysical value of the number of atoms even in first coordination sphere (e.g. 1.8 instead of 4) as in papers [43, 49] is resulted from the fitting.

To overcome the problem with two kinds of inclusions with different magnetic properties the MnSb nanoclusters were considered in [50]. The main advantage of MnSb compound is T_C well above the RT (587 K) while for MnAs it is only 318 K. Additional advantage is a possibility to form the hexagonal MnSb precipitates directly during the MBE process without subsequent annealing. It was shown in [50] from EXAFS, that the properly elaborated MBE process allows to gather all Mn atoms into MnSb hexagonal inclusions and to prevent the Mn atoms from being incorporated into GaSb or GaAs matrix. Moreover, the formation of the inclusions does not prevent the two-dimensional layer by layer growth in contrast to the GaMnAs system where MnAs segregation leads to the formation of the nanowires as demonstrated in [51].

In the InMnAs samples grown by organometallic vapor phase epitaxy method a very different atomic order around Mn atoms was proposed in [52]. The investigated films show ferromagnetism above RT. The authors declare that as the concentration of Mn increases, the average local environment surrounding Mn changes from that of a Mn–Mn interstitial pair to a dimer, trimer, or cubic MnAs structure and then to the hexagonal MnAs structure/interstitial pair. In contrast to random substitution, the Mn impurity atoms in these dimer, trimer, or cubic MnAs structure occupy adjacent sites in the In sublattice. Nevertheless, no reports confirming these have been published.

Majority of examples of applying EXAFS analysis to control the nano-inclusions formation were devoted to the MBE growing samples, but even more difficult is get information about atomic order around implanted magnetic ions. In contrast to MnAs, forming the MnSb inclusions by an implantation process encounters many difficulties. Detailed EXAFS analysis presented in papers [53, 54] has applied comparison with simulated and measured reference spectra, to estimate the local atomic order around implanted atoms without and with post implantation processing. It was shown that implantation tends to remove Sb atoms from the neighborhood of the Mn atoms even with the implantation temperature kept close to liquid nitrogen temperature. It was possible to reintroduce Sb atoms by the annealing in Sb vapor but together with oxygen which is more reactive and was bounded to the Mn atoms. Keeping the high temperature during the implantation prevented the migration of oxygen but still the chemical affinity of Mn to Sb is lower than to Ga and the Mn-Ga clusters were formed.

Attempts to dope GaAs and GaN semiconductor nanowires with Mn have been also reported and are presented in Chap. 13. It was found in [55] that the wires had a wurtzite structure and both EXAFS and transport measurements suggested that Mn atoms actually diffuse into the nanowires where they act as dopants. A further study was carried out in [56] on Mn-induced into InAs wires in which morphological investigations showed the effectiveness of this metal in forming wires. In this case, the structural study of the Mn environment by EXAFS evidenced the formation of MnAs precipitates. Again, a characterization and a physical understanding of the local structure of Mn in (GaMn)As nanowires are of great importance. Two main questions should be answered: the degree of substitutional incorporation of Mn in the nanowire lattice and/or whether Mn-defective structures are formed, and as demonstrated in [57] just only EXAFS offers decisive answer.

As was discussed above, the aggregation of magnetic impurities resulting at the nanoscale in formation of chemical and/or crystallographic inhomogeneities has a dramatic influence on the magnetic, electronic and optical properties of the systems. The possibility to reach a control over the formation of inhomogeneities and its effect on the chemical and physical behavior of the host material, can have outstanding consequences in the design and functionality of the next generation devices. In the [58] the efficiency of an approach based on the controlled formation of complexes, involving one magnetic impurity and one or more electrically active dopant was recently demonstrated. With EXAFS methods supported by ab initio computations, have been shown that co-doping of GaN with Mn and Mg resulted in the formation of cation complexes Mn–Mg_k. Depending on the number of ligands k predefined by fabrication conditions, a strong electron-phonon interaction and consequently an efficient and broadband infrared photoluminescence were promoted. Control over the formation of Mn–Mg_k cation complexes allowed to extend towards the infrared the already remarkable optical capabilities of nitrides, and represent a new perspective for magnetic semiconductors.

In summary, an analysis of EXAFS is a unique technique to study the non homogenous materials. Due to its element selectivity and sensitivity to the local atomic order in the case of element being bonded in two different crystal structures it provides full information about abundance of both structures. In the case of element occupying different positions in the crystal structure with different atomic order, one can find the fraction of atoms located in each position. Advantages of EXAFS method were demonstrated in the case of GaMnAs layers with different Mn content and after different thermal treatment. Such considerations allowed to propose procedures to control the formation of magnetic atoms nanoclusters in DMSs. The problem is becoming more and more important as the composite materials are very attractive for commercial use.

15.4 Electronic Structure of Magnetic Ions in Semiconductors—XANES Studies

15.4.1 Substitutional and Interstitial Positions of the Magnetic Ions

An understanding of the local environment of magnetic ion in semiconductors and its valence state, has an important impact on the understanding and the optimization of its properties. As was discussed in the Sect. 15.3.1 and shown in the case of several examples in previous chapters XANES is sensitive for long range order and chemical bonding and even small differences in the scattering amplitude of neighbours might lead to appreciable differences in the spectra. In this chapter the power of this technique which is even more sensitive than EXAFS when strengthened by the use of advanced theory, will be demonstrated. The simulation of XANES, like it was described in Sect. 1.2.4, is based on ab initio or multiple scattering approach. Both approaches need the assumption of model of atoms potential and geometrical arrangement. Considering of all possible arrangements can be arduous and time demanding but finally provide decisive answer. In some cases as in [59] the core-hole effects should be additionally implemented to achieve agreement with the experimental spectra.

Majority of XANES analysis is devoted to determination the valence state of magnetic ions and to confirmation the cation substitution. The shape and energy position of XANES K-edge is very characteristic for Mn atomic local structure and together with prediction of simulation in many papers is used to confirm Mn substitutional location. Particularly many studies are reported for wurtzite GaMnN (e.g. in [55, 60–62]) and cubic GaMnAs (e.g. in [33, 34]). There is a common agreement that valence state of magnetic ions in GaAs and GaSb matrices is Mn^{2+} but regarding matrices with nitrides discussion is still going on. The studies of the d-electrons reflected in $L_{2,3}$ edges presented in [62] indicate the Mn^{2+} but most of K-edge studies which reflect mostly p-states and in the pre-peak the states hybridized with d-states indicate the Mn^{3+} valence state (e.g. in [36, 61]). Moreover, the coexistence of Mn^{2+}/Mn^{3+} valence resulting in p-type conductivity via d-electron hopping is suggested in [63] to play a very important role in the ferromagnetism. Although Mn at different hosts shows different fingerprints in general, it is very difficult to distinguish the difference in valence between Mn^{2+} and Mn^{3+} in GaMnN only by the XANES studies at the K-edge if both Mn ions are present at the substitutional sites.

The very interesting opportunity of exploiting XANES spectroscopy was proposed by Rovezzi and applied in paper by Bonanni et al. [64]. The high resolving power of a wavelength dispersive spectrometer coupled with a high brilliance synchrotron beamline permits to collect fluorescence K_β emission lines even on dilute species in strong absorbing matrix, as DMSs. Due to the intrinsic spin-polarization of d electrons and strong 3p3d intra-atomic exchange interaction, the spin momentum is determined by the non-resonant X-ray emission spectrum that probes the transitions from the spin-up ($K_{\beta\prime}$) and the spin down ($K_{\beta 1,3}$) levels. This effect was used to record spin-selective XANES . The pre-edge features, enhanced by the sharpening effect of the high resolution selective collection mode, quantify the $t_2^{\uparrow}$ -$t_2^{\downarrow}$ states splitting.

Let us consider now in more details the sensitivity of XANES to distinguish between different Mn locations on the example of GaAs matrix. The energy distribution of unoccupied density of states calculated for different interstitial position, using FEFF8.4 codes described in [65], significantly differs and compared with the experimental spectra allowed distinguishing between different interstitial positions

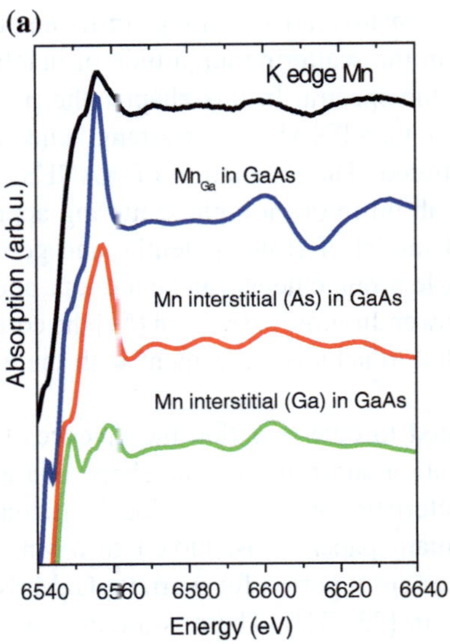

Fig. 15.6 **a** Experimental spectrum for $Ga_{1-x}Mn_xAs$ with $x = 0.06$ versus FEFF8.4 calculations for different Mn positions. **b** Experimental spectrum for $Ga_{1-x}Mn_xAs$ with $x = 0.06$ compared with results of fittings different models

in GaAs matrix as shown e.g. in [33, 43, 66]. This was not possible by other techniques. In Fig. 15.6a the FEFF8.4 calculations of Mn spectra for substitutional Mn_{Ga}, interstitial (As) and interstitial (Ga) are shown under assumption that all Mn atoms are located in chosen positions only. The XANES spectra differ significantly for each Mn position, but it is clear that none of the theoretical spectra is close to the experimental XANES. Obviously, Mn atoms occupy more than one position in the crystal lattice. Three possible combinations of Mn locations in the GaAs lattice sites were considered. The results of the best fitting are shown in Fig. 15.6b assuming the given model. In curve marked by (a) interstitial (As) and Mn_{Ga} were fitted, in (b) interstitial (As) and interstitial (Ga), in (c) interstitial (Ga) and Mn_{Ga}, in (d) interstitial (As), interstitial (Ga) and Mn_{Ga}.

The fits closest to experiment were obtained in the case of model (c) and (d). This suggests that Mn atoms prefer to be located in the substitutional positions (43(2) %) and the interstitial (Ga) (57(2) %). However, it cannot be excluded that a small part of them is located in the interstitial (As) position as was assumed in the model d). The performed analysis of K-edge XANES resulted in even smaller amount of Mn located in the substitutional positions than estimated from EXAFS analysis discussed in Sect. 15.3.1. The details of analysis are presented in [66].

In the K-edge spectra the density of unoccupied p states are reflected and it was proved above that these states are sensitive for Mn location in the crystal structure. Let us check now if much more localised d states are sensitive to Mn positions in the crystal lattice. In Fig. 15.7 the results of similar as in the case of K-edge analysis are presented for L_3-edge.

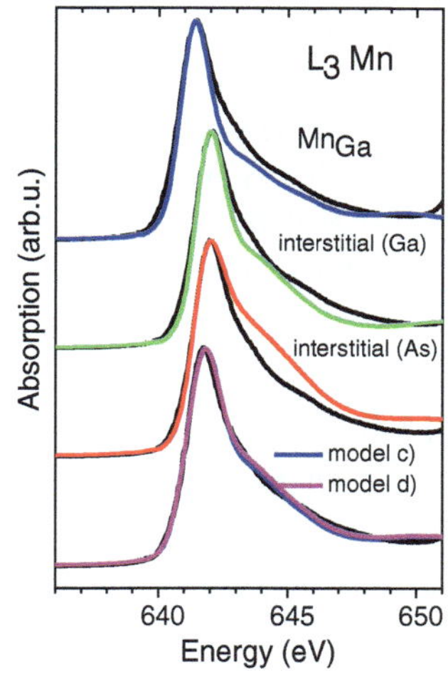

Fig. 15.7 XANES spectra for $Ga_{1-x}Mn_xAs$ with x = 0.06 at Mn L_3-edge compared with FEFF8.4 calculations for different Mn positions in the GaAs matrix and models

It can be seen that the calculated L_3 spectra for the Mn_{Ga} and interstitial (Ga) models are close to the experimental one, but there is a substantial difference in the results for the interstitial (As). For this model the full width at half maximum is significantly wider than for others. Therefore, from results of L_3 spectra simulation, it can be also concluded that the interstitial Mn atoms prefer a position with Ga atoms as the first neighbors. Moreover, the weighted sum of the theoretical L_3 spectra performed with the values obtained from Mn K-edge fitting maintains a good agreement for (c) and (d) models thus proving that the states of -d symmetry are very sensitive to the location of Mn in the crystal lattice. Similar effect was reported for GaMnN in the case of Mn valence in [62]. Therefore, performed analysis for the first time indicated that Mn atoms prefer interstitial positions with Ga atoms as the first neighbors and can form with Ga shorter bond than with As atoms. This can explain the limited solubility of the Mn in GaAs matrix where Mn has the As neighbors.

The possibility of Mn to be in the significant amount located at the As antisites was discussed in [43] and ruled out by comparison of the calculated XANES with the experimental spectra. The As antisite positions lead to a strong resonance in the middle of the edge, which is reduced to only a bump in the Mn_{Ga} site. This enhancement can be tentatively explained by the fact that neighboring Ga atoms possess more unoccupied states in the 4*p* band than As.

15.4.2 Formation of Nanoinclusions

The formation of nano-inclusions can be not only detected by EXAFS analysis but due to the well established fact that XANES is a fingerprint of atom bonding in the given compound, XANES is frequently used for inclusions of other phase identification and the abundance of given phases estimation. There can be found in the literature several examples of this kind analysis in the case of III–V semiconductors. In GaCrN studies presented in [39, 40] the direct comparison with the reference of CrN sample were used to indicate to formation of CrN nanoclusters. The formation of MnSb clusters during the MBE growth was evidenced by comparing with XANES of MnSb compound powder in [50]. The identification of the Mn location in the GaSb implanted with the Mn was performed with this kind of approach in [53, 54].

15.5 Magnetic Structure of the Magnetic Ions in Semiconductors—XMCD Studies

The X-ray magnetic circular dichroism (XMCD) is defined as the difference in the absorption cross section between left and right circularly polarized X-ray. Since its experimental evidence by Schutz et al. [67] XMCD has developed in to widely used technique for the element specific characterization of magnetic materials. Moreover, the quantitative evaluation of magnetic moments, separated into spin and orbital contributions, is possible from integrals over XMCD spectra by applying the so-called sum rules, proposed in 1992 and 1993 in [68, 69].

To demonstrate the possibilities of XMCD, let us consider the L spectra of the transition metals. The p states in TM are splited into the $2p_{3/2}$ and $2p_{1/2}$ levels by the spin orbit interaction. At the $2p_{3/2}$ and the $2p_{1/2}$ levels the orbital angular momentum $\boldsymbol{l}$ and the spin angular momentum $\boldsymbol{s}$ are coupled parallel and anti-parallel, respectively. The spin-orbit coupling allows to excite electrons of the 2p shell by circular polarized light spin selectively into the empty valence shell. The spin polarization arises from the selection rule for the orbital magnetic moment depending on the polarization of the absorbed X-ray photon. Because of the parallel coupling of $\boldsymbol{l}$ and $\boldsymbol{s}$ at the $2p_{3/2}$ level and the anti-parallel at the $2p_{1/2}$ level, transitions from the $2p_{3/2}$ and the $2p_{1/2}$ levels into the empty valence shell occur with different spin polarization. Possible final states for the photo-excitation are the unoccupied 3d and 4s states above the Fermi level. The dipole transition is spin conservative which means that a spin up electron can only be promoted in to a spin up empty state and vice versa.

In the XMCD experiment the chosen helicity of the X-ray are usually used and measurement is carried out in magnetic field for opposite magnetization directions. The magnetic field is ordering the election states in given temperature. The difference between two spectra μ^+ and μ^-, recorded at different directions of magnetic field, provides XMCD. Using integrated intensity of the $L_{2,3}$-edge XAS and XMCD spectra of a magnetic atom, one can separately estimate the orbital (M_{orb}) and spin (M_{spin}) magnetic moments applying XMCD formulas described in [68, 69]:

$$M_{\mathrm{orb}} = -\frac{4\int_{L_3+L_2}\left(\mu^+ - \mu^-\right)dw}{3\int_{L_3+L_2}\left(\mu^+ + \mu^-\right)dw}(10 - N_d),$$

$$M_{\text{spin}} + 7M_T = -\frac{6\int_{L_3}(\mu^+ - \mu^-)\,dw - 4\int_{L_3+L_2}(\mu^+ - \mu^-)\,dw}{\int_{L_3+L_2}(\mu^+ + \mu^-)\,dw}(10 - N_d),$$

where, $\mu^+(\mu-)$ is the absorption intensity for the different direction of magnetic field (or helicity of radiation), N_d is the d electron occupation number of the specific atom. The L_3 and L_2 denote the integration range, which can be either both edges or only L_3 edge. M_T is the expectation value of the magnetic dipole operator, which is small when the local symmetry of the atomic site is high and is usually neglected for TM atoms.

In the numerical evaluation of the orbital and spin moments in the case of TM and particularly Mn atoms two main sources of errors should be kept in mind. The coupling between L_2 and L_3 edges (so called branching ratio) which influenced spin moment. It was predicted in [70] that the intermixing of the L_2 and L_3 absorption edges is mainly present in the early TM where the electron core hole interaction increases while the spin-orbit splitting decreases. For such metals (e.g. Mn) the application of the spin sum rule can produce an error up to 30 %. The second important source of error in applying the spin sum rule as discussed in [71] is the energy dependent radial matrix elements. Though the sum rules are normalized to the isotropic absorption cross-section which is proportional to the number of holes in the d-shell, into the measured absorption cross-section also continuum states contribute. To eliminate this contribution a double step function is fitted to the absorption spectrum and subtracted. This procedure is likely to introduce a systematic error in the determination of the number of holes in the final state. Additional corrections for degree of X-ray circular polarity and photon incident angle should be also introduced to the experimental data.

The experimental confirmation of the XMCD sum rules for Fe and Co was presented by Chen et al. [72]. The presented measurements were performed with high precision in transmission with in situ grown thin films. This procedure eliminated experimental artifacts and proved that the evaluated spin and orbital moments are in good agreement with those obtained from Einstein-de Haas gyromagnetic ratio measurements.

Although, the best mode of XMCD detection is the transmission for majority of reported in the literature experiments for DMSs the total electron yield (TEY) and total fluorescence yield (TFY) were applied because most of investigated samples are the thin films. These methods suffer from saturation and self-absorption effects that are very difficult to correct for. Moreover, the TEY method can be sensitive to varying applied magnetic field and changing the electron detecting efficiency (photocurrent). The TFY is insensitive to the applied field, but the yield is intrinsically not proportional to the absorption cross section, because the radiative to non-radiative relative core-hole decay probability depends strongly on the symmetry and spin polarization of the XAS final states. These two detection modes differ also in the examined depth and can provide different result in the case of not homogenous samples, as shown in Fig. 15.8b for sample containing the cubic GaMnAs and hexagonal MnAs nano-clusters. The XMCD signal characteristic for the tetragonal and octahedral

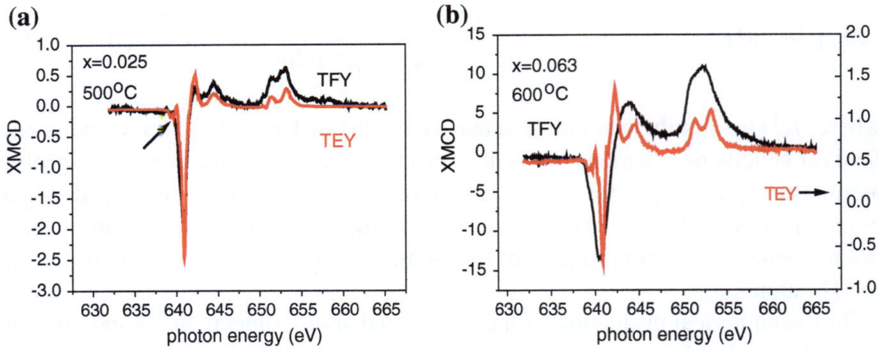

Fig. 15.8 The XMCD signal for TEY mode of detection (*red/gray*) and TFY (*black*) measured at the ID08 station (ESRF) for samples with Mn content **a** x = 0.025 annealed at 500 °C with only cubic nano-clusters, and **b** x = 0.06 annealed at 600 °C with 50 % of cubic GaMnAs and 50 % of hexagonal MnAs nano-clusters. *Arrow* in (**a**) indicated the feature sensitive on the hole density as discussed in [73]

symmetry of atomic arrangement is reflected for TEY and TFY detection mode, respectively. This indicated that at the surface Mn atoms have cubic neighboring and can be located either in the GaAs matrix or cubic clusters and in bulk the hexagonal MnAs clusters dominated. In the case of sample with only cubic nano-clusters not significant difference between both modes of detection was observed in Fig. 15.8a.

The orbital and spin magnetic moments are fundamental quantities for understanding the macroscopic magnetic properties of materials, therefore in spite of the listed problems significant experimental efforts have been performed to find them for DMSs, particularly for GaMnAs. Already in the first XMCD experiments at the Mn L edges it was shown that only small fraction of Mn atoms is participating in the ferromagnetic order (e.g. 13 % in the sample with 2 % of Mn reported in [74]) most of the atoms are paramagnetic. The reason for this was examined during the subsequent years.

Calculations of the GaMnAs band structure and element, orbital, and spin resolved densities of states, based on model of Mn substituted for Ga in the GaAs lattice, predict that the mediating states are primarily derived from As 4*p* valence band states [e.g. [75] and reference therein]. These imply a substantial induced moment on the As atoms neighboring the Mn ions that is antiparallel to the Mn 3*d* moments, and a small parallel Ga moment. Taking advantage of element selectivity of XMCD it was possible to confirm these predictions in [76]. Induced Ga and As moments in ferromagnetic $Ga_{1-x}Mn_xAs$ were detected using XMCD at the Mn, Ga, and As $L_{3,2}$ edges.

Applying the density functional full-potential linearized augmented plane wave approach, the XMCD spectra of $Ga_{1-x}Mn_xAs$ have been calculated in [77] taking into account many different possibilities of atom arrangements around Mn, such as antisite defects, interstitial Mn, as well as various Mn dimers. Intriguingly, the spectroscopic features of Mn-XMCD appeared to be very sensitive to the change in environment around Mn. This promises a potential to use XMCD for unambiguous determination of impurity distribution in DMS through strong interplay between experiment and theory. As an example the comparison of the calculations with the experimental results for sample with x = 0.07 is shown in [77]. The best agreement was achieved under assumption of the 50 % of Mn atoms in substitional position and 50 % in interstitial which were forming a dimers. Moreover, the presence of Mn_{Ga}–Mn_I dimers was predicted to drastically reduce the effective magnetic moment, as observed in most experiments. In the paper [35] from EXAFS analysis of as grown sample with similar content of Mn (x = 0.063) also so many Mn atoms were found in interstitial positions. Nevertheless, no second Mn atom was detected in the distance up to 4 Å. Consequently, the presence of Mn dimers was not confirmed by EXAFS studies.

The influence of the surface with the MnO on the value of XMCD signal registered by TEY was demonstrated by Edmonds et al. [78]. After removing MnO surface layer by chemical etching a large XMCD, approaching the calculated maximum value of ~58 %, was observed at the Mn $L_{3,2}$ edge. This indicates that, after etching, most of the measured Mn atoms contributing to the ferromagnetic order. The sum rules were applied to set of etched samples in [79]. The small orbital moment $M_{\mathrm{orb}} = 0.16(2)\mu_B$ and spin moment $M_{\mathrm{spin}} = 4.3(3)\mu_B$, evaluated under assumption of $N_d = 5.1$ and with the correction factor for branching effect 1.47 were found for annealed and etched samples independent of concentration. An orbital-to-spin moment ratio was estimated to be 0.037. In as grown sample (without annealing), a clear signature of antiferromagnetic coupling was observed and the Mn moment per atom was significantly reduced compared to the low temperature annealed films, and increases slowly and linearly on increasing the magnetic field from 2 to 5 T at temperatures well below T_C. Tc explain these authors applied a modified Brillouin function and got an antiferromagnetic coupling temperature of 22 K. This antiferromagnetic coupling was assigned to the presence of interstitial-substitutional pairs (dimers), which break upon annealing. The deconvolution of XMCD spectra into ferromagnetic and paramagnetic component was also discussed in [80, 81].

The ability to separately resolve localized and hybridized *d* states in GaMnAs was demonstrated by performing angle dependent XMCD and have been showed in [73] that XMCD can be a powerful method for determining the electronic structure of magnetic semiconductors. The cubic features are shown to be atomic-like, with no net anisotropy in the magnetic moments, and no significant Mn or hole concentration dependence. In contrast, the small uniaxial feature indicated by arrow in Fig. 15.8a, has a systematic dependence on hole density, indicating that this feature corresponds to states close to the Fermi energy.

The temperature, magnetic field and Mn concentration dependences of the ferromagnetism in as grown $Ga_{1-x}Mn_xAs$ samples were investigated by XMCD to extract the intrinsic magnetic component from TEY measurements in [82]. It was demonstrated that short range ferromagnetic correlations develop significantly above T_C and that antiferromagnetic interaction between the Mn_{Ga} and Mn_I is important to understand the magnetic properties of $Ga_{1-x}Mn_xAs$. Therefore, the amount of the Mn_i ions is strongly related with T_C.

The magnetic coupling between metal and semiconductor layers in a series of Fe/GaMnAs bilayer films was recently studied in [83] exploiting the XMCD. Bulk sensitive SQUID magnetometry measurements, as well as measurements of the magnetic response of the individual layers using XMCD, demonstrate an exchange bias effect in the GaMnAs induced by the antiferromagnetic coupling to the neighboring Fe layer. Comparison of the magnetic signal to that of a control film indicated that the bias field affects the whole of the GaMnAs layer, such that the layer is homogeneously magnetized antiparallel to the Fe layer at low magnetic fields. Furthermore, taking advantage of the different penetration depths of TFY and TEY detection modes, it was found that while the bulk of the layer reorients to parallel alignment when an external field large enough to overcome the bias field is applied, the Mn moments within a subnanometer interface layer remain aligned antiparallel to the Fe layer even for the largest external fields investigated.

Although, the most frequently studied by XMCD was GaMnAs, several papers devoted to other III–V semiconductors were published. In [84] the XMCD studies for GaMnP with different Mn content are presented. The Mn XAS and XMCD spectra from GaP matrix are nearly identical in shape to those from GaAs suggesting that the charge state and local environment of Mn atoms in these two materials are very similar. The very different XMCD was reported for GaCrN in [85]. The temperature dependence of the Cr XMCD main peak intensity was described by the Curie–Weiss law, and ferromagnetic interaction between Cr ions suggested. Multiple heterostructure of GaN/$Ga_{1-x}Mn_xN$ of four periods was investigated in [86] and theoretical analysis based on configuration-interaction theory was adopted to understand the local electronic structure. Performed XMCD measurements proved that Mn ions were responsible for the ferromagnetism of the investigated sample.

Due to problems with reaching the solubility of magnetic ions sufficiently high to get magnetic properties at room temperature the attempt to produce the magnetic inclusions immersed into semiconducting matrix was performed as discussed in Sects. 15.3.2 and 15.4.2. Magnetic properties of the hexagonal MnAs nano-clusters segregated by high-temperature annealing of $Ga_{1-x}Mn_xAs$ using XAS and XMCD were reported in [87]. The XAS spectrum drastically changed from the as-grown samples and showed the shape characteristic for octahedral coordination.

The detailed studies of the samples with cubic GaMnAs and hexagonal MnAs inclusions were recently performed by K. Lawniczak-Jablonska and co-workers at ID08 beamline of ESRF. The orbital and spin magnetic moments in function of magnetic field were examined. The particular interest was devoted to changes of magnetic moments with formation of nanoclusters. The as grown and annealed in 500 and 600 °C samples were investigated. This set of samples was fully characterized by several techniques as reported in [25, 35, 47]. In the as grown samples similar as in other reports (e.g. in [79]), the small orbital moment was detected under magnetic field 5T (0.006 μ_B) at T=150 K. Orbital moment was increased when clusters were formed and for cubic GaMnAs cluster was 0.014 μ_B but for small hexagonal MnAs nanoclusters reach already 0.25 μ_B. The orbital moment was found not depend significantly on the magnetic field and change of the temperature as can be seen in Fig. 15.9.

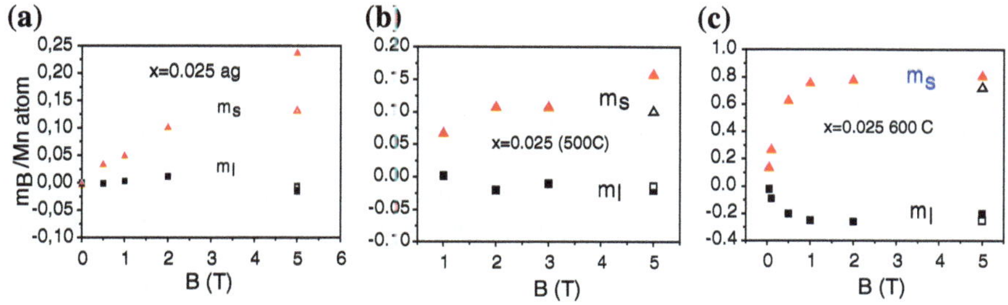

Fig. 15.9 The field dependency of the spin (*triangles*) and orbital (*squares*) of the magnetic moments estimated from the XMCD measurements for $Ga_{1-x}Mn_xAs$ for x = 0.025 **a** as grown samples; **b** after annealing in 500 °C; **c** after annealing in 600 °C. The measurements performed at 150 K (full symbols) and *open symbols* at 300 K

The value of spin moment changes with the magnetic field differently depending on the location of Mn in the sample. In the as grown sample, similar as it was reported in [79] for solid solution of Mn in GaAs, the spin moment increases linearly with the magnetic field reaching 0.24 μ_B at 5T in Fig. 15.9a. In sample with small (~5 nm) cubic GaMnAs nanoclusters spin moment at 1T is larger (0.07 μ_B) than for as grown sample (0.05 μ_B), and does not change much with the magnetic field indicating on ferromagnetic coupling of Mn ions in Fig. 15.9b. The effect is more pronounced in the sample with still small but already hexagonal MnAs nanoclusters in Fig. 15.9c. In this sample spin moment saturate at 1T (0.8 μ_B).

The magnetic properties of MnSb inclusions formed in GaSb matrix directlyduring the molecular beam epitaxial growth have been studied by XMCD and macroscopic methods in [46]. The shape of XMCD signal was typical for octahedral coordination. Moreover, one can change the direction of easy and hard axis of formed nanoclusios by proper chose of substrate crystallographic direction.

15.6 Summary

This chapter has reviewed the application of XAS to III–V semiconductors doped with magnetic ions. The theoretical prediction of the possibility to achieve the RT magnetic semiconductor with randomly distributed magnetic ions in III–V semiconductors, was and still is an inspiration for the extensive technological work. This resulted with the tremendous number of papers presenting the results of technological attempts to achieve RT magnetic semiconductors. It is not possible to review all. Therefore, only these which underlined the unique properties of XAS and XMCD techniques were briefly overviewed. Many technological problems can be solved or prediction of theory verified only by XAS. Thanks to element selectivity and sensitivity on the local atomic order in many cases only this unique technique can provide decisive answer.

References

1. R.R. Galazka, in *Semimagnetic Semiconductors*, ed. by B.L.H. Wilson Proceedings of 14th International Conferences on Physics of Semiconductors (Edinburgh 1978), vol. 43 (Institute of Physics Publishing, Bristol, 1979), p. 133
2. T. Story, R.R. Galazka, R.B. Frankel, P.A. Wolff, Phys. Rev. Lett. **56**, 777 (1986)
3. H. Ohno, H. Munekata, T. Penney, S. von Moln'ar, L.L. Chang, Phys. Rev. Lett. **68**, 2664 (1992)
4. H. Ohno, A. Shen, F. Matsukura, A. Oiwa, A. Endo, S. Katsumoto, Y. Iye, Appl. Phys. Lett. **69**, 363 (1996)
5. T. Dietl, H. Ohno, F. Matsukura, J. Cibert, D. Ferrand, Science **287**, 1019 (2000)
6. N.W. Ashcroft, N.D. Mermin, *Solid State Physics* (Saunders, Philadelphia, 1976)
7. H.A. Kramers, Physica **1**, 182 (1934)
8. P.W. Anderson, Phys. Rev. **79**, 350 (1950)
9. C. Zener, Phys. Rev. **82**, 403 (1951)
10. C. Zener, Phys. Rev. **81**, 440 (1951)
11. T. Jungwirth, K.Y. Wang, J. Masek, K.W. Edmonds, J. König, J. Sinova, M. Polini, N.A. Goncharuk, A.H. MacDonald, M. Sawicki, A.W. Rushforth, R.P. Campion, L.X. Zhao, C.T. Foxon, B.L. Gallagher, Phys. Rev. B **72**, 165204 (2005)
12. K. Sato, L Bergqvist, J. Kudrnovsky, P.H. Dederichs, O. Eriksson, I. Turek, B. Sanyal, G. Bouzerar, H. Katayama-Yoshida, V. A. Dinh, T. Fukushima, H. Kizaki, R. Zeller. Rev. Mod. Phys. **82**, 1633–1690 (2010)
13. T. Jungwirth, J. Sinova, J. Masek, J. Kucera, A.H. MacDonald, Rev. Mod. Phys. **78**, 809–864 (2006)
14. A. Zunger, in *Solid State Physics*, vol. 39 ed., by F. Zeits, H. Ehrenreich, D. Turnbull (Academic, New York, 1986), p. 275
15. A. Oswald, R. Zeller, P.J. Braspenning, P.H. Dederichs, J. Phys. F: Met. Phys. **15**, 193 (1985)
16. A.I. Liechtenstein, M.I. Katsnelson, V.P. Antropov, V.A. Gubanov, J. Magn. Magn. Mater. **67**, 65 (1987)
17. K. Sato, P.H. Dederichs, H. Katayama-Yoshida, J. Kudrnovský, J. Phys.: Condens. Matter **16**, S5491 (2004)
18. K. Sato, P.H. Dederichs, H. Katayama-Yoshida, Europhys. Lett. **61**, 403 (2003)
19. K. Sato, W. Schweika, P.H. Dederichs, H. Katayama-Yoshida, Phys. Rev. B **70**, 201202(R) (2004)
20. K.W. Edmonds et al., Appl. Phys. Lett. **81**, 3010 (2002)
21. K.W. Edmonds, P. Boguslawski, K.Y. Wang, R.P. Campion, S.N. Novikov, N.R.S. Farley, B.L. Gallagher, C.T. Foxon, M. Sawicki, T. Dietl, M. Buongiorno Nardelli, J. Bernholc Phys. Rev. Lett. **92**, 037201 (2004)
22. K.W. Edmonds et al., Appl. Phys. Lett. **81**, 4991 (2002)
23. F. Matsukura et al., Phys. Rev. B **57**, R2037 (1998)
24. D. Chiba et al., Appl. Phys. Lett. **82**, 3020 (2003)
25. K. Lawniczak-Jablonska, J. Libera, A. Wolska, M.T. Klepka, P. Dluzewski, J. Sadowski, D. Wasik, A. Twardowski, A. Kwiatkowski, K. Sato, Phys. Status Solidi RRL **5**, 62 (2011)
26. G. Martinez-Criado, A. Somogyi, S. Ramos, J. Campo, R. Tucoulou, M. Salome, J. Susini, M. Hermann, M. Eickhoff, M. Stutzmann, Appl. Phys. Lett. **86**, 131927 (2005)
27. M. van Schilfgaarde, O.N. Mryasov, Phys. Rev. B **63**, 233205 (2001)
28. K. Sato, H. Katayama-Yoshida, P.H. Dederichs, Jpn. J. Appl. Phys. **44**, L948–951 (2005)
29. T. Dietl, J. Phys.: Condens. Matter **19**, 165204 (2007)
30. S. Ye, P.J. Klar, Th. Hartmann, W. Heimbrodt, M. Lampalzer et al. Appl. Phys. Lett. **83**, 3927 (2003)
31. H. Akinaga, S. Miyanishi, K. Tanaka, W. Van Roy, K. Onodera, Appl. Phys. Lett. **76**, 97 (2000)
32. R. Shioda, K. Ando, T. Hayashi, M. Tanaka, Phys. Rev. B **58**, 1100 (1998)
33. R. Bacewicz et al., J. Phys. Chem. Solids **66**, 2004 (2005)

34. I.N. Demchenko et al., Phys. Condens. Matter **19**, 496205 (2007)
35. K. Lawniczak-Jablonska et al., Rad. Phys. Chem. **78**, S80 (2009)
36. W. Stefanowicz, D. Sztenkiel, B. Faina, A. Grois, M. Rovezzi, T. Devillers, F. d'Acapito, A. Navarro-Quezada, T. Li, R. Jakiea, M. Sawicki, T. Dietl, A. Bonanni, Phys. Rev. B **81**, 235210 (2010)
37. M. Rovezzi, F. d'Acapito, A. Navarro-Quezada, B. Faina, T. Li, A. Bonanni, F. Filippone, A.A. Bonapasta, T. Dietl, Phys. Rev. B **79**, 195209 (2009)
38. M. Hashimoto, H. Tanaka, S. Emura, M.S. Kim, T. Honma, N. Umesaki, Y.K. Zhou, S. Hasegawa, H. Asahi, J. Cryst. Growth **273**, 149 (2004)
39. S. Kimura, S. Emura, H. Ofuchi, Y.K. Zhou, S. Hasegawa, H. Asahi, J. Cryst. Growth **301**, 651 (2007)
40. M. Hashimoto, S. Emura, H. Tanaka, T. Honma, N. Umesaki, S. Hasegawa, H. Asahi, J. Appl. Phys. **100**, 103907 (2006)
41. S. Kimura, S. Emura, K. Tokuda, Y.K. Zhou, S. Hasegawa, H. Asahi, J. Cryst. Growth **311**, 2046 (2009)
42. K.C. Ku et al., Appl. Phys. Lett. **82**, 2302 (2003)
43. F. d'Acapito, G. Smolentsev, F. Boscherini, M. Piccin, G. Bais, S. Rubini, F. Martelli, A. Franciosi, Phys. Rev. B **73**, 035314 (2006)
44. Y.L. Soo, S. Wang, S. Kim, G. Kim, M. Cheon, X. Chen, H. Luo, Y.H. Kao, Y. Sasaki, X. Lui, J.K. Furdyna, Appl. Phys. Lett. **83**, 2354 (2003)
45. K. Ganesan, H.L. Bhat, J. Supercond. Nov. Magn. **21**, 391 (2008)
46. K. Lawniczak-Jablonska, A. Wolska, M.T. Klepka, S. Kret, J. Gosk, A. Twardowski, D. Wasik, A. Kwiatkowski, B. Kurowska, B.J. Kowalski, J. Sadowski, J. Appl. Phys. **109**, 074308 (2011)
47. K. Lawniczak-Jablonska et al., J. Solid State Chem. **184**, 1530 (2011)
48. K. Lawniczak-Jablonska et al., Phys. Status Solidi B **248**, 1609 (2011)
49. Y.L. Soo, G. Kioseoglu, X. Chen, H. Luo, Y.H. Kao, Y. Sasaki, X. Lui, J.K. Furdyna, Appl. Phys. Lett. **80**, 2654 (2002)
50. Wolska, M.T. Klepka, K. Lawniczak-Jablonska, J. Sadowski, A. Reszka and B.J. Kowalski, Rad. Phys. Chem. **80**, 1026 (2011)
51. J. Sadowski, P. Dluzewski, S. Kret, E. Janik, E. Łusakowska, J. Kanski, A. Presz, F. Terki, S. Charar, D. Tang, Nano Letters **7**, 2724 (2007)
52. Y.L. Soo, S. Kim, Y.H. Kao, A.J. Blattner, B.W. Wessels, S. Khalid, C. Sanchez Hanke, C.-C. Kao, Appl. Phys. Lett. **84**, 481 (2004)
53. A. Wolska, K. Lawniczak-Jablonska, M.T. Klepka, A. Barcz, A. Hallen, D. Arvanitis, Acta Phys. Pol. A **117**, 286 (2010)
54. A. Wolska in *Ion Implantation*, ed. by M. Goorsky, InTech, p. 105 (2012), http://www.intechopen.com/books/ion-implantation
55. F. Martelli et al., Nano Letters **6**, 2130 (2006)
56. F. Jabeen, M. Piccin, L. Felisari, V. Grillo, G. Bais, S. Rubini, F. Martelli, F. d'Acapito, M. Rovezzi, F. Boscherini, J. Vac. Sci. Technol. B **28**, 478 (2010)
57. F. d'Acapito, M. Rovezz, F. Boscherini, F. Jabeen, G. Bais, M. Piccin, S. Rubini, F. Martelli, Semicond. Sci. Technol. **27**, 085001 (2012)
58. T. Devillers, M. Rovezzi, N. Gonzalez Szwacki, S. Dobkowska, W. Stefanowicz, D. Sztenkiel, A. Groil, J. Suffczynski, A. Navarro-Quezada, B. Faina, T. Li, P. Glatzel, F. d'Acapito, R. Jakiela, M.j Sawicki, J. A. Majewski, T. Dietl. A. Bonanni, Sci. Rep. **2**, 722 (2012)
59. I.N. Demchenko, J.D. Denlinger, M. Chernyshova, K.M. Yu, D.T. Speaks, P. Olalde-Velasco, O. Hemmers, W. Walukiewicz, A. Derkachova, K. Lawniczak-Jablonska, Phys. Rev. B **82**, 075107 (2010)
60. R. Bacewicz, J. Filipowicz, S. Podsiado, T. Szyszko, M. Kaminski, J. Phys. Chem. Sol. **64**, 1469 (2003)
61. A. Titov, E. Kulatov, Yu.A. Uspenskii, X. Biquard, D. Halley, S. Kuroda, E. Bellet-Amalric, H. Mariette, J. Cibert, J. Magn. Magn. Mater. **300**, 144 (2006)
62. J.I. Hwang, Y. Ishida, M. Kobayashi, H. Hirata, K. Takubo, T. Mizokawa, A. Fujimori, J. Okamoto, K. Mamiya, Y. Saito, Y. Muramatsu, H. Ott, A. Tanaka, T. Kondo, H. Munekata, Phys. Rev. B **72**, 085216 (2005)

63. S. Sonoda, I. Tanaka, H. Ikeno, T. Yamamoto, F. Oba, T. Araki, Y. Yamamoto, K. Suga, Y. Nanishi, Y. Akasaka, K. Kindo, H. Hori, J. Phys.: Condens. Matter **18**, 4615 (2006)
64. A. Bonanni, M. Sawicki, T. Devillers, W. Stefanowicz, B. Faina, T. Li, T.E. Winkler, D. Sztenkiel, A. Navarro-Quezada, M. Rovezzi, R. Jakiea, A. Grois, M. Wegscheider, W. Jantsch, J. Suffczynski, F. D'Acapito, A. Meingast, G. Kothleitner, T. Dietl, Phys. Rev. B **84**, 035206 (2011)
65. A.L. Ankundinov et al., Phys. Rev. B **58**, 7565 (1998)
66. A. Wolska et al., Acta Phys. Pol. **114**, 357 (2008)
67. G. Schutz et al., Phys. Rev. Lett. **58**, 737 (1987)
68. B.T. Thole, P. Carra, F. Sette, G. van der Laan, Phys. Rev. Lett. **68**, 1943 (1992)
69. P. Carra, B.T. Thole, M. Altarelli, X. Wang, Phys. Rev. Lett. **70**, 694 (1993)
70. J. Schwitalla, H. Ebert, Phys. Rev. Lett. **80**, 4586 (2005)
71. R. Wu, D. Wang, A.J. Freeman, Phys. Rev. Lett. **71**, 3581 (1993)
72. C.T. Chen, Y.U. Idzerda, H.-J. Lin, N.V. Smith, G. Meigs, E. Chaban, G.H. Ho, E. Pellegrin, F. Satte, Phys. Rev. Lett. **75**, 152 (1995)
73. K.W. Edmonds, G. van der Laan, A.A. Freeman, N.R.S. Farley, T.K. Johal, R.P. Campion, C.T. Foxon, B.L. Gallagher, E. Arenholz, Phys. Rev. Lett. **96**, 117207 (2006)
74. H. Ohldag, V. Solinus, F.U. Hillebrecht, J.B. Goedkoop, M. Finazzi, F. Matsukura, H. Ohno, Appl. Phys. Lett. **76**, 2928 (2000)
75. E. Kulatov, H. Nakayama, H. Mariette, H. Ohta, Yu.A. Uspenskii Phys. Rev. B **66**, 045203 (2002)
76. D.J. Keavney, D. Wu, J.W. Freeland, E. Johnston-Halperin, D.D. Awschalom, J. Shi, Phys. Rev. Lett. **91**, 187203 (2003)
77. R. Wu, Phys. Rev. Lett. **94**, 207201 (2005)
78. K.W. Edmonds, N.R.S. Farley, R.P. Campion, C.T. Foxon, B.L. Gallagher, T.K. Johal, G. van der Laan, M. MacKenzie, J.N. Chapman, E. Arenholz, Appl. Phys. Lett. **84**, 4065 (2004)
79. K.W. Edmonds, N.R.S. Farley, T.K. Johal, G. van der Laan, R.P. Campion, B.L. Gallagher, C.T. Foxon, Phys. Rev. B **71**, 064418 (2005)
80. Y. Ishiwata, M. Watanabe, R. Eguchi, T. Takeuchi, Y. Harada, A. Chainani, S. Shin, T. Hayashi, Y. Hashimoto, S. Katsumoto, Y. Iye, Phys. Rev. B **65**, 233201 (2002)
81. O. Rader, K. Fauth, C. Gould, C. Ruster, G.M. Schott, G. Schmidt, K. Brunner, L.W. Molenkamp, G. Schutz, F. Kronast, H.A. Durr, W. Eberhardt, W. Gudat, J. Electron Spectrosc. Relat. Phenom. **144–147**, 789 (2005)
82. Y. Takeda, M. Kobayashi, T. Okane, T. Ohkochi, J. Okamoto, Y. Saitoh, K. Kobayashi, H. Yamagami, A. Fujimori, A. Tanaka, J. Okabayashi, M. Oshima, S. Ohya, P.N. Hai, M. Tanaka, Phys. Rev. Lett. **100**, 247202 (2008)
83. K. Olejnik, P. Wadley, J.A. Haigh, K.W. Edmonds, R.P. Campion, A.W. Rushforth, B.L. Gallagher, C.T. Foxon, T. Jungwirth, J. Wunderlich, S.S. Dhesi, S.A. Cavill, G. van der Laan, E. Arenholz, Phys. Rev. B **81**, 104402 (2010)
84. P.R. Stone, M.A. Scarpulla, R. Farshchi, I.D. Sharp, E.E. Haller, O.D. Dubona, K.M. Yu, J.W. Beeman, E. Arenholz, J.D. Denlinger, H. Ohldag, Appl. Phys. Lett. **89**, 012504 (2006)
85. H. Makino, J.J. Kim, T. Nakamura, T. Muro, K. Kobayashi, T. Yao, Curr. Appl. Phys. **4**, 615 (2004)
86. J.I. Hwang, M. Kobayashi, G.S. Song, A. Fujimori, A. Tanaka, Z.S. Yang, H.J. Lin, D.J. Huang, C.T. Chen, H.C. Jeon, T.W. Kang, Appl. Phys. Lett. **91**, 072507 (2007)
87. J. Okabayashi, M. Mizuguchi, M. Oshima, H. Shimizu, M. Tanaka, M. Yuri, C.T. Chen, Appl. Phys. Lett. **83**, 5487 (2003)

Chapter 16
Magnetic Ions in Group II–VI Semiconductors

Steve M. Heald

Abstract The group II–VI materials include a number of technologically important systems ranging from IR detection materials ($Hg_{1-x}Cd_xTe$), to X-ray detectors ($Cd_{1-x}Zn_xTe$), and the versatile ZnO. With the advent of interest in magnetic semiconductors for possible spintronics application, doping the II–VI compounds with transition metals was seen as potential route to such materials. EXAFS and XANES have proven very useful in characterizing such materials. They can determine the location and valence of the dopants, and are sensitive to potential second phase formation. This chapter looks at the application of XAS methods to a variety of magnetically doped II–VI systems over the last 30 years.

16.1 Introduction

There is a long history of interest in group II–VI materials for their widespread applications and interesting properties. For example, the ternary $Hg_{1-x}Cd_xTe$ system has long been used in IR detectors since its bandgap can be easily tuned by varying x, and the similarity in lattice constants of HgTe and CdTe is convenient for epitaxial growth of the mixed system [1]. The important ZnO based materials also have many applications [2]. In the area of magnetic semiconductors it was realized early on that the II–VI materials are excellent hosts for magnetic ions [3]. Interest in new materials became especially strong as prospects for spintronics devices improved. Many spintronics devices need magnetic semiconductors, and the popular prototype material, Mn doped GaAs, is only magnetic at low temperatures [4]. In 2000, Dietl et al. showed that a simple Zener model can explain the Curie temperature in Mn doped GaAs and ZnTe [5]. This same paper also predicted that ZnO would be a promising candidate for a high transition temperature, and led to a renewed interest in doping related materials with magnetic ions [6].

S.M. Heald (✉)
X-ray Science Divison, Advanced Photon Source, Argonne National Lab,
Lemont, IL 60439, USA
e-mail: heald@aps.anl.gov

C.S. Schnohr and M.C. Ridgway (eds.), *X-Ray Absorption Spectroscopy of Semiconductors*, Springer Series in Optical Sciences 190,
DOI 10.1007/978-3-662-44362-0_16

Spintronics are devices that take advantage of the electron spin in addition to or in place of the charge [7]. This offers the potential for devices with improved properties such as non-volatile memory or reduced power consumption. If coherence and quantum interactions between the spins can be preserved then a whole new class of quantum devices could be enabled. Early devices such as spin-valves were based on ferromagnetic metals. However, for integration into standard semiconductor processing, magnetic semiconductors offer many advantages. Because of mismatch at the interface, ferromagnetic metals are much inferior to semiconductors for injecting polarized spins into a semiconductor. What is needed is a semiconductor that is magnetic at room temperature, and, preferably, can also be doped into p or n types. As discussed above the II–VI materials are attractive semiconducting hosts for magnetic doping, and much of the work described in this chapter is a result.

In this chapter several examples are discussed that illustrate the above points for a number of II–VI systems. It starts with a discussion of the application of XAS methods to magnetic semiconductors, including some experimental subtleties of working with single crystals and thin films. Early work concentrated on Mn doping of non-oxide II–VI materials. This is summarized along with another interesting system: Cr doped ZnTe. The chapter ends with a look at work on ZnO based materials, an area where interest has exploded in the last decade.

16.2 Application of XAFS to Magnetic Semiconductors

When studying the magnetic properties of doped semiconductors it is important to characterize the location and chemical state of the magnetic dopant. Many of the dopants can form second phases that are magnetic, complicating the interpretation of the magnetic measurements. Generally second phases are unwanted since they do not contribute to the spin currents of the semiconductor host. However, it is possible that the second phase could enhance the magnetic properties of the doped semiconductor if it is also magnetic [8]. There is also the possibility that the observed magnetism may not be due to the doping, but rather from defects that are caused by the doping. For example, doping of an oxide with a dopant that has a different valence from the host atom could result in oxygen vacancies or interstitials that are needed to maintain charge neutrality. EXAFS and XANES are ideal techniques for characterizing the fate of the dopants. EXAFS can determine if the dopant is taking up the expected substitutional sites or is forming a second phase. Often these second phases are in the form of small disordered nanoparticles that are difficult to detect using diffraction techniques. The XANES can determine the valence of the dopant and the symmetry of its site. This can be an independent measure of the amount of second phase if, as is often the case, its valence is different from the substitutional atoms. For example, a common second phase is the formation of zero-valence metallic nanoparticles that are easily seen in the XANES. For cases where all of the atoms take up substitutional sites the XANES can verify that the valence is the expected value.

Doped semiconductor samples for XAFS measurements can be either in the form of powders or as thin films typically grown on single crystal substrates. Since the ultimate goal for many of the materials is device applications, many samples are in the form of thin films. Thin film deposition methods can also allow the production of materials that are not stable in bulk form. The use of thin films on single crystals can complicate the measurements. Fluorescence detection will usually be needed since the substrate absorption does not allow for transmission. The single crystal substrate or sample can produce Bragg diffraction peaks as the energy is scanned, adding spurious structure to the measured spectra. The most serious are reflections that hit the detector, but even reflections that miss the detector will change the fluorescence signal since additional fluorescence can be excited by the exiting X-rays.

There are several approaches for measuring such samples. The first is to search for a sample orientation that is free from Bragg reflections. This can be difficult and time-consuming. It is also possible to measure the sample at two or more angles and splice together the regions of the spectra free from Bragg peaks. A better approach, that can be generally applied, is to spin the sample about the surface normal, averaging out the Bragg reflections. This removes most Bragg interference, but there still might be occasional interference if a strong reflection is briefly excited that hits the detector. However, these residual peaks can be removed by using an energy resolving detector as described in Chap. 1. The Bragg reflection photons are at the incident energy and can be separated from the lower energy fluorescence photons.

Fluorescence detection can be used for another purpose, the determination of the dopant concentration. By comparing the fluorescence signals from the dopant atom with that from the film constituents, it is possible to accurately determine the composition. While other methods can be used for measuring composition, the fluorescence measurement essentially comes for free when fluorescence XAFS measurements are being made. These composition measurements are particularly important for materials like doped ZnO where thin films are often found to have different compositions from the starting materials. This is thought to be related to the high vapor pressure of Zn at typical deposition conditions, often resulting in enhancement of the dopant concentration. An example of a fluorescence spectrum is shown for Mn:ZnO in Fig. 16.1. In this case the deposition target material had 2 % Mn while the resulting film was measured to have 3.8 % [9].

Although beyond the scope of this chapter, it should be mentioned that another useful XAS technique for studying magnetic materials is X-ray magnetic circular dichroism (XMCD). XMCD measures the difference in the absorption as the relationship between the helicity of the circular polarized X-rays and the magnetization direction is varied to be parallel or antiparallel [10]. By tuning the X-rays to the appropriate edges the element specific magnetism can be measured. This can help localize the origin of the magnetism. For transition metals, it is especially useful when applied to the L-edge to directly probe the important 3d states that are involved in the magnetism. Some XMCD studies of transition metal ions incorporated into group III–V semiconductors are discussed in Chap. 15.

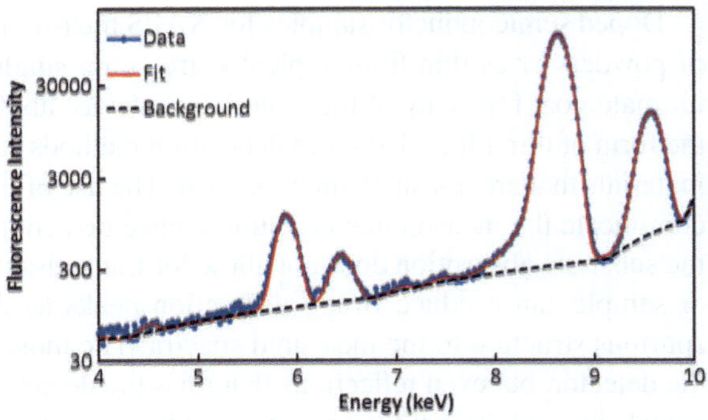

Fig. 16.1 Fluorescence spectrum for a Mn doped ZnO sample grown on Al_2O_3. The fit determined that the Mn content is 3.8 %

16.3 Search for Dilute Magnetic Semiconductors in II–VI Systems

16.3.1 Mn Doping

A good review of early work on II–VI and III–V based dilute magnetic semiconductors has been given by Furdyna [3]. There was an emphasis on Mn doping of II–VI materials. These materials were of interest for their electroluminescence properties as well as for potential magnetic semiconductors. The II–VI semiconductors typically have the zinc blende or similar wurtzite crystal structures. Both have tetrahedral near-neighbor symmetry based on sp3 bonding. Mn has a half filled d-band with two 4s valence electrons that can contribute to the sp3 bonding if it substitutes for the group II atom. As pointed out by Furdyna, the half-filled d-band in Mn has all 5 spins aligned, and it would cost considerable energy to add another electron. This aspect of Mn makes it unique among the transition metals in resembling the filled d shell of the group II atoms. The result is that large concentrations of Mn are often stable as it substitutes for the group II atoms in the II–VI materials. Thus, these materials were important tests for theories of mixed composition lattices.

The lattice parameters of Mn doped II–VI compounds generally follow Vegard's law with a linear dependence on the concentration of Mn. An important result of early EXAFS studies was the realization that on the local scale Vegard's law does not apply to the individual bond lengths [11]. Balzarotti et al. [12, 13] carried out and summarized much of the early work. Figure 16.2 shows an example for the Mn doped CdTe system. This system can be stabilized over a wide range of Mn concentrations. The X-ray diffraction lattice parameter varies smoothly between the CdTe and MnTe endpoints. However, EXAFS measurements found that the Mn-Te and Cd-Te bond lengths were nearly independent of the concentration. This results in a bimodal distribution of Te-Te distances depending on whether the intervening atom is Cd or Mn. As a result, many of the Te atoms must be shifted from their ideal lattice positions. A similar result was also found in the pioneering EXAFS study of $In_{1-x}Ga_xAs$ by Mikkelson and Boyce [11] as discussed in Chap. 2. A variety of

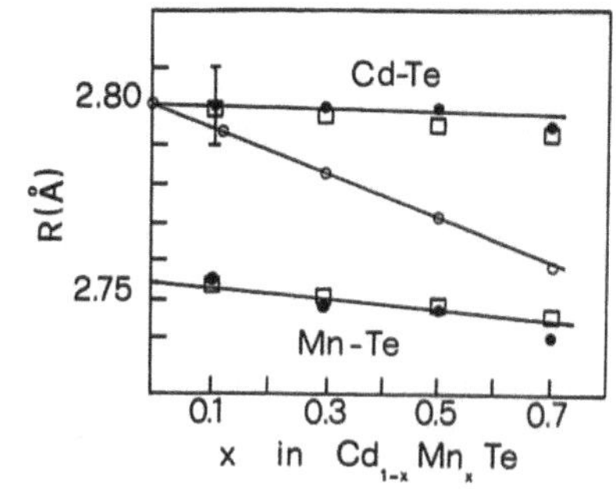

Fig. 16.2 Comparison of the X-ray lattice parameter (*open circles*) with the near neighbor bond lengths (*solid points*) determined by EXAFS for $Cd_{1-x}Mn_xTe$. Reprinted Fig. 11 with permission from [12]. Copyright 1985 by the American Physical Society

EXAFS studies have shown this behavior to be common in many other II-VI systems including $Cd_{1-x}Zn_xTe$ [13], $Hg_{1-x}Se_xTe$ [13], $Hg_{1-x}Cd_xTe$ [14], $Hg_{1-x}Mn_xTe$ [14], and $Zn_{1-x}Mn_xS$ [15] (see also Chap. 2).

The local distortions necessary to accommodate the microscopic picture provided by XAFS can have important implications for understanding these materials. The ideal symmetry of the lattice is broken, possibly allowing previously forbidden optical transitions. For theories of magnetism in these materials the superexchange interaction between Mn atoms is thought to be a major contributor. Locally there can be large variations in the Mn-Mn distance and bond angles with the bonding anion. These variations arise since individual atoms have different environments depending on the composition of the next nearest neighbors (see Chap. 2). For a true random alloy this next nearest neighbor composition is expected to have a range of values. This is discussed in detail by Robouch et al. [16] who develop a strained tetrahedra model for the site occupancy probabilities for zinc-blende ternary alloys, and compare them to EXAFS results. Both EXAFS distances and coordination numbers can be used to determine site preferences, but obtaining accurate coordination results is more difficult. In general, it was found that the distributions of atoms are often far from random, with many systems showing a preference for certain types of bonds. Of course, the implications of this for theoretical calculations are obvious. Typically, in the absence of other information, theorists will assume random occupations with the full symmetry of the host lattice.

16.3.2 Cr Doped ZnTe

As described above, Mn stands out among the transition metals in that it can reach high levels of doping in II–VI materials without second phase formation. As an example of a system where this is not the case, consider Cr doping of ZnTe. At low concentration ($x < 0.04$), $Zn_{1-x}Cr_xTe$ is found to be magnetic with a transition temperature reaching 15 K for $x = 0.035$ [17]. This seemed to be a true dilute magnetic semiconductor, and the transition temperature was higher than other Mn doped II–VI systems. Thus, attempts were made to increase the transition temperature by increasing the doping level, and room temperature magnetism was soon reported for $x = 0.20$ [18]. These results were bolstered by theoretical calculations that suggested the potential for half-metallic ferromagnetism that could result in high Curie temperature [19].

However, there were reasons to question the high concentration results. Vegard's law seemed to be followed at low concentration, but not above $x = 0.04$ [20]. This suggests that some other phase may be involved. Indeed, an XAFS study of the Cr environment with concentration found changes in both the XANES and EXAFS that could be interpreted as the formation of Cr_2Te_3 or CrTe second phases. Both of these materials are magnetic at room temperature. An issue with this work is the similarity in the Cr EXAFS expected from the substitutional site and the second phases. However, the XANES also showed changes consistent with second phase formation. Recent work [8, 21, 22] for high concentrations of Cr with Iodine codoping has found that the Cr concentration was highly inhomogeneous. I doping was used since small amounts of I result in dramatic changes in the transition temperature. The I doping is found to simultaneously increase T_c while inducing an inhomogeneous Cr concentration. The high Cr regions were found to have a CrTe like structure using both TEM and XRD.

This system demonstrates a common problem with studies of these types of materials. Samples are produced by different groups using different preparation conditions, and are analyzed by different techniques. Some of the common characterization methods such as XRD have been shown to be relatively insensitive to second phase formation. Thus, a single experimental result is often not conclusive evidence of a true dilute magnetic semiconductor. For Cr doped ZnTe there are a number of results supporting both possible sources of ferromagnetism, intrinsic or second phase. In these cases XAFS is a powerful method in that it is an atomic structural probe whose signal weights all of the atoms equally, be they in substitutional sites or in a second phase.

It should also be emphasized that XAFS results while extremely useful may not be the final answer. In the case of $Zn_{1-x}Cr_xTe$, the XAFS results pointed the way to the correct interpretation of the data, but were not conclusive. It is also possible that two samples could have fully substitutional doping, but have differing magnetic properties where the differences come from factors to which XAFS has low sensitivity. For example, in the case of Co doped TiO_2 anatase a careful combined XAFS and diffraction study [23] showed that the magnetism had a strong inverse correlation to the structural perfection of the sample. The structural defects, presumably grain boundaries and O vacancies had little effect on the XAFS. XAFS, however, was still important in ruling out second phase formation. XAS studies of doped II–VI semiconductors are also presented in Chaps. 4, 12, and 13.

16.4 Doped ZnO

16.4.1 Co Doping

Transition metal doped ZnO has been one of the hot topics in materials science over the last decade. A review in 2008 identified over a 1,000 papers on the topic [6], and research has continued at a high pace since then. One of the most studied systems is Co doped ZnO (Co:ZnO). This activity was sparked by theoretical predictions, [5, 24] and subsequent observations of high temperature magnetism [25]. There have been numerous publications that find magnetism at room temperature in Co:ZnO. However, the experimental results span a range of results that call into question an intrinsic origin of the magnetism [26–28]. Many results have not included structural studies that conclusively rule out second phases. When careful structural characterization has been carried out, ferromagnetic second phases have often been observed [29–32]. In addition, cases of well characterized single phase materials exist that

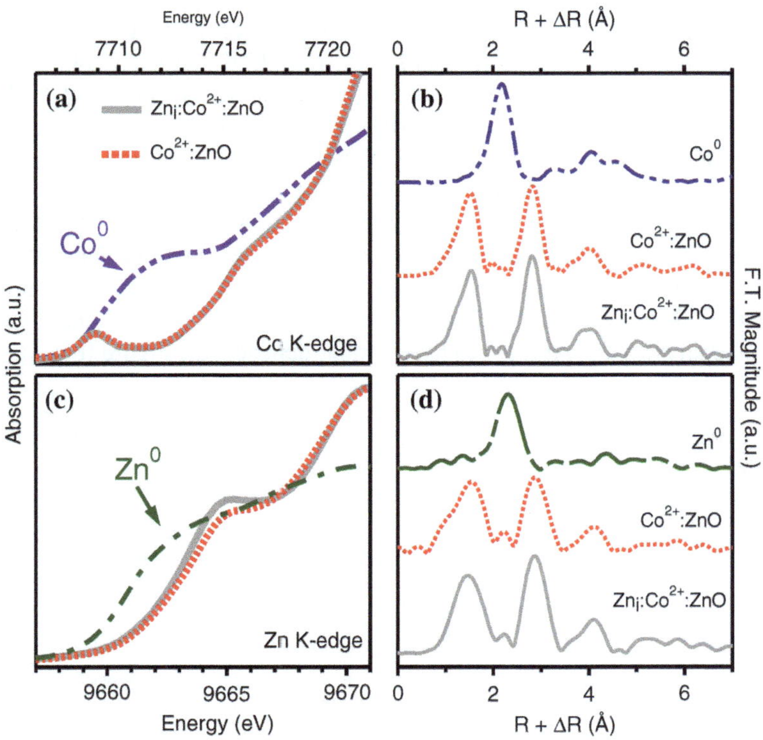

Fig. 16.3 Comparison of the XANES and EXAFS from a sample that was cycled between a magnetic state (Zni:Co^{2+}:ZnO) and a non-magnetic state (Co^{2+}:ZnO) by annealing in Zn vapor and air respectively. The sample was grown by MOCVD on r-plane Al_2O_3 and contained 9 % Co. Figure 16.2 from [38]. Copyright 2006 by the American Physical Society

do not show magnetism [33, 34]. Debate continues since there have been reports of magnetic moments too large to be explained by metallic cobalt second phases [35–37]. This is especially the case for samples that were also doped with Al. Al doping is used to prepare n-type highly conductive films.

A possible source of the confusion is the role of defects in the samples [6]. As mentioned earlier there are several important possibilities for defects. These include vacant lattice sites or atoms at interstitial locations. XAS studies of cobalt—oxygen vacancy complexes are discussed in Chap. 5. In another interesting result, [38] it was shown that a Co:ZnO sample could be repetitively made to be magnetic by annealing in Zn vapor and demagnetized by annealing in air. These results were explained by the incorporation of a small amount of Zn interstitials ($<1\,\%$). XAFS measurements shown in Fig. 16.3 were used to rule out second phase formation and found all of the Co was in substitutional sites with no change upon annealing. There was a small but significant change in the Zn XANES upon annealing, but no indication of additional Zn phases in the Zn EXAFS.

For thin film samples a common structural method checking for phase purity is X-ray diffraction. However, diffraction techniques have reduced sensitivity in thin film samples, and the second phases can be present as small nanoparticles. Such small particles can result in broad difficult-to-detect diffraction peaks that can be further reduced by structural disorder. A study [39] that used both X-ray and neutron diffraction demonstrated the lack of sensitivity to second phases in Co:ZnO for commonly used Cu Kα radiation XRD. Element specific probes such as XAS methods are much better to detect second phases. The Co XAFS equally weights the signals from all of the Co atoms in the samples. The XANES determination of valence is unaffected by disorder allowing easy characterization of atoms in disordered sites.

The polarization dependence of the XAS signal can be used to enhance the sensitivity to second phases. In spite of its relatively symmetric structure, Zn atoms in the ZnO lattice show strong polarization dependence in their XANES, and dopants in substitutional sites are expected to have similar dependence. Figure 16.4 shows the

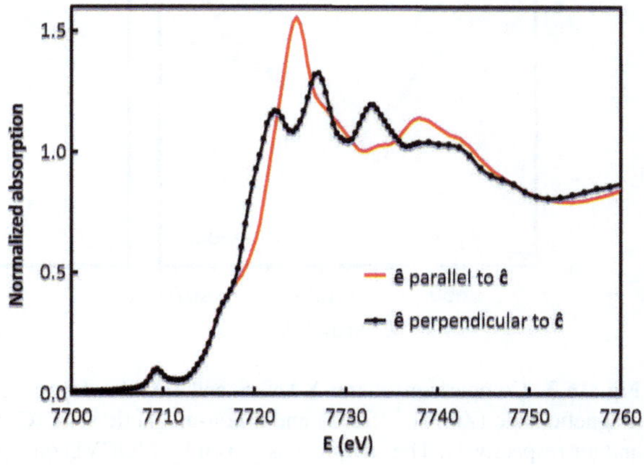

Fig. 16.4 Polarization dependence of the Co XANES in a Co doped ZnO sample grown on c-plane Al_2O_3

orientation dependent XANES for Co in ZnO. Films are often grown in the (0001) orientation, and the XANES is easily measured with the polarization perpendicular to the plane of the film ($\hat{\mathbf{e}}$ parallel to $\hat{\mathbf{c}}$) and the polarization in the plane of the film ($\hat{\mathbf{e}}$ perpendicular to $\hat{\mathbf{c}}$). Second phases can be expected to have a much different polarization dependence or the in case of metallic particles, very little polarization dependence. Thus measurement of the polarization dependent XANES can enhance the sensitivity to second phases. This has been discussed in detail by Ney et al. [40] for the case of Co:ZnO. In this paper a procedure is described for separating the signal of the true substitutional atoms from the second phases using the difference spectra from the two polarizations. The part of the signal without the expected polarization dependence is isolated for more detailed analysis. The difference signal or X-ray linear dichroism (XLD) can also be used as a criteria for separating 'good' samples (undetectable second phases) from contaminated samples. Figure 16.5 taken from this work shows examples for a variety of samples. Metallic Co has an edge shifted to lower energy. However, as seen in Fig. 16.5, Co^{2+} in a non-centrosymmetric environment has a fairly strong pre-edge signal at energies similar to the initial rise for Co metal. This can reduce the sensitivity to small amounts of metallic Co phases. The XLD can highlight the differences in the samples and is also sensitive to non-metallic second phases. The data for sample SPM-RMS shows a Co^{2+} like pre-edge, but the XLD does not have the expected polarization dependence. This indicates the formation of Co^{2+} that is not in substitutional sites.

While Co metal is the most common magnetic second phase, it is also possible to have other metallic second phases. For example, the compound CoZn is also magnetic and might be expected to form under similar reducing conditions as is conducive to Co metal formation. Indeed this was seen in a sample grown on an r-sapphire substrate and annealed in Zn vapor [41]. In this case the polarization dependent EXAFS signal was crucial in its identification. Figure 16.6 shows the polarization dependent EXAFS signal and the inset shows the structure responsible for the strong polarization dependence. The primary Co site in CoZn has a nearly planar arrangement of Co neighbors and a more symmetric arrangement of Zn atoms at a slightly larger distance. When the polarization is in the plane of the Co neighbors, this contribution is enhanced and interferes strongly with the Zn signal. This gives rise to the strong observed polarization dependence. Fits to the data based on an oriented CoZn second phase give a good explanation of the polarization dependence. This same model also explains the shift to higher R from the expected Co-Co peak in Co metal. It turns out that CoZn is also magnetic, and could explain the magnetism seen in this sample.

Varying the X-ray incidence angle can also be used to study the depth dependence of the signal. Figure 16.7 shows the angle dependent signal from a sample grown on r-plane Al_2O_3. This sample was annealed in Zn vapor, which resulted in a significant metal contribution. By reducing the incident angle to below the critical angle of reflection (in this case about 0.2°) the X-rays are confined to the near surface region. The enhancement of the metallic component at small angles confirms that it is located near the surface. XAS studies of Co doped ZnO nanowires are discussed in detail in Chap. 13.

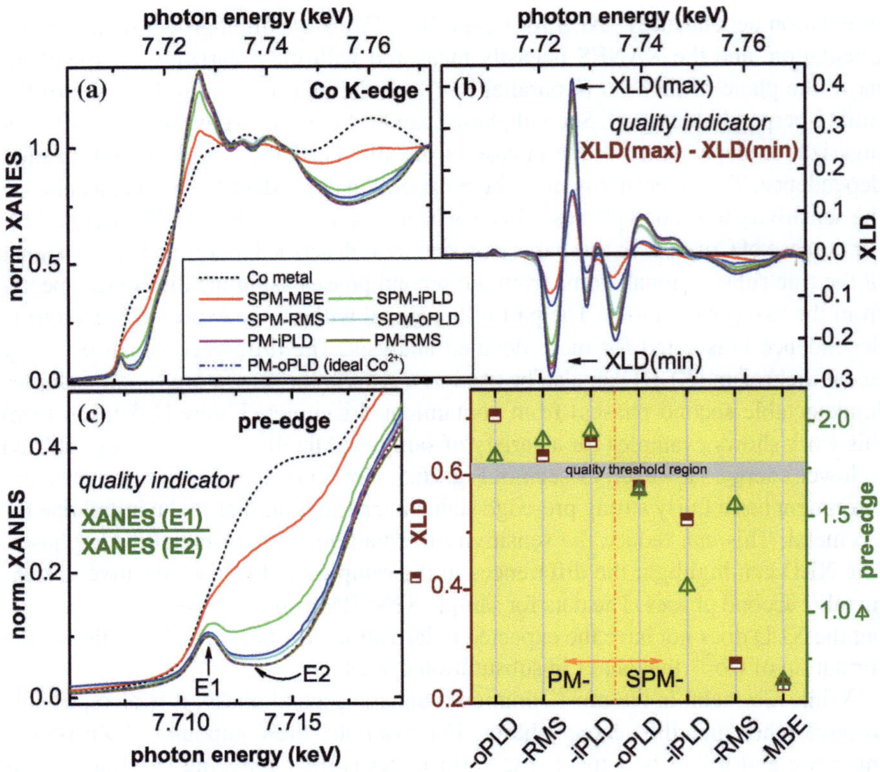

Fig. 16.5 (From [40]) Two methods for determining the quality of the doping in Co doped ZnO. The XANES method looks at the height of the pre-edge peak E1 relative to the background E2, and the linear dichroism (XLD) method looks at the difference between the max and min of the dichroism in the pre-edge region. As shown in the *yellow figure* both can see metallic Co second phases, but the XLD can also detect non-substitutional Co^{2+} impurities. See [40] for a description of the samples. Reprinted from [40] © IOP Publishing & Deutsche Physikalische Gesellschaft. CC BY-NC-SA

16.4.2 Doping of ZnO by Other Transition Metals

Magnetism near room temperature has been observed in ZnO doped with many different transition metals [6, 42]. These include Sc, Ti, V, Cr, Mn, Fe, Ni, and Cu in addition to Co. As was the case for Co, a second dopant is often added to improve the magnetism or other properties such as the conductivity. Thus, the range of possible materials is large and XAFS measurements can and have been applied to many of them. In this section we briefly consider the cases of Mn:ZnO and Fe:ZnO.

Based on the success of Mn doped GaAs, Mn:ZnO has been the focus of a lot of effort. Mn has a high solubility in ZnO, and samples with a wide range of concentrations have been studied. The overall situation is similar to the story for Co:ZnO. Early work for films with concentrations exceeding $x = 0.35$ [43] found a strong room

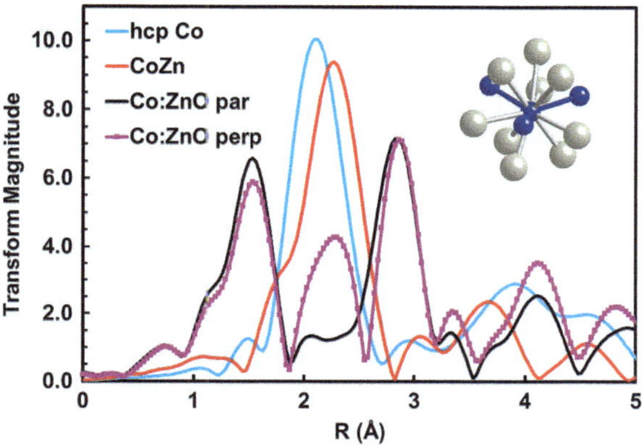

Fig. 16.6 Polarization dependence of the Fourier Transform of the EXAFS from a Co doped ZnO sample grown on r-plane Al_2O_3. The *inset* shows the Co near neighbors (Co in *blue*, Zn in *gray*) in CoZn. Data from [41]

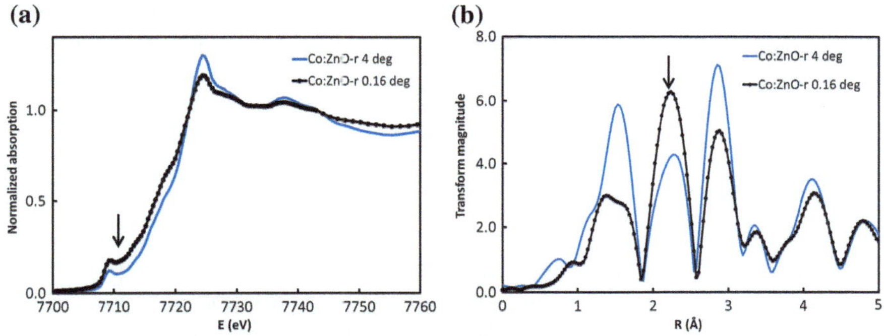

Fig. 16.7 Angle dependent **a** XANES and **b** Fourier transformed EXAFS data for a Co:ZnO sample that was annealed in Zn vapor. The *arrows* point out the metal signals that are enhanced at glancing angles, indicating the metal is located near the surface of the film

temperature magnetoresistance signal. However, later magnetic measurements on similar samples could be interpreted as spin-glass behavior [44]. Just as for Co:ZnO there have been a wide variety of findings some suggesting magnetism and other finding non-magnetic behavior. Lower concentration samples of high structural quality verified by various techniques including XAFS were found to be paramagnetic [45, 46], although based on XMCD it was suggested that the Mn concentration was inhomogeneous with local regions of higher concentrations [46]. It is expected that two Mn atoms separated by an intervening O will be antiferromagnetically coupled. Thus, clustering of the Mn atoms would enhance the antiferromagnetic coupling reducing the possibilities for ferromagnetism. The case for intrinsic ferromagnetism arising from substitutional Mn^{2+} atoms is still unresolved. As was found for Co:ZnO there are indications that defects may play an important role in many of the observed magnetic signals found so far [6, 47]. A strong magnetic signal from a low concentration of defects could explain the typically small magnetic signal per Mn atom found in many samples.

Early work on Fe:ZnO powders [48] found that doping these samples with small amounts of Cu significantly enhanced the magnetic properties, raising the strength of the magnetic signal and the transition temperature. In this case XANES measurements found the Fe to be predominantly Fe^{2+}, and the Cu in the Cu^{1+} valence. However, neutron scattering on similar powder samples suggested that the magnetism was due to $ZnFe_2O_4$ second phase [49].

Later work was also plagued by second phase formation. On $LiNbO_3$ substrates it was found that the maximum Fe concentration for substitutional Fe was about 4 % [50]. Again XANES was used in part to determine that at higher concentrations magnetite was the dominant phase. Being magnetic this phase resulted in a strong increase in the magnetic signal. At lower concentrations where a second phase could not be detected the magnetic signal was about 0.1 μ_B per Fe atom. Later work including XAFS have confirmed that low concentration films can be prepared without detectable second phases [51]. As for Mn:ZnO and Co:ZnO, the small magnetic signal coupled with the wide variation of results suggests that defects may be playing an important role. XAS studies of Mn and Cu doped ZnO nanowires are reviewed in Chap. 13.

16.5 Summary

This chapter has reviewed the application of XAFS to doped II–VI materials. There has been a tremendous amount of work on these materials motivated primarily by prospects for obtaining dilute magnetic semiconductors. There has been no attempt to comprehensively summarize the vast literature, but rather to indicate the opportunities for applying XAFS methods. XAFS is one of the best techniques for characterizing the structural quality of the films, since it is a bulk technique that treats all atoms equally. This means it can be used to detect second phase formation even in cases where these second phases are present in small disordered nanoparticles nearly invisible to diffraction methods. It can also confirm the dopant valence, and determine any local structural distortions about the dopant atom.

References

1. P. Norton, HgCdTe infrared detectors. Opto-electron. Rev. **10**, 159–174 (2002)
2. U. Ozgur, Y.I. Alivov, C. Liu, A. Teke, M.A. Reshchikov, S. Dogan, V. Avrutin, S.J. Cho, H. Morkoc, A comprehensive review of ZnO materials and devices. J. Appl. Phys. **98**, 041103–041301 (2005)
3. J.K. Furdyna, Diluted magnetic semiconductors. J. Appl. Phys. **64**, R29–R64 (1988)
4. A.H. MacDonald, P. Schiffer, N. Samarth, Ferromagnetic semiconductors: moving beyond (Ga, Mn)As. Nat. Mater. **4**, 195–202 (2005)

5. T. Dietl, H. Ohno, F. Matsukura, J. Cibert, D. Ferrand, Zener model description of ferromagnetism in zinc-blende magnetic semiconductors. Science **287**, 1019–1022 (2000)
6. F. Pan, C. Song, X.J. Liu, Y.C. Yang, F. Zeng, Mater. Sci. Eng. **R62**, 1–35 (2008)
7. S.A. Wolf, D.D. Awschalom, R.A. Buhrman, J.M. Daughton, S. von Molnár, M.L. Roukes, A.Y. Chtchelkanova, D.M. Treger, Spintronics: a spin-based electronics vision for the future. Science **294**, 1488–1495 (2001)
8. N. Samarth, Ferromagnetic semiconductors: ruled by a magnetic-rich minority. Nat. Mater. **6**, 403–404 (2007)
9. K.A. Yates, A.J. Behan, J.R. Neal, D.S. Score, H.J. Blythe, G.A. Gehring, S.M. Heald, W.R. Branford, L.F. Cohen, Spin-polarized transport current in n-type codoped ZnO thin films measured by Andreev spectroscopy. Phys. Rev. B **80**, 245207 (2009)
10. F. Baudelet, *X-ray Magnetic Circular Dichroism in Neutron and X-ray Spectroscopy*, ed. by F. Hippert et al. (Springer, Dordrecht, 2006), pp. 103–130
11. J.C. Mikkelsen Jr., J.B. Boyce, Atomic-scale structure of random solid solutions: extended X-ray-absorption fine-structure study of $Ga_{1-x}In_xAs$. Phys. Rev. Lett. **49**, 1412–1415 (1982)
12. A. Balzarotti, N. Motta, A. Kisiel, M. Zimnal-Starnawska, M.T. Czyyk, M. Podgórny, Model of the local structure of random ternary alloys: experiment versus theory. Phys. Rev. B **31**, 7526–7539 (1985)
13. A. Balzarotti, Lattice distortions around atomic substitutions in II–VI alloys. Phys. B+C **146**, 150–175 (1987)
14. R.A. Mayanovic, W.F. Pong, B.A. Bunker, X-ray-absorption fine-structure studies of $Hg_{1-x}Cd_xTe$ and $Hg_{1-x}Mn_xTe$ bond lengths: bond relaxation and structural stability of ternary alloys. Phys. Rev. B **42**, 11174–11182 (1990)
15. R.J. Iwanowski, K. ławniczak-Jabłońska, I. Winter, J. Hormes, EXAFS studies of local atomic structure in $Zn_{1-x}Mn_xS$. Solid State Commun. **97**, 879–885 (1996)
16. B.V. Robouch, A. Kisiel, J. Konior, Statistical model for atomic distances and site occupation in zinc-blende diluted magnetic semiconductors (DMSs). J. Alloy Compd. **340**, 13–26 (2002)
17. H. Saito, W. Zaets, S. Yamagata, Y. Suzuki, K. Ando, Ferromagnetism in II-VI diluted magnetic semiconductor $Zn_{1-x}Cr_xTe$. J. Appl. Phys. **91**, 8085–8087 (2002)
18. H. Saito, V. Zayets, S. Yamagata, K. Ando, Room-temperature ferromagnetism in a II-VI diluted magnetic semiconductor $Zn_{1-x}Cr_xTe$. Phys. Rev. Lett. **90**, 207202 (2003)
19. W.-H. Xie, B.-G. Liu, Half-metallic ferromagnetism in ternary transition-metal compounds based on ZnTe and CdTe semiconductors. J. Appl. Phys. **96**, 3559–3561 (2004)
20. K. Shinji, O. Nobuhiko, N. Nozomi, K. Tatsumi, M. Stéphane, T. Kôki, Growth and magnetic properties of novel ferromagnetic semiconductor (Zn, Cr) Te. Sci. Technol. Adv. Mater. **6**, 558 (2005)
21. H. Kobayashi, Y. Nishio, K. Kanazawa, S. Kuroda, M. Mitome, Y. Bando, Structural analysis of the phase separation in magnetic semiconductor (Zn, Cr)Te. Phys. B: Condens. Matter **407**, 2947–2949 (2012)
22. S. Kuroda, N. Nishizawa, K. Takita, M. Mitome, Y. Bando, K. Osuch, T. Dietl, Origin and control of high-temperature ferromagnetism in semiconductors. Nat. Mater. **6**, 440–446 (2007)
23. T.C. Kaspar, S.M. Heald, C.M. Wang, J.D. Bryan, T. Droubay, V. Shutthanandan, S. Thevuthasan, D.E. McCready, A.J. Kellock, D.R. Gamelin, S.A. Chambers, Negligible magnetism in excellent structural quality $Cr_xTi_{1-x}O_2$ anatase: contrast with high-T_C ferromagnetism in structurally defective $Cr_xTi_{1-x}O_2$. Phys. Rev. Lett. **95**, 217203 (2005)
24. K. Sato, H. Katayama-Yoshida, Material design for transparent ferromagnets with ZnO-based magnetic semiconductors. Jpn. J. Appl. Phys. **39** L555 (2000)
25. K. Ueda, H. Tabata, T. Kawai, Magnetic and electric properties of transition-metal-doped ZnO films. Appl. Phys. Lett. **79**, 988–990 (2001)
26. S.A. Chambers, Ferromagnetism in doped thin-film oxide and nitride semiconductors and dielectrics. Surf. Sci. Rep. **61**, 345–381 (2006)
27. J.M.D. Coey, Dilute magnetic oxides. Curr. Opin. Solid State Mater. **10**, 83 (2006)
28. S.J. Pearton, D.P. Norton, M.P. Ivill, A.F. Hebard, J.M. Zavada, W.M. Chen, I.A. Buyanova, Ferromagnetism in transition-metal doped ZnO. J. Electron. Mater. **36**, 462–471 (2007)

29. J.H. Park, M.G. Kim, H.M. Jang, S. Ryu, Y.M. Kim, Co-metal clustering as the origin of ferromagnetism in Co-doped ZnO thin films. Appl. Phys. Lett. **84**, 1338 (2004)
30. S. Deka, R. Pasricha, P.A. Joy, Experimental comparison of the structural, magnetic, electronic, and optical properties of ferromagnetic and paramagnetic polycrystalline Zn $_{1-x}Co_xO$ (x=0, 0.05, 0.1). Phys. Rev. B **74**, 033201 (2006)
31. T.C. Kaspar, T. Droubay, S.M. Heald, M.H. Engelhard, P. Nachimuthu, S.A. Chambers, Hidden ferromagnetic secondary phases in cobalt-doped ZnO epitaxial thin films. Phys. Rev. B **77**, 201303(R) (2008)
32. K. Rode, R. Mattan, A. Anane, V. Cros, E. Jacquet, J.-P. Contour, F. Petroff, A. Fert, M.-A. Arrio, P. Sainctavit, P. Bencok, F. Wilhelm, N.B. Brookes, A. Rogalev, Magnetism of (Zn, Co)O thin films probed by x-ray absorption spectroscopies. Appl. Phys. Lett. **92**, 012509 (2008)
33. A.R. Han, S.-J. Hwang, Y. Zhao, Y.-U. Kwon, X-ray absorption spectroscopic and magnetic characterization of cobalt-doped zinc oxide nanocrystals prepared by the molten-salt method. J. Magn. Magn. Mater. **320**, 1591 (2008)
34. T.C. Kaspar, T. Droubay, Y. Li, S.M. Heald, P. Nachimuthu, C.M. Wang, V. Shutthanandan, C.A. Johnson, D.R. Gamelin, S.A. Chambers, Lack of ferromagnetism in n-type cobalt-doped ZnO epitaxial thin films. New J. Phys. **10**, 055010 (2008)
35. M. Venkatesan, C.B. Fitzgerald, J.G. Lunney, J.M.D. Coey, Phys. Rev. Lett. **93**, 177206 (2004)
36. C. Song, K.W. Geng, F. Zeng, X.B. Wang, Y.X. Shen, F. Pan, Y.N. Xie, T. Liu, H.T. Zhou, Z. Fan, Phys. Rev. B **73**, 024405 (2006)
37. P. Sati, R. Hayn, R. Kuzian, S. Regnier, S. Schafer, A. Stepanov, C. Morhain, C. Deparis, M. Laugt, M. Goiran, Z. Golacki, Phys. Rev. Lett. **96**, 017203 (2006)
38. K.R. Kittilstved, D.A. Schwartz, A.C. Tuan, S.M. Heald, S.A. Chambers, D.R. Gamelin, Direct kinetic correlation of carriers and ferromagnetism in Co2+: ZnO. Phys. Rev. Lett. **97**, 037203 (2006)
39. H.-J. Lee, S.H. Choi, C.R. Cho, H.K. Kim, S.-Y. Jeong, The formation of precipitates in the ZnCoO system. Europhys. Lett. **72**, 76 (2005)
40. A. Ney, M. Opel, T.C. Kaspar, V. Ney, S. Ye, K. Ollefs, T. Kammermeier, S. Bauer, K.W. Nielsen, S.T.B. Goennenwein, M.H. Engelhard, S. Zhou, K. Potzger, J. Simon, W. Mader, S.M. Heald, J.C. Cezar, F. Wilhelm, A. Rogalev, R. Gross, S.A. Chambers, Advanced spectroscopic synchrotron techniques to unravel the intrinsic properties of dilute magnetic oxides: the case of Co:ZnO. New J. Phys. **12**, 013020 (2010)
41. S.M. Heald, T. Kaspar, T. Droubay, V. Shutthanandan, S. Chambers, A. Mokhtari, A.J. Behan, H.J. Blythe, J.R. Neal, A.M. Fox, G.A. Gehring, X-ray absorption fine structure and magnetization characterization of the metallic Co component in Co-doped ZnO thin films. Phys. Rev. B (Condens. Matter Mater. Phys.) **79**, 075202–075211 (2009)
42. J.M.D. Coey, M. Venkatesan, C.B. Fitzgerald, Donor impurity band exchange in dilute ferromagnetic oxides. Nat. Mater. **4**, 173–179 (2005)
43. T. Fukumura, Z. Jin, A. Ohtomo, H. Koinuma, M. Kawasaki, An oxide-diluted magnetic semiconductor: Mn-doped ZnO. Appl. Phys. Lett. **75**, 3366–3368 (1999)
44. T. Fukumura, Z. Jin, M. Kawasaki, T. Shono, T. Hasegawa, S. Koshihara, H. Koinuma, Magnetic properties of Mn-doped ZnO. Appl. Phys. Lett. **78**, 958–960 (2001)
45. C.N.R. Rao, F.L. Deepak, Absence of ferromagnetism in Mn- and Co-doped ZnO. J. Mater. Chem. **15**, 573–578 (2005)
46. T.C. Droubay, D.J. Keavney, T.C. Kaspar, S.M. Heald, C.M. Wang, C.A. Johnson, K.M. Whitaker, D.R. Gamelin, S.A. Chambers, Correlated substitution in paramagnetic Mn2+-doped ZnO epitaxial films. Phys. Rev. B **79**, 155203 (2009)
47. W. Yan, Z. Sun, Q. Liu, Z. Li, Z. Pan, J. Wang, S. Wei, D. Wang, Y. Zhou, X. Zhang, Zn vacancy induced room-temperature ferromagnetism in Mn-doped ZnO. Appl. Phys. Lett. **91**, 062113–062113 (2007)
48. S.J. Han, J.W. Song, C.H. Yang, S.H. Park, J.H. Park, Y.H. Jeong, K.W. Rhie, A key to room-temperature ferromagnetism in Fe-doped ZnO: Cu. Appl. Phys. Lett. **81**, 4212–4214 (2002)
49. J.H. Shim, T. Hwang, S. Lee, J.H. Park, S.-J. Han, Y.H. Jeong, Origin of ferromagnetism in Fe- and Cu-codoped ZnO. Appl. Phys. Lett. **86**, 082503–082503 (2005)

50. X.X. Wei, C. Song, K.W. Geng, F. Zeng, B. He, F. Pan, Local Fe structure and ferromagnetism in Fe-doped ZnO films. J. Phys.: Condens. Matter **18**, 7471 (2006)
51. H. Liu, J. Yang, Z. Hua, Y. Liu, L. Yang, Y. Zhang, J. Cao, Cu-doping effect on structure and magnetic properties of Fe-doped ZnO powders. Mater. Chem. Phys. **125**, 656–659 (2011)

Index

C.S. Schnohr and M.C. Ridgway (eds.), *X-Ray Absorption Spectroscopy of Semiconductors*, Springer Series in Optical Sciences 190,
DOI 10.1007/978-3-662-44362-0

Z

MIX
Papier aus verantwortungsvollen Quellen
Paper from responsible sources
FSC® C105338

If you have any concerns about our products,
you can contact us on
ProductSafety@springernature.com

In case Publisher is established outside the EU,
the EU authorized representative is:
Springer Nature Customer Service Center GmbH
Europaplatz 3, 69115 Heidelberg, Germany

Printed by Libri Plureos GmbH
in Hamburg, Germany